OZONE DEPLETION, CHEMISTRY AND IMPACTS

OZONE DEPLETION, CHEMISTRY AND IMPACTS

SEM H. BAKKER
EDITOR

Nova Science Publishers, Inc.
New York

For permission to use material from this book please contact us:
Telephone 631-231-7269; Fax 631-231-8175
Web Site: http://www.novapublishers.com

LIBRARY OF CONGRESS CATALOGING-IN-PUBLICATION DATA

Ozone depletion, chemistry, and impacts / Sem H. Bakker (editor).
p. cm.
ISBN 978-1-60692-007-7 (hardcover)
1. Ozone layer depletion. 2. Atmospheric ozone--Reduction. 3. Air--Pollution. 4. Ozone-depleting substances. I. Bakker, Sem H.
QC879.7.O936 2008
363.738'75--dc22 2008027831

Published by Nova Science Publishers, Inc. ✦ New York

CONTENTS

Preface **vii**

Chapter 1 Ozone Precursor Monitoring—New Problems and Solutions **1**
Liming Zhou

Chapter 2 Ozone Decomposition by Catalysts and Its Application in Water Treatment: An Overview **17**
J. Rivera-Utrilla, M. Sánchez-Polo and J.D. Méndez-Díaz

Chapter 3 Decolorization Efficiency of Ozone and Ozone Derivatives for Several Kinds of Dye in Aqueous Solution **53**
Hanjin Luo

Chapter 4 Ozone in the Degradation of Phenols and Xenobiotics **97**
Maurizio D'Auria, Lucia Emanuele and Rocco Racioppi

Chapter 5 Airway Inflammation and Hyper-Responsiveness Induced by Ozone Exposure **117**
M.D. Yang Xiang and Xiaoqun Qin

Chapter 6 Ozone History and Ecosystems: A Goliath from Impacts to Advance Industrial Benefits and Interests, to Environmental and Therapeutical Strategies **135**
Eugenia Bezirtzoglou and Athanasios Alexopoulos

Chapter 7 Catalytic Ozonization: A New Approach to the Treatment of Wastewater **147**
Luciana Serra Soeira and Renato S. Freire

Chapter 8 Bivariate Stochastic Volatility Models Applied to Mexico City Ozone Pollution Data **163**
Jorge A. Achcar, Henrique C. Zozolotto and Eliane R. Rodrigues

Chapter 9 Rate Constants of the Gas-Phase Reaction of Ozone with Organosulfides at Room Temperature **187**
Maofa Ge, Lin Du and Kun Wang

Chapter 10 A Model-Based Warning System for Air Pollution Monitoring (Application to Ground-Level Ozone and PM10 Time Series Measured in Bordeaux, France) **195**
Ali Zolghadri

Chapter 11 The Quasi-biennial Oscillations in the Equatorial Stratosphere: Seasonal Regularities, Dependence on the Solar UV Flux, and Relation to Ozone Depletion in Antarctica **239**
I. Gabis and O. Troshichev

Chapter 12 Ozone/Activated Carbon: A New Advanced Oxidation Process to Remove Water **269**
J. Rivera-Utrilla and M. Sánchez-Polo

Index **315**

PREFACE

Ground-level ozone is an air pollutant with harmful effects on the respiratory systems of animals. Ozone in the upper atmosphere filters potentially damaging ultraviolet light from reaching the Earth's surface. It is present in low concentrations throughout the Earth's atmosphere. It has many industrial and consumer applications.This new book presents the latest research in the field from around the globe.

As explained in Chapter 1, tropospherical ozone is regulated by US EPA (Environmental Protection Agency) as a criterion atmospheric pollutant because of their harmful effects on human's health. VOCs (Volatile Organic Compounds) are monitored in many places as ozone precursor gases such as the PAMS (Photochemical Assessment Monitoring Stations) network by EPA. Online GCs (Gas Chromatograph) and GC/MS are common methods that have been used to analyze speciated VOCs. Usually, these methods can give continuous measurement of speciated VOCs with hourly temporal resolution. Recently, new problems arise in the ozone precursor management and new monitoring techniques can provide the solutions.

Studies have shown that highly reactive VOCs from very local sources, especially fugitive sources, can cause rapid ozone events at very high concentrations at receptor sites, especially in the areas with dense point sources such as refineries and other petrochemical facilities. In order to monitor VOCs at the fenceline of major point sources, traditional continuous GCs are not able to capture these short-lived events by the rapid variations in either emission rate or the meteorological conditions. Running the online GC in trigger mode proved to be very effective in capture these events and hence reducing this type of VOC emissions. Analytical instruments with fast response can also help to address this type of issues. For example, PTR-MS (Transfer Reaction Mass Spectrometry), FTIR (Fourier transform infrared spectroscopy, including extractive mode and open path mode), and fast micro GCs can provide speciated VOC concentrations in seconds to minutes, and may be used for special monitoring purpose. Each of these instruments has its own limitations and need more field test before they can be widely used for ambient air quality monitoring.

Nearly half of the VOC emissions from the chemical industries are fugitive. Industrial operators have been using FID based detectors to manually inspect each component for leaks in their facilities as described in the LDAR (Leak Detection and Repair) program. Most of the time, given the huge number of possible leaking components (~millions for a typical mid scale refinery), this turns out to be a highly labor intensive and time consuming process. IR imaging techniques are developed as an efficient tool for the LDAR purposes. The VOC plumes from the leaking components are visualized by the IR cameras. The detection efficiency can be dramatically improved, and previous unexpected source categories could also be found by using an IR camera. With AI (Artificial Intelligent) tools developed for machine to identify VOCs from IR videos, IR cameras can be operated continuously and in an unattended way to further reduce VOC emissions and improve the industrial process reliabilities.

Comprehensive two dimensional GC techniques have proved to be very powerful in analyzing complex systems such as VOC in ambient air. With a modulator between two independent columns based on different separation mechanisms, two dimensional GCs can better separate background and noises from signal peaks, resulting 2 to 4 fold improvement in sensitivity than traditional one dimensional GC. More importantly, two dimensional GC makes it possible to separate peaks that are previously lumped in one column. For ambient air samples, the number of peaks identified by a two dimensional GC can identify is ten times of a traditional GC. The application of two dimensional GC for ozone precursor monitoring may provide more information to identify the sources and better understanding the ozone formation processes. As a common analytical tool, two dimensional GC is getting mature but methods need to be developed and standardized for their specific applications in ambient air monitoring area such as speciated VOCs.

Ozone has recently received much attention in water treatment technology for its high oxidation and disinfection potential. The use of ozone brings several benefits but has a few disadvantages that limit its application in water treatment, including: i) low solubility and stability in water, ii) low reactivity with some organic compounds and iii) failure to produce a complete transformation of organic compounds into CO_2, generating degradation by-products that sometimes have higher toxicity than the raw micropollutant. To improve the effectiveness of ozonation process efficiency, advanced oxidation processes (AOPs) have recently been developed (O_3/H_2O_2, O_3/UV, O_3/catalysts). AOPs are based on ozone decomposition into hydroxyl radicals ($HO^{\cdot}$), which are high powerful oxidants. Chapter 2 offers an overview of AOPs, focusing on the role of solid catalysts in enhancing ozone transformation into $HO^{\cdot}$ radicals. Catalytic ozonation is a new way to remove organic micropollutants from drinking water and wastewater. The application of several homo- and heterogeneous ozonation catalysts is reviewed, describing their activity and identifying the parameters that influence the effectiveness of catalytic systems. Although catalytic ozonation has largely been limited to laboratory applications, the good results obtained have led to investigations now under way by researchers worldwide. It is therefore timely to provide a summary of achievements to date in the use of solid materials to enhance ozone transformation into $HO^{\cdot}$ radicals.

In Chapter 3, eleven kinds of dyes from the three dye categories of azo, anthraquinone and lush were selected to investigate the decolorization efficiency of dyes by ozone in aqueous solutions. Changes in pH, total organic carbon (TOC), and electronic conductivity (EC) in solutions over reaction time were examined. The major components of dye derivatives during the degradation process were quantified. The results indicated that except for disperse dyes, the decolorization efficiencies of ozone for all soluble dyes exceeded 79% in 20 min. The sequence of the decolorization rate is: reactive > acid > direct > cationic > disperse. In the same type of soluble dyes, the decolorization rates were quicker for dyes with smaller molecular mass. With increasing reaction time, some acidic materials were produced, resulting in a decrease in pH. The final pH for all treatments was less than 4.5. During the reaction process, the decolorization efficiencies increased, while the pH and TOC content decreased. EC increased with the increasing of decolorization efficiency. Concentrations of the related ions (NO_3^-, NO_2^-, NH_4^+, Cl^-, $H_2PO_4^-$, and SO_4^{2-}) and organic derivatives during the reaction process were monitored. The results showed that complicated dye molecules were degraded to simpler organic compounds. Almost all the substituents such as –Cl and PO_4^{3-} in the dye molecules were transformed into Cl^-, $H_2PO_4^-$. Almost all $-SO_3H$ is converted to SO_4^{2-}. Nitrogen was partially degraded to NH_4^+ or NO_3^- due to the types of groups in dye molecules, and NO_2^- was not detected in the degraded solution. Other organics were detected and the organic acids were identified in the solutions. Based on the intermediates produced and the variation of their concentrations, some tentative degradation pathways were proposed.

The irradiation of lignin from pine from steam explosion process in the presence of oxygen, in conditions described for the formation of superoxide ion, for different irradiation time was followed isolating the lignin and determining the average molecular weight. The experiments showed that, until eight hours irradiation, M_n decreases, while M_w and M_z increases. After eight hours irradiation an inverse behaviour was observed, with an increase of M_n and a decrease of M_w and M_z. These results in Chapter 4 are in agreement with an initial polymerization process followed by a photoinduced degradation. Ozonization was carried out in acetonitrile – methanol solution. The reaction showed a zero order kinetics. After 50 min. the average molecular weight of lignin is the half. The reaction mixture was analyzed by using GC-MS. Oxalic acid was determined. The treatment of diluted olive oil mill waste water with Fenton's reagent reduces COD. The reaction followed a zero order kinetics. The reaction needs to use large amounts of reagents to have an appreciable reduction of COD. Treatment of olive oil mill waste water with ozone reduces COD. The reaction followed a first order kinetics. The uv spectrum of olive oil mill waste water after treatment with ozone did not show absorptions. Different degradation methods have been applied to assess the suitability of advanced oxidation process (AOPs) to promote mineralization of imazethapyr, (*RS*)-5-ethyl-2-(4-isopropyl-4-methyl-5-oxo-2-imidazolin-2-yl)nicotinic acid, a widely used imidazolinone class herbicide, the persistence of which has been demonstrated in surface and ground waters destined to human uses. Independently of the oxidation process assessed, the decomposition of imazethapyr always followed a pseudo-first order kinetic. The direct UV-irradiation of the herbicide as its oxidation with O_3 and H_2O_2/UV-oxidation were sufficiently slow to permit the identification of intermediate products, the formation pathway of which has been proposed. O_3/UV, O_3/UV+TiO_2, TiO_2/UV, and TiO_2/UV+H_2O_2 treatments were characterized by a faster degradation and rapid formation of a lot of little molecules, which were quickly destroyed.

As explained in Chapter 5, acute ozone exposure is known to decrease pulmonary function. We have successfully constructed an animal airway hyper-responsiveness (AHR) model by ozone stress, mimicking the airway obstruction, airway inflammatory response, and increased airway responsiveness observed in human AHR disease. The mechanisms leading to the increased AHR are not clear, but epithelial injury is involved. The epithelium is not merely a passive barrier but can generate a range of mediators that may play a role in the inflammatory and remodeling responses. Damage of the bronchial epithelium associated with leukocyte infiltration and increased airway responsiveness are consistent features of asthma. It is reasonably hypothesized that disruption of these functional processes or defects in airway epithelium integrity may be the initial steps leading to airway hyper-responsiveness. Therefore, we damaged the airway epithelium with ozone stress in cultured BEC and animal model, focusing in particular on the roles of airway epithelium in airway inflammation and hyper-responsiveness.

Years of research in our laboratory showed that, after repeated stimulation of bronchial epithelial cells by ozone, a serious of events are programmed to occur: defect in function (e.g., anti-oxidation or secretion) or structural integrity (e.g., imbalance in adhesive molecules expression) weaken the protective ability of BEC against exogenous factors or antigens, such that BEC are easily stressed, damaged and even denuded; and thus the sensitivity of the epithelium and sensory nerve ends are enhanced aberrantly. Subsequently the inflammatory mediators are released; antigen-presenting activities are increased, recruiting and activating immune or inflammatory responsive cells with enhanced airway inflammation and hyper-responsiveness. Pulmonary peptidergic innervation remodeling and airway remodeling continued to increase the airway resistance. These events may be involved in the pathogenesis of AHR.

Ozone is known to be formed naturally in the atmosphere, as a colorless gas having a pungent odor and associated with the ability to guard against the sun's harmful ultraviolet radiation. Historically[1], the characteristic odor of ozone was noticed by Van Marum at 1785 in his electrostatic machine during passage of electric sparks. Later at 1801, Cruickshank notice this odor at the anode electrode during water electrolysis. Finally, in 1840 Shonbein named this chemical substance as ozone from the Greek word ozein, which means to smell. Siemens 's invention followed the early ozone method production by a corona discharge which is based on passing an electric discharge through dry oxygen or air.

From the side of chemistry, ozone is characterized as a triatomic allotropic form of oxygen (O_3) having a molecular weight of 48.

Chapter 6 has focused on the general analysis and evaluation of the industry with the introduction of the ozone generating equipment from the perspective of the evolution of its application and usefulness, as well as its interaction with the environmental and social variables, together with the problems and challenges to be faced contributing in the industry's development. Moreover, ozone application during the last decades brought together great advances in the industry contributing to an effective and sustainable production policy based in a main and capital tool, which is a clean technology with environmental benefits.

Chapter 7 is a review about catalytic ozonation, which is a new way of contaminants removal from wastewater. Despite its current application is mainly limited to laboratory use, the results obtained employing this approach in pollutant degradation showed to be promissory to scale large applications, however further investigation about its efficiency and drawback must be carried out. The aim of this chapter is to give a survey of the application of several homo- and heterogeneous catalysts, with special attention given to the last one. In this context, some metals (for example, Fe, Co, Mn, Zn and Ni) under various forms (salt of reduced metal, solid oxide or deposited metal on support) can be used to enhance the efficiency of ozone towards the removal and/or conversion of different organic compounds in aqueous solution. Moreover, the activity and the parameters that influence the efficiency of these two catalytic systems will be present as a short overview.

In Chapter 8, we consider recently introduced bivariate stochastic volatility models commonly used to analyse financial time series, to study problems related to air pollution data. Such models are used here to estimate the volatility of weekly averaged ozone measurements taking into account two different sets of data provided by the monitoring network of Mexico City. A Bayesian analysis is developed using Markov Chain Monte Carlo (MCMC) methods to simulate samples from the joint posterior distributions and perform the estimates of interest.

The atmospheric sulfur cycle has been the subject of intensive investigation for several decades because of the need to assess the contribution of anthropogenically produced sulfur to such problems as acid rain, visibility reduction, and climate modification. The atmospheric chemistry of sulfur-containing compounds is directly relevant to the formation of sulfur aerosol in marine air. Reduced organic sulfur compounds have been estimated to account for approximately 25% of the total global gaseous sulfur budget. Besides the predominant CH_3SCH_3 (dimethyl sulfide, DMS), other reduced sulfur compounds should also be estimated, such as $C_2H_5SCH_3$ (ethylmethyl sulfide, EMS), n-$C_3H_7SCH_3$ (n-propylmethyl sulfide, PMS), and $C_2H_5SC_2H_5$ (diethyl sulfide, DES), of which there are only a few kinetics investigations. Based on our previous work of DMS and DES, we have measured the rate constants of the gas-phase reactions of ozone with EMS and PMS at room temperature in our self-made smog chamber. Experiments were conducted under supposedly pseudo-first-order decay conditions, keeping $[\text{sulfide}]_0 > 50[O_3]_0$, but having different combinations of $[\text{sulfide}]_0$ and $[O_3]_0$. Cyclohexane was added into the reactor to eliminate the effect of OH radicals. The rate constants of the gas-phase reactions of ozone with EMS and PMS were determined to be $(1.12\pm0.18)\times10^{-19}$ and $(1.24\pm0.15)\times10^{-19}$ cm^3 $molecule^{-1}$ s^{-1}, respectively. The results in Chapter 9 will enrich the kinetics data of atmospheric chemistry, and provide some useful information for evaluating the loss processes of reduced organic sulfur compounds.

Chapter 10 describes a model based approach to develop an operational public warning system for air pollution monitoring. The proposed methodology is based on hard and soft computing techniques and combines an adaptive nonlinear state space-based prediction mechanism, a gain scheduling strategy and neural network techniques to develop an integrated operational warning system. The overall method was applied to ground-level ozone and PM_{10} time-series data measured in Bordeaux, France over four years (1998 to 2001). The aim of model building was to develop predictive models in order to provide forecasts of the maximal daily ground-level ozone and the daily mean PM_{10} concentrations. The goals of the forecast are to provide information in order to satisfy needs for public information and to further reduction and prevention of exposure, in cases where pollutant limit values are exceeded over a specified length of time. A key characteristic of such a system is that it is constantly fed with new available data, so its behaviour can be adapted to the short term changes of air pollution. Moreover, the warning system provides additional information regarding the extent of a smog episode. This is an important aspect for planning of counter actions and for assessments of human health hazards and negative environmental effects. Finally, he chapter discuss some inherent shortcomings associated with the commonly used statistical techniques for air pollution modelling, and an alternative solution based on l_∞ optimization techniques is proposed.

The term Quasi-Biennial Oscillation (QBO) was designated to describe the regular reversals of zonal winds in the equatorial stratosphere occurred with periodicity about 28 months. The phase of QBO cycle, East (E) or West (W) is determined by direction of the zonal winds. Although the wind QBO is mainly a tropical circulation feature, its effects are displayed well beyond the equatorial region. Indeed, the year-to-year fluctuations in the extratropical atmosphere are controlled by QBO, which influence is observed in such parameters as temperature, the total ozone distribution, and others. The most important thing is the well known at present QBO property to modulate the solar activity impact on the stratosphere and troposphere.

Today the dissipation of upward propagating equatorial waves usually examined as a theoretical mechanism for QBO generation. So, it is generally accepted that the proximity of the average QBO period to the double annual one being merely statistical result, whereas the actual period varies from 19 to 36 months. As a result of the high variability and elusive relation to the seasonal cycles, the QBO phase duration seemed to be unpredictable.

A new method for the investigation of the QBO cycle evolution is presented in Chapter 11. The analysis of the height wind profiles makes it possible to reveal new regularities in the equatorial stratosphere wind reversals. The stagnation stage of descending easterlies always starts in solstice, either in December–January or in June–July, and always completes by the equinox being of quantized duration in different QBO-cycles (about 3, 9, or 15 months). So the length of the complete QBO-cycle, if defined as a period between the inceptions of the successive stagnation stages, depends on their durations and may be equal to 24, 30, or 36 months. Consequently, the full QBO cycle length can be predicted as soon as end of the stagnation stage is found while observing the zonal wind transformation.

In turn the duration of the stagnation stage depends on the intensity of solar UV flux. The short stalling period (3–4 months) is observed under condition of the high level or steady increase of the UV irradiance during the first equinox in course of the partial QBO cycle. If the UV flux is low or decreases during the proper equinox, the easterly winds do not descend from the upper layer to the lower stratosphere, and the stalling period length at about 20-40 hPa increases to 9–10 or 15–16 months. The intensity of the spring ozone depletion within Antarctic polar vortex is thought to be also dependent on the phase of the equatorial QBO cycle. It is shown the relation of the season regularities of the equatorial QBO-circulation to the interannual fluctuations and the intensity of total ozone (TOZ) destruction over Antarctica during "ozone hole" phenomenon.

Rising concerns about the quality of drinking water have led both public and private bodies to invest considerable human and economic resources in the development of novel water treatment processes to more effectively remove organic micropollutants, highly toxic for human health (pesticides, herbicides, microtoxins) and sometimes responsible for altering organoleptic properties of the water. Processes based on the simultaneous use of ozone and activated carbon have proven to be very effective for removing contaminants from drinking waters. The results obtained by using naphthalenesulphonic acids as model pollutants have shown that O_3/activated carbon systems have a great efficiency in drinking water treatment because i) it is possible to remove micropollutants due to the high adsorption capacity of activated carbon, and/or ii) they oxidise polar micropollutants, characterised by a low reactivity with ozone, due to the enhancement of ozone transformation into OH· radicals catalysed by activated carbon. Moreover the presence of activated carbon during water ozonation processes reduces the concentration of dissolved organic carbon (TOC) increasing the benefits of the process. Two processes are involved in TOC decrease: i) adsorption of oxidation by-products on activated carbon and ii) mineralization of organic matter by hydroxyl radicals generated in the interaction between ozone and activated carbon. New carbon materials have been developed in our lab in order to potentiate ozone transformation into OH· radicals (activated coke, nitrogen enriched activated carbons and metal doped carbon aerogels) obtaining interesting results especially with metal doped carbon aerogels. The mechanism and the influence of operational parameters in the efficacy of O_3/activated carbon system have also been studied in Chapter 12.

In: Ozone Depletion, Chemistry and Impacts
Editor: Sem H. Bakker, pp. 1-16
ISBN: 978-1-60692-007-7

Chapter 1

OZONE PRECURSOR MONITORING—NEW PROBLEMS AND SOLUTIONS

Liming Zhou
Providence Engineering and Environmental LLC, Baton Rouge, LA

Abstract

Tropospherical ozone is regulated by US EPA (Environmental Protection Agency) as a criterion atmospheric pollutant because of their harmful effects on human's health. VOCs (Volatile Organic Compounds) are monitored in many places as ozone precursor gases such as the PAMS (Photochemical Assessment Monitoring Stations) network by EPA. Online GCs (Gas Chromatograph) and GC/MS are common methods that have been used to analyze speciated VOCs. Usually, these methods can give continuous measurement of speciated VOCs with hourly temporal resolution. Recently, new problems arise in the ozone precursor management and new monitoring techniques can provide the solutions.

Studies have shown that highly reactive VOCs from very local sources, especially fugitive sources, can cause rapid ozone events at very high concentrations at receptor sites, especially in the areas with dense point sources such as refineries and other petrochemical facilities. In order to monitor VOCs at the fenceline of major point sources, traditional continuous GCs are not able to capture these short-lived events by the rapid variations in either emission rate or the meteorological conditions. Running the online GC in trigger mode proved to be very effective in capture these events and hence reducing this type of VOC emissions. Analytical instruments with fast response can also help to address this type of issues. For example, PTR-MS (Transfer Reaction Mass Spectrometry), FTIR (Fourier transform infrared spectroscopy, including extractive mode and open path mode), and fast micro GCs can provide speciated VOC concentrations in seconds to minutes, and may be used for special monitoring purpose. Each of these instruments has its own limitations and need more field test before they can be widely used for ambient air quality monitoring.

Nearly half of the VOC emissions from the chemical industries are fugitive. Industrial operators have been using FID based detectors to manually inspect each component for leaks in their facilities as described in the LDAR (Leak Detection and Repair) program. Most of the time, given the huge number of possible leaking components (~millions for a typical mid scale refinery), this turns out to be a highly labor intensive and time consuming process. IR imaging techniques are developed as an efficient tool for the LDAR purposes. The VOC plumes from the leaking components are visualized by the IR cameras. The detection efficiency can be dramatically improved, and previous unexpected source categories could also be found by using an IR camera. With AI (Artificial Intelligent) tools developed for machine to identify VOCs from IR videos, IR cameras can be operated continuously and in an unattended way to further reduce VOC emissions and improve the industrial process reliabilities.

Comprehensive two dimensional GC techniques have proved to be very powerful in analyzing complex systems such as VOC in ambient air. With a modulator between two independent columns based on different separation mechanisms, two dimensional GCs can better separate background and noises from signal peaks, resulting 2 to 4 fold improvement in sensitivity than traditional one dimensional GC. More importantly, two dimensional GC makes it possible to separate peaks that are previously lumped in one column. For ambient air samples, the number of peaks identified by a two dimensional GC can identify is ten times of a traditional GC. The application of two dimensional GC for ozone precursor monitoring may provide more information to identify the sources and better understanding the ozone formation processes. As a common analytical tool, two dimensional GC is getting mature but methods need to be developed and standardized for their specific applications in ambient air monitoring area such as speciated VOCs.

1. Introduction

Ground level ozone or tropospheric ozone is regulated by US EPA (Environmental Protection Agency) as one of the criterion pollutant gases because of their harmful health effects. Tropospheric ozone is formed through photochemical reactions by the precursor gases including Volatile Organic Compounds (VOCs) and NO_x[1]. Most NO_x are produced by combustion at high temperatures where oxygen and nitrogen in the air directly react. NO_x itself is also regulated by US EPA as a criterion pollutant through ambient monitoring and CEMS (Continuous Emission Monitoring System).

VOCs come from various sources, including both natural and anthropogenic sources. The VOC emissions experience a number of physical process as well as chemical reactions. When meeting NO_x during the transport, photochemical reactions may happen at correct conditions and secondary ozone is produced. Since ozone formation is directly associated with VOC emission, VOCs are monitored to provide information for ozone management purposes.

Usually, there are three sampling methods to gather VOC samples from the ambient air for successive analysis, dynamic enrichment, denudation enrichment, and passive enrichment[2]. The dynamic method uses a pump to draw air samples directly into either a container or the analytical instrument. The denudation method uses the dynamic sampling method and a tube where the analytes diffuse to the coated wall. In passive methods, analytes diffuse into the container or the tube without extra devices to provide driving force. Because of its slow sampling rate, passive methods are used for long term sampling and monitoring. In this article, only a dynamic or a denudation method is used for any mentioned VOC monitoring activities.

2. Conventional Gas Chromatography Techniques

Gas Chromatography (GC) Techniques are the major methods in VOC monitoring [3,4]. Based on sampling methods, this type of techniques can be classified into online GC and canister sampling followed by a GC analysis in an analytical laboratory. When running in an online mode, pre-concentrators based on thermal absorption and desorption are usually used to enrich the analyte. Modern pre-concentrating techniques can improve the sensitivity up to

1000 times. These methods are standardized by EPA as described in EPA methods TO14 and TO15.

Two types of detectors are commonly used in combination with GC, FID (Flame Ionized Detector) and MSD (Mass Spectrum Detector). FID detectors rely on the retention time on the column to identify VOC species and high resolution column is needed for a sufficient separation of VOC species.

Figure 1 illustrates a GC-FID system for online VOC sampling and analyses. Two columns separate VOC species on boiling point. The BP column is used for heavy components and the PLOT column is for light components. Samples first go through the BP column. Light components elute first on the BP column and are switched to PLOT column for better separation and then go to an FID detector. Heavy components are directly sent to another FID since they may poison the PLOT column. Typically, this configuration needs 40 to 50 min to complete the GC analytical cycle and can identify over 50 VOC species. When running at a continuous manner, a careful coordination with the pre-concentrator, with usually a 40 min sampling period, can provide hourly speciated VOC concentrations. This system is used in EPA's PAMS (Photochemical Assessment Monitoring Stations) network (http://www.epa.gov/oar/oaqps/pams/). The VOC species monitored in PAMS include alkanes, alkenes and some mono-ring aromatic hydrocarbons.

GC-FID may suffer from co-elution issues, that different may have similar retention time. Using an MSD as detector may partially resolve this issue. Certain mass charge ratios are used to identify a component in addition to retention time (EPA Method TO14). Because of its relatively simple operation and reliability, GC-FID is more often used in ambient air monitoring stations in an unattended way.

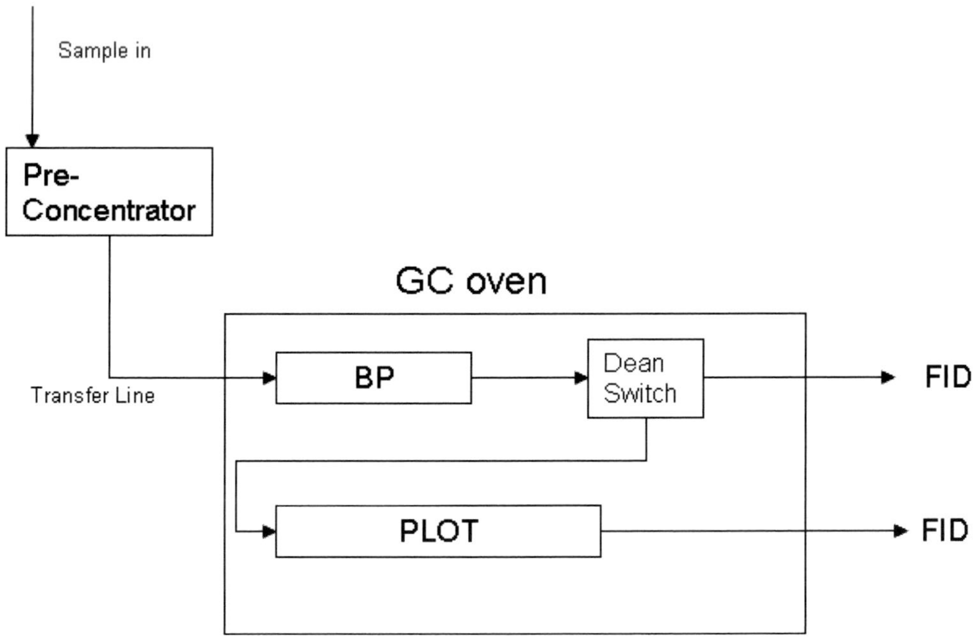

Figure 1. Dean switch dual column TD-GC system for ozone precursor measurement.

Other detectors used include PID (PhotoIonization Detector) and ELCD (Electrolytic Conductivity Detector). PID is used to detect aromatic and olefinic hydrocarbons in the presence of alkanes and other saturated hydrocarbons. ELCD is used for a selective detection of halogen-containing compounds (Allen AE on Chlorine compound on ozone).

3. Fast Response Monitoring Techniques

The contributions from very local sources changes rapidly with the micro meteorological conditions or the emission itself does not last enough long time, especially in an area with a dense distribution of industrial facilities. Recent research has shown that half of the VOC emission is fugitive and concentrated in a very short period rather than evenly distributed across a year [5]. The time scale of these changes is usually at sub hour level [6]. Thus, the one hour resolution provided by the online GC in the previous section becomes in sufficient for identifying these very local sources.

Although current total non-methane VOC analyzer provides a time resolution of 10 min, it cannot provide speciated information. This type of analyzer is usually used in combination with canister samplers to collect sample of elevated total hydrocarbon events for a successive lab analysis with GC or GC-MS.

3.1. Online GC in a Trigger Mode

In order to provide a prompt response for rapid and short life VOC plumes, a TNMOC analyzer has been used with an online GC for fenceline VOC monitoring[6].

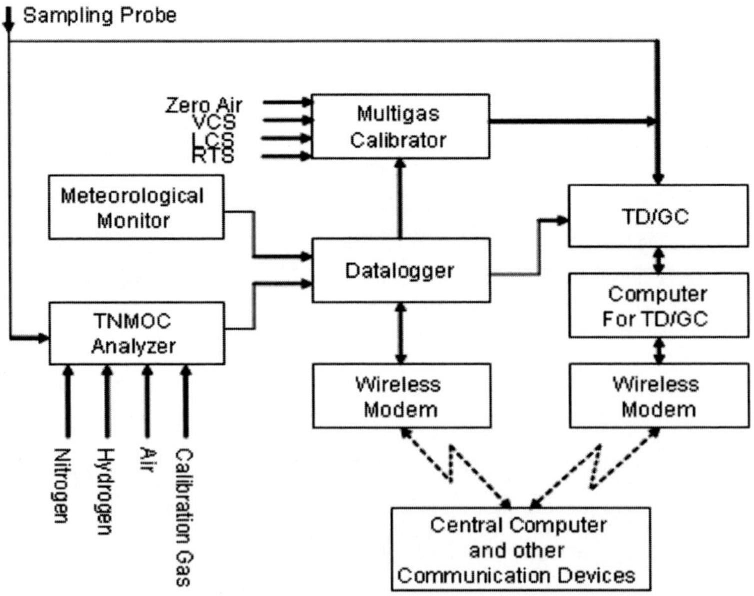

Figure 2. Schematic of a triggered GC system.

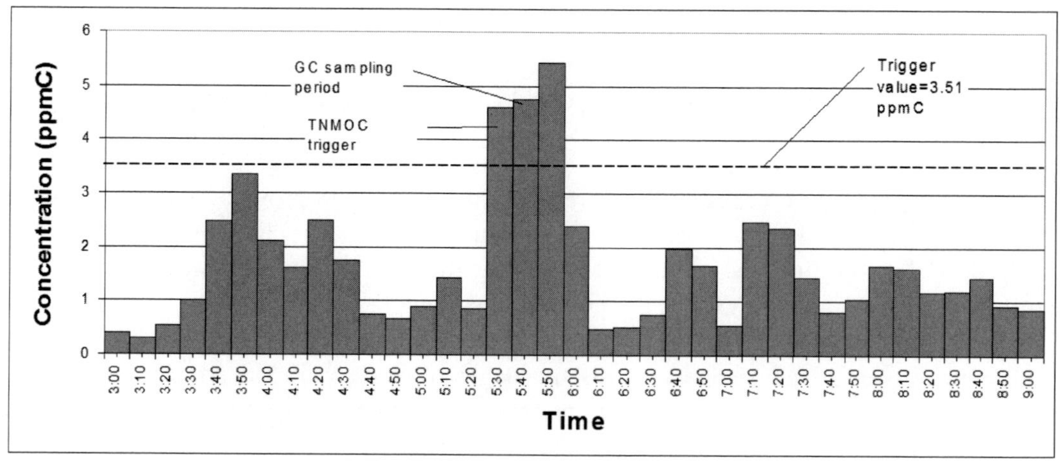

Figure 3. TNMOC events and a GC trigger sample.

A schematic of such a system is presented in Figure 2. A datalogger is used to record total non-methane organic carbon (TNMOC) concentration and meteorological data. When the TNMOC exceeds a certain threshold, the datalogger triggered the online GC to start sampling and analysis. A shorter sampling period is used to better capture the VOC plumes from nearby sources. In Figure 3, an example of TNMOC concentration and GC sample are illustrated, where TNMOC events lasts for twenty minutes. If a continuous GC is used, the sample capture may not be as good as the triggered GC and the speciated VOC concentration may not be less representative. Also, since the sample is performed in a dynamic way, the calibrations are also performed dynamically controlled by the datalogger in accordance with the trigger samples. With this measure, the triggered GC system has good performance comparable to continuous online GCs.

This method proves effective in providing prompt analysis results for both environmental and industrial operators to locate and fix VOC sources. This practice has proved to be very successful in reducing expected VOC release and has contributed to reducing local ozone levels [7]

3.2. PTR (Transfer Reaction Mass Spectrometry) –MS

Online VOC measurement with an MS using direct sampling method is popular for vehicle based mobile monitoring stations[8]. For these common mass spectrometers, thermal electrons are used as the ionization source [9]. The electrons bombard the target molecules to generate ions. When used for analysis of organic molecules, instead of forming a single ionized species, most molecules break down into smaller fragment ions. When several compounds enter the MS detector simultaneously, the final mass spectrum may be too complex to interpret or to quantify. Although species can be separated by using a GC in front, the slow GC cycles prevent fast response. An inter-comparison of PTR-MS and GC-PTR-MS was performed and the validation of using a GC in front verified that the PTR-MS gives accurate measurement [10] for the ambient air samples collected during the study.

PTR-MS uses a “soft ionization” method to ionize the organic molecules, where the VOC molecules are reacted in a drift tube with charged ions, with hydroxonium ions (H_3O^+)

produced in an external glow discharge ion source. The ratio of the electric field and the vacuum of in the drift tube has significant effect on the performance. Since the molecules are not broken, the mass charge ratios can reflect the molecular weight, and the compounds will be easy to identify.

PTR-MS can provide ppt (part per trillion) level sensitivity within a second of measurement time, and because of its fast response time, PTR-MS provides the opportunity to observe compounds that cannot survive pre-concentration and chromatography, and is especially ideal for the measurement on an aircraft or other online measurements. The application of PTR-MS on VOC measurement is summarized in PTR-MS conferences[11]. The application of PTR-MS for VOC measurement in ambient air is reviewed by Hewitt et al.[12] where both the advantages and disadvantages of using PTR-MS are discussed. PTR-MS has been used in Texas for aircraft measurement of highly reactive VOCs [13]. The speciated VOC data from continuous PTR-MS measurement is also used for source apportionment of VOCs [14].

PTR-MS has been successfully used in characterizing the VOC emissions from biosphere. Warneke et al. observed emissions of acetone, methanol, acetaldehyde and other VOCs from decaying vegetation [15]. Substantial quantities of partially oxygenated VOCs, including hexenal family of compounds, alcohols, aldehydes, ketones, esters and acids, are also observed when the vegetation is wounded from such as grazing of grassland, harvesting of crops, freezing and thawing of plant tissue [16,17,18].

Problems may arise during the measurement of complex and qualitatively unknown gaseous mixtures such as the VOCs in the atmosphere. Although PTR-MS is designed to minimize fragmentation, fragmentation is not completely eliminated for all VOC species. For example, Tani et al. observed the fragmentation of some C10 monoterpene compounds at high concentrations and the mass fragments may be misidentified as toluene, isoprene or other lower molecular mass species [19]. Since more than one VOC species may contribute to the same observed ion signal, PTR-MS is not a truly species-specific method and a complimentary method is needed for VOC measurement in the atmosphere in order to obtain comprehensive and accurate information of VOC composition[12].

3.3. FTIR (Fourier Transform Infrared Spectroscopy)

FTIR is a common analytical tool. When used for VOC monitoring, it can be running at two modes, extracted and open-path. In the extracted mode, the samples are introduced into a long cell for measurement. The overall measurement time is determined by the sampling time since the instrumentation measurement time is negligible compared with the sampling time. FTIR can measure a wide range of VOCs with detection limits for most species at ppb levels. EPA method TO16 describes how an FITR technology should be used for VOC monitoring. The theories and applications of FTIR are recently summarized by Bacsik et al.[20,21].

FTIR analyzers can work in two modes, extracted and open path. In the extracted mode, samples are drawn into a cell for analysis. Based on the Lambert-Bill's law, longer cell length will produce higher sensitivity. A 100 m cell length can provide sub ppb levels of sensitivities. In an open path mode, the open air is treated as a cell for the infrared light to pass. The open path mode provides a much faster response since the sampling procedure is avoided. However, ambient interferences such as water vapor, CO_2 and metrological conditions may also defect the accuracy of the measurement. Hong et al.[22] developed novel software called COL1SB to handle the chemical interference in FTIR spectrum and successfully applied it for VOC and ozone open-path measurement.

In the extracted mode, FTIR usually uses a light source to illuminate the analyte; while in the open-path mode, both active and passive mode can be used for remote sensing purposes. In an active mode for open-path FITR, a light source is placed to the other side of the path or the same side with a reflection mirror. Hu et al.[23] discussed the progress in using FITR for remote sensing the atmospheric pollutants. Their review included chemometrics, computed tomography, FTIR spectra interpretation, and spatial distribution of air pollutant concentration. A particular advantage of using FTIR is the spatial distribution of VOCs can be reconstructed by using the Computer Tomography (CT) technology, which is useful to identify and quantify the point sources and source contributions. A protocol for using open-path FTIRs to map plume locations and distributions is released by UP EPA [24].

When running in a passive mode, natural light source or the light emitted or reflected by the analyte itself is used. The detection of the pollutants is based on the temperature difference in the pollutant plume and the background [25]. The passive open path FTIR has the advantages of the ability to detect any directions, fast response, and convenient operations. However, at the same time, the spectrum processing also becomes more difficult since the signal from the natural light source is weak and the radiation from the background needs also to take into account [26].

3.4. Fast MICROGC

In order to obtain enough separation effects, GCs usually use long columns up to 50 m. Although it improved the resolution of peaks, the retention time becomes long. Fast microGCs reduce the overall instrumental sizes and also the column lengths with also a fast temperature program [27,28]. Therefore, the chromatography retention time can be reduced to within a minute, with a sacrifice of some sensitivities and peak resolution. This type of instrument is small in volume and portable. It consumes much less gases and power than traditional GCs and is ideal for field applications. The application of microGCs in VOC monitoring is described by Politzer A.R.[29]. At present, there are still very few applications of using microGCs for VOC monitoring purposes.

4. IR (Infrared) Imaging

Air monitoring samples are obtained from a point to represent a certain area. This is usually true for ambient monitoring stations. However, in the areas with concentrated point sources, the spatial distribution of VOCs becomes not that uniform. Fresh plumes usually concentrate within a very limited space and become difficult to detect by point sampling methods.

4.1. IR Cameras

An example is the VOC leaking detection in chemical and petrochemical facilities. EPA required an LDAR (Leak Detection and Repair) program for these industrial facilities to periodically check and fix leaking components. In a typical LDAR process, operators use handheld FID detectors to "sniff" around any possible leaking components. The process is labor-intensive and time consuming, given the millions of possible leaking components. Sometimes the results may not be highly reliable if the operator misses any space around the leaking components where the VOC plume may be released into.

IR imaging techniques have been used recently to identify leaks in petrochemical industry. The principle of operation of the active system is the production of an optical image by reflected (backscattered) laser light, where the laser wavelength is such that it is strongly absorbed by the gas of interest. The system illuminates the scene with infrared light and a video camera-type scanner picks up the backscattered infrared light. The camera captures the backscattered infrared light to form images and video clips. Since the scanner is only sensitive to illumination from the infrared light source and not the sun, the camera is capable of displaying an image in either day or night conditions. The passive camera uses natural lights. The camera only detects the light within the VOC absorption band. By superimposing the filtered light (at a frequency that displays VOC gas) on a normal video screen, the camera can display the VOC cloud in real time in relationship to the surrounding process equipment. The operator can "see" a plume of VOC gas emanating from a leak[30].

Figure 4. VOC leak at a tank roof.

Currently, various cameras have been tested for fugitive VOC leak detection purposes. Passive cameras are more convenient to use in industrial facilities and long distance monitoring. IR cameras have different mechanisms for wavelength selectivity. The simplest way is to allow a fixed band width to pass to the detector. Usually this band width is selected where most VOC species have absorption. Some cameras have tunable filter so the camera can work at different wavelengths and an optimal wavelength can be used for a certain VOC species. If multi-wavelengths are used, the image becomes hyperspectral and more VOC species can be observed at the same time with better sensitivity. Hyperspectral IR cameras include dispersive cameras and FTIR cameras.

Dispersive cameras uses a lens to disperse the light vertically so a line of hyperspectral pixels can be detected and the camera then scans line by line to finally obtain a hyperspectral data cube (http://www.specim.fi/). A FTIR camera provides a full spectrum through Fourier transformations for each pixel in the image (http://www.telops.com/).

The IR cameras provide plenty of information in images or videos that can be used for both qualitative and quantitative purposes. Epperson et al.[31] discussed the method to quantify the emission rate from the IR images. An IR camera can also be operated in a continuous mode without an operator as a video surveillance system to minimize the time to find and fix leaks. Zhou et al.[32] developed a method to process IR videos for automatic VOC plume identifications. The upper image in Figure 4 shows a frame of an IR video with VOC leaks from a tank roof. The pixels of all frames at the same location form a time series. Fourier transform of the pixel time series is a common method to detect fire or plumes in the video surveillance area. The spectral power at certain frequencies reflects the flickering motions caused by atmospheric turbulences. However, the vibrations of the camera may cause the video frames not aligned, and processing the unaligned frames will lead to meaningless results as indicated in the middle image in Figure 4. After the alignment through a series of spatial Fourier transformations[32], the spectral power of the flickering frequencies in the lower image in Figure 4 can represent the leaked VOC plume with the stationary background removed. This alignment method will be useful for automatic and quantitatively video processing purposes.

4.2. Infrared Tunable Laser Spectroscopy

The development of tunable infrared laser sources is contributing to the development of infrared spectroscopy, including trace atmospheric pollutants measurement. Past intensive development efforts have resulted in extremely reliable and room temperature semiconductor lasers in the visible and near-infrared. These lasers are now commercially available.

Curl and Tittel[33] reviewed the theories of IR laser and the applications in different areas. Differential Absorption Lidar (DIAL) systems are used to identify methane leaks in pipelines from an aircraft. The DIAL system runs at two wavelengths, one with strong methane absorption and the other with no methane absorption [34]. The differential signal gives sensitive detection of methane plumes. Although each measurement only gives one point, the instrument can rapidly scan a number of points to form a contour that can overlay onto an aerial photo by a common camera.

Tunable semiconductor lasers with multiple wavelengths are also now available and a spectroscopic system with such a laser source could be used for speciated VOC remote detection. The application of semiconductor laser spectroscopy in atmospheric pollutants is just started and further development and more contributions are expected for spatial and remote VOC detections.

5. Two-Dimensional (2d) GC

For a complex system like the atmospheric pollutants, GC method may fail since any column has limited peak capacity. Analytes peaks become broadened as it travels across the column, and only finite number of peaks can be resolved at the column outlet. The only solution is to provide additional separation based on a different mechanism. The two dimensional GC is developed for this purpose, with a modulator between two independent columns to make two dimensional chromatograms. Two dimensional GC techniques, invented in the early 90s in the last century, have proved to be one of the most powerful analytical tools for complex systems such as food, oil, metabolic and environmental analyses[35,36,37,38].

5.1. Orthogonal Theory of Two Dimensional GC

The best separation results are obtained if the two columns have independent separation mechanisms [39]. The independence of the two columns is also called orthogonality. Orthogonality is a critical concept in multidimensional separations and determines the magnitude of supranational space that is used. The correlation of the retention time of the two columns will reduce the maximum peak capacity. A high correlation may reduce a multidimensional separation to a one-dimensional separation with peaks distributed along a diagonal[39]. Usually, the primary column is a boiling point column and the secondary column is based on polarity. Ryan et al.[40] investigated how the separation space was affected by varying the polarity of the first column, and concluded that the maximum separation on the second polarity column was achieved with the maximum differentiation between the two columns.

5.2. Modulation

The modulator is the heart of a two-dimensional GC system. The modulator collects the effluence from the primary column and holds them, and at the end of the modulation period, the accumulated analyte will be injected into the secondary column for further separation. The resolution of the primary column will be reduced to the modulation period and the maximum retention time of the secondary column should not exceed the modulation period. In order to keep the peak definition, the modulation period should be no shorter than one third of the peak width. The peak width on the first is usually 0.5 to 1 min and the modulation period is usually chosen at several seconds.

Various modulators have been developed in the past. Cryogenic and valve-based modulators are most widely used. As a cryogenic modulator, the dual-jet modulator with liquid carbon dioxide or nitrogen for cooling is believed ideal for essentially all applications Adahchour et al.[36]. A dual stage jet modulator uses a cold jet working continuously and a hot jet working periodically as shown in Figure 5. Analyte from the primary column is trapped twice. The first trap serves as a buffer and the second trap refocus the analyte. When the first trap is heated, analyte flows to the second trap. At the same time, the second trap is heated to inject the analyte into the secondary column. When the analyte from the primary trap arrives at the second trap, the second trap has completed desorption and becomes cold again. This type of modulator is commercialized by Zoex [41]. By using liquid nitrogen, this modulator can be used for analyzing C_3-C_{40} compounds.

Valve-based modulator can be used for very fast second-dimension separations of 1 sec or less, but they provide a rather low upper temperature limit and also lose some of the analytes from the first column [42]. A total-transfer valve is developed for eliminate mass loss from the first column by blocking certain ports of a high speed six-port diaphragm valve [43].

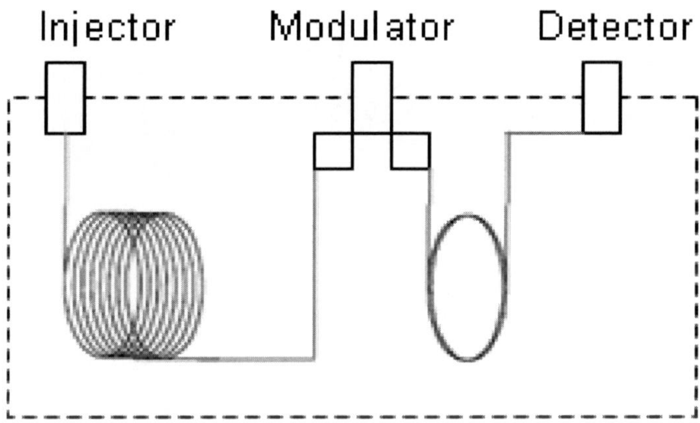

Figure 5. Illustration of a two-dimensional GC structure.

In this configuration, the temperature limitation of the valve-based modulator is addressed by placing the valve partially inside the GC oven [44]. Based on Seeley's original concept[45], a differential flow modulation 2D GC is recently commercialized by using a switch type modulator through micro-fabrication techniques [46].

5.3. VOC Measurement Using 2D GC

Current known VOC species may not be sufficient to explain the ozone formations. Unidentified peaks in one dimensional GC chromatograms are found important for ozone production especially transient ozone formations [47].

Lewis et al.[48] used 2D GC for urban VOC measurements and found over 500 individual VOC species, including a wide variety of multisubstituted compounds, were not identified previously. Xu et al [49] performed in situ measurements of C7–C14 compounds in urban air in Crete (Greece) and revealed ~650 identifiable compounds as well as similar amount of unidentifiable peaks.

Analysis of ambient air in an urban environment (Leeds, U.K.) detected the presence of 147 monoaromatic species with up to 8 carbon atoms added to the aromatic ring. State-of-the-art, single-column measurements typically reveal only 8–15 of these compounds [50]. These previously unmeasured monoaromatic compounds may be of particular significance in urban areas because of their high and variable ozone production potential.

VOC monitoring data by two dimensional GCs are also used for the study of VOC oxidation reactions. Hamilton et al.[51] also used 2D GC for chamber study of VOC oxidation and found the current reaction mechanism significantly over-predicted the concentrations of some products. Bartenbach et al.[52] compared 2D GC-FID with GC-MS for continuously measuring VOCs and found a good agreement between the two methods. The VOC monitoring data are then used to estimate the concentration of OHs.

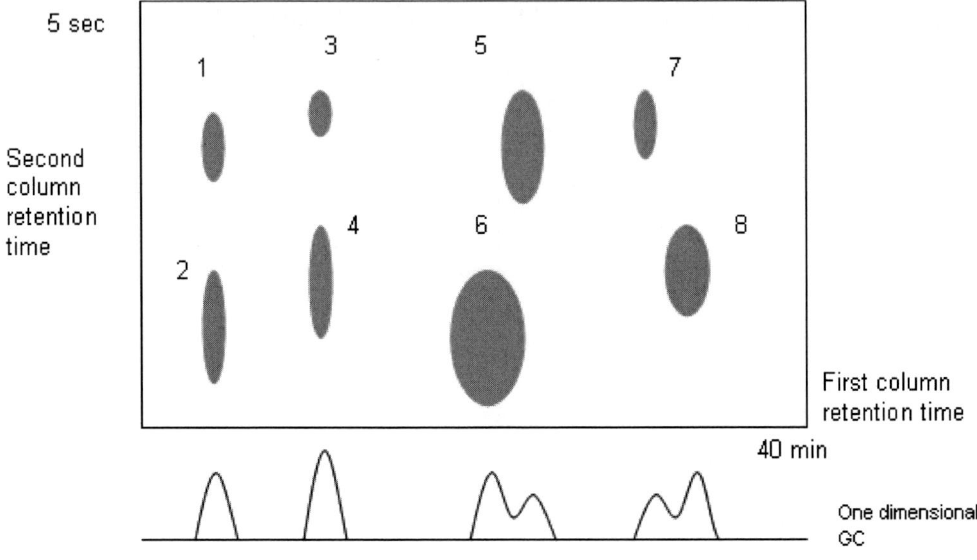

Figure 6. Illustration of two-dimensional peaks versus one dimensional peaks.

Seeley et al.[42] used 2DGC with dual secondary columns for measuring VOC samples from outdoor, indoor and exhaled breath. By using a specially designed differential modulator, more than 100 VOCs can be measured in less than 10 min, including oxygenated and aromatic VOCs. His method could also be used for local source characterization because of its relatively fast response.

5.4. Remaining Issues for VOC Measurement Using 2D GC

For the hardware side, 2D GC has become mature and commercial instruments are now available. However, the data analysis techniques have not been finalized in most application areas of 2D GC. Chemometric methods can provide an efficient use of the data but may fail when a temperature program is used [53]. The identification of a VOC species is through the retention indices at both columns and when an MS is used as the detector, the full mass spectrum will be compared with the standard spectra in a mass spectrum library. Vogt[54] et al. developed an automated cluster method to classify the measured mass spectra with retention time of both GC columns.

So far, 2D GC measurement of VOCs or atmospheric aerosols cannot identify all GC peaks and the number of unidentified peaks may be up to half of all peaks. This issue is not prominent in one dimensional GC applications since many unknowns are buried under the known peaks. Some methods are needed to excavate the information in these unidentified peaks to further understand the ozone issue.

When used for online field measurement, similar to one dimensional GC[6] , 2D GC may also suffer from retention time shift issues. Automated retention time alignment method may be needed for long time monitoring activities.

Finally, for VOC monitoring purposes, standard data processing methods are necessary to compare the results at different locations.

6. Conclusion

As one of the major atmospheric pollutants, ozone, because of its health effects on human lungs and respiratory tracts, has been of public concerns. As US EPA tightens the national standard of ozone, many places will feel difficult to be in compliance for ozone. As one of the two precursors, VOC emissions have to be reduced to lower the ozone formations. Unfortunately, current VOC emission inventories are insufficient to explain the observed VOC concentrations for the areas with significant ozone issues [55], with an underestimation up to an order of magnitude. In order to assess the effects of ozone control strategies through photochemical modeling, reliable VOC emission information needs to be obtained first. Thus, enhanced VOC monitoring is indispensable to address the ozone challenges with novel technical progresses in the areas such as gas chromatography, mass spectrometry, IR spectroscopy.

In many heavy industrialized areas with concentrated VOC point sources, the VOC concentrations vary fast both spatially and temporally. Dynamic or fast VOC monitoring techniques are necessary to capture the emission from very local sources. With high temporal meteorological measurement and detailed source profiles, very local point sources could be identified or separated. These techniques include dynamically operated GC, micro-fast GCs, PTR-MS and FTIR. Further field experiment and assessment is needed for using these techniques.

Conventional monitoring techniques only measure a point of samples and cannot provide spatial distribution of the pollutant plumes. Open path FTIRs and IR imaging techniques provide the spatial information, and are especially ideal for detecting unexpected industrial sources. IR cameras can scan a large area and has been tested in petrochemical industries in order to be used in routine leak detections. Automatic and continuous operation of IR cameras with certain pattern recognition method could greatly reduce the time interval to detect and fix leaks in an industrial environment. Infrared LIDAR techniques have also started to contribute to the methane remote sensing field. In the future, its development into multi-wavelengths for other VOC species will be able to contribute to ozone precursor monitoring purposes.

Comprehensive two dimensional GC techniques have a great potential to address ozone issues. With an additional orthogonal column, the column capacities are increased by an order of magnitude and the sensitivity also is increased several folds. Current field tests have shown that more than half of the peaks in a two dimensional GC chromatograph cannot be identified, which implies more unknown VOC species, sources or chemical processes. As the hardware has become mature, to develop effective methods for data analysis for two dimensional GCs for VOC monitoring will be of more importance, and finally a standard method is needed to widely use this technique. During this process, chemometrics will be anticipated to have important contributions.

References:

[1] Liang J., B. Jackson, A. Kaduwela, *Atmos. Environ.* 2006, 40, 5156–5166.

[2] Partyka M., Zabiegala B., Namiesnik J., Przyjazny A., *Critical Rev. in Anal. Chem.* 2007, 37, 51-78.

[3] Helmig D, *J. Chromatogr.* 1999, 843, 129–146.

[4] Dewulf J., van Langenhove H., *J. Chromatogr.* 1999, 843, 163–177.

[5] Nam J., Kimura Y., Vizuete W., Murphy C., Allen D.T., *Atmos. Environ.* 2006, 40, 5329-5341.

[6] Zhou L., Zeng Y., Hazlett P.D., Matherne V., *Anal. Chem. Act.* 2007, 596, 156–163

[7] Oubre M. HRVOC AOC Final Report, Louisiana Department of Environmental Quality, April 2007.

[8] Badjagbo K., Moore S., Sauve S., *Trends in Anal. Chem.*, 2007, 26, 931-940.

[9] Lindinger., Hansel A. and Jordan A., *Inte.. J. Mass Spec. Ion. Process.*, 173, 191-241.

[10] Warneke C., de Gouw J.A., Kuster W.C.., Goldan P.D., and Fall R., *Envrion Sci. Tech.*, 2003, 37, 2494-2501.

[11] The 3rd International Conference on Proton Transfer Reaction Mass Spectrometry and Its Applications, Innsbruck Austria, 2007. Edited by: Hansel A., Märk T. D..

[12] Hewitt C. N., Hayward S. and Tani A., J. Environ. Monit, 2003, 5, 1-7.

[13] Zhang R. and Fortner E., Aircraft Measurements of Highly Reactive Volatile Organic Compounds Using Proton Transfer Reaction Mass Spectrometry (PTRMS) During TexAQS II, 2005

[14] Zhao W., Hopke P.K., Karl T., *Environ. Sci. Tech.* 2004, 38, 1338-1347.

[15] Warneke C., Karl T., Judmaier H., Hansel A., Jordan A., Lindinger W. and Crutzen P.J., *Global. Biogeochem. Cycles,* 1999, 13, 9.

[16] De Gouw J. A., Howard C. J., Custer T. G., Baker B. M. and Fall R., *Environ. Sci. Technol.*, 2000, 34, 2640.

[17] Karl T., Fall R., Jordan A., and Lindinger W., Environ. Sci. Technol., 2001, 35, 2926.

[18] Karl T., Guenther A., Jordan A., Fall R. and Lindinger W., *Atmos. Environ.*, 2001, 35, 491.

[19] Tani A., Hayward S., Hewitt C. N. *Inte. J. Mass. Spec.*, 2002, 223-224, 561-578.

[20] Bacsik Z., Mink J. Keresztury G., *App. Spectr. Rev.*, 2004, 39, 295 – 363.

[21] Bacsik Z., Mink J. Keresztury G., *App. Spectr. Rev.*, 2005, 40, 327 – 390.

[22] Hong DW., Heo GS., Han JS., Cho SY., *Atmos. Environ.* 2004, 38, 5567-5576.

[23] Hu L., Li Y., Zhang L. , Zhang LM., Wang JD., *Spectr. Spec. Anal.*, 2006, 26, 1863-1866.

[24] EPA protocol OTM-10 Radial Plume Mapping, 2006.

[25] Harig R. and Matz G., *Field Anal. Chem. Tech.* 2001, 5, 75-90.

[26] Harig, R.; Matz, G.; Rusch, P., Scanning Infrared Remote Sensing System for Identification, Visualization, and Quantification of Airborne Pollutants, Instrumentation for Air Pollution and Global Atmospheric Monitoring, Proceedings of SPIE Vol. 4574, p83-94, James O. Jensen; Robert L. Spellicy; Eds.

[27] Overton EB. , Carney KR., Roques N, Dharmasena HP. *Field Anal. Chem. & Tech.* 2001, 5, 97 – 105.

[28] Eiceman, GC., Gardea-Torresdey J., Overton E., Carney K., Dorman F., *Anal. Chem.* 2004, 76, 3387-3394.

[29] Politzer A.R. An examination of the relationship between environmental science and law due to emerging micro-scale gas chromatography technology, thesis, The Department of Environmental Studies, Louisiana State University, 2004

[30] ENVIRON, 2004, "Development of Emissions Factors and/or Correlation Equations for Gas Leak Detection, and the Development of an EPA Protocol for the Use of a Gas-imaging Device as an Alternative or Supplement to Current Leak Detection and Evaluation Methods", Final Report to Texas Council on Environmental Technology and the Texas Commission on Environmental Quality, October, 2004.

[31] Epperson D., Lev-On M., Taback H., Siegell J., Ritter K., Equivalent, *J. Air. Waste. Manag. Assos.*,2007, 9, 1050-1060.

[32] Zhou L., Zeng Y., Automatic *Anal. Chim. Act.,* 2007, 584, 223-227.

[33] Curl R. F. and Tittel F. K., *Annu. Rep. Prog. Chem., Sect. C*, 2002, 98, 219–272.

[34] EPA, Airborne Leak Detection Hits National News, *Natural Gas STAR Partner Update*, Spring 2006,1.

[35] Adahchour M., Beens J., Vreuls R.J.J., Brinkman U.A.Th., *Trends. in Anal. Chem.*, 2006, 25, 438-454.

[36] Adahchour M., Beens J., Vreuls R.J.J., Brinkman U.A.Th., *Trends. in Anal. Chem.*, 2006, 25, 540-553.

[37] Adahchour M., Beens J., Vreuls R.J.J., Brinkman U.A.Th., *Trends. in Anal. Chem.*, 2006, 25, 726-741.

[38] Adahchour M., Beens J., Vreuls R.J.J., Brinkman U.A.Th., *Trends. in Anal. Chem.*, 2006, 25, 821-840.

[39] Giddings J.C., *Anal. Chem.* 1984, 56, 1258A-1270A.

[40] Ryan D., Morrison P., Marriott P., *J. Chromatography A*, 2005, 1071,47-53.

[41] Ledford E.B. et al. Zoex Corporation Technical Note KT030606-1: What is Loop Modulation, 2003.

[42] Seeley J., Kramp F.J., Sharpe K.S., Seeley S.Y., *J. Sep. Sci.*, 2002, 25, 53-59.

[43] Mohler R.E., Prazen B.J., Synovec R.E., *Anal. Chem. Act.* 2006, 1, 68-74.

[44] Sinha A.E., Johnson K.J., Prazen B.J., Lucas S.V., Fraga C.G., Synovec R.E., *J. Chromat.* 2002, 1-2, 195-204.

[45] Seeley, J.V., Kramp F., Hicks C.J., *Anal. Chem.* 2000, 72, 4346-5352.

[46] Agilent, 2008, http://www.chem.agilent.com/scripts/pds.asp?lpage=62367.

[47] Gan F., Hopke P.K., *Anal. Chim. Acta.* 2003, 490: 153-158.

[48] Lewis, A. C.; et al. *Nature*, 2000, 405, 778–781.

[49] Xu, X.; et al. *Atmos. Chem. Phys.* 2003, 3, 665–682.

[50] Hamilton, J. F.; Lewis, A.C. *Atmos. Environ.* 2003, 37, 589–602.

[51] Hamilton J. F., Lewis A. C., Bloss C., Wagner V., Henderson A. P., Golding B. T., Wirtz K., Martin-Reviejo M., and Pilling M. J., *Atmos. Chem. Phys.*,2003, 3, 1999–2014.

[52] Bartenbach S., Williams J., Plass-Dülmer C., Berresheim H., and Lelieveld J., *Atmos. Chem. Phy.*, 2007, 7, 1–14.

[53] Sinha A.E, Fraga C.G., Prazen B.J., Synovec R.E., , *J. Chromat.* 2004, 1027, 269-277.

[54] Vogt L., Gröger T., Ziemmermann R., *J. Chromat.* 2007, 1150, 2-12.

[55] Cowling EB., Furiness C., Dimitriades, Parrish D. Final Rapid Science Synthesis Report: Findinigs from the Second Texas Air Quality Study (TexAQSII), 2007, available at: http://www.tceq.state.tx.us/assets/public/implementation/air/texaqs/doc/rsst_final_report.pdf.

In: Ozone Depletion, Chemistry and Impacts
Editor: Sem H. Bakker, pp. 17-51

ISBN: 978-1-60692-007-7

Chapter 2

OZONE DECOMPOSITION BY CATALYSTS AND ITS APPLICATION IN WATER TREATMENT: AN OVERVIEW

J. Rivera-Utrilla, M. Sánchez-Polo and J.D. Méndez-Díaz
Inorganic Chemistry Department, Faculty of Science
University of Granada, 18071 Granada, Spain

Abstract

Ozone has recently received much attention in water treatment technology for its high oxidation and disinfection potential. The use of ozone brings several benefits but has a few disadvantages that limit its application in water treatment, including: i) low solubility and stability in water, ii) low reactivity with some organic compounds and iii) failure to produce a complete transformation of organic compounds into CO_2, generating degradation by-products that sometimes have higher toxicity than the raw micropollutant. To improve the effectiveness of ozonation process efficiency, advanced oxidation processes (AOPs) have recently been developed (O_3/H_2O_2, O_3/UV, O_3/catalysts). AOPs are based on ozone decomposition into hydroxyl radicals ($HO^{\cdot}$), which are high powerful oxidants. This chapter offers an overview of AOPs, focusing on the role of solid catalysts in enhancing ozone transformation into $HO^{\cdot}$ radicals. Catalytic ozonation is a new way to remove organic micropollutants from drinking water and wastewater. The application of several homo- and heterogeneous ozonation catalysts is reviewed, describing their activity and identifying the parameters that influence the effectiveness of catalytic systems. Although catalytic ozonation has largely been limited to laboratory applications, the good results obtained have led to investigations now under way by researchers worldwide. It is therefore timely to provide a summary of achievements to date in the use of solid materials to enhance ozone transformation into $HO^{\cdot}$ radicals.

1. Introduction

Demographic development and an exponential increase in industrial activity have largely been responsible for the large rise in the demand for water for domestic, public, and industrial use, and for the high volume of effluents discharged into waters. Legislation has been introduced to reduce permissible contamination levels. However, conventional treatment systems cannot completely remove a large amount of organic and inorganic contaminants present in waters, because most cannot be metabolized by microorganisms as carbon source and can even inhibit the activity of these microorganisms, leading to their bioaccumulation in the food chain. Hence, there is an increasing demand for more effective treatments to reduce the potential environmental impact of effluents and to comply with increasingly strict legislations. Successful water treatment requires the use of more sophisticated methods, including:

- *Biological systems for nitrogen removal*: Ammonium can be transformed into nitrate by using nitrifier microorganisms in aerobic medium. Nitrate can be removed in a subsequent stage under anaerobic conditions, when it is transformed by denitrifying bacteria into molecular nitrogen.
- *Advanced oxidation processes for removal of toxic organic compounds, chromophore compounds, and other non-biodegradable organic compounds.* These are based on the use of highly oxidizing agents, e.g., ozone or hydrogen peroxide. A greater effectiveness is observed when these oxidizing agents are used in the presence of UV radiation.
- *Ionic exchange for ion removal.* This is highly effective for removing cations and anions from the aqueous phase but transfers the problem to the solid phase by concentrating the pollutant in the adsorbent medium.
- *Adsorption on activated carbon for removal of metals, organic compounds, etc.* This has the same drawback as described above for ionic exchange.
- *Chemical precipitation for phosphorus removal:* Chemical agents ($Al_2(SO_4)_3$, $Fe_2(SO_4)_3$ or $FeCl_3$) are used to precipitate phosphorus.
- *Distillation for removal of volatile organic compounds.* This is only appropriate when there are high concentrations of the contaminant and its recovery brings economic benefits.
- *Liquid-liquid extraction.* This is also only useful under the above conditions.

Chemical oxidation processes currently play a very important role in water treatment. Table 1 lists the most common chemical oxidants used in water treatments and their corresponding reduction potentials[1]. Oxidants can be used to remove both inorganic pollutants and toxic organic compounds (pesticides, hydrocarbons, toxins, etc.)[2,3]. In addition, oxidants are widely used to degrade compounds responsible for odor, color or taste[4-6.

Ozone, due to its high oxidation and disinfection potential, has recently received much attention in water treatment technology. Despite several advantages of using ozone, it has a few disadvantages that limit its application in water treatment, including: i) low solubility and stability in water, ii) low reactivity with some organic compounds, and iii) failure to produce a complete transformation of organic compounds into CO_2, generating degradation by-products that sometimes have a higher toxicity than the raw micropollutant. To improve the effectiveness of ozonation, advanced oxidation processes (AOPs) have recently been developed (O_3/H_2O_2, O_3/UV, O_3/catalysts). AOPs are based on the decomposition of ozone into hydroxyl radicals ($HO^{\cdot}$), a very powerful oxidant.

This chapter offers an overview of the role of different solid catalysts in enhancing ozone transformation into $HO^{\cdot}$ radicals. Catalytic ozonation is a new way to remove organic micropollutants from drinking water and wastewater. The application of several homo- and heterogeneous ozonation catalysts is reviewed, describing their activity and identifying the parameters that influence the effectiveness of catalytic systems. Although catalytic ozonation has largely been limited to laboratory applications, the good results obtained have led to investigations now under way by researchers worldwide. It is therefore timely to provide a summary of achievements to date in the use of solid materials to enhance ozone transformation into $HO^{\cdot}$ radicals.

Table 1. Reduction potentials of different oxidants used in water treatments.

Oxidant	Reduction semireaction	E^0 (V)
Ozone	$O_3 (aq) + 2H^+ + 2e^- \rightarrow O_2 (aq) + H_2O$	2.08
Permanganate	$2MnO_4^- + 8H^+ + 6e^- \rightarrow 2MnO_2 (s) + 4H_2O$	1.68
Chlorine dioxide	$ClO_2 + e^- \rightarrow ClO_2^-$	0.95
Hypochlorous acid	$HOCl + H^+ + 2e^- \rightarrow Cl^- + H_2O$	1.48
Hypochlorite ion	$ClO^- + 2H^+ + 2e^- \rightarrow Cl^- + H_2O$	1.64
Dichloramine	$NHCl_2 + 3 H^+ + 4e^- \rightarrow 2Cl^- + NH_4^+$	1.34
Oxygen	$O_2 (aq) + 4H^+ + 4e^- \rightarrow 2H_2O$	1.23
Hydroxyl radical	$HO^{\cdot} + H^+ + e^- \rightarrow H_2O$	2.85
Hydrogen peroxide	$H_2O_2 + 2H^+ + 2e^- \rightarrow 2H_2O$	1.78

2. Ozone

Ozone, O_3, discovered by Schönbein in 1840[7], is an allotrope of oxygen consisting of three atoms. It is a diamagnetic compound that is an unstable gas at room temperature with a characteristic sharp odor. Experimental results show a bond angle of 116.8 ± 0.5° and an interatomic distance of 127.8 pm between central oxygen and each terminal[**8**]. Fig 1 depicts the two resonant forms of the ozone molecule according to the valence bond theory:

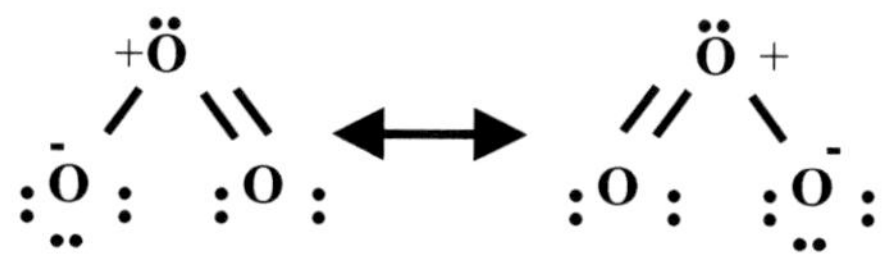

Figure 1. Chemical structure of ozone.

Due to its configuration, ozone is a highly oxidant compound ($E^\circ = +2.08$ V), and its natural tendency is to transfer an oxygen atom and release O_2. Applications of ozone applications in water treatment can be grouped into three categories: i) as disinfectant or biocide, ii) as oxidant for the removal of organic pollutants and, iii) as pre- or post-treatment in another procedure (coagulation, flocculation, sedimentation, biological oxidation, adsorption on activated carbon, etc.).

Ozone began to be used as bactericide agent to treat drinking waters in Nice (France) at the beginning of the 20th century. As a result of the large amount of resources invested in the study of ozone, other advantages of its use in treatment plants[9] are now widely known, including: i) removal of compounds that produce odor, taste or color in water, ii) oxidation of inorganic chemical compounds, e.g., iron and manganese, iii) removal of algae and other aquatic microorganisms, iv) oxidation of organic micropollutants, v) absence of the increase in the presence of organochlorinated compounds found when chlorine is used for the treatment, and vi) enhanced performance of adsorption and coagulation processes. Nevertheless, its large-scale application to treat industrial liquid waste did not become widespread due to the high economic costs involved and the chemical complexity of industrial effluents. However, ozone has now become an attractive option for the treatment of these effluents because conventional systems are inadequate to reduce the toxic organic compounds they contain to levels required by new environmental legislation.

Start

$$O_3 + OH^- \longrightarrow O_2^{\cdot-} + HO_2^{\cdot} \quad (1)$$

$$O_3 + H_2O \longrightarrow 2HO_2^{\cdot} \quad (2)$$

Propagation

$$HO_2^{\cdot} \longrightarrow H^+ + O_2^{\cdot-} \quad (3)$$

$$O_2^{\cdot-} + O_3 \longrightarrow O_2 + O_3^{\cdot-} \quad (4)$$

$$O_3^{\cdot-} + H^+ \longrightarrow HO^{\cdot} + O_2 \quad (5)$$

$$O_3^{\cdot-} \longrightarrow O^{\cdot-} + O_2 \quad (6)$$

$$HO^{\cdot} \longrightarrow O^{\cdot-} + H^+ \quad (7)$$

$$HO^{\cdot} + O_3 \longrightarrow HO_2^{\cdot} + O_2 \quad (8)$$

End

$$HO_2^{\cdot} + HO_2^{\cdot} \longrightarrow H_2O_2 + O_2 \quad (9)$$

$$HO_2^{\cdot} + O_2^{\cdot-} \longrightarrow HO_2^{-} + O_2 \quad (10)$$

$$O_3^{\cdot-} + O_2^{\cdot-} + H_2O \longrightarrow 2HO^{-} + 2O_2 \quad (11)$$

Figure 2. Mechanism of ozone decomposition in aqueous medium.

Over the past two decades, there has been a notable increase in research into the reaction between ozone and numerous organic and inorganic compounds, especially aromatic compounds[10-14. Because it is highly reactive, ozone can interact with a large number of organic and inorganic substances by direct oxidation/reduction reaction, cycloaddition-substitution, or nucleophilic addition[15-17. The direct reaction between ozone and a given compound is not the only pathway by which ozone can act to degrade pollutants, since ozone is very unstable in aqueous solution and spontaneously decomposes by a complex chain mechanism (Fig 2) in which different radical species participate[18]. The radicals generated are of great interest for water treatment because some of them, e.g., the hydroxyl radical $HO^{\cdot}$, are even more reactive than ozone and play an essential role in removing pollutants in solution[19]. These systems will be described in greater depth in next section, which is devoted to AOPs.

Despite its high efficacy in some systems, ozone has inadequate capacity to degrade surfactants and tensioactive substances, among other pollutants. Narkis et al.[20] observed that ozone treatment favored the biodegradation of non-ionic surfactants but was unable to remove them. Other authors also reported the lack of reactivity of saturated cationic surfactants[21]. Regarding anionic surfactants, although some researchers achieved a high degradation of alkylbenzene-sulphonates with ozone[22], the reaction rate constant values were subsequently determined to be low[23], suggesting that an indirect mechanism was largely responsible for the degradation.

3. Advanced Oxidation Processes Based on Ozone

Polluted waters can generally be effectively treated by biological, adsorbent or conventional chemical approaches (chlorination, ozonation or oxidation with permanganate). However, as stated earlier, these methods are sometimes inadequate to degrade pollutants to levels required by law or necessary for subsequent utilization of the effluent. AOPs have proven highly effective in the oxidation of numerous organic and inorganic compounds and are based on the generation of free radicals, notably hydroxyl radical $HO^{\cdot}$. These free radicals are highly reactive species that can successfully attack most organic and inorganic compounds with very high reaction rate constants (10^6-10^9 $M^{-1}s^{-1}$). The numerous systems that can be produced by these radicals (Table 2) account for the high versatility of AOPs.

Table 2. Water treatment technologies based on advanced oxidation processes.

Non-photochemical processes	Photochemical processes
• Oxidation in sub/supercritical water	• Photolysis of water with vacuum UV (VUV)
• Fenton's reagent (Fe^{2+}/H_2O_2)	• UV/hydrogen peroxide
• Electrochemical oxidation	• UV/ozone
• Radiolysis	• Photo-Fenton
• Non-thermal plasma	• Heterogeneous photocatalysis
• Ultrasound	•
• Ozonation in alkaline medium (O_3/OH^-)	•
• Ozonation in the presence of hydrogen peroxide (O_3/H_2O_2)	•
• Catalytic ozonation	•

Advanced oxidation processes based on the use of ozone are briefly described below, highlighting catalytic ozonation (also see section 4).

3.1. Ozonation in Alkaline Medium

Based on studies by Hoigné et al.[10-12], Aieta et al.[24] published an illustrative diagram in 1988 describing ozone in aqueous solution. Figure 3 summarizes the two reaction pathways of molecular ozone: i) direct reaction of the substrate with molecular ozone, which is selective but slow or null with some species, and ii) decomposition and generation of $HO^{\cdot}$ hydroxyl radicals, which attack rapidly but not selectively.

Error!

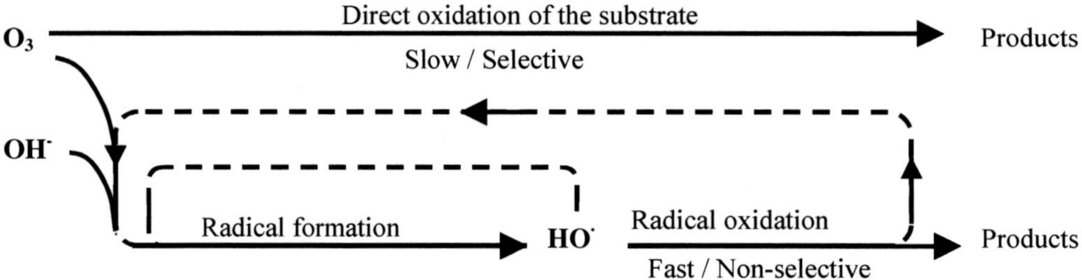

Figure 3. Reaction pathways of ozone.

Moreover, hydroxyl radicals react rapidly with molecular ozone, contributing towards autocatalytic decomposition of the ozone. Several researchers have studied the mechanism and reaction kinetics involved in ozone decomposition in aqueous phase[18,25-33. Ozone stability largely depends on the aqueous matrix, especially the pH, type of organic matter present, and alkalinity[34]. The water pH is especially important, because hydroxyl ions considerably increase the ozone decomposition rate[18]. Thus, according to equations 12 and 13, ozone decomposition spontaneously accelerates with an increase in the solution pH, leading to an AOP. However, it must be taken into account that a high pH increase can have a negative effect on the degradation, depending on the composition of the water, because of the inhibiting action of $HO^{\cdot}$ radical scavengers, e.g., bicarbonate or carbonate ions[35].

$$O_3 + OH^- \rightarrow HO_2^- + O_2 \quad k = 70\ M^{-1}s^{-1} \tag{12}$$

$$O_3 + HO_2^- \rightarrow HO^{\cdot} + O_2^{\cdot -} + O_2 \quad k = 2.8 \cdot 10^6\ M^{-1}\ s^{-1} \tag{13}$$

3.2. Ozonation in the Presence of Hydrogen Peroxide

The addition of hydrogen peroxide favors the ozonation of organic compounds in the medium. The combination of ozone and hydrogen peroxide is largely used to oxidize pollutants that require high ozone consumption. Hydrogen peroxide is a weak acid (pKa =

11.6), a powerful oxidant (See Table 1), and an unstable compound that easily dismutes[8] (equations 14-16).

$$H_2O_2 \rightarrow HO_2^- + H^+ \quad (14)$$

$$H_2O_2 + 2e^- + 2H^+ \rightarrow 2\,H_2O \quad (15)$$

$$H_2O_2 + HO_2^- \rightarrow H_2O + O_2 + HO^- \quad (16)$$

The mechanism by which H_2O_2 favors free radical generation was described by Forni et al.[25], who demonstrated that the conjugated base of hydrogen peroxide initiates ozone decomposition in aqueous phase *via* an electronic transference reaction.

$$HO_2^- + O_3 \rightarrow HO_2 + O_3^- \quad (17)$$

Taking advantage of the capacity of H_2O_2 to initiate ozone decomposition in aqueous phase (Figure 4), numerous researchers have used this process for a faster and more effective oxidation of organic matter[36-43. The presence of H_2O_2 in the system favors oxidation, although the possibility that added H_2O_2 is consumed in reactions with other contaminants has hampered application of this method to treat industrial effluents. Fernández et al.[44] compared the efficacy of the O_3/H_2O_2 system and photolysis to remove linear alkylbenzene-sulphonates (LAS), observing the complete degradation of the mixture of surfactants after 20 min of ozonation.

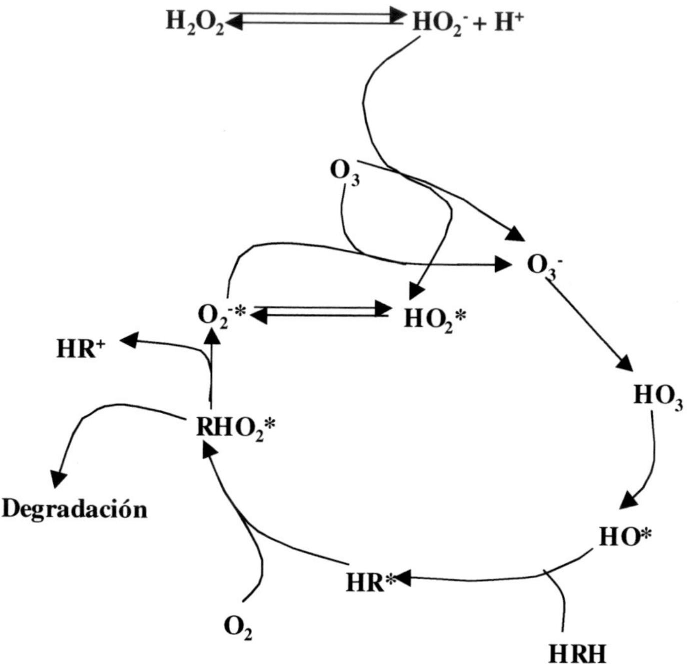

Figure 4. Mechanism of oxidation of an organic compound (HRH) by means of O_3/H_2O_2.

Table 3. Application of systems based on O_3, UV radiation, and catalysts for removing pollutants from waters.

System	Catalyst	Pollutants	Experimental conditions	Observations	Ref.
O_3, H_2O_2, UV, O_3/UV, O_3/H_2O_2, UV/H_2O_2, O_3/UV/H_2O_2		Phenolic compounds	Semi-continuous; pH 8; 17°C; flow 1-1.5 L/min; low-pressure lamps (254nm)	Treatments without ozone are less effective in pollutant removal	[45]
UV, UV/TiO_2, UV/Fenton	TiO_2 (Degussa P25)	Different wastewaters	Semi-continuous; 25°C; medium-pressure Hg lamp 400 W; 2 g/L of catalyst;	Better performance of catalyzed experiments; Fenton process more cost-effective than UV	[46]
O_2/UV/cat	Ag/ZnO	Textile effluent	Semi-continuous; pH 3-7; medium-pressure Hg lamp 125 W; flow 10 mL/min	Total removal of toxicity of certain effluents, but little reduction in TOC	[47]
O_2/UV/cat	TiO_2 (Degussa P25), ZnO	Colorants	Semi-continuous; pH 5; 20°C; medium-pressure Hg lamp 125 W; gas flow10 mL/min; 0.8 g/L catalyst	High mineralization; better catalytic activity of ZnO; synergic effect between TiO_2 and ZnO	[48]
UV/TiO_2	TiO_2	Formic acid	Discontinuous with recirculation; pH 3.8; 20°C; Lamp 6W	Better performance in combined use; determination of reaction constants	[49]
UV/H_2O_2,		2-methyl benzimidazole carbamate	Discontinuous; low-pressure Hg lamp	Determination of radical reaction rate constants (hydroxyl and carbonate); identification of degradation products	[50]
O_3/UV		Oxalic acid	Semi-continuous; pH 7; 1.5 L/min; UV lamps 3-9·10-6 E/L-s	Kinetic model proposed; study of effect of operational variables	[51]
O_2/UV/TiO_2	TiO_2 (Degussa P25)	Tetracyclines	Semi-continuous; flow 0.2L/min; 3 lamps of different power; 0.5-1 g/L catalyst	Degradation is not observed in the absence of TiO_2; significant mineralization with use of catalyst	[52]
O_3, H_2O_2, UV, O_3/UV, O_3/H_2O_2, UV/H_2O_2,		Diazinon	Discontinuous; pH 2-9; 20-40°C; low-pressure Hg lamp 15W	Kinetic study and comparison between systems; determination of quantum performances and reaction constants	[53]
UV, O_3/UV, O_3/TiO_2, O_3/UV/TiO_2	TiO_2	Acetic acid, monochloroacetic acid, phenol, dimethyl-2,2,2-trichlorine-1- hydroxyethyl-phosphate	Semi-continuous; low-pressure Hg lamp 6 W	Proposal of the reaction mechanism; detection of formic, acetic, glyoxylic, and glycolic acids as reaction intermediates	[54]

Table 3. Continued

System	Catalyst	Pollutants	Experimental conditions	Observations	Ref.
O_3, UV/TiO_2, $O_3/UV/TiO_2$	TiO_2	2,4-dichlorine phenoxyacetic acid	Semi-continuous; pH 3-11; 18-22ºC; flow 1-220 mg O_3/L; lamp 20W	Study of kinetics and operational variables; O_3/UV improves treatment efficiency; identification of reaction intermediates	[55]
$O_3/UV/TiO_2$	TiO_2	Aniline	Semi-continuous; pH 3; 25ºC; flow 5·10_{-4} mol O_3/min; medium-pressure Hg lamp 125 W; 2 g/L of catalyst	Proposal of reaction mechanism; simultaneous use of $O_3/UV/TiO_2$ is more effective versus O_3 and UV/TiO_2 in parallel	[56]
O_3, O_3/UV, UV/TiO_2, $O_3/UV/TiO_2$	TiO_2 (Degussa P25)	Alkylamines, alkanolamines, nitrogenated heterocyclic and aromatic compounds	Semi-continuous; pH 6.5; 20ºC; flow 35 L/h; Xe Lamp 450 W; 0-3 g/L of catalyst	Proposal of reaction mechanism; identification of intermediate reaction products; study of effect of structure and concentrations on reaction rate	[57]
UV/TiO_2, O_3/UV, $O_3/UV/TiO_2$	TiO_2 (Degussa P25)	Pyridine; monochloroacetic acid	Semi-continuous; pH 3; 20ºC; flow 20 L/h; UV Lamp; 0.5g/L catalyst	Proposal of reaction mechanism; much higher performance of system based on $O_3/UV/TiO_2$	[58]
O_3, UV/TiO_2, $O_3/UV/TiO_2$	TiO_2 (Degussa P25)	Textile effluent	Semi-continuous; pH 11; flow 6 L/h; high-pressure Hg lamp 125 W; 0.1 g/L catalyst	Determination of toxicity and TOC and color of effluent as a function of treatment time	[59]
$O_3/UV/TiO_2$	TiO_2	4- chlorobenzaldehyde	Semi-continuous; pH 6; flow 1 L/min; low-pressure Hg lamp 10 W; 1 mg/mL of catalyst	Higher efficiency of the combined system; detection of degradation by-products	[60]
UV/TiO_2, $O_3/UV/TiO_2$, UV/Fenton	TiO_2(Degussa P25)	2-dimethyl-pyrazine, monochloroacetic acid	Semi-continuous; high-pressure Hg lamp 125W	Proposal of disjunctive mechanism: UV promotes electrons in TiO_2 or radical reaction generation	[61]
$O_3/UV/TiO_2$	TiO_2 (Degussa P25 y BDH)	Cyanides	Semi-continuous; pH 11.3; 20ºC; medium-pressure Hg lamp 400 W	Kinetic study and mechanism; formation of carbonate, cyanate, and nitrate; strong O_3/TiO_2 interaction	[62]
O_3, O_3/H_2O_2, O_3/H_2O_2/UV $O_3/UV/TiO_2$, $O_3/H_2O_2/UV/TiO_2$	TiO_2	Wastewater	Semi-continuous; pH 4.4; flow 500 L/h; Hg Lamp (200-280 nm); 2 g/L of catalyst	$O_3/H_2O_2/UV/TiO_2$ is the only one able to appreciably reduce TOC; after treatment, anaerobic digestion improves CH_4 production	[63]

Table 3. Continued

System	Catalyst	Pollutants	Experimental conditions	Observations	Ref.
UV/TiO_2, O_3, $O_3/UV/TiO_2$	TiO_2	Formic acid	Semi-continuous; pH 2-12; 10-50ºC; gas flow $5 \cdot 10^{-3}$ m^3/min; 6W Lamp	Diffusion effects worsen performance	[64]
$O_3/UV/TiO_2$	TiO_2	Cyanides	Continuous; pH 10; gas flow 4 L/min; low-pressure lamp 40 W	Mechanisms is proposed and products identified; effluent is reutilized to obtain deionized water	[65]
$O_3/UV/TiO_2$	TiO_2 (Degussa P25 , SG)	Phenol	Semi-continuous; 25ºC; flow 2.1 mg O_3/L; Hg Lamp (220-380nm); 15 g/L of catalyst	Detection of intermediates; improvement of mineralization with use of combined system	[66]
O_2, O_3, O_3/UV, UV/TiO_2, $O_3/UV/TiO_2$	Carbon-TiO_2	Catechol	Semi-continuous; low-pressure lamp 15 W	Determination of rate constant and reaction order; higher efficacy of $O3/UV/TiO_2$ system to remove TOC	[67]
$O_3/UV/TiO_2$	TiO_2 (Degussa P25)	Humic acid	Semi-continuous; flow 4.8 mg O_3/min; 125 W Lamp; 0.25-1 g/L of catalyst	Kinetic study; UV compensates for low capacity of O_3 to remove organic matter; adsorption study of pollutant on catalyst	[68,69, 70]
O_3, O_3/UV, UV/TiO_2, $O_2/UV/TiO_2$, $O_3/UV/TiO_2$	TiO_2 (Degussa P25)	Acetic acid, formic acid, propionic acid	Semi-continuous; pH 2; 25ºC; flow 0.9 L/min; medium-pressure Hg lamp 6 W; 1 g/L catalyst	Identifies intermediates; better performance of combined process versus ozone and photocatalysis separately	[71]
O_3/Foto fenton, $O_3/UV/TiO_2$	TiO_2 (Degussa P25)	Alachlor, atrazine, diuron, isoproturon, pentachlorophenol	Semi-continuous; pH 3, 7; 25ºC; flow 1.6 g O_3/h; UVA Lamp 6W; 0.25 g/L of catalyst	Photo-fenton system (order 1) shows better results than $O_3/UV/TiO_2$ (order 0)	[72]
O_3/UV, O_3/TiO_2, O_3/V-O/TiO_2	TiO_2, V-O/TiO_2	Sulphosalicylic acid	Semi-continuous; pH 4, 7, 9; 20ºC; Lamp (254 nm) 14 W; 1.6 g/L of catalyst	Higher removal efficacy of O_3/V-O/TiO_2 system; maintains activity even in the presence of carbonates	[73]
O3, $O3/TiO_2$, $O_3/UV/TiO_2$, O_3/UV, UV, UV/TiO_2	TiO_2 (Degussa P25)	Phenol, p-chlorophenol, p-nitrophenol	Semi-continuous; pH 5-6; flow 51 L/h; high-pressure Hg lamp 700 W; 1.5 g/L catalyst	Kinetic study and identification of intermediates; determination of reaction constants	[74]
O_3, UV/TiO_2, $O_3/UV/TiO_2$	TiO_2	Water with fungicide	Semi-continuous; UV Lamp 6W	Synergic effect; $O_3/UV/TiO_2$ system removes organic compounds and inhibits germination of fungi	[75]

3.3. Ozonation in the Presence of UV Radiation

Ozonation coupled with UV radiation is one of the most effective chemical oxidation techniques to treat polluted waters. This process is capable of oxidizing organic substances at room temperature and generates products that are innocuous to the environment. As in the O_3/H_2O_2 system, the UV radiation of O_3 generates hydroxyl radicals in solution[76]. The reactions involved in this process are:

$$O_3 + h\nu\ (\lambda < 310\ nm) \longrightarrow O_2 + O \quad (18)$$

$$O + H_2O \longrightarrow 2HO^{\cdot} \quad (19)$$

Prengle et al.[77] of "Houston Research Inc." were the first researchers to describe the application of an O_3/UV-based system for water treatment. Glaze et al.[78,79] subsequently analyzed the effectiveness of the O_3/UV system to remove microcontaminants and determined the reactions involved in the mechanisms of this process[80]. They demonstrated that compounds not degraded by UV light behave very similarly in response to O_3/UV and O_3/H_2O_2 systems[81].

The O_3/UV process has been mainly applied in the oxidation of aromatic derivatives[82,83] but it has also shown high efficacy to remove surfactants[84-86, pesticides[87], and other organic compounds[88]. Taking advantage of the high photocatalytic capacity of TiO_2, a new treatment process has been developed based on the simultaneous use of O_3/UV/TiO_2. Table 3 displays information on the type of pollutant, experimental conditions and the main conclusions of publications on the application of the O_3/UV system to remove organic pollutants from waters; in the majority of these studies, TiO_2 was used as catalyst.

4. Catalytic Ozonation

Research into novel alternatives to conventional advanced oxidation processes has significantly increased over the past few decades. These new treatments are based on the addition of reagents to the system, generally heavy metals that increase the effectiveness of ozone as oxidizing agent. These substances, called catalysts, participate in the reaction mechanism but are not consumed in the process. Depending on the state of the species that acts as catalyst, two types can be distinguished: homogeneous catalysis (catalyst dissolved in aqueous phase) and heterogeneous catalysis (solid or supported catalyst). Each of these catalytic processes is studied in depth below.

4.1. Homogeneous Catalytic Ozonation

One of the first investigations in this field was by Hewes and Davison[89], who observed that addition of certain salts during ozonation of phenolic compounds increased the mineralization of organic matter. Subsequently, other researchers[90,91] found that these catalysts could improve the decoloration of effluents from textile industries and accelerate the oxidation of compounds at acid pH values.

Table 4. Most relevant data from publications on homogeneous catalyzed ozonation.

System	Catalyst	Contaminants	Experimental Conditions	Observations	Ref.
Ozone/H_2O_2	Fe^{2+}, Co^{2+}, Ni^{2+}, Cu^{2+} Sulfates	*	Discontinuous; acid pH	Study of the catalytic effect of several cations; reaction mechanism is proposed	[92]
Ozone	Co^{2+}	*	Discontinuous; acid pH; 0ºC	Study of mechanism and kinetics; acid pH inhibits the process	[93]
Ozone	Co^{2+}	Acetic acid	Discontinuous; acid pH; 0ºC	Study of mechanism and kinetics	[94]
Ozone	Co^{2+}, Ti^{2+}, Mn^{2+}	Secondary effluent	Discontinuous and recirculation; pH 5-10, 30-60ºC	Improvement of DQO removal	[89]
Ozone	Fe^{2+}	*	Discontinuous; acid pH; [O_3] $3\cdot10^{-4}$M	Determination of reaction stoichiometry; mechanism proposal; influence of several radicals	[95]
Ozone	Cu^{2+}, Zn^{2+}, Cr_2O_3	Azoic colorants	Semi-continuous; flow 100 L/h	Improvement in discoloration rate	[90]
Ozone	Fe^{2+}, Mn^{2+}	*	Discontinuous, pH 5-8, 25ºC, 1mg/L O_3	Determination of reaction stoichiometry and rate constant	[96]
Ozone	NOM	Mn^{2+}	Discontinuous, 0.5-2 mg/L	Significant removal of Mn^{2+}; influence of O_3 dose	[97]
Ozone	Mn^{2+}	Oxalic Acid	Discontinuous and semi-continuous (flow 36 L/h); acid pH; 20ºC	Kinetic study and determination of rate constant; proposal of reaction mechanism	[91]
Ozone	Fe^{2+}	*	Reactor with flow stop; pH 0-2; 25ºC; $1\cdot10^{-4}$-$6\cdot10^{-5}$ M O_3	Kinetic study and determination of rate constant at the beginning of the process; mechanism is proposed	[98]
Ozone	NOM	Mn^{2+}, DQO	Continuous; diffusion plate	Efficacy is tested in different waters	[99]
Ozone	Mn^{2+}	*	Discontinuous; pH 2-8; 4-$5\cdot10^{-5}$ M O_3	Reaction mechanism is proposed	[100]
Ozone	Mn^{2+}	Pyrazine, oxalic acid	Semi-continuous; pH 4; flow 36 L/h	Oxalic- Mn complex as catalyst	[101]
Ozone	Ag^{+}, Fe^{2+}, Fe^{3+}, Co^{2+}, Cu^{2+}, Mn^{2+}, Zn^{2+}, Cd^{2+}	TOC	Discontinuous; pH 7; flow 1.5g O_3/h	Intermediates are identified; TOC reduction improves with catalyst	[102]

Table 4. Continued

System	Catalyst	Contaminants	Experimental Conditions	Observations	Ref.
Ozone	Mn^{2+}	Atrazine	Semi-continuous and recirculation; pH; 20ºC	Considerably increases atrazine removal with Mn^{2+}	[103]
Ozone	Mn^{2+}	TOC	Semi-continuous; pH 8; flow 1.6g/ O_3/h	TOC removal increases with the catalyst; intermediates are identified	[104]
Ozone	Mn^{2+}	Pyruvic acid	Semi-continuous; pH 2-4; 25ºC; flow 36 L/h	Only removes pyruvic acid if catalyzed; negative effect of pH	[105]
Ozone	Mn^{2+}	Atrazine, DQO	Semi-continuous and recirculation; pH 7; 20ºC; 1 mg/L Mn^{2+}	Humic acids improve atrazine removal; reaction mechanism is proposed	[106]
Ozone	Fe^{2+}, Fe^{3+}, Mn^{2+}	Chlorobenzenes	Semi-continuous; pH 7	Presence of the catalyst improves contaminant removal	[107]
Ozone	Mn^{2+}	Glyoxylic acid	Semi-continuous; pH 2-5.4; flow 36 L/h	Reaction intermediates are identified; mechanism is proposed	[108]
Ozone	Mn^{2+}	Pyruvic acid	Semi-continuous; pH 1-3; flow 36 L/h	Kinetic study; similar behavior of Mn^{2+} and Mn^{4+}	[109]
Ozone	Mn^{2+}	p-Nitrotoluene in acetic anhydride	Semi-continuous; acid pH; flow 10^{-2} L/s	Determination of the kinetic constant and reaction intermediates	[110]
Ozone/H_2O_2	Fe^{2+}	Acids and colorants	Semi-continuous; 4-10.5; flow 29mg O_3/min	Kinetic study	[111]
Ozone	Pb^{+}, Cu^{2+}, Zn^{2+}, Fe^{2+}, Mn^{2+} Nitrates	Ortho-chlorophenol	Semi-continuous; pH 3; flow mg O_3/min	Mechanism is proposed; order of metal catalyst activity is established	[112]
Ozone	Co^{2+}	Oxalic acid	Semi-continuous; acid pH; 20ºC; flow 24 L/h	Kinetic study and reaction stoichiometry determination	[113]
Ozone	Co^{2+}	*	Discontinuous; pH 1.6-8.4; 10^{-5}-$2{\cdot}10^{-4}$ M	Influence of pH, O_3 dose and catalyst; mechanism is proposed	[114]
Ozone	Fe^{3+}	Oxalic acid	Semi-continuous; pH 2.5; 20ºC, flow 24 L/h	Improvement in oxalic acid removal; reaction mechanism is proposed	[115]
Ozone	Mn^{2+}, Cu^{2+}	Phenol	Semi-continuous; pH 3 and 10	Phenol and TOC removal improved with catalyst	[116]

*Publications on ozone decomposition in the absence of pollutants

Unlike for other AOPs, a general reaction mechanism cannot be established for homogeneous catalytic ozonation because the system can behave very differently according to the compounds involved. Whereas some researchers indicated that the presence of transition metals does not promote the generation of free radicals[117], explaining that the higher oxidation of the substrate is due to the formation of a complex that is highly reactive to ozone, others explained the enhancement of substrate degradation by the generation of hydroxyl radicals[118]. A review of the literature shows that metals (either in cation or anion form) that are susceptible to oxidization in the ozonation process can act as initiators/promoters of the transformation of ozone into $HO^{\cdot}$ radicals. Despite the improvement in performance, the need to add metal species to the system increases the difficulties in applying homogeneous catalytic ozonation in liquid waste treatments, because it introduces highly contaminant species into the system. Table 4 summarizes data from the main published studies on the use of homogeneous catalytic ozonation to remove organic microcontaminants from waters.

4.2. Heterogeneous Catalytic Ozonation

Heterogeneous catalysis takes place in systems with two or more phases, most frequently with catalyst in solid phase and reagents in gas or liquid phase. Most publications in this field have centered on the study of reactions in gas phase, although increasing numbers of researchers have studied reactions in liquid phase in recent years. Since Chen et al.[119] published results on heterogeneous catalytic ozonation of some compounds using Fe_2O_3 as catalyst, numerous catalysts have been tested, including metal oxides[120,121], supported metal oxides[122,123], supported metals[124], mesoporous materials[125], and activated carbons[126]. In general, the catalytic activity of these catalysts is mainly based on the generation of radical species such as hydroxyl; therefore, the efficacy of heterogeneous catalysis in ozone decomposition will be largely influenced by solution characteristics (pH, temperature, ionic strength, etc.) and by chemical and textural properties of the catalyst. Among the numerous materials that have been used as heterogeneous catalysts, there are four major types:

1) Supported metals.
2) Metal oxides.
3) Supported metal oxides.
4) Activated carbons.

4.2.1. Supported Metals

After the first results published by Heinig et al.[127], the use of catalysts based on metals deposited on different supports began to be studied in the late 1990s. At present there is considerable uncertainty about the mechanisms involved and the influence of different operational variables on this process. Results obtained by Karpel Vel Leitner et al.[128] indicated that adsorption of dissolved humic matter on the surface of the catalyst reduces the catalytic activity of metal catalysts supported on alumina, whereas catalysts supported on TiO_2, where there is little such adsorption, show a better performance.

Table 5. Ozonation catalyzed by supported metals.

System	Catalyst	Contaminants	Experimental Conditions	Observations	Ref.
O_3, O_2	Ag/Al_2O_3	Microorganisms	Fixed bed; 500 g of catalyst	Bacteria and virus removal; strong catalytic interaction with dissolved O_2	[127]
O_3	Cu supported on alumina, anatase, and attapulgite, Ru/CeO_2	Organic compounds (TOC), succinic acid	Discontinuous; and semi-continuous (flow 1.25 g O_3/h); 0.2-1.2 g/L catalyst	Demonstrates the catalytic activity of supported metals; proposes mechanism by adsorption and/or hydroxyl radical attack	[129]
O_3	Metals supported on Al_2O_3, TiO_2 and clay	Salicylic, peptide acid, humic substances	Continuous (6.5 mg O_3/L) and semi-continuous; acid pH	TOC removal increases with certain contaminants; low adsorption on catalyst surface	[128]
O_3	Ru/CeO_2	Succinic acid	Semi-continuous; pH 3.4; flow $2.75 \cdot 10^{-4}$ mol O_3/h; 0.8 g/L of catalyst	High TOC mineralization in catalysts prepared by impregnation and reduction	[129]
O_3	Ru/CeO_2	Succinic acid	Semi-continuous; pH 3.4; flow 1.25 g O_3/h; 0.8-3.2 g/L of catalyst	Effect of the catalyst structure on its efficacy; organic molecules produce metal sintering	[130]
O_3	Ru/CeO_2-TiO_2	Chloroacetic acid, chlorosuccinic acid	Semi-continuous; pH 2.6-3.6; flow 15.6 L/h; 0.8 g/L of catalyst	High contaminant removal rates	[130]
O_3	Cu/Al_2O_3	Alachlor	Semi-continuous; flow 0.5 mg O_3/min	Use of the catalyst increases TOC, chloride and nitrate formation	[131]
O_3	Mg and Al on Fe, Co, Ni and Cu oxides	Phenol, oxalic acid	Semi-continuous; pH 2; 20ºC; flow 200 mL/min; 1 g/L of catalyst	Identifies oxalic acid as product of phenol ozonation; oxalic acid appears to cause losses of Cu and Ni	[132]

Although the value of solution pH and the adsorption process are important, they do not always determine the activity of the catalyst in the ozonation of organic contaminants. Thus, the catalyst synthesis method and consequently its structure are critical for its capacity to transform ozone into $HO^{\cdot}$ radicals[133,134]. Lin et al.[135] recently studied the catalytic activity of more than 20 catalysts (alumina, silica, activated carbons impregnated with metals, etc.) in the ozonation of formic acid, concluding that the transformation of ozone into $HO^{\cdot}$ radicals is mainly influenced by its diffusion rate on the catalyst surface. Table 5 lists the catalysts tested in some of the most important research studies in this field, including data on the pollutants studied, the experimental conditions, and some observations.

4.2.2. Metal Oxides

Besides its chemical stability, the most important parameter determining the ability to use a metal oxide as catalyst of ozone decomposition is its acidity or basicity. Thus, active centers of the surface of a catalyst are located in groups that act as Brönsted or Lewis acids, which is largely conditioned by the medium pH. Adsorption of organic compounds on the catalyst surface plays an essential role in the catalytic activity of the oxide. This is because: i) it is the first step in the catalytic mechanism of ozone decomposition, and ii) the affinity of the metal oxide for certain inorganic species in solution (carbonates, phosphates, etc) can block active sites and reduce the efficacy of the catalyst. Researchers have used numerous metal oxides to study ozone decomposition and/or the ozonation of organic contaminants in solution. Thus, different groups successfully degraded oxalic acid by using catalysts such as TiO_2, MnO_2 and Al_2O_3[120,136,137]. Several authors have studied the optimization of experimental conditions and have investigated reaction mechanisms (Table 6), proposing catalysts of different origin to achieve or improve the degradation of organic contaminants and mineralize dissolved organic matter in waters. Table 6 shows the most recent investigations into the simultaneous use of ozone and metal oxides to remove pollutants from waters, including the main study conclusions.

4.2.3. Supported Metal Oxides

Attempts have been made to enhance the catalytic activity of metal oxides in the transformation of ozone into $HO^{\cdot}$ radicals by increasing their surface area, supporting them on alumina, clays, silica, or zeolites. However; results obtained were not completely satisfactory because of leaching. According to Copper et al.[138], the presence of a heterogeneous surface facilitates the transference of ozone in solution and its decomposition into radical species. For this reason, many organic compounds were more effectively degraded when ozonation was carried out in the presence of these supported catalysts. Table 7 summarizes the main conclusions of different research groups, and describes the experimental conditions for the simultaneous use of ozone and supported metal oxides to remove organic contaminants from waters.

Table 6. Most recent research on catalyzed ozonation with metal oxides.

System	Catalyst	Contaminants	Experimental conditions	Observations	Ref.
Ozone	Fe_2O_3	Phenol, DQO, TOC	Continuous; pH 4.3-6.3	High efficacy to remove phenol and DQO; mechanisms is proposed	[119]
Ozone	CuO, Fe_2O_3, NiO, Cr_2O_3, Co_2O_3	Aniline, DQO	Semi-continuous; flow 12 L/h; 55 g/L of catalyst	Good catalytic activity; catalyst leaching	[139]
Ozone	Mo, Zn, Sn, V, Ni, Co, Cu, Pb, Mn, Fe, Ce, Ag and Bi oxides	*	Continuous; 25°C; flow 0.15 L/min; 0.5 g of catalyst	Deactivation of certain oxides due to possible adsorption of H_2O on active sites	[140]
Ozone	α-FeOOH	Chlorobenzene, bicarbonates	Semi-continuous; flow 0.2 L/min; 0.05-0.2 g/L of catalyst	Increased chlorobenzene removal in the presence of catalyst	[141]
Ozone	MnO_2	Oxalic acid	Semi-continuous; pH 3-7, flow 36 L/h; 0.25 g/L of catalyst	Improved performance with lower pH; reaction on surface	[120]
Ozone	γ-Al_2O_3	2-chlorophenol	Semi-continuous; pH 3-9; flow 18 mg O_3/h; 2 g/L of catalyst	Kinetic study; toxicity and TOC behaviors	[142]
Ozone	TiO_2	Oxalic acid	Semi-continuous; pH 2.5, 5; 30°C; 5 g/L catalyst	Total removal of TOC; zero-order reaction mechanism is proposed	[136]
Ozone	Tierra, Fe(O), α-FeOOH	*	Continuous; 5-50 g/L catalyst	Study of operational variables; effect of metal oxides and NOM on the process	[143]
Ozone	FeOOH	NOM (TOC)	Semi-continuous, pH 7, flow 0.5 L/min; 10 g/L catalyst	Proposal of radical mechanism; improved TOC removal	[144]
Ozone	Mn^{2+}, β-MnO_2, α-MnO_2, $MnSO_4$	Sulfosalicylic and propionic acid	Semi-continuous, pH 1-8.5	Study of effects of pH and state of Mn on catalytic activity	[145]
Ozone	Mixed metal oxides, metal oxides	p-chlorobenzoic acid	Discontinuous (pH 7 and deionized water); semi-continuous (natural water); 43 mg/L of catalyst	No catalysis in deionized water; Co and Cu-Zn-Al oxides catalyze in natural water	[146]
Ozone	CuO-Al_2O_3	Phenol, chlorophenol, nitrophenol	Reactor tubular; pH 3-11; 21°C; flow 20 L/h; 0.25-2 g/L catalyst	Intermediates are identified; the catalyst reduces ozone consumption and increases rate	[147]

Table 6. Continued

System	Catalyst	Contaminants	Experimental conditions	Observations	Ref.
Ozone	Perfluorinated alumina	Methyl tert-butyl ether, isopropyl ether, ethyl tert-butyl ether, tert-amyl methyl ether	Semi-continuous with and without recirculation; pH 5; 10-60 g/L of catalyst	Alumina does not catalyze; kinetic constant is determined and reaction by-products are identified	[148]
Ozone	Al_2O_3	Oxalic, acetic, salicylic, and succinic acids	Discontinuous and semi-continuous (flow 100 L/h); pH 5.5; 20-50 g/L of catalyst	Removal increase depends on type of contaminant	[137]
Ozone	Al_2O_3	NOM (DQO)	Semi-continuous; pH 7-8; flow 0.4 mg O_3/L; 75 g/L of catalyst	Increased NOM removal and decreased formation of reaction by-products	[149]
Ozone	TOCCATA®	Wastewaters (DQO)	Semi-continuous; pH 7-8; flow 40 mg O_3/L	Improvement in DQO removal by oxidation and precipitation	[150]
Ozone	Sludge with metals and metal oxides	*	Semi-continuous; pH 3; 25°C	Feasibility study on the use of sludge as catalyst of ozone decomposition	[151]

*Publications on ozone decomposition in absence of contaminants

Table 7. Removal of organic contaminants by systems based on the simultaneous use of ozone and supported metal oxides.

System	Catalyst	Contaminants	Experimental conditions	Observations	Ref.
Ozone	CuO, MnO_2, Pd, supported on Al_2O_3	Phenol	Semi-continuous; pH 6-9; 20°C; flow 1.43 L/min	Without catalyst activity	[152]
Ozone	Fe_2O_3/Al_2O_3	Phenol, hydroquinone, carboxylic acids, aldehydes	Semi-continuous; acid pH	Catalyst activity removes more TOC; it does not oxidize hydroquinone and phenol	[153,154]
Ozone	Ag y Ni supported on zeolite	*	Continuous; 25°C; flow 0.15 L/min; 0.5 g of catalyst	High efficacy of Ag to decompose O_3; it does not appear to affect adsorbed H_2O	[139]
Ozone	TiO_2, TiO_2/Al_2O_3, TiO_2/arcilla	Fulvic acid, proteins, disaccharides	Semi-continuous; pH 8; 30 g of catalyst	High removal of TOC and pollutants in the presence of the catalyst	[155]
Ozone	MnO_2/Al_2O_3	*	Semi-continuous	Decomposition of O_3, in the presence of the catalyst is zero order with respect to water; mechanism proposed with superoxide and peroxide intermediates	[156]

Table 7. Continued

System	Catalyst	Contaminants	Experimental conditions	Observations	Ref.
Ozone	TiO_2/Al_2O_3	Fulvic acid	Discontinuous; pH 7.5; 20°C; 10 g/L de catalyst	Use of the catalyst increases mineralization	[157]
Ozone	Al_2O_3, Fe_2O_3/Al_2O_3, TiO_2/Al_2O_3	Chlorophenol, oxalic acid, chloroethanol	Semi-continuous; flow 24 mg O_3/(L·h)	Use of the catalyst increases solubility and decomposition of O_3,improving pollutant removal	[151]
Ozone	TiO_2/Al_2O_3	Humic acid, natural river water (TOC)	Semi-continuous; pH 7.2; flow 0.4 g O_3/h; 2.5-10 g/L catalyst	By-products are identified; TOC removal increases at low humic acid concentration	[123, 158]
Ozone	V-O/SiO_2, V-O/TiO_2, MnO_2	Sulfosalicylic acid	Semi-continuous, pH 3, flow 0.67 L/min	Formation of oxalic acid; improved TOC removal and reaction rate with catalysts; MnO_2 does not catalyze	[159]
Ozone	Metal oxides on Al_2O_3, SiO_2, TiO_2 y $SiO_2{\cdot}Al_2O_3$	*	Fixed-bed; flow 0.175 L/min of aqueous solution with O_3; 1.5 g of catalyst	Proposes a mechanism of ozone decomposition; catalytic efficacy of noble metals; catalyst effect	[160]
Ozone	MnO_x/Al_2O_3, MnO_x/SiO_2, MnO_x/TiO_2, MnO_x/ZrO	Benzene	Fixed-bed; gas flow of 0.25-1L/min; 20-75 mg/L of catalyst	Catalyst deactivation and subsequent regeneration at high temperatures	[161]
Ozone	TiO_2/Al_2O_3	Oxalic acid	Semi-continuous; pH 2.5; 10-40°C; flow 12-36 L/h; 1.25-3.75 g/L of catalyst	The catalyst notably improves pollutant removal; Eley-Rideal mechanism; reaction order and activation energy determined	[162]
Ozone	MnO_x/activated carbon	Nitrobenzene	Continuous (flow 0.075 L/h) and semi-continuous; pH 2-9	Higher catalytic activity at acid pH; pollutant oxidation and adsorption is detected on catalyst surface	[163]
Ozone	MnO_x/Al_2O_3	Cyclohexane	Semi-continuous; flow 0.5 L/min; 0.1 g/L of catalyst	Indicates formation of intermediate compounds on catalyst surface	[164]
Ozone	Co oxides supported on alumina and hydrotalcite	Naphthol blue black	Semi-continuous; 25°C; 30 L/h; 0.3 g/L of catalyst	Co leaching; dispersion is studied in catalysts and reutilization in 3 cycles	[165]

*Publications on ozone decomposition in the absence of contaminants

4.2.4. Activated Carbons

The porous texture and chemical nature of activated carbons makes them ideal materials for use as adsorbents and also as catalysts or catalyst supports[166]. Although there has been far less study on the use of activated carbons as catalyst or metal catalyst support for reactions in aqueous solution than on their role as adsorbent, research interest has been increasing over the past 10 years. In the 1990s, it was observed that the presence of activated carbon catalyses ligand substitution between complex compounds[167] and can produce oxidation of sulfide into sulfate[168]. These studies found a relationship between the catalytic activity of activated carbon and its porous structure, although they did not consider the influence of the surface chemistry of the carbon, which was subsequently addressed by other authors[169,170].

The catalytic activity of activated carbon in aqueous solution has been studied in the synthesis[171,172] and degradation of ammonic nitrate[173], phenolic compounds[174,175] or chlorinated compounds[176,177], among others. The main interest in this behavior of carbon is related to the oxidation of organic matter. Thus, it has been observed that activated carbon can decompose ozone in aqueous solution into radicals with a high oxidizing power, increasing the extent of ozonation[178]. Therefore, the combined use of ozone and activated carbon in a single process was proposed as an attractive option to destroy toxic organic compounds.

Until recently, the major advantage of adding activated carbon to the treatment system was considered its high adsorption capacity[179]. However, Jans and Hoigné[178] showed that both carbon black and activated carbon catalyze the decomposition of ozone in aqueous phase. These authors indicated that both types of carbon initiate the radical-type chain reaction of ozone, which continues in aqueous phase and accelerates the transformation of ozone into free hydroxyl radicals. Zaror et al.[179] reported that the presence of activated carbon drastically reduced the stability of ozone in aqueous solution, which they attributed to a combination of the decomposition catalyzed by the surface and the participation of activated carbon surface groups. Subsequently, in-depth research into ozone/activated carbon systems by Rivera-Utrilla et al.[180-183 demonstrated that the surface chemistry and textural characteristics of activated carbon play a major role in its behavior as catalyst of ozone decomposition into hydroxyl radicals. These authors ascribed this effect to basic carbon groups, e.g., pyrone and chromene[180]. Zaror et al. found that addition of activated carbon to the system favored the oxidation rate of 1,2-dihydroxybenzene with ozone[179]. Hu et al.[184] reported a more effective removal of organic matter from textile industry effluents by using reduced Cu supported on activated carbon. Table 8 summarizes the results of some of the most relevant studies on the simultaneous use of ozone and activated carbon in water treatments.

It is necessary to study the mechanisms underlying this catalytic process in order to achieve its economic and effective large-scale optimization and implementation. Various researchers have investigated the mechanisms involved in heterogeneous catalytic ozonation[185,186], but these have not yet been fully elucidated because of the complexity of these simultaneous pollutant oxidation and adsorption processes.

Table 8. Relevant publications on catalyzed ozonation with activated carbons.

System	Catalyst	Contaminants	Experimental conditions	Observations	Ref.
Ozone	Granular activated carbon	Phenols	Semi-continuous; pH 2; 15-35°C	Lower consumption of O_3 with AC; AC improves selectivity but not the reaction constant	179
Ozone	Activated carbon	Chlorinated hydrocarbons, COD	Fixed bed "ECCOCLEAR"	Combines catalyzed ozonation with biological treatment and nano filtration; low influence of carbonates as scavengers	[187]
Ozone	Activated carbon	Subterranean waters, domestic and industrial wastewaters	Fixed bed "ECCOCLEAR"	Results of industrial plants; catalyzed ozonation requires less ozone and is effective over a wide pH range	[188]
Ozone	Activated carbon, carbon black	p-Chlorobenzoate, acetate, methanol	Discontinuous; 20 mg/L of catalyst	Radical pathway process; Inhibitor and promoter effects of ozone decomposition	[178]
Ozone	Activated carbon	Textile wastewaters	Semi-continuous; flow 60-360 L/h; 200 g of activated carbon column	Kinetic and stability study and catalyst activity; removes color	[189]
Ozone	Granular activated carbon	1,2 dihydroxybenzene	Fixed bed; pH 2, 8; 25°C; flow 5 L/min; 4 g/L of catalyst	Solid-gas catalytic reaction; increases % of oxidized groups after activated carbon ozonation	[190]
Ozone	Granular activated carbon	*	Discontinuous; pH 2-9; 5-30°C; 2 g/L de CA	Proposes mechanism; heterogeneous and homogeneous processes simultaneously; influence of variables and kinetic study	[191]
Ozone	Granular activated carbon	1,3,6 naphthalene-trisulfonic acid	Discontinuous; pH 2.3	Higher catalytic activity of basic activated carbon with/without metals in mineral matter	[180]
Ozone	Granular activated carbon	(mono,di,tri) naphthalene-trisulfonic acids	Semi-continuous; pH 2-12; flow 76 mg O_3 /min; 0.5-2 g/L of activated carbon	Study of adsorption capacity and surface chemical modification of ozonated activated carbon	[181]
Ozone	Granular activated carbon	(mono,di,tri) naphthalene-trisulfonic acids	Semi-continuous; pH 2; flow 76 mg O_3/min	Reduces genotoxicity of ozonation products when catalyzed with activated carbon	[192]
Ozone	Granular activated carbon	1,3,6 naphthalene-trisulfonic acid	Discontinuous; pH 2.3; 2 g/L of catalyst	Identification of intermediates; reduction in catalytic activity of activated carbon with ozonation	[193]
Ozone	Activated carbon	Pyruvic acid	Semi-continuous; pH 7.5; 20°C; 2.5 g/L CA	Catalytic and non-catalytic reaction rate constants determined; transfer of O_3 mass is limiting step at high carbon doses	[194]

Table 8. Continued

System	Catalyst	Contaminants	Experimental conditions	Observations	Ref.
Ozone	Granular activated carbon	*	Discontinuous; 22ºC; 1 g/L of catalyst	Higher catalytic activity of basic activated carbon and larger surface area; greater ozone decomposition with higher concentration	[183]
Ozone	Granular activated carbon, pellets	Succinic acid	Semi-continuous; pH 7; 5, 15, 25ºC; flow 20-60 L/h; 10-20 g/L CA	Use of activated carbon permits almost complete mineralization; reaction *via* 2 pathways: water and activated carbon surface	[195]
Ozone	Activated carbon (0.1-0.3 mm)	*	Semi-continuous; pH 3-9; 25ºC; 4-15 mg/L O_3; 0.2 g/L CA	Because of activated carbon oxidation, the surface chemistry is only important in the first cycles, after which the texture is more influential	[196]
Ozone	Activated carbon	Rainwater (DQO)	Columns in line; 100 g CA; 3 mg/L O_3	Optimization of operational variables; synergic effect in the O3/AC /CAB system	[197]
Ozone, ozone/H_2O_2	Granular and powdered activated carbon	Sodium dodecyl-benzenesulfonate	Discontinuous; pH 7; 25ºC; $2{\cdot}10^{-5}$M; 0.1 g/L of activated carbon	Effect of operational variables (activated carbon size and dose, O_3 dose, etc.); activated carbon reduces effect of radical scavengers	[198]
Ozone, ozone/H_2O_2	Granular activated carbon	Organic and inorganic compounds in natural lake water	Discontinuous; pH 7, 9; 25ºC; $2{\cdot}10^{-5}$M O3; 0.5 g/L CA	Removal of TOC and microcontaminants; similar performance of O_3/GAC and O_3/H_2O_2 in generation of $HO^{\cdot}$ radicals	[199]
Ozone	Carbon aerogels with Mn^{2+}, Co^{2+}y Ti^{4+}	Para-chlorobenzoic acid	Discontinuous; pH 7; 25ºC; $2{\cdot}10^{-5}$M O_3; 2.5-10 mg/L aerogel	Metals susceptible to oxidation with O_3 improve the generation of $HO^{\cdot}$ radicals; activated (more basic) samples are more effective	[200]
Ozone	Granular activated carbon	Benzothiazole	Semi-continuous; pH 2,7, 11; 15ºC; flow 54 L/h; 1 g/L CA	Proposes mechanism and determines kinetic constants; high radical contribution to contaminant removal	[201]
Ozone	Activated carbon fiber	Phenol	Semi-continuous; 25ºC; flow 1.5 L/min; 4 g/L activated carbon fiber	Notable catalytic activity in phenol removal; *in situ* regeneration of activated carbon fiber during ozonation, although its chemical and textural properties are modified	[202]
Ozone	Granular activated carbon	Polyphenols	Semi-continuous; pH 2.5, 5; 25ºC; flow 30 L/h; 10 g/L CA	Identification of reaction by-products; improves TOC removal; generation of H_2O_2	[203]
Ozone	Activated carbon (0.1-0.3 mm), oxidized with HNO_3	Oxalic acid, oxamic acid	Semi-continuous; pH 3, 7; 25ºC; flow 0.15 L/min; 0.5 g/L CA	Study of the effect of pH, presence of radical scavengers and surface chemistry; proposes mechanism; better result with basic activated carbons	[204]

5. Conclusions

There is now a considerable body of knowledge on advanced oxidation processes (O_3/H_2O_2, O_3/UV) in the literature, including data on the mechanisms and reactions involved and on the influence of operational variables in the transformation of ozone into $HO^{\cdot}$ radicals. This information has allowed the large-scale application of these processes to remove a wide variety of organic microcontaminants.

The use of dissolved or supported metals to transform ozone into $HO^{\cdot}$ radicals is not very effective. Studies show that i) the simultaneous use of ozone and dissolved metals must be approached in a more systematic manner; and ii) there is considerably controversy about the reactions involved in the process and the metal characteristics that promote the transformation of ozone into $HO^{\cdot}$ radicals. Research is required into the mechanisms that underlie this catalytic process in order to achieve its economical and effective optimization and implementation on a large scale. Because of the complexity of simultaneous pollutant oxidation and adsorption processes, the mechanisms involved in heterogeneous catalytic ozonation have not yet been fully elucidated.

Finally, activated carbon appears to be the most appropriate solid catalyst to use for enhancing micropollutant ozonation. Besides considerably accelerating the process of ozone transformation into $HO^{\cdot}$ radicals, thereby increasing the rate of organic contaminant removal, activated carbon also has a high adsorbent capacity, further contributing towards the removal of microcontaminants from waters.

References

[1] Singer P.C., Reckhow D.A., *Water quality & treatment: A handbook of community water supplies,* McGraw-Hill, New York (1999).

[2] Zou L.Y., Li Y., Hung Y.S., Wet air oxidation for waste treatment, *Handbook of Environ. Eng.,* **5**, 575-610 (2007).

[3] von Sonntag C., The basics of oxidants in water treatment. Part A: OH radical reactions, *Wat. Sci. & Tech.*, **55**, 19-23 (2007).

[4] Pitochelli A., Biocides: Useful applications for the use of chlorine dioxide in water treatment, *Ultrapure Water*, **23**, 40-42 (2006).

[5] Westerhoff P., Nalinakumari B., Pei P., Kinetics of MIB and geosmin oxidation during ozonation, *Ozone: Sci. & Eng*., **28**, 277-286 (2006).

[6] Peter A., Von Gunten U., Oxidation kinetics of selected taste and odor compounds during ozonation of drinking water, *Env. Sci. & Tech.*, **41**, 626-631 (2007).

[7] Rubin M.B., The history of ozone. The Schönbein period, 1839-1868, *Bull. Hist. Chem.*, **26** (1), 40-56 (2001).

[8] Greenwood N.N., Earnshaw A., *Chemistry of the elements*, Butterworth-Heinemann, Oxford (1997).

[9] Bermond A., Camel V., The use of ozone and associated oxidation processes in drinking water treatment, *Wat. Res.*, **32**, 3208-3222 (1998).

[10] Hoigné J., Bader H., Rate constants of reactions of ozone with organic and inorganic compounds in water. 1. Non-dissociated organic compounds, *Wat. Res.*, **17**, 173-183 (1983).
[11] Hoigné J., Bader H., Rate constants of reactions of ozone with organic and inorganic compounds in water. 2. Dissociated organic compounds, *Wat. Res.*, **17**, 185-194 (1983).
[12] Hoigné J., Bader H., Haag W., Staehelin J., Rate constants of reaction of ozone with organic and inorganic compounds in water. 3.- Inorganic compounds and radicals, *Wat. Res.*, **19**, 993-1004 (1985).
[13] Gurol, M.D., Vatistas R., Kinetic behaviour of ozone in aqueous solution of substituted phenols, *Ind. Eng. Chem. Fundam.*, **23**, 54-60 (1984).
[14] Yao C.C.D., Haag W.R., Rate constants for direct reactions of ozone with several drinking water contaminants, *Wat. Res.* **25**, 761-773 (1991).
[15] Vollhardt K.P.C., Schore N.E., *Química Orgánica*, Omega, Barcelona (1996).
[16] Eisenhauer H.R., Ozonation of phenolic wastes, *J. Wat. Poll. Cont. Fed.*, **40** (11), 1877-1899 (1968).
[17] Riebel A.H., Erikson R.E., Abshire C.J., Bailey P.S. , Ozonation of carbon-nitrogen double bonds I. Nucleophilic attack of ozone, *J. of the Amer. Chem. Soc.*, **82**, 1801-1807 (1960).
[18] Stahelin J., Hoigné J., Decomposition of ozone in water: rate of initiation by hydroxyde ions and hydrogen peroxide, *Env. Sci. Tech*, **16** (10), 676-681 (1982).
[19] Hoigné J., Bader H., The role of hydroxyl radical reactions in ozonation processes in aqueous solutions, *Wat. Res.* 10**,** 377-386 (1976).
[20] Narkis N., Ben-David B., Schneider-Rotel M., Non ionic surfactants interactions with ozone, *Tens. Surf. & Det.*, **24**, 200-210 (1987).
[21] Corless C., Reynolds G., Graham N., Gibson T.M., Haley J., Aqueous ozonation of a quaternary cationic surfactant, *Wat. Res.*, **23**, 1367-1371 (1989).
[22] Evans F.L., Ryckman D.W., Ozonated treatment of wastes containing ABS, *Eng. Ext. Series (Purdue Univ.),* **115**, 141-157 (1963).
[23] Beltrán F.J., García-Araya J.F., Álvarez P.M., Sodium dodecylbenzenesulfonate renoval from water and wastewater 1.Kinetics of decomposition by ozonation, *Ind. Eng. Chem. Res.*, **39**, 2221-2227 (2000).
[24] Aieta E.M., Reagan K.M., Lang J.S., McReynolds L., Kang J.W., Glaze W.H., Advanced oxidation processes for treating groundwater contaminated with TCE and PCE: Pilot-scale evaluations, *Journ. AWWA*, **80** (5), 64-72 (1988).
[25] Forni L., Bahnemann D., Hart E.J., Mechanism of hydroxide ion initiated decomposition of ozone in aqueous solution, *J. Phys. Chem.*, **86** (2), 255-259 (1982).
[26] Sehested K., Holcman J., Hart E.J., Rate constants and products of the reactions of eq^-, dioxide (1-) ($O2^-$) and H with ozone in aqueous solutions, *J. Phys. Chem.*, **87** (11), 1951-1954 (1983).
[27] Bühler R.E., Staehelin J., Hoigné J., Ozone decomposition in water studied by pulse radiolysis 1. HO2/$O2^-$ and OH3/$O3^-$ as intermediates, *J. Phys. Chem.*, **88** (12), 2560-2564 (1984).

[28] Sehested K., Holcman J., Bjergbakker E., Hart E.J., A pulse radiolytic study of the reaction hydroxyl + ozone in aqueous medium, *J. Phys. Chem.*, **88**(18), 4144-4147 (1984).

[29] Staehelin J., Buhler R., Hoigné J., Ozone decomposition in water studied by pulse radiolysis 2. Hydroxyl and hydrogen tetroxide (HO4) as chain intermediates, *J. Phys. Chem.*, **88** (24), 5999-6004 (1984).

[30] Staehelin J., Hoigné J., Decomposition of ozone in water in presence of organic solutes acting as promoters and inhibitors of radical chain reactions, *Env. Sci., Tech.*, **19** (12), 1206-1213 (1985).

[31] Tomiyasu H., Fukutomi H., Gordon G., Kinetics and mechanism of ozone decomposition in Basic aqueous solution, *Inorg. Chem.*, **24** (19), 2962-2966 (1985).

[32] Sehested K., Corfitzen H., Holcman J., Fischer C.H., Hart E.J., The primary reaction in the decomposition of ozone in acidic aqueous solutions, *Env. Sci. Tech.*, **25** (9), 1589-1596 (1991).

[33] Sehested K., Corfitzen H., Holcman J., Hart E.J., On the mechanism of the decomposition of acidic O3 solutions, thermally or H2O2-initiated, *J. Phys. Chem. A*, **102** (16), 2667-2672 (1998).

[34] Hoigné J., Chemistry of aqueous ozone, and transformation of pollutants by ozonation and advanced oxitadion processes, *The handbook of environmental chemistry quality and treatment of drinking water*, Springer, Berlin (1998).

[35] Buxton G., Wilmarth W.K., Aqueous chemistry of inorganic free radicals V. Carbon monoxide as scavenger for hydroxil radicals generated by the photolysis of hydrogen peroxide, *J. Phys. Chem.*, **67** (12), 2835-2841 (1963).

[36] Bruny R., Bourbigot M.M., Doré M., Oxidation of organic compounds through the combination ozone-hydrogen peroxide, *Ozone Science & Engineering*, **6**, 163-183 (1984).

[37] Duguy J.P., Brodard E., Dussert B., Mallevialle J., Improvement in the effectiveness of ozonation of drinking water through the use of hydrogen peroxide, *Ozone Science & Engineering*, **7**, 241-257 (1985).

[38] Paillard H., Bruny R., Doré M., Optimal conditions for applying an ozone-hydrogenperoxide oxiding systems, *Wat. Res.*, **22**, 91-103 (1988).

[39] Glaze W.H., Kang J.W., Advanced oxidation processes. Description of a kinetic model for the oxidation of hazardous materials in aqueous media with ozone and hydrogen peroxide in a semibatch reactor, *Ind. Eng. Chem. Res.*, **28,** 1573-1580 (1989).

[40] Glaze W.H. y Kang J.W., Advanced oxidation processes. Test of a kinetic model for the oxidation of organic compounds with ozone and hydrogen peroxide in a semibatch reactor, *Ind. Eng. Chem. Res.*, **25**, 1580-1587 (1989).

[41] Volk C., Renner C., Roche P., Paillard P., Roche H., Joret J.C., Effects of ozone on the production of biodegradable dissolved organic carbon (BDOC) during water treatment, *Ozone: Sci. & Eng.*, **15**, 389-404 (1993).

[42] Duguy J.P., Application of combined ozone-hydrogen peroxide for the removal of aromatic compounds from groundwater, *Ozone Science & Engineering*, 12, 281-293 (1990).

[43] Beltrán F.J., Ovejero G., Rivas J., Oxidation of polynuclear aromatic hydrocarbons in water. 4. Ozone combined with hydrogen peroxide, *Ind. Eng. Chem. Res.*, **35**, 891-898 (1996).

[44] Fernández J., Riu J., García-Calvo E., Rodríguez A., Fernández-Alba A.R., Barceló D., Determination of photodegradation and ozonation by products of linear alkylbenzene sulfonates by liquid chromatography and ion chromatography under controlled laboratory experiments, *Talanta*, **64**, 69-79 (2004).

[45] Kamenev S., Kallas J., Munter R., Trapido M., Chemical oxidation of biologically treated phenolic effluents, *Waste Manag*., **15**, 203-208 (1995).

[46] Bauer R., Waldner G., Fallmann H., Hager, S., Klare M., Krutzler T., Malato S., Maletzky P., The photo-fenton reaction and the TiO2/UV process for waste water treatment - novel developments, *Catalysis Today*, **53**, 131-144 (1999).

[47] Gouvea C.A, Wypych F., Moraes S.G., Duran N., Peralta-Zamora P., Semiconductor-assisted photodegradation of lignin, dye, and kraft effluent by Ag-doped ZnO, *Chemosphere*, **40**, 427-432 (2000).

[48] Gouvea C.A, Wypych F., Moraes S.G., Duran N., Peralta-Zamora P., Semiconductor-assisted photocatalytic degradation of reactive dyes in aqueous solution, *Chemosphere*, **40**, 433-440 (2000).

[49] Wang S., Shiraishi F., Nakano K., Decomposition of formic acid in a photocatalytic reactor with a parallel array of four light sources, *J. of Chem. Tech. & Biotech*., **77**, 805-810 (2002).

[50] Mazellier P., Leroy E., De Laat J., Legube B., Degradation of carbendazim by UV/H2O2 investigated by kinetic modelling, *Environ. Chem. Letters*, **1**, 68-72 (2003).

[51] Garoma T., Gurol M.D., Modeling Aqueous Ozone /UV Process Using Oxalic Acid as Probe Chemical, *Environ. Sci. & Tech*., **39**, 7964-7969 (2005).

[52] Reyes C., Fernandez J., Freer J., Mondaca M.A., Zaror C., Malato S., Mansilla H.D., Degradation and inactivation of tetracycline by TiO2 photocatalysis, *J. of Photochem. & Photobiol. A: Chem*., **184**, 141-146 (2006).

[53] Gomez C., Rodriguez J., Freer J., Lizama C., Zaror C., Mansilla H.D., Coupling of photocatalytic and biological reactors to remove EDTA-Fe from aqueous solution, *Env. Tech*., **28**, 123-127 (2007).

[54] Tanaka K., Abe K., Hisanaga T., Photocalytic water treatment on inmobilized TiO2 combined with ozonation, *J. Photochem. Photobiol. A*, **101**, 85-87 (1996).

[55] Muller T.S, Sun Z., Kumar G., Itoh K., Murabayashi M., The combination of photocatalysis and ozonolysis as a new approach for cleaning 2,4-dichlorophenoxyaceticacid polluted water, *Chemosphere*, **36**, 2043-55 (1998).

[56] Sánchez L., Peral J., Doménech X., Aniline degradation by combined photocatalysis and ozonation, *Appl. Catal. B. Environ*., **19**, 59-65 (1998).

[57] Klare M., Waldner G., Bauer R., Jacobs H., Broekaert J.A., Degradation of nitrogen containing organic compounds by combined photocatalysis and ozonation. *Chemosphere*, **38**, 2013-2027 (1999).

[58] Kopf P., Gilbert E., Eberle S.H., TiO2 photocatalytic oxidation of monochloroacetic acid and pyridine: influence of ozone, *J. of Photochem. & Photobiology A: Chem*., **136**, 163-168 (2000).

[59] De Moraes, S.G., Freire R.S., Duran N., Degradation and toxicity reduction of textile effluent by combined photocatalytic and ozonation processes, *Chemosphere*, **40**, 369-373 (2000).

[60] Balcioglu, I. A.; Getoff, N.; Bekbolet, M. A comparative study for the synergistic effect of ozone on the γ -irradiated and photocatalytic reaction of 4-chlorobenzaldehyde. *Journal of Photochemistry and Photobiology, A: Chemistry*, **135**, 229-233 (2000).

[61] Pichat P., Cermenati L., Albini A., Mas D., Delprat H., Guillard C., Degradation processes of organic compounds over UV-irradiated TiO2. Effect of ozone, *Res. on Chem. Intermediates*, **26**, 161-170 (2000).

[62] Hernandez-Alonso M.D., Coronado J.M., Javier Maira A., Soria J., Loddo V., Augugliaro V., Ozone enhanced activity of aqueous titanium dioxide suspensions for photocatalytic oxidation of free cyanide ions, *Applied Catalysis, B: Environmental*, **39**, 257-267(2002).

[63] Martin M.A., Raposo. F., Borja R., Martin A., Kinetic study of the anaerobic digestion of vinasse pretreated with ozone, ozone plus ultraviolet light, and ozone plus ultraviolet light in the presence of titanium dioxide, *Process Biochemistry*, **37**, 699-706 (2002).

[64] Wang S., Shiraishi F., Nakano K., A synergistic effect of photocatalysis and ozonation on decomposition of formic acid in an aqueous solution, *Chemical Engineering Journal*, **87**, 261-271 (2002).

[65] Wada H., Murayama T., Kuroda Y., Recycling of cyanide wastewater applying combined UV-ozone oxidation with a titanium dioxide catalyst and ion exchange resin method, *Bulletin of the Chemical Society of Japan*, **75**, 1399-1405 (2002).

[66] Villaseñor J., Reyes P., Pecchi G., Catalytic and photocatalytic ozonation of phenol on MnO2 supported catalysts, *Catalysis Today*, **76**, 121-131 (2002).

[67] Li L., Zhu W., Zhang P., Chen Z., Han W., Photocatalytic oxidation and ozonation of catechol over carbon-black-modified nano-TiO2 thin films supported on Al sheet, *Wat. Res.*, **37**, 3646-3651 (2003).

[68] Kerc A., Bekbolet M., Saatci A.M., Sequential oxidation of humic acids by ozonation and photocatalysis, *Ozone Sci. & Eng.*, **25**, 497-504 (2003).

[69] Kerc A., Bekbolet M., Saatci A.M., Effect of partial oxidation by ozonation on the photocatalytic degradation of humic acids, *Int. Journal of Photoenergy*, **5**, 75-80 (2003).

[70] Kerc A., Bekbolet M., Saatci A.M., Effects of oxidative treatment techniques on molecular size distribution of humic acids, *Water sci. & tech. a journal of the Int. Assoc. on Water Poll. Res.*, **49**, 7-12 (2004).

[71] Ilisz I., Bokros A., Dombi A., TiO2-based heterogeneous photocatalytic water treatment combined with ozonation, *Ozone: Sci. & Eng.*, **26**, 585-594 (2004).

[72] Farre M.J., Franch M.I., Malato S., Ayllon J.A., Peral J., Doménech X., Degradation of some biorecalcitrant pesticides by homogeneous and heterogeneous photocatalytic ozonation, *Chemosphere*, **58**, 1127-1133 (2005).

[73] Tong S., Xie D., Wei H., Liu W., Degradation of sulfosalicylic acid by O3/UV, O3/TiO2/UV, and O3/V-O/TiO2: A comparative study, *Ozone: Sci. & Eng.*, **27**, 233-238 (2005).

[74] Beltran F.J., Rivas F.J., Gimeno O. Comparison between photocatalytic ozonation and other oxidation processes for the removal of phenols from water, *J. of Chem. Tech. & Biotech.*, **80**, 973-984 (2005).

[75] Hur J.,Oh S., Lim K., Jung J., Kim J., Koh Y.J., Novel effects of TiO2 photocatalytic ozonation on control of postharvest fungal spoilage of kiwifruit, *Postharvest Biology and Technology,* **35**, 109-113 (2005).

[76] Okabe H., Photochemistry of small molecules, New York, NY: Wiley Interscience (1978).

[77] Prengle H.W., Hewes C.G., Mauk C.E., Oxidation of refractory materials by ozone with ultraviolet radiation, in *Proc. Second International Symposium on Ozone Technology,* R.G. Rice, P. Pichy & M.A. Vincent Eds., Norwalk (1976).

[78] Peyton G.R., Huang F.Y., Burleson J.L., Glaze W.H., Destruction of pollutants in water with ozone in combination with ultraviolet radiation. 1. General principles and oxidation of tetrachloroethylene, *Environ. Sci. Technol.* **16**, 448-454 (1982).

[79] Glaze W.H., Peyton G.R., Lin S., Huang F.Y., Burleson J.L., Destruction of pollutants in water with ozone in combination with ultraviolet radiation. 2. Natural trihalomethane precursors, *Env.. Sci. Tech.*, **16**, 454-462 (1982).

[80] Peyton G.R., Glaze W.H., Mechanism of photolytic ozonation, en Photochemistry of Environmental Aquatic Systems, R.G. Zika & W.J. Cooper, Eds., ACS Symposium Series 327, (Washington, DC: Am. Chem. Soc.), pp. 76-88 (1986).

[81] Glaze W.H., Kang J.W., Chapin D.H., The chemistry of water treatment processes involving ozone, hydrogen peroxide and ultraviolet radiation, *Ozone Science & Engineering*, **9**, 335-342, (1987).

[82] Gurol M.D. y Vatistas R., Oxidation of phenolic compounds by ozone and ozone+UV radiation: a comparative study, *Wat. Res.* **21**, 895-900, (1987).

[83] Ku Y., Su W.J. y Shen Y.S., Decomposition of phenols in aqueous solution by the O3/UV process, *Ozone Sci. & Eng.*, **18**, 417-442, (1996).

[84] Matsuo T., Nishi T., Matsuda M., Izumida T., Compatibility of the ultraviolet Light ozone system for laundry wastewater treatment in nuclear power plants, *Nuclear Tech.*, **119** (2), 149-157 (1997).

[85] Goncharuk V.V., Vakulenko V.F., Sova A.N., Svadchina Y.O., Effect of UV radiation on the kinetics of oxidation of sodium alkylbenzene sulfonate by ozone in water, *Khim. Thekn. Vody*, **24** (2), 99-108 (2002).

[86] Amat A.M., Arques A., Miranda M.A., Vicente R., Segui S., Degradations of two commercial anionic surfactants by means of ozone and/or UV irradiation, *Env. Eng. Sci.,* **24** (6), 790-794 (2007).

[87] Beltrán F.J., García Araya, J.F. y Acedo B., Advanced oxidation of atrazine in water. II Ozonation combined with ultraviolet radiation, *Wat. Res.*, **28**, 2165-2174 (1994).

[88] Takahashi N., Decomposition of low molecular weight organic compounds by combination of ozonation and other methods, *Kogai to Taisaku*, **25**, 1500-1504 (1989).

[89] Hewes C.G., Davison R.R., Renovation of waste water by ozonation, *AIChE Symposium Series,* **69** (129), 71-80 (1973).

[90] Abdo M.S.E., Shaban H. y Bader M.S.H., Decolorization by ozone of direct dyes in presence of some catalysts, *J. Environ. As. Health*, **A23** (7), 697-710 (1988).

[91] Andreozzi R., Insola A., Caprio V. y D´Amore M.G., The kinetics of Mn(II)-catalysed ozonation of oxalic acid in aqueous solution, *Wat. Res.*, **26** (7), 917-921 (1992).

[92] Taube H., Bray W.C., Chain reactions in aqueous solutions containing ozone, hydrogen peroxide and acid, *J. of the American Chem. Soc.*, **62** 3357-3373 (1940).

[93] Hill G.R., Kinetics, mechanism, and activation energy of the cobaltous-ion-catalyzed decomposition of ozone, *J. of the American Chem. Soc.*, **70** 1306-1307 (1948).

[94] Hill G.R., Kinetics of the oxidation of cobaltous ion by ozone, *J. of the American Chem. Soc.*, **71** 2434-2435 (1949).

[95] Yang T.C., Neely W.C., Relative stoichiometry of the oxidation of ferrous ion by ozone in aqueous solution, *Analytical Chem.*, **58**, 1551-1555 (1986).

[96] Reckhow D.A., Knocke W.R., Kearney M.J., Parks C.A., Oxidation of iron and manganese by ozone, *Ozone: Sci. & Eng.*, **13**, 675-695 (1991).

[97] Toui S., The oxidation of manganese and disinfection by ozonation in water purification processing, *Ozone: Sci. & Eng.*, **13**, 623-637 (1991).

[98] Loegager T., Holcman J., Sehested K., Pedersen T., Oxidation of ferrous ions by ozone in acidic solutions, *Inorganic Chem.*, **31**, 3523-3529 (1992).

[99] McKnight K.F., Carlson M., Fortin P., Ziesemer C., Comparison of ozone efficiency for manganese oxidation between raw and settled water, *Ozone: Sci. & Eng.*, **15**, 331-341 (1993).

[100] Seby F., Potin-Gautier M., Castelbon A., Astruc M., Study of the ozone-manganese reaction and the interactions of disulfonate indigo carmin/oxidized manganese forms as a function of pH, *Ozone: Sci. & Eng.*, **17**, 135-147 (1995).

[101] Andreozzi R., Caprio V., D'amore G., Insola A., Manganese catalysis in water pollutants abatement by ozone, *Environ. Tech.*, **16**, 885-891 (1995).

[102] Gracia R., Argues J.L., Ovelleiro J.L., Study of the catalytic ozonation of humic substances in water and their ozonation byproducts, *Ozone: Sci. & Eng.*, **18**, 195-208 (1996).

[103] Ma Jun, Graham N.J.D., Preliminary investigation of manganese-catalyzed ozonation for the destruction of atrazine, *Ozone: Sci. & Eng.*, **19**, 227-240 (1997).

[104] Gracia R., Aragues J.L., Ovelleiro J.L., Mn(II)-catalyzed ozonation of raw Ebro River water and its ozonation byproducts, *Wat. Res.*, **32**, 57-62 (1998).

[105] Andreozzi R., Caprio V., Insola A., Marotta R., Tufano V., The ozonization of pyruvic acid in aqueous solutions catalyzed by suspended and dissolved manganese, *Wat. Res.*, **32,** 1492-1496 (1998).

[106] Ma J., Graham N.J.D., Degradation of atrazine by manganese-catalyzed ozonation: influence of humic substances, *Wat. Res.*, **33**, 785-793 (1998).

[107] Cortes S., Sarasa J., Ormad P., Gracia R., Ovelleiro J.L., Comparative efficiency of the systems O3/high pH and O3/catalyst for the oxidation of chlorobenzenes in water, *Ozone: Sci. & Eng.*, **22**, 415-426 (2000).

[108] Andreozzi R., Marotta R., Sanchirico R., Manganese-catalyzed ozonation of glyoxalic acid in aqueous solutions, *J. of Chem. Tech. & Biotech.*, **75,** 59-65 (2000).

[109] Andreozzi R, Caprio V, Marotta R, Tufano V, Kinetic modeling of pyruvic acid ozonation in aqueous solutions catalyzed by Mn(II) and Mn(IV) ions, *Wat. Res.*, **35**, 109-120 (2001).

[110] Potapenko E.V., Galstyan G.A., Galstyan A.G., Kudyukov Y.P., Oxidation of 4-nitrotoluene by ozone in acetic anhydride in the presence of manganese sulphate, *Kinetics & Catal.*, **42**, 796-799 (2001).

[111] Hassan M.M., Hawkyard C.J., Decolourisation of aqueous dyes by sequential oxidation treatment with ozone and Fenton's reagent, *J. of Chem. Tech. & Biotech.*, **77**, 834-841 (2002).

[112] Ni C.H., Chen J., Yang P., Catalytic ozonation of 2-dichlorophenol [2-chlorophenol] by metallic ions, *Wat. Sci. & Tech.*, **47**, 77-82 (2003).

[113] Beltrán F.J., Rivas F.J., Montero-de-Espinosa R., Ozone-enhanced oxidation of oxalic acid in water with cobalt catalysts. 2. Heterogeneous catalytic ozonation, *Ind. & Eng. Chem. Res.*, **42**, 3218-3224 (2003).

[114] Rivas F.J., Beltrán F.J., Carbajo M., Gimeno O., Homogeneous catalyzed ozone decomposition in the presence of Co(II), *Ozone: Sci. & Eng.*, **25**, 261-271 (2003).

[115] Beltrán F.J., Rivas F.J., Montero-de-Espinosa R., Iron type catalysts for the ozonation of oxalic acid in water, *Wat. Res.*, **39**, 3553-3564 (2005).

[116] Assalin M.R., Lima da Silva P., Comparison of the efficiency of ozonation and catalytic ozonation (Mn(II) and Cu(II)) in phenol degradation, *Quimica Nova*, **29**, 24-27 (2006).

[117] Nowell L.H. y Hoigné J., Interaction of iron (II) and other transition metals with aqueous ozone, 8th Ozone World Congress, Zurich, September (1987).

[118] Pines D.S., Reckhow D.A., Effect of dissolved cobalt(II) on the ozonation of oxalic acid, *Env. Sci. Tech.*, **36** (19), 4046-4051 (2002).

[119] Chen J.W., Hui C., Keller T., Smith G., Catalytic ozonation in aqueous system, *AIChE Symposium Series,* **73**, 206-212 (1977).

[120] Andreozzi R., Insola A., Caprio V., Marotta R. y Tufano V., The use of manganese dioxide as a heterogeneous catalyst for oxalic ozonation in aqueous solution, *App. Catal. A: General,* **138**, 75-81 (1996).

[121] Ma J., Graham N.J.D., Preliminary investigation of manganese-catalyzed ozonation for the destruction of atrazine, *Ozone Science & Engineering*, **19**, 227-236 (1997).

[122] Al Hayek N., Legube B. y Doré M., Ozonation catalytique (Fe(III)/Al2O3) du phénol y de ses produits d´ozonation, *Env. Tech. Letters*, **10**, 415-422 (1992).

[123] Gracia R., Cortés S., Sarasa J., Ormad P., Ovelleiro J.L., Catalytic ozonation with supported titanium dioxide. The stability of catalyst in water, *Ozone Sci. Eng.,* **22** (2), 185-193 (2000).

[124] Karpel Vel Leitner N., Fu H., pH effects on catalytic ozonation of carboxylic acids with metal on metal oxides catalysts, *Topics in Catal.*, **33** (1-4), 249-256 (2005).

[125] Cooper C., Burch R., Mesoporous materials for water treatment processes, *Wat.Res.*, **33** (18), 3689-3694 (1999).

[126] McKay G., McAleavey G., Ozonation and carbon adsorption in a three-phase fluidised bed for colour removal from peat water, *Chem. Eng. Res. Des.*, **66**, 532-536 (1988).

[127] Heinig C.F.Jr., Ozone or O2 and Ag: a new catalyst technology for aqueous phase sanitation, *Ozone: Sci. & Eng.*, **15**, 533-546 (1993).

[128] Karpel Vel Leitner N., Delouane B., Legube B., Luck F., Effects of catalysts during ozonation of salicylic acid, peptides and humic substances in aqueous solution, *Ozone: Sci. & Eng.*, **21**, 261-276 (1999).

[129] Legube B., Karpel Vel Leitner N., Catalytic ozonation. A promising advanced oxidation technology for water treatment, *Catalysis Today*, **53**, 61-72 (1999).

[130] Fu H., Karpel Vel Leitner N., Legube B., Catalytic ozonation of chlorinated carboxylic acids with Ru/CeO2-TiO2 catalyst in the aqueous system, *New Journal of Chem.*, **26**, 1662-1666 (2002).

[131] Qu J., Li H., Liu H., He H., Ozonation of alachlor catalyzed by Cu/Al2O3 in water, *Catalysis Today*, **90**, 291-296 (2004).

[132] Shiraga M., Kawabata T., Li D., Shishido T., Komaguchi K., Sano T., Takehira K., Memory effect-enhanced catalytic ozonation of aqueous phenol and oxalic acid over supported Cu catalysts derived from hydrotalcite, *Appl. Clay Sci.*, **33**, 247-259 (2006).

[133] Karpel Vel Leitner N., Delanoe F., Acedo B., Legube B., Reactivity of various Ru/CeO2 catalysts during ozonation of succinic acid aqueous solutions, *New Journal of Chem.*, **24**, 229-233 (2000).

[134] Delanoë F., Acedo B., Karpel Vel Leitner N., Legube B., Relationship between the structure of Ru/CeO2 catalysts and their activity in the catalytic ozonation of succinic acid aqueous solutions, *Appl. Catal. B: Environ.*, **29**, 315-325 (2001).

[135] Lin J., Nakajima T., Jomoto T., Hiraiwa K., Effective catalysts for wet oxidation of formic acid by oxygen and ozone, *Ozone: Sci. & Eng.*, **22**, 241-247 (2000).

[136] Beltran F.J., Rivas F.J., Montero-de-Espinosa R., Catalytic ozonation of oxalic acid in an aqueous TiO2 slurry reactor, *Appl.Catal. B: Environ.*, **39**, 221-231 (2002).

[137] Ernst M., Lurot F., Schrotter J.C., Catalytic ozonation of refractory organic model compounds in aqueous solution by aluminum oxide, *Appl. Catal. B: Environ.*, **47**, 15-25 (2004).

[138] Cooper C., Burch R., An investigation of catalytic ozonation for the oxidation of halocarbons in drinking water preparation, *Wat. Res.*, **33**, 3695-3700 (1999).

[139] Munter R., Kamenev S.B., Preis S.V., Siirde E., Khudak V.I., Catalytic wastewater treatment with ozone, *Khimiya i Tekhnologiya Vody*, **7**, 17-19 (1985).

[140] Imamura S., Ikebata M., Ito T., Ogita T., Decomposition of ozone on a silver catalyst, *Ind. & Eng. Chem. Res.*, **30,** 217-221 (1991).

[141] Bhat N.N., Gurol M.D., Oxidation of chlorobenzene by ozone and heterogeneous catalytic ozonation, *Hazard. & Ind. Wastes*, **27**, 371-382 (1995).

[142] Ni C.H, Chen J.N., Heterogeneous catalytic ozonation of 2-chlorophenol aqueous solution with alumina as a catalyst, *Wat. Sci. & Tech.*: A journal of the Int. Assoc. on *Water Poll. Res.*, **43**, 213-220 (2001).

[143] Lim H., Choi H., Hwang T., Kang J., Characterization of ozone decomposition in a soil slurry: kinetics and mechanism, *Wat. Res.*, **36**, 219-229 (2002).

[144] Park J.S., Choi H., Ahn K.H., The reaction mechanism of catalytic oxidation with hydrogen peroxide and ozone in aqueous solution, *Wat. Sci. & Tech.*, **47**, 179-184 (2003).

[145] Tong S., Liu W., Leng W., Zhang Q., Characteristics of MnO2 catalytic ozonation of sulfosalicylic acid and propionic acid in water, *Chemosphere*, **50**, 1359-1364 (2003).

[146] Pines D.S., Reckhow D.A., Solid phase catalytic ozonation process for the destruction of a model pollutant, *Ozone: Sci. & Eng.*, **25**, 25-39 (2003).

[147] Udrea I., Bradu Corina, Ozonation of substituted phenols in aqueous solutions over CuO-Al2O3 catalyst, *Ozone: Sci. & Eng.*, **25**, 335-343 (2003).
[148] Kasprzyk-Hordern B., Andrzejewski P., Dabrowska A., Czaczyk K., Nawrocki J., MTBE, DIPE, ETBE and TAME degradation in water using perfluorinated phases as catalysts for ozonation process, *Appl. Catal. B: Environ.*, **51**, 51-66 (2004).
[149] Kasprzyk-Hordern B., Raczyk-Stanislawiak U., Swietlik J., Nawrocki J., Catalytic ozonation of natural organic matter on alumina, *Appl. Catal. B: Environ.*, **62**, 345-358 (2006).
[150] Fontanier V., Farines V., Albet J., Baig S., Molinier J., Study of catalyzed ozonation for advanced treatment of pulp and paper mill effluents, *Wat. Res.*, **40**, 303-310 (2006).
[151] Muruganandham M., Chen S.H., Wu J.J., Evaluation of water treatment sludge as a catalyst for aqueous ozone decomposition, *Catal. Communications*, **8**, 1609-1614 (2007).
[152] Joshi M.G., Shambaugh R.L., The kinetics of ozone-phenol reaction in aqueous solutions, *Wat. Res.*, **16**, 933-938 (1982).
[153] Al-Hayek N., Legube B., Dore M., Catalytic ozonization (iron(III)/aluminum oxide (Al2O3)) of phenol and its ozonization by-products, *Environ. Tech.* Letters, **10**, 415-426 (1989).
[154] Al-Hayek N., Dore M., Oxidation of phenols in water by hydrogen peroxide on alumina-supported iron, *Wat. Res.*, **24**, 973-982 (1990).
[155] Allemane H., Delouane B., Paillard H., Legube B., Comparative efficiency of three systems (O3, O3/H2O2 and O3/TiO2) for the oxidation of natural organic matter in water, *Ozone: Sci. & Eng.*, **15**, 419-432 (1993).
[156] Dhandapani B., Oyama S.T., Kinetics and mechanism of ozone decomposition on a manganese oxide catalyst, *Chem. Letters*, **(6)**, 413-414 (1995).
[157] Volk C., Roche P., Joret J.C., Paillard H., Comparison of the effect of ozone, ozone-hydrogen peroxide system and catalytic ozone on the biodegradable organic matter of a fulvic acid solution, *Wat. Res.*, **31**, 650-656 (1997).
[158] Gracia, R., Cortes S., Sarasa J., Ormad P., Ovelleiro J.L., TiO2-catalyzed ozonation of raw Ebro River water, *Wat. Res.*, **34**, 1525-1532 (2000).
[159] Ping T.S., Hua L.W., Qing Z.J., Nan C.C., Catalytic ozonation of sulfosalicylic acid, *Ozone: Sci. & Eng.*, **24**, 117-122 (2002).
[160] Lin J., Kawai A., Nakajima T., Effective catalysts for decomposition of aqueous ozone, *Appl. Catal. B: Environ.*, **39**, 157-165 (2002).
[161] Einaga H., Futamura S., Catalytic oxidation of benzene with ozone over alumina-supported manganese oxides, *J. of Catal.*, **227**, 304-312 (2004).
[162] Beltran F.J., Rivas F.J., Montero-de-Espinosa R., A TiO2/Al2O3 catalyst to improve the ozonation of oxalic acid in water, *Appl. Catal. B: Environ.*, **47**, 101-109 (2004).
[163] Ma J., Sui M., Zhang T., Guan C., Effect of pH on MnOx/GAC catalyzed ozonation for degradation of nitrobenzene, *Wat. Res.*, **39**, 779-786 (2005).
[164] Einaga H., Futamura S., Oxidation behavior of cyclohexane on alumina-supported manganese oxides with ozone, *Appl. Catal. B: Environ.*, **60**, 49-55 (2005).

[165] Gruttadauria M., Liotta L.F., Di Carlo G., Pantaleo G., Deganello G., Lo Meo P., Aprile C., Noto R., Oxidative degradation properties of Co-based catalysts in the presence of ozone, *Appl. Catal. B: Environ.*, **75**, 281-289 (2007).

[166] Radovic L.R. y Rodríguez-Reinoso F., *Carbon materials in catalysis, en Chemistry and Physics of Carbon,* Vol. 25, Ed. Peter Thrower, Marcel Dekker: New York, 243-358 (1997).

[167] Kimura M., Miyamoto I., Fujita S., Kumashiro S., Difference in the kinetic behaviour on the AC-catalyzed decomposition of [Co(C2O4)3]3- and $[Co(NH3)5(H2O)]^{3+}$, and on the AC-catalyzed ligand-substitution reactions between each complex and EDTA in an aqueous solution, *Bull. Chem. Soc. Jpn.,* **69**, 2897-2900 (1996).

[168] Goncharuk V.V., Sergeev V.P., Taranukhina L.D., Baglei L.D., Gorokhovatskaya N.V. y Plygan E., Activated fibrous carbon materials as catalyst in liquid-phase oxidation of sulfide-containing aqueous solutions, *Ukr. Khim. Zh.*, **61**, 27-31 (1995).

[169] Rivera-Utrilla J., Sanchez-Polo M., The role of dispersive and electrostatic interactions in the aqueous phase adsorption of naphthalenesulphonic acids on ozone-treated activated carbons, *Carbon*, **40**, 2685-2691 (2002).

[170] Alvarez P.M., García-Araya J.F., Beltrán F.J., Giradlez I., Jaramillo J., Goméz-Serrano V., The influence of various factors on aqueous ozone decomposition by granular activated *carbons* and the development of a mechanistic approach, *Carbon*, **44** (14), 3102-3112 (2006).

[171] Díaz E., Mohedano A.F., Calvo L., Gilarranz M.A., Casas J.A., Rodríguez J.J., Hydrogenation of phenol in aqueous phase with palladium on activated *carbon* catalysts, *Chem. Eng. Journal*, **131**, 65-71 (2007).

[172] Zhong Y., Li G., Zhu L., Yan Y., Wu G., Hu C., Low temperature hydroxylation of benzene to phenol by hydrogen peroxide over Fe/activated *carbon* catalyst, *J. Molec. Cat. A: Chemical,* **272**, 169-173 (2007).

[173] Inazu K., Kitahara M., Aika K., Decomposition of ammonium nitrate in aqueous solution using supported platinum catalysts, *Catalysis Today*, **93-95,** 263-271 (2004).

[174] Alvarez P.M., McLurgh D., Plucinski P., Copper oxide mounted on activated *carbon* as catalyst for wet air oxidation of aqueous phenol 1. Kinetic and mechanistic approaches., *Ind. & Eng. Chem. Res.*, **41** (9), 2147-2152 (2002).

[175] Rubalcaba A., Suarez-Ojeda M.E., Carrera J., Font J., Stueber F., Bengoa C., Fortuna A., Fabregat A., Biodegradability enhancement of phenolic compounds by hydrogen peroxide promoted catalytic wet air oxidation, *Catalysis Today*, **124**, 191-197 (2007).

[176] Mackenzie K., Battke J., Khoeler R., Kopinke F.D., Catalytic effects of activated *carbon* on hydrolysis reactions of chlorinated organic compounds, *Applied Catalysis B: Environmental,* **59**, 171-179 (2005).

[177] Baker F.S., Catalytic activated *carbon* for removal of chloramines from water, U.S. Pat. Appl. Publ. (2004).

[178] Jans U. y Hoigné J., "Activated *carbon* and *carbon* black catalyzed transformation of aqueous ozone into OH-radicals", *Ozone Science & Engineering*, **20**, 67-90 (1998).

[179] Zaror C.A., "Enhanced oxidation of toxic effluents using simultaneous ozonation and activated *carbon* treatment", *J. Chem. Tech and Biotech.*, **70**, 21-28 (1997).

[180] Rivera-Utrilla J., Sánchez-Polo M., Ozonation of 1,3,6-naphthalenesulphonic acid catalysed by activated *carbon* in aqueous phase, *Appl. Catal. B: Environ.*, **39** (4), 319-329 (2002).

[181] Rivera-Utrilla J., Sánchez-Polo M., The role of dispersive and electrostatic interactions in the aqueous phase adsorption of naphthalenesulphonic acids on ozone-treated activated *carbon*, *Carbon* **40** (14), 2685-2691 (2002).

[182] Sánchez-Polo M., Rivera-Utrilla J., Zaror C.A., Advanced oxidation with ozone of 1,3,6 naphthalenetrisulfonic acid in aqueous solution, *J. Chem. Tech. & Biotech.*, **77**, 148-154 (2002).

[183] Sánchez-Polo M., von Gunten U., Rivera-Utrilla J., Efficiency of activated *carbon* to transform ozone into $HO^{\cdot}$ radicals: Influence of operational parameters, *Wat. Res.*, **39** (14), 3189-3198 (2005).

[184] Hu X., Lei L., Ping Chu H., Lock Yue P., Copper/activated *carbon* as catalyst for organic wastewater treatment, *Carbon*, **37**, 631-637, (1999).

[185] Legube B., Leitner N.K.V., Catalytic ozonation. A promising advanced oxidation technology for water treatment, *Catal. Today*, **53** (1), 61-72 (1999).

[186]Kasprzyk-Hordern B., Ziolek M., Nawrocki J., Catalytic ozonation and methods of enhancing molecular ozone reactions in water treatment, *App. Catal. B: Environ.,* **46** (4), 639-669 (2003).

[187] Kaptijn J.P., The Ecoclear process. Results from full-scale installations, *Ozone: Sci. & Eng.*, **19**, 297-305 (1997).

[188] Logemann F.P., Annee J.H.J., Water treatment with a fixed bed catalytic ozonation process, *Water Sci. & Tech.*, **35**, 353-360 (1997).

[189] Lin S.H., Lai C.L., Kinetic characteristics of textile wastewater ozonation in fluidized and fixed activated *carbon* beds, *Wat. Res.*, **34**, 763-772 (2000).

[190] Zaror C., Soto G., Valdes H., Mansilla H., Ozonation of 1,2-dihydroxybenzene in the presence of activated *carbon*, *Water Sci. & Tech.*, **44**, 125-130 (2001).

[191] Beltran F.J., Rivas J., Alvarez P., Montero-de-Espinosa R., Kinetics of heterogeneous catalytic ozone decomposition in water on an activated *carbon*, *Ozone: Sci. & Eng.*, **24**, 227-237 (2002).

[192] Rivera-Utrilla J., Sanchez-Polo M., Mondaca M. A., Zaror C. A., Effect of ozone and ozone/activated *carbon* treatments on genotoxic activity of naphthalenesulfonic acids, *J. of Chem. Tech. & Biotech.*, **77**, 883-890 (2002).

[193] Sanchez-Polo M., Rivera-Utrilla J., Effect of the ozone-*carbon* reaction on the catalytic activity of activated *carbon* during the degradation of 1,3,6-naphthalenetrisulphonic acid with ozone, *Carbon*, **41**, 303-307 (2003).

[194] Beltran F.J., Acedo B., Rivas F.J., Gimeno O., Pyruvic acid removal from water by the simultaneous action of ozone and activated *carbon*, *Ozone: Sci. & Eng.*, **27**, 159-169 (2005).

[195] Beltran F.J., Garcia-Araya J.F., Giraldez I., Masa F.J., Kinetics of activated *carbon* promoted ozonation of succinic acid in water, *Ind. & Eng. Chem. Res.*, **45**, 3015-3021 (2006).

[196] Faria P.C.C., Orfao J.J.M., Pereira M.F.R., Ozone decomposition in water catalyzed by activated *carbon* : Influence of chemical and textural properties, *Ind. & Eng. Chem. Res.*, **45**, 2715-2721 (2006).

[197] Li L., Zhu W., Zhang P., Zhang Q., Zhang Z., AC/O3-BAC processes for removing refractory and hazardous pollutants in raw water, *J. of Hazard. Materials*, **135**, 129-133 (2006).

[198] Rivera-Utrilla J., Mendez-Diaz J., Sanchez-Polo M., Ferro-Garcia M.A., Bautista-Toledo I., Removal of the surfactant sodium dodecylbenzenesulphonate from water by simultaneous use of ozone and powdered activated *carbon* : Comparison with systems based on O3 and O3/H2O2, *Wat. Res.*, **40**, 1717-1725 (2006).

[199] Sanchez-Polo M., Salhi E., Rivera-Utrilla J., von Gunten U., Combination of ozone with activated *carbon* as an alternative to conventional advanced oxidation processes, *Ozone: Sci. & Eng.*, **28**, 237-245 (2006).

[200] Sanchez-Polo M., Rivera-Utrilla J., von Gunten U., Metal-doped *carbon* aerogels as catalysts during ozonation processes in aqueous solutions, *Wat. Res.*, **40,** 3375-3384 (2006).

[201] Valdes H., Zaror C.A., Heterogeneous and homogeneous catalytic ozonation of benzothiazole promoted by activated *carbon* : Kinetic approach, *Chemosphere*, 65, 1131-1136 (2006).

[202] Qu X., Zheng J., Zhang Y., Catalytic ozonation of phenolic wastewater with activated *carbon* fiber in a fluid bed reactor, *J. of Colloid & Interf. Sci.*, **309**, 429-434 (2007).

[203] Giraldez I., Garcia-Araya J.F., Beltran F.J., Activated *Carbon* Promoted Ozonation of Polyphenol Mixtures in Water : Comparison with Single Ozonation, *Ind. & Eng. Chem. Res.*, **46**, 8241-8247 (2007).

[204] Faria P.C.C., Orfao J.J.M., Pereira M.F.R., Activated *carbon* catalytic ozonation of oxamic and oxalic acids, *Appl. Catal. B: Environ.*, **79**, 237-243 (2008).

In: Ozone Depletion, Chemistry and Impacts
Editor: Sem H. Bakker, pp. 53-95

ISBN: 978-1-60692-007-7

Chapter 3

DECOLORIZATION EFFICIENCY OF OZONE AND OZONE DERIVATIVES FOR SEVERAL KINDS OF DYE IN AQUEOUS SOLUTION

Hanjin Luo[1]
College of Environmental Science and Engineering, South China University of Technology, Guangzhou 510006, People's Republic of China

Abstract

Eleven kinds of dyes from the three dye categories of azo, anthraquinone and lush were selected to investigate the decolorization efficiency of dyes by ozone in aqueous solutions. Changes in pH, total organic carbon (TOC), and electronic conductivity (EC) in solutions over reaction time were examined. The major components of dye derivatives during the degradation process were quantified. The results indicated that except for disperse dyes, the decolorization efficiencies of ozone for all soluble dyes exceeded 79% in 20 min. The sequence of the decolorization rate is: reactive > acid > direct > cationic > disperse. In the same type of soluble dyes, the decolorization rates were quicker for dyes with smaller molecular mass. With increasing reaction time, some acidic materials were produced, resulting in a decrease in pH. The final pH for all treatments was less than 4.5. During the reaction process, the decolorization efficiencies increased, while the pH and TOC content decreased. EC increased with the increasing of decolorization efficiency. Concentrations of the related ions (NO_3^-, NO_2^-, NH_4^+, Cl^-, $H_2PO_4^-$, and SO_4^{2-}) and organic derivatives during the reaction process were monitored. The results showed that complicated dye molecules were degraded to simpler organic compounds. Almost all the substituents such as –Cl and PO_4^{3-} in the dye molecules were transformed into Cl^-, $H_2PO_4^-$. Almost all $-SO_3H$ is converted to SO_4^{2-}. Nitrogen was partially degraded to NH_4^+ or NO_3^- due to the types of groups in dye molecules, and NO_2^- was not detected in the degraded solution. Other organics were detected and the organic acids were identified in the solutions. Based on the intermediates produced and the variation of their concentrations, some tentative degradation pathways were proposed.

Keywords: Dye; Decolorization;Ozone;Derivative;Organic acids; Degradation mechanism

---

[1] E-mail address: luohj@scut.edu.cn. 86 20 88375960 (Phone), 86 20 87110517 (Fax), Corresponding Author: Hanjin Luo

1. Introduction

Wastewater derived from the dye industry is characterized by high color and COD content, and pH varying from 2 to 12, which is very toxic, resistant to physicochemical treatments and not easily biodegradable [1-8]. The color of dye results from conjugated chains or rings that can absorb light of various wavelengths. The chromophores of dyes are usually composed of carbon carbon double bonds, diazo bonds, carbon nitrogen double bonds, and aromatic and heterocyclic rings containing oxygen, nitrogen or sulfur [9, 10]. Conventional biochemical oxidative treatment of wastewaters containing dye often results in colored water unfit for reuse, while other methods such as coagulation, absorption, ultra filtration, photo decomposition, etc., alone or in combination, are found to be ineffectual due to their cost, regeneration requirement or reusability, and secondary pollution [11-16]. Chemicals such as hypochlorite, ozone, and hydrogen peroxide, in the absence of and in the presence of UV light and hydrogen peroxide with ferrous ions have been used for pretreatment of dye-bearing wastewater [17-18]. Ozonation is an important wastewater chemical process. Ozone is very effective for decolorizing dye wastewaters because it attacks conjugated double bonds which are often associated with color [19-22]. Ozone reacts with aqueous compounds in two ways: directly corresponding to the action of molecular ozone, and an indirect path resulting from the decomposition of ozone to radicals, which is favored by basic pH in which the reaction is initiated by hydroxyl ions (OH^-) [23-25]. Muthukumar et al. (2004) reported that the decolorization efficiency of ozone was high either in acidic or alkaline conditions. Lower salt concentration gave faster decolorization of the effluent when maximum COD removal of 64% was obtained at lower salt concentration in alkaline condition. Some literatures [26-31] suggested that ozonation and UV enhanced ozonation were both very powerful for degrading dyes, but in comparison of ozonation alone, the photolytic ozonation showed little enhancement on degradation of dyes.

Only a few comparative studies have been carried out on the decolorization efficiency of ozone for various dyes, and on the composition of the derivatives. The objective of the present work was to compare the ozonation decolorization efficiency of ozonation for dyes of various chemical structures and types, and the variation of pH and TOC during the reaction. The derivatives in the degradation solutions, as well as the reasons for pH and TOC changes were also studied. The derivatives in the degradation solutions were analyzed by using an infrared spectrometric analyzer, ion chromatography and GC-MS.

2. Materials And Methods

2.1. Materials

The chemical structures and properties of dyes [32] are shown in Table 1. The dyes considered contain azo, anthraquinone or lush. The types of dyes include acidic, direct, dispersed, cationic and reactive. The experimental dyes were all analytically pure.

Table 1 Dye structure and property

No.	Name of dye	Dye structure	Type	λ_{max} /nm
1	Acid Light Yellow G	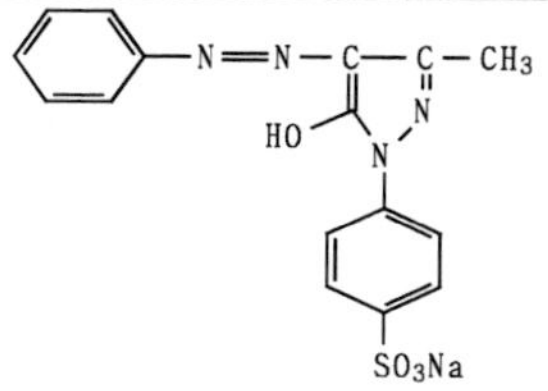	Azo	389
2	Acid Blue R	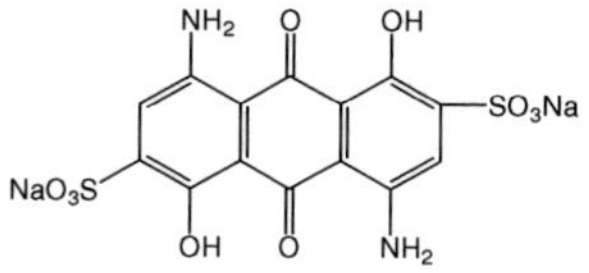	Anthraquinone	635
3	Reactive Light Yellow X-7G	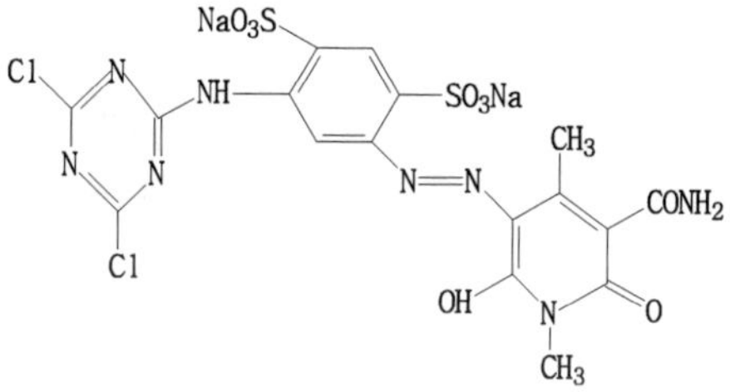	Azo	417
4	Reactive Brilliant Blue X-BR	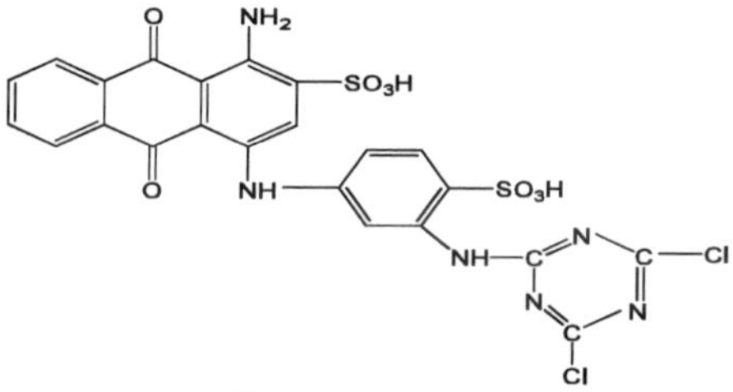	Anthraquinone	580
5	Direct Scarlet 4BS	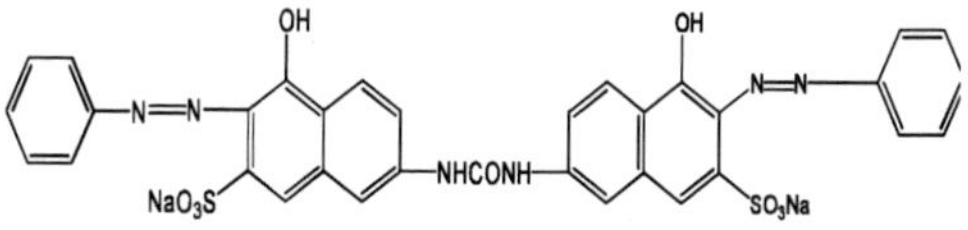	Azo	505
6	Direct Fast Blue B2RL	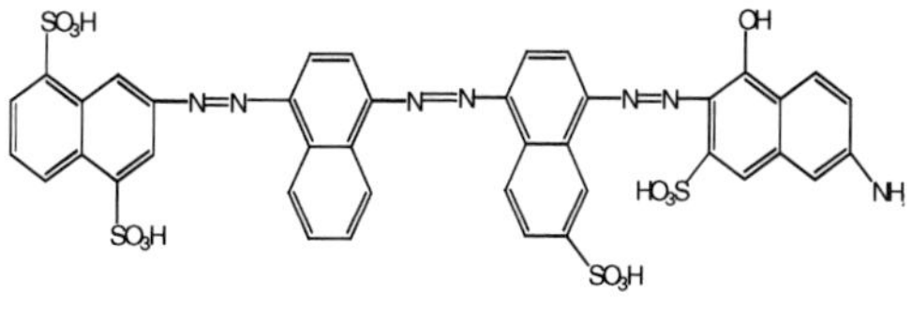	Anthraquinone	568
7	Cationic Red GTL	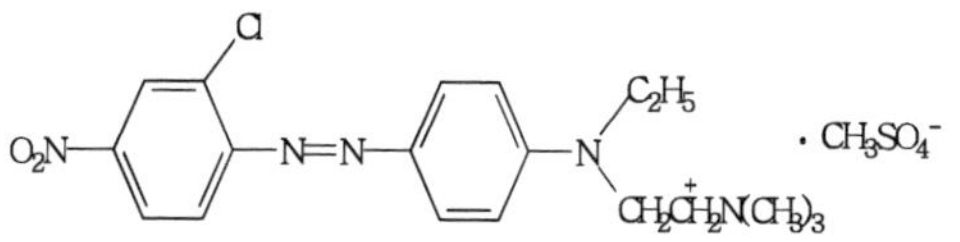	Azo	530
8	Cationic Blue FGL	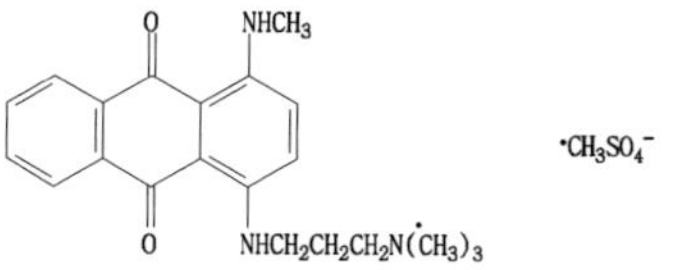	Anthraquinone	606

Table 1. Continued

No.	Name of dye	Dye structure	Type	λ_{max} /nm
9	Cationic Pink FG	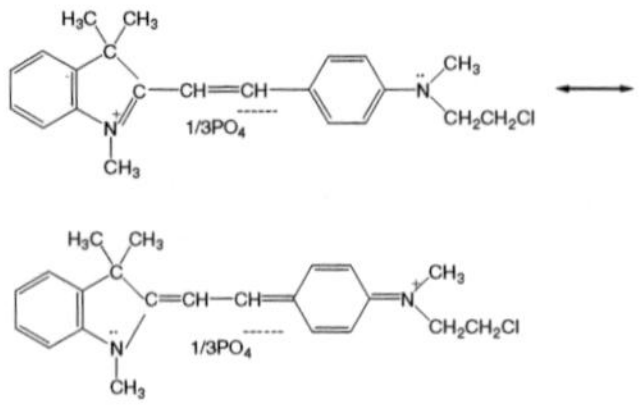	Lush	523
10	Dispersed Orange G	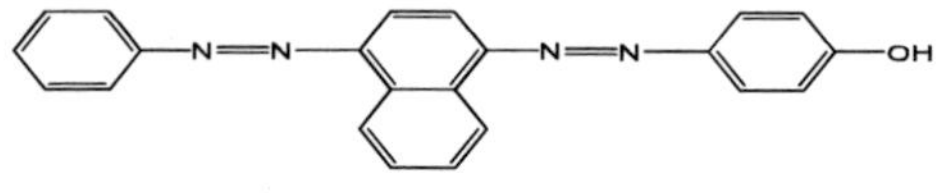	Azo	227
11	Dispersed Red 3B	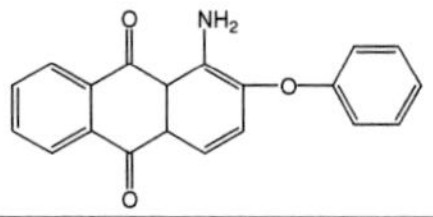	Anthraquinone	519

2.2. Experimental Set-up

The experimental set-up included an oxygen concentrator (Sim O_2 plus, China), a HF-3 model ozone generator (Ozonetek Ltd., Shanghai, China) which was used to produce a maximum of 3g ozone h^{-1}, an ozonation chamber with a capacity of 800 ml, and two gas absorption bottles. The oxygen flow rate to the generator was monitored with a rotameter incorporated into the ozone generator. Ozone gas was supplied to the bottom of the reactor. Excess ozone was passed into two gas absorption bottles containing 2% KI solution. All tubes from the ozone generator to the chamber and the gas absorption bottles were made of Neoprene and the fittings were made of Teflon.

2.3. Experimental Methods

500 ml of dye solution at a concentration of 300 mg L^{-1} was prepared and exposed to ozone in the ozonation chamber. 5 ml of the solution was sampled at 0, 1, 3, 5, 10, 15, 20, 25, 30 and 40 min.

For the preparation of GC-MS samples, the degraded aqueous solutions were extracted three times by 30 ml trichloromethane, at a pH of 2 and 11, adjusted by HCl and NaOH, respectively. The extracted phase was purged to 1 ml using pure nitrogen gas at 25 □, and then adjusted to 5 ml by adding acetone for GC-MS analysis.

2.4. Analytic Methods

The samples were determined through absorbance at maximum wavelength (λ_{max}) of the dyes by using a Hitachi UV–vis spectrophotometer (U-3210, Japan). The decolorization efficiency is calculated by the equation:

$$\eta = \frac{[(OD)_A - (OD)_B] \times 100\%}{(OD)_A} \tag{1}$$

Where η is the decolorization efficiency of dye, $(OD)_A$ is the absorbance for the initial aqueous solution at the maximum wavelength for the dye, and $(OD)_B$ is the absorbance for the sample of a particular time at the maximum wavelength for the dye.

The pH value was determined using a Thermo pH meter (Orion 210A+, USA). An O-I-Analytical TOC analyzer (1020A, USA) was used to measure TOC concentration to characterize the mineralization of dyes.

A Bruker infrared spectrometric analyzer (Tensor 27, Germany) was used to analyze the change in molecuar composition. After ozonation, the solutions were measured by using Dionex ion chromatography (ICS-2000, USA). 20 μl of the sample was injected into the Metrosep a Supp 4 anion column for analysis. The mobile phase for anion was a mixture of 1.8 mM Na_2CO_3, 1.7 mM $NaHCO_3$ and 5% acetone. The mobile phase for cationic was 11 mM H_2SO_4.

A Shimadzu GC-MS spectrometer (QP2010, Japan) was used to identify the degradation products. The capillary column used was an HP-5 (cross linked 5% phenyl methyl siloxane, 30.0 m×0.25 mm×0.25 μm). One μl of the solution was chromatographed under the following conditions: injector temperature was 280 °C, the initial column temperature was held constant at 40 °C for 2 min, ramped at 10 °C min^{-1} to 150 °C and held constant for 2 min^{-1}, then ramped further at 5 °C min^{-1} to 250 °C and held constant for 2 min, and finally ramped at 10 °C min^{-1} to 280 °C and held constant for 3 min.

3. Results and Discussion

3.1. Decolorization Efficiency

Figure 1 shows the decolorization efficiencies for the various dyes over time. In spite of the different structures, colors, and types, the decolorization efficiencies for all soluble dyes investigated, except dispersed dyes, exceeded 79% within 20 min. This indicates that ozone is very effective for decolourizing soluble dyes. The decolorization rate is high in the first 20 min, but then diminishes in the following 20 min. According to chemical reaction dynamics, the chemical reaction rate is directly proportional to the dye concentration. Hence the reaction is initially fast in the first 20 min (when the dye concentration is high), but slows in the following 20 min with depletion of the dye). As shown in Figure 1, decolorization rates for the dyes investigated are varied in the same time. The decolorization efficiencies in 20 min are shown in Table 2.

The sequence of the decolorization reaction rate is: Reactive Brilliant Blue X-BR > Acid Blue R Reactive Light Yellow X-7G > Acid Light Yellow G > Direct Fast Blue B2RL > Direct Scarlet 4BS >> Cationic Pink FG > Cationic Blue FGL > Cationic Red GTL > Dispersed Orange G > Dispersed Red 3B. Thus it can be seen that ozone has different oxidative effects on the various types and structures of dye. The sequence of the decolorization reaction rate is: Reactive > Acid > Direct > Cationic > Disperse. For the same type of soluble dyes, the decolorization rates were faster for dyes with smaller molecular mass.

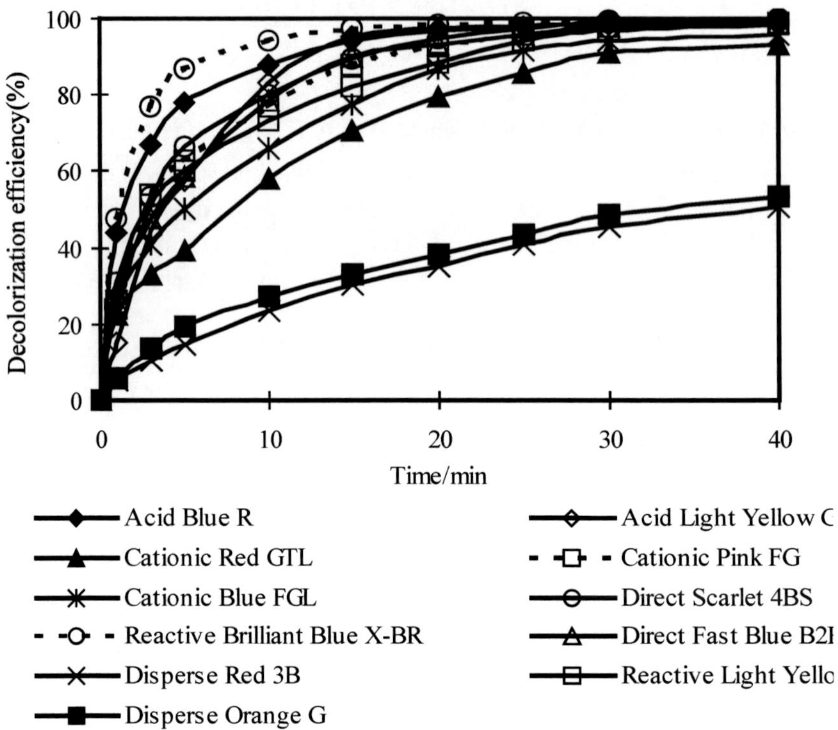

Figure 1. Decolorization efficiency vs. time.

Table 2. The decolorization efficiency in 20 min and 40min

No.	Decolorization efficiency（20min, %）	Decolorization Efficiency（40min, %）
1 Acid Light Yellow G	97.1	98.1
2 Acid Blue R	98.3	99.3
3 Reactive Light Yellow X-7G	97.7	98.7
4 Reactive Brilliant Blue X-BR	98.8	99.8
5 Direct Scarlet 4BS	94.3	96.8
6 Direct Fast Blue B2RL	95.8	97.6
7 Cationic Red GTL	79.5	93.3
8 Cationic Blue FGL	87.1	96.0
9 Cationic Pink FG	92.8	96.0
10 Dispersed Orange G	38.2	53.3
11 Dispersed Red 3B	35.2	50.6

The dyes of Acid Light Yellow G and Reactive Brilliant Blue X-BR have the easily ruptured imine (−NH−). However, because the imine in Acid Light Yellow G is far away from chromophores diazo bonds (e.g., −N=N−), the rupture of the imine by ozone does not accelerate decolorization. Hence, the decolorizration rate of ozone for Acid Light Yellow G is less than that of Reactive Brilliant Blue X-BR. Since Acidic dyes have only one diazo bond (−N=N−), they are degraded faster than Direct dyes, which contain three diazo bonds. Since the molecules in Cationic dyes carry charges, their reaction rates with O_3 are reduced. Dispersed dyes have the lowest degradation efficiencies due to their low solubility.

3.2. pH

In order to study the changes of solution properties, pH was measured in the experimental process. The changes in pH over time is shown in Figure 2, which indicates a decrease of pH value with increasing reaction time, and that the final pH values are less than 4.5. The final pH value of Reactive Light Yellow is the lowest of 3.45.

As shown in Figure 1 and Figure 2, the decolorization rates of the dyes increase significantly along with the evident decrease of the pH value in the first 20 min. The pH value has a mild decreasing trend with the reduction of decolorization rate 20 min after reaction. It indicates that the faster the decolorization rate, the quicker the decrease of pH. The former has the consistent changing trend with the latter.

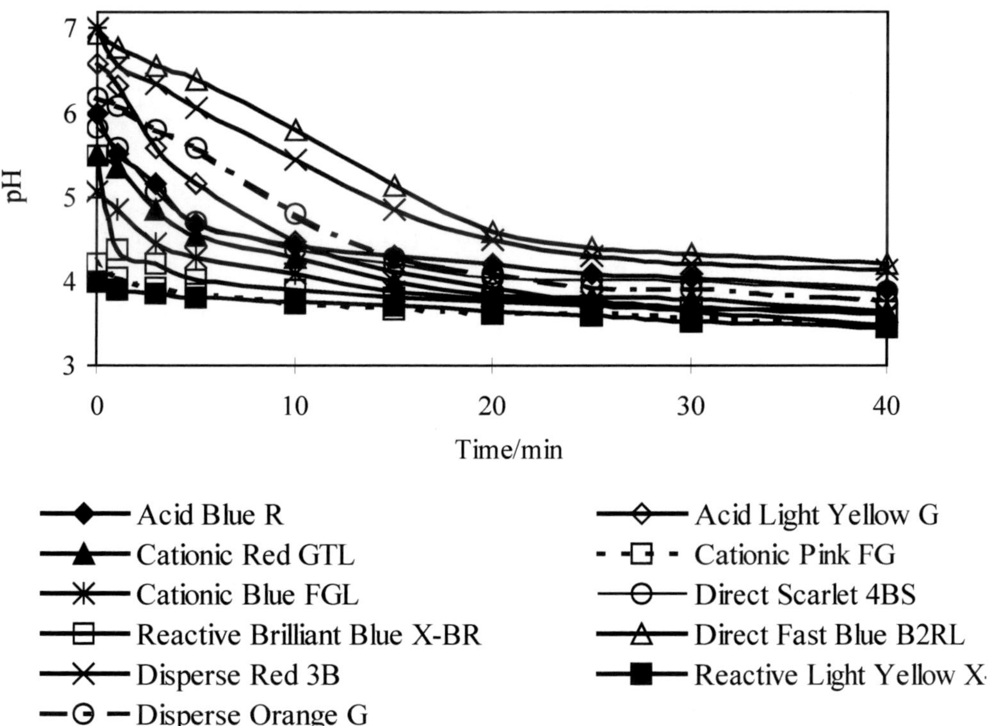

Figure 2. pH vs. time.

Hence, it is deduced that H^+ was released in the reaction and the dye molecules could possibly be decomposed to organic acids or inorganic acids. It can be concluded that new acidic materials are produced in the dye degradation process. The main reasons may be: (1)

O_3 reacts with H_2O and then releases H^+; (2) Dye molecules are degraded into low molecular weight organic acids and inorganic acids.

Among the eleven kinds of dyes being used in the experiment, Acid Light Yellow has the largest drop in pH, with the ΔpH of 3.12. The pH of Reactive Light Yellow X-7G is always low over reaction time. Therefore, the drop of its pH during ozonation is the smallest.

At the same time, by comparing the pH value of the original solutions and the ΔpH value among eleven kinds of dyes, Table 6 shows that the higher the pH of the original solutions, the greater the drop in pH. The reasons may be because a plenty of new materials are produced in the solutions during the reaction with the ozone, the H^+ and other acidic materials would inhibit the reaction of ozone with dyes and intermediate products. In the dye solutions with minor acidic materials (comparatively large pH value), the reaction between ozone and solutions proceeded more quickly than in those with low initial pH, with a large drop in pH.

3.3. Variation of UV-vis Spectrum and Its Dynamics in Dye Ozonation

Decolorization efficiencies of eleven kinds of dyes were reported in 3.1. This showed that the breakage of the main chromophores is important, but the study reported there can not display the reaction status of other chromophores or the production of the new chromophores. Spectrum at the ultraviolet and visible wavelengths has been widely used to study the dynamics of dye reaction.

3.3.1. UV-Vis Spectrum during Dye Ozonation

According to the analysis of decolorization of dyes, 0~20 min period belongs to fast reaction phase with high decolorization rate. The variation of UV-vis spectrum and its mechanisms in this period are reported in this chapter. Figure 3(1) - (11) shows the variation of UV-vis spectrum for 11 kinds of dyes during the first 20 min of the ozonation treatments.

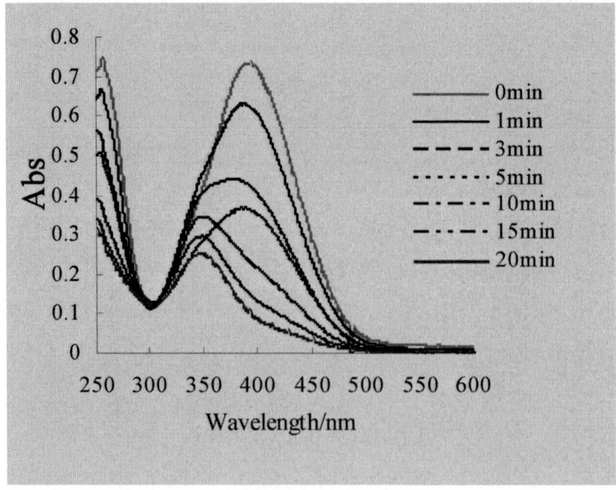

Figure 3(1). The UV-vis spectrum of Acid Light Yellow G.

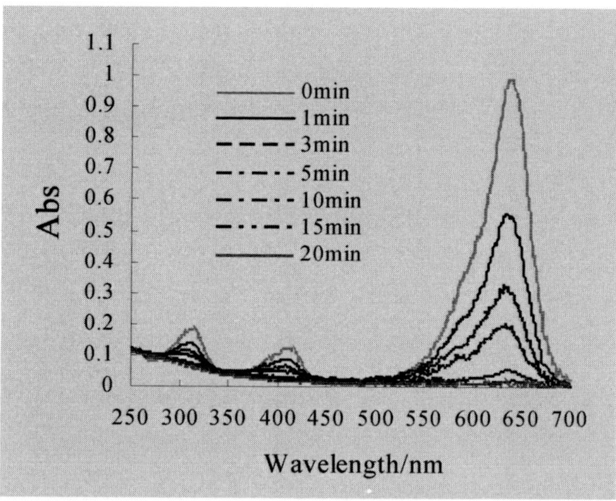

Figure 3(2). The UV-vis spectrum of Acid Blue R.

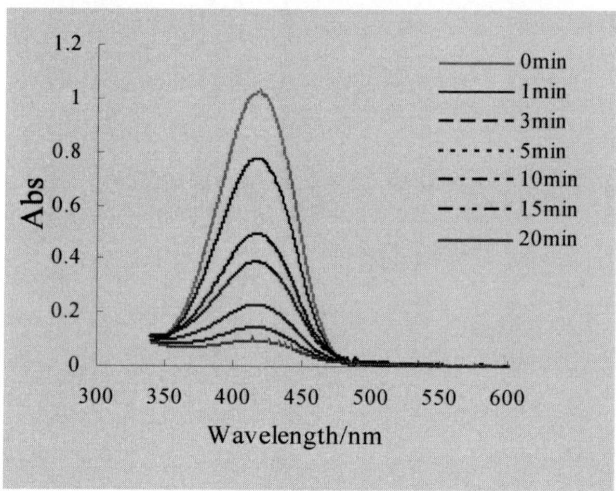

Figure 3(3). The UV-vis spectrum of Reactive Light Yellow X-7G.

As shown in Figure 3(1) - 3(11), because of the strong oxidizability of ozone, the absorbance of the 11 dyes decreased not just in a certain wave band, but over the whole ultraviolet-visible spectrum. It indicates that ozone oxidizes and degrades the whole dye molecule, not just for some certain functional groups. For example, Acid Blue R has two chromophores : a quinoid group and an acetophenone group.

In Figure 3(2), the absorption peak (635 nm) in the long wave band is caused by the quinoid group, which belongs to $n \rightarrow \pi^*$ transition; the absorption peak (420nm) in the short wave band is caused by the acetophenone group, which belongs to superposition of $n \rightarrow \pi^*$ transition and benzene absorption.

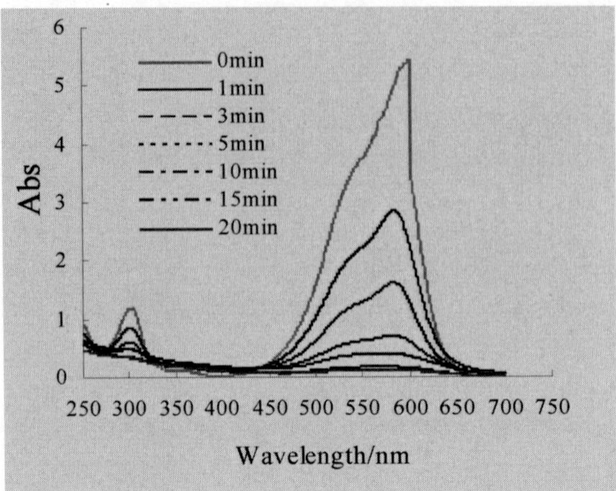

Figure 3(4). The UV-vis spectrum of Reactive Brillant X-BR.

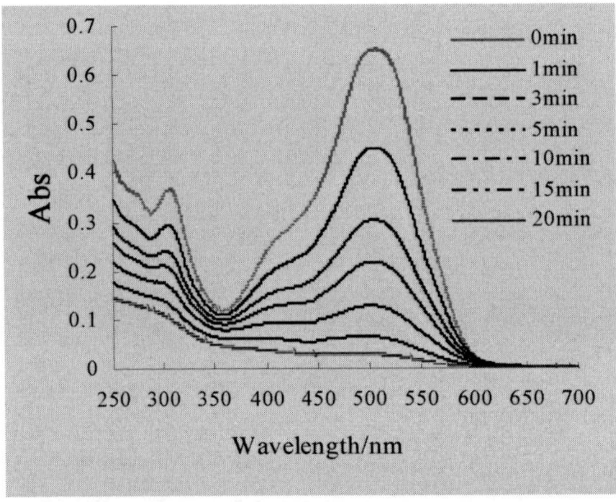

Figure 3(5). The UV-vis spectrum of Direct Scarlet 4BS.

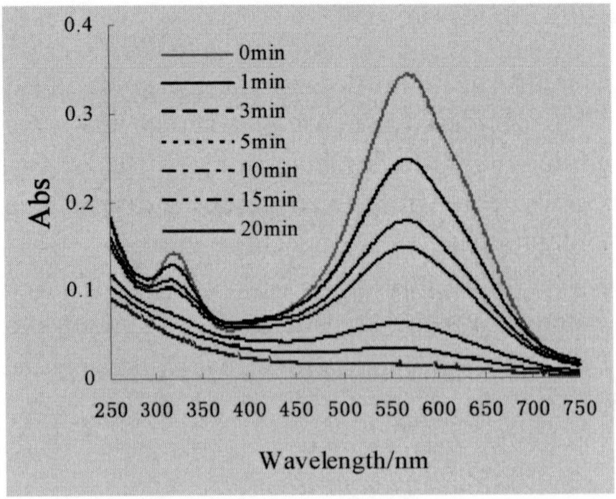

Figure 3(6). The UV-vis spectrum of Direct Fast Blue B2RL.

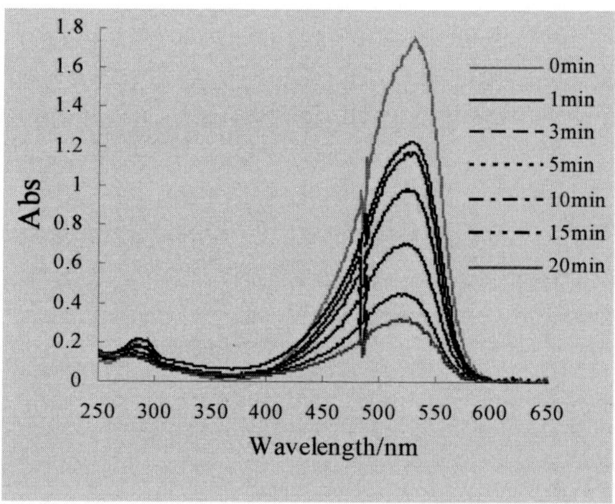

Figure 3(7). The UV-vis spectrum of Cationic Red GTL.

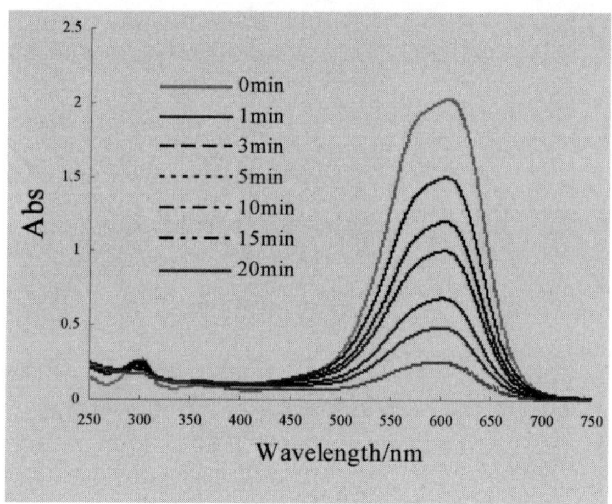

Figure 3(8). The UV-vis spectrum of Cationic Blue FGL.

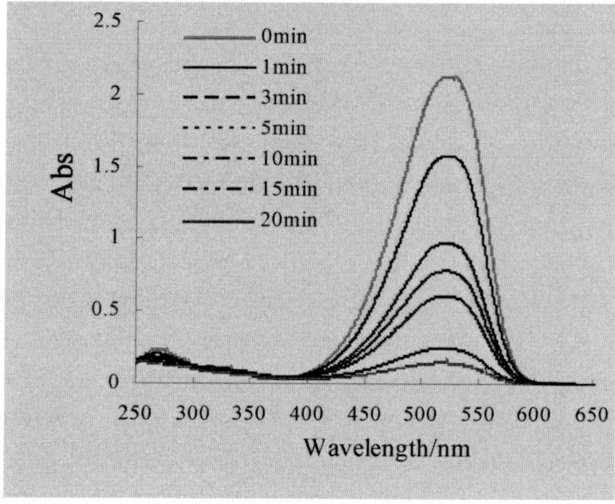

Figure 3 (9). The UV-vis spectrum of Cationic Pink FG.

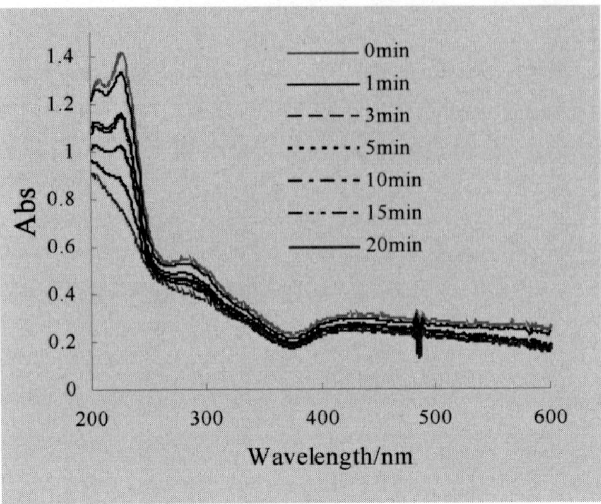

Figure 3 (10). The UV-vis spectrum of Dispersed Orange G.

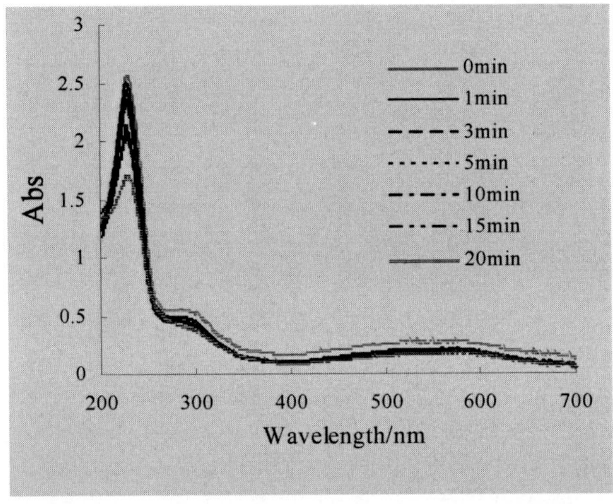

Figure 3 (11). The UV-vis spectrum of Dispersed Red 3B.

The two absorption peaks decreased in wavelengths gradually with increasing reaction time. In addition, the reaction rate of the quinoid group is higher than that of the acetophenone group. The quinoid group is more active than the acetophenone group and can react with ozone more easily with consequent dye degradation.

Figure 3(2) indicated that the maximum absorption peaks transited to shorter wave band in the UV-vis spectrum curve, which was a violet-shift. This phenomenon was seen in Acid Light Yellow G, Acid Blue R, Active Brilliant Blue X-BR, Direct Fast Blue B2RL, Cationic Red GTL, Cationic Blue FGL, and Cationic Pink FG. It indicated that new materials with smaller absorption intensity were produced in ozonation processes. Electron-donating groups were broken from dye molecules, which reduced the density of the electronic cloud of the color base and caused a violet-shift. In some case, the maximum absorption peaks shifted to the longer wave band in the UV-vis spectrum curve, which was a red-shift. This existed in

Reactive Light Yellow X-7G, Direct Scarlet Red 4BS, Dispersed Orange G and Dispersed Red 3B.

It indicated that substances with larger absorption were produced in the degradation processes. The electron-losing groups were broken from the dye molecules, which increased density of electron cloud of the color base and caused a red-shift.

3.3.2. Dynamics of Dye Ozonation

The direct reaction of ozone is assumed to be first order with respect to both dye and ozone concentration. When the amount of ozone is in excess or the ozone concentration is assumed to reach a stationary stage at the interface, the oxidation rate follows pseudo-first-order kinetics with respect to the concentration of the organic substances. In this study, the kinetics of dyes ozonation was evaluated by plotting ln (C_0/C_t) values versus reaction time following the equation:

$$\ln (C_0/C_t) = k_d \cdot t \tag{2}$$

Where C_t and C_0 are dye concentrations at reaction time t = t and t = 0 min, respectively. k_d stands for the pseudo-first-order reaction rate constant.

For Reactive Light Yellow X-7G, the following curve shows decoloration process can be described by the pseudo-first-order kinetics. The coefficient (R^2) of the determination is 0.986 and the slope of the log-linear curve was $k_d = 0.140$ min^{-1}.

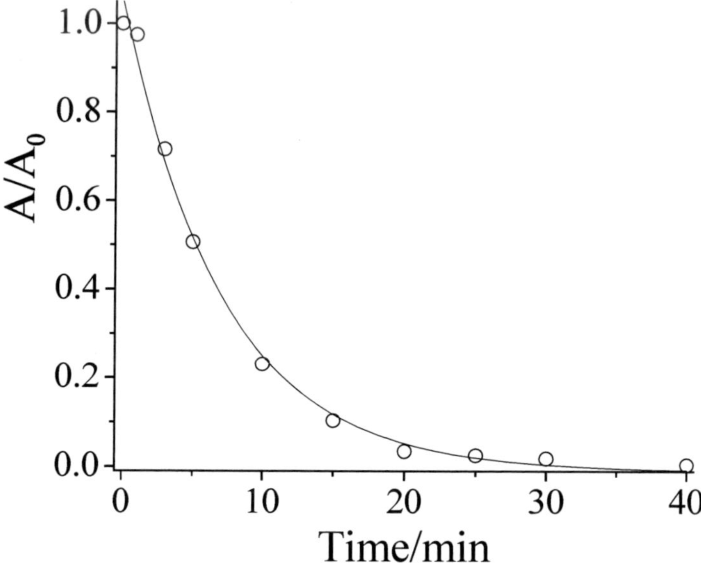

Figure 4. A/A_0 vs. time for Recative Light Yellow X-7G.

It was found out that the pseudo first-order behaviors were observed during ozonation of for all the other 10 kinds of dyes. The results are shown in Table 4. A similar result has been widely reported in many literatures, such as López-López et al. [33].

Table 4. Kinetic equation of ozonation for 11 dyes

Dyes Name	Regression Equation	k_d / min^{-1}	R^2
Acid Light Yellow G	$\ln(C_0/C_t)=0.135t+0.109$	0.135	0.985
Acid Blue R	$\ln(C_0/C_t)=0.166t+0.4378$	0.166	0.946
Reactive Light Yellow X-7G	$\ln(C_0/C_t)=0.141t+0.243$	0.141	0.986
Reactive Brilliant Blue X-BR	$\ln(C_0/C_t)=0.177t+0.554$	0.177	0.971
Direct Scarlet Red 4BS	$\ln(C_0/C_t)=0.121t+0.219$	0.121	0.987
Direct Fast Blue B2RL	$\ln(C_0/C_t)=0.132t+0.127$	0.132	0.995
Cationic Red GTL	$\ln(C_0/C_t)=0.078t+0.146$	0.078	0.982
Cationic Blue FGL	$\ln(C_0/C_t)=0.092t+0.158$	0.092	0.980
Cationic Pink FG	$\ln(C_0/C_t)=0.113t+0.203$	0.113	0.996
Dispersed Orange G	$\ln(C_0/C_t)=0.031t+0.022$	0.031	0.970
Dispersed Red 3B	$\ln(C_0/C_t)=0.019t-0.002$	0.019	0.995

3.4. Mineralization of Sample Solutions and Release of the Ions

As shown in Figure 5, electrical conductivity increases with increasing reaction time. The electrical conductivity increment in the first 20 min is larger than in the next 20 min. The graphs of electrical conductivity, decolorization efficiency and pH over time have consistent variation trends. The changing trend of first 20 min is larger than second 20 min. The electrical conductivity of eleven kinds of dyes increased 20 min after reaction, and Cationic Blue FGL solution took the largest increment, 280 μs/cm. Acid Light Yellow G had the smallest increase, 46 μs/cm. 40 min later, Cationic Blue FGL had the highest increase of electrical conductivity 380 μs/cm. On the contrary, with a 72 μs/cm increment in electrical conductivity, Direct Fast Blue B2RL occupied the smallest increase among 11 kinds of dyes. The increase of the electrical conductivity demonstrated that ions were produced in the reaction process. That means the mineralization in the dye solutions increased and dye molecules were decomposed gradually. Dye molecules were decomposed to many kinds of ions or other substances during reaction with ozone.

During the reaction process, the increasing of the electrical conductivity in sample solutions indicates that dye molecules were decomposed to many kinds of ions or other substances after reaction with ozone. The variation trend of electrical conductivity is consistent with decolorization efficiency and pH.

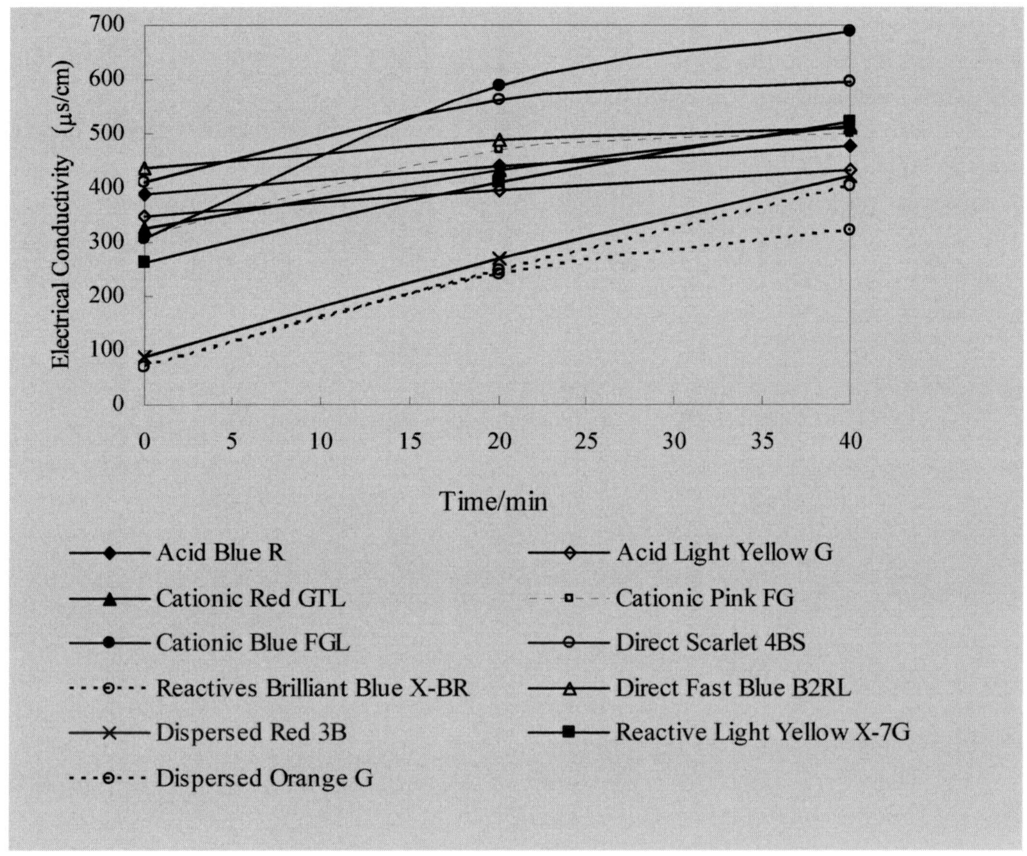

Figure 5. Electric conductivity of samples vs time.

3.5. Formation of Ions in Ozonation of Dyes

Mineralization usually implies the production of inorganic products, since hetero-atoms are converted into anions, in which they are at their highest oxidation state [34].

For eleven kinds of dyes investigated, the elements like S, N, Cl and P can be transformed to inorganic substances. The concentrations of SO_4^{2-}, Cl^-, NO_2^-, NO_3^- and $H_2PO_4^-$ were measured. Figures 6-15 showed the changes of the concentrations of SO_4^{2-}, Cl^- , NO_2^-, NO_3^- and $H_2PO_4^-$ during the ozonation. As shown in these figures, 40 min after the reaction between dyes and ozone, except for the production rates of SO_4^{2-} in dye No.3, No.4 are comparatively low, which are 78% and 81%, respectively, other dyes have reached up to 95% (the production rate is the ratio of S molar concentration in the SO_4^{2-} to its content in original dye molecule).

3.5.1. Comparative Study on SO_4^{2-} Roduction in Azo and Anthraquinone Dyes

Generally speaking, the formation of SO_4^{2-} during ozonation of dye solution is large. A similar phenomenon was reported by He et al. [35]. Figure 6 shows the formation of SO_4^{2-} from acid blue solution is larger than from acid light yellow. One of the reasons may be that acid blue molecule has two $-SO_3H$, but acid light yellow molecule has only one. In addition, both the $-SO_3H$ link to the anthraquinone ring in the acid blue molecule, but $-SO_3H$ of acid light yellow molecule links to a benzene ring, so the former is more active in ozonation. The formation rate of SO_4^{2-} in reactive yellow solution is much greater than in reactive blue.The production rate of SO_4^{2-} in Reactive Yellow is similar to that of Reactive Blue.

As shown in Figure 7, dyes No. 5 and 6 have a similar formation rate of SO_4^{2-}. The reason is $-SO_3H$ of two dyes link to naphthalene ring and thus have similar activity in ozonation. In the dyes No. 7 and 8, the formation rates of SO_4^{2-} in cationic red and cationic blue solution are similar too. Both dyes are ionic solution, so free $CH_3SO_4^-$ ions in the two dyes can be oxidized to SO_4^{2-} at similar rates.

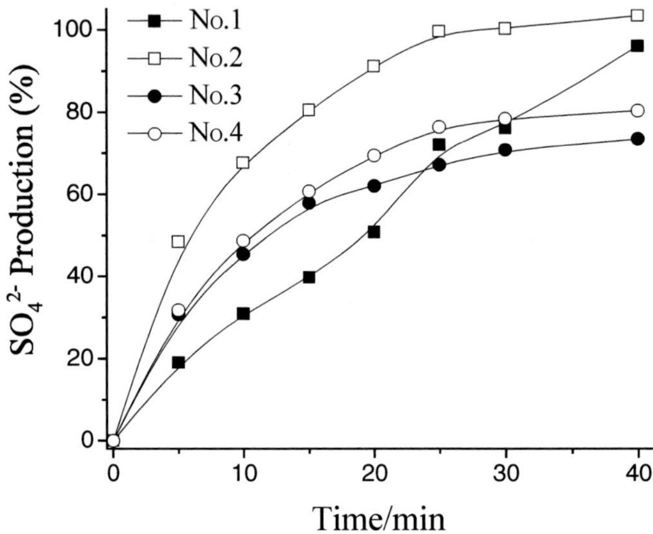

Figure 6. SO_4^{2-} production (%) of Acid and Reactive dyes.

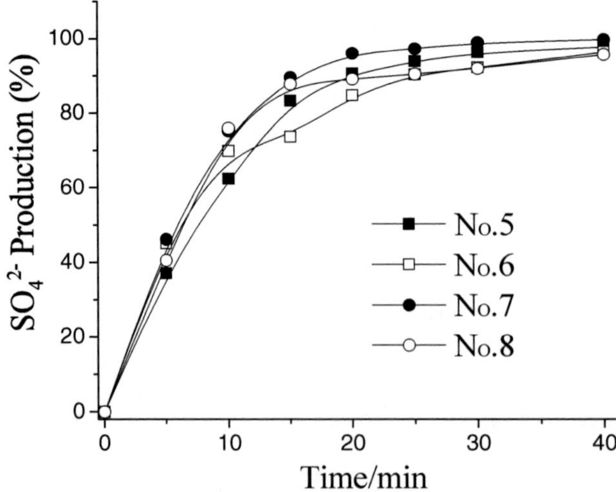

Figure 7. SO_4^{2-} production (%) of Direct and Cantionic dyes.

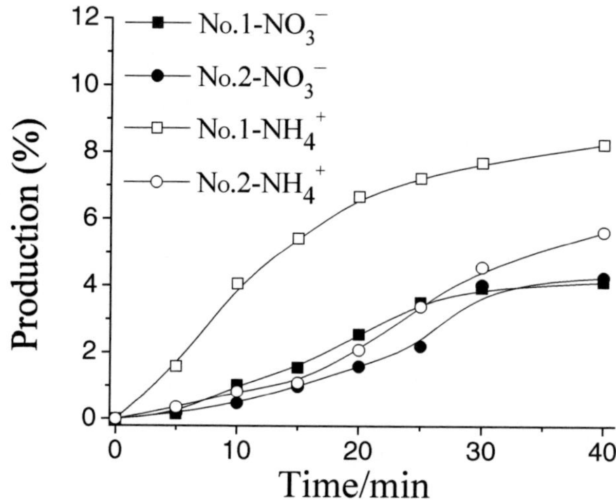

Figure 8. NH_4^+ and NO_3^- production (%) of Acid dyes.

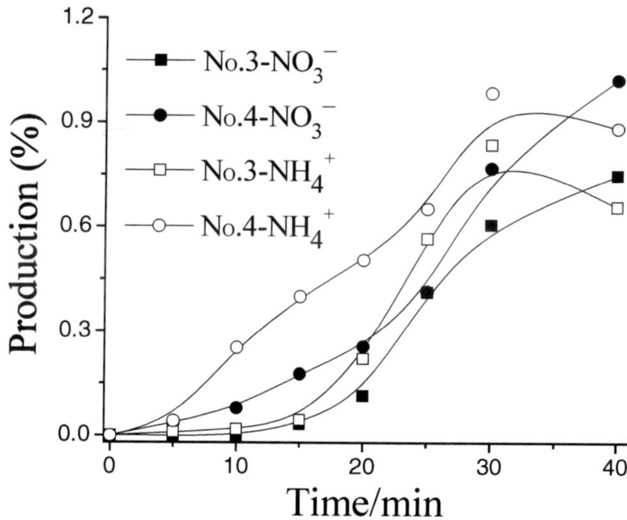

Figure 9. NH_4^+ and NO_3^- production (%) of Reactive dyes.

From the above analysis, among the same dye types, the production rate of SO_4^{2-} depends on the place where $-SO_3H$ links in the dye structure. But there are no other obvious differences among different dye structures.

3.5.2. Comparative Study on NH_4^+ and NO_3^- Production in Azo and Anthraquinone Dyes

Nitrogen is the main element of dye color chromophores. Studying the effect of ozone on the transformation of the dye structure is of great importance to clarify degradation mechanisms. As shown in Figures 8 – 13, after degradation, the production rates of NO_3^-, NH_4^+ were low.

The production rates of NO_3^-, NH_4^+ after partial degradation were also investigated for two dyes with similar structure and different types (anthraquinone and azo).

As shown in Figures 8 – 11, the production of NH_4^+ is larger than for NO_3^- in dye solution in the first 20 min. The concentrations of NH_4^+ in Direct Red, Reactive Yellow, Reactive Blue and Cantionic Red solutions began to decrease from about 25th or 30th min. The formation of NH_4^+ occurred rapidly and the formation of NO_3^- occurred slowly during ozonation.

The reasons need further elaboration. As shown in Figure 8, the production rates of NH_4^+ and NO_3^- in dye No.2 were higher than in No.1. However, in dye No.2, the $–NH_2$ group linking to the benzene is more active than the nitrogen in the ring. The production mechanisms cannot be explained completely from analysis of the production rates of NH_4^+ and NO_3^-.

The three nitrogen atoms in dye No.8 combine in the form –C–N–. Dye No.7 has only two nitrogen atoms linking in the form –C–N–. Through, dye No.7 has more nitrogen than No.8 (five vs. three); the former has nitrogen-nitrogen double bonds which are released in the form of N_2 and another nitrogen atom exists in the form of NO_2^- which was converted to NO_3^- after degradation. The production rate of NH_4^+ in dye No.7 is smaller than in No.8 and dropped 25 min after degradation. However, the production rate of NO_3^- in dye No.7 is larger than in No.8.

As depicted in Figure 9, the NH_4^+ and NO_3^- production in Reactive Blue solution is greater than in Reactive Red solution. This can be explained from the molecular structure. There are three N atoms exist in triazinyl in the reactive blue molecule, which can hardly be oxidized. Half of all N atoms in Reactive Blue molecules can be transformed into NH_4^+ and NO_3^- ions. But there are only three N atoms which can be transformed into NH_4^+ and NO_3^- ion from the eight N atoms in Reactive Red Molecules. One N atom of the three N atoms exists in a heterocycle, which is less reactive than the substituents. The NH_4^+ and NO_3^- production in Reactive Blue solution are larger than Reactive Red solution.

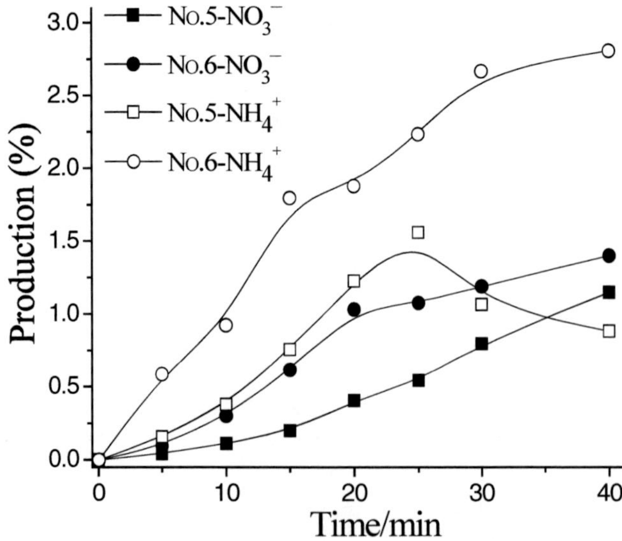

Figure 10. NH_4^+ and NO_3^- production (%) of Direct dyes.

As observed in Figure 10, NH_4^+ and NO_3^- production in Direct Blue solution is larger than in Direct Red solution. There are two azo bonds and three N atoms in other forms in Direct Red molecules, but there is only one N atom in other forms among the seven N atoms in Direct Blue molecule. All the N atoms in Direct Red molecules are in the interior of conjugation groups, and no N exists in subtituents. Direct blue molecules contain a $-NH_2$ substituent, which is active in ozonation, N atoms of Direct Blue molecules can be transformed into NH_4^+ and NO_3^- ions to a greater percentage.

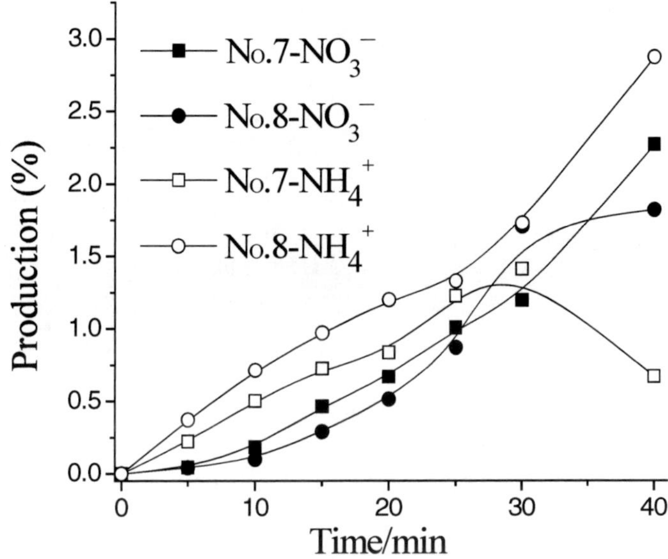

Figure 11. NH_4^+ and NO_3^- production (%) of Cantionic dyes.

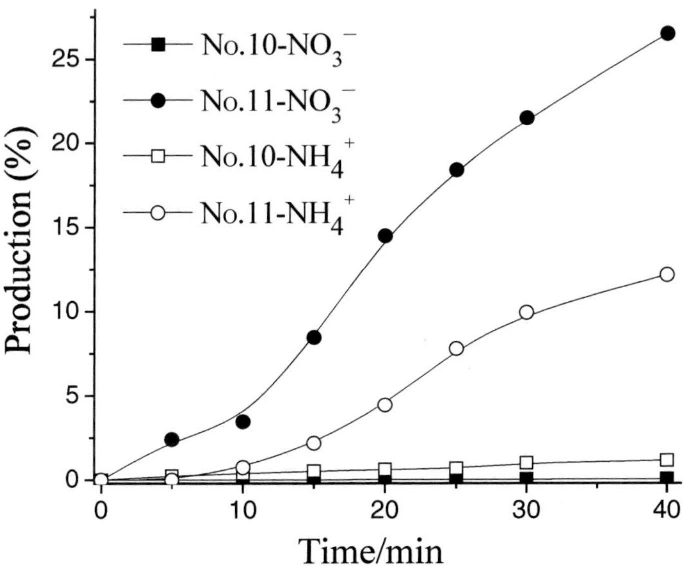

Figure 12. NH_4^+ and NO_3^- production (%) of Dispersed dyes.

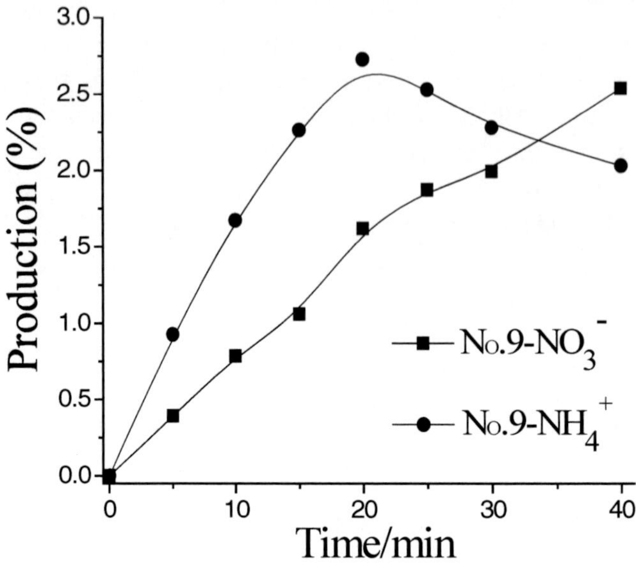

Figure 13. NH_4^+ and NO_3^- production (%) of Cationic Pink.

Figure 12 illustrates that the production of NH_4^+ and NO_3^- in Dispersed Red solution reached 12% and 26.4%. The only N in Dispersed Red molecules exists in the form of $-NH_2$ which means that the $-NH_2$ can be protonated to NH_4^+ and be further oxidized to NO_3^-. The NH_4^+ and NO_3^- production in Dispersed Orange are nearly zero. The reason is that Dispersed Orange molecules contain two azo bonds and no N atom in other forms, which proves that azo bond is oxidized to nitrogen gas in the ozonation process [36].

Figure 13 shows the production rate of NO_3^- and NH_4^+ in the cyanine type of dyes after degradation. Form curves for the same type but different chemical structure, taking the dye No. 7 and 8 as example, the production rate of NO_3^- and NH_4^+ in dye No.9, No.7 are similar. The two curves show the same downward trends 25 min after degradation. The Nitrogen group in the dye No.9 is more active than the Nitrogen in the dye No.7. The production rates of NO_3^- and NH_4^+ in dye No.9 are greater than in No.7.

According to the above analysis, the form, chemical structure, and chemical charge of nitrogen in the dye structure varied. So did the reaction paths with ozone.

3.5.3. Comparative Study on Cl^- Production of Azo, Anthraquinone and Lush Dyes

Figure 14 shows the production rate of Cl^- in Reactive Yellow (No.3) is a little larger than in Reactive Blue (No.4). The Cl^- groups in both two dyes linked to the triazine and have the same reaction capabilities. The Reactive Yellow structure has two Cl^- groups linking to the triazine group. Due to the repulsion effects between the two Cl^- groups in the Reactive Yellow, the reaction capability of Cl^- group in Reactive Yellow is larger than in Reactive Blue and accelerates the formation of Cl^- in Reactive Yellow.

Dyes No.7 and No.9 have different types of chemical structure, but have similar production rates for Cl^-. Based on the above analysis, no matter what kind of groups link with the –Cl group, they are transformed and converted to equivalent amounts of Cl^- after degradation.

As shown in Figure 15, the PO_4^{3-} group in dye No.9 was transformed to $H_2PO_4^-$ after degradation in the solutions. During the ozonation process, the results indicated that NO_2^- didn't exist in any of the eleven dyes. NO_2^- is rapidly oxidized to NO_3^-.

From the above analysis, it can be concluded that during ozonation of dye solutions, the formed ions are related to the molecule structure of the dye and the number and position of the reactive groups. Because there are many reactive groups, not only SO_4^{2-}, NH_4^+, NO_3^-, Cl^-, PO_4^{3-} can be produced during ozonation, but also acid and volatile organic compounds. So the degradation rate of dye solution cannot be completely specified by the concentration and production rate of ions in solution. The degradation rate of dye solutions is a complex result of ozonation.

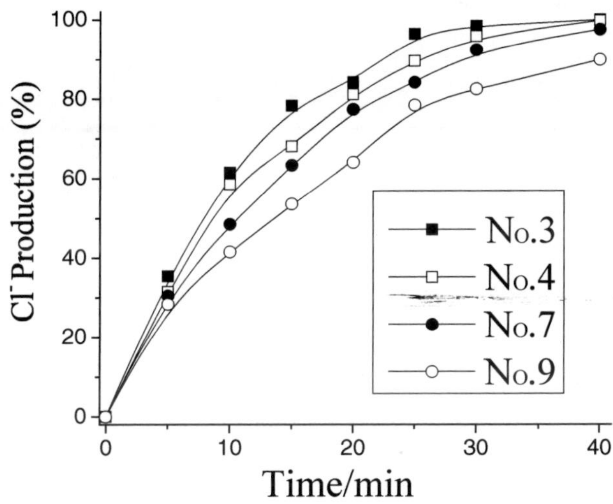

Figure 14. Cl^- production (%) of 4 dyes.

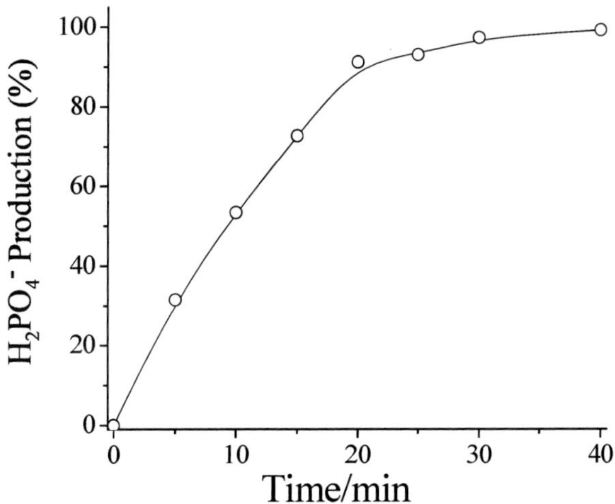

Figure 15. $H_2PO_4^-$ production (%) of Cationic Pink.

The concentrations of various kinds of cations and anions in the degradation solutions were measured by using ion chromatography, and were also calculated by stoicheiometry. The results are showed in the Table 5. As shown in Table 5, the theoretical concentration of N (calculated by stoichiometry) and the actual concentration of NH_4^+, NO_2^- and NO_3^- (measured using ion chromatography) were compared. It is observed that the actual concentration is much less than the theoretical one, indicating that N in the dye molecule is not completely decomposed to NH_4^+, NO_2^- or NO_3^-, and some other compounds containing N are produced. At the same time, the actual concentrations of NO_2^- in the degradation solutions of all the dyes are zero, indicating that NO_2^- is oxidized by ozone to NO_3^-.

Table 5 Comparison of stoichiometric and measured concentration for N, S, Cl, P

No.	Theoretical molar concentration by stoichiometry ($mmol\cdot L^{-1}$)				Actual molar concentration by ion chromatography ($mmol\cdot L^{-1}$)					
	N	S	P	Cl	NO_2^-	NO_3^-	NH_4^+	SO_4^{2-}	$H_2PO_4^-$	Cl^-
1	3.158	0.789	0	0	0	0.130	0.261	0.767	0	0
2	1.266	1.304	0	0	0	0.054	0.071	1.256	0	0
3	3.687	0.922	0	0.922	0	0.028	0.024	0.722	0	0.922
4	2.826	0.942	0	0.942	0	0.029	0.025	0.757	0	0.940
5	2.583	0.738	0	0	0	0.029	0.023	0.721	0	0
6	2.232	1.275	0	0	0	0.031	0.063	1.124	0	0
7	2.991	0.598	0	0.598	0	0.068	0.020	0.591	0	0.583
8	1.944	0.650	0	0	0	0.035	0.056	0.621	0	0
9	1.562	0	0.260	0.781	0	0.041	0.032	0	0.258	0.702
10	3.409	0	0	0	0	0.002	0.039	0	0	0
11	0.946	0	0	0	0	0.251	0.115	0	0	0

3.6. Formation of Ions in Degradation of Dyes Ozonation by Ozone

It is deduced that the degradation solutions of Dispersed Orange G did not contain NO_3^-. Little NO_3^- exists in Acid Light Yellow G, Reactive Light Yellow X-7G, Reactive Brilliant Blue X-BR, Direct Scarlet 4BS, Direct Fast Blue B2RL, Cationic Red GTL, Cationic Blue FGL, and Cationic Pink FG. The reasons are that the compounds contain diazo bonds or aniline are oxidized to N_2 by ozone. N, N'-R_1,R_2-substituted arylamines are oxidized easily to N, N'-R_1,R_2-substituted arylamino oxides. Due to space constrains in R_1, R_2 and Ar, N, N'-R_1, R_2-substituted amino oxides break off from the benzene ring easily and are then partially oxidized to NO_3^-.

It is notable that aniline exists in both Acid Blue R (No.2) and Dispersed Red 3B (No.11) dye molecules.

Although NO_3^- was produced in the degradation solutions of many kinds of dyes, their production rates are different, which signifies the relation between production rate of NO_3^-

and the dye molecular structure. For example, for Dispersed Red 3B (No.11), the calculated ratio of NO_3^- concentration in the solutions after degradation to the total Nitrogen content is 26.4%, while the value for Acid Blue R is 4.27%.

The anthracene ring in Dispersed Red 3B (No.11) has a —O—(benzene ring) phenoxy group. The ether bond in the anthracene ring can be easily broken, followed by the rupture of the anthracene ring. Along with the rupture of the anthracene ring, the $Ar\text{-}NH_2$ is oxidized to NO_3^-. However, Acid Blue R (No.2) has a more stable structure than Dispersed Red 3B (No.11); its $Ar\text{-}NH_2$ is oxidized to $Ar\text{-}NO_2$, rather hard be transformed to NO_3^-.

NH_4^+ existed in the degradation aqueous solutions for eleven kinds of dyes. The formation mechanisms have been reported already. In the molecule of Dispersed Orange G, nitrogen exists in the form of nitrogen-nitrogen double bonds. Generally, it's believed that nitrogen is transformed to N_2 during ozonation. NH_4^+ is detected in the degradation solutions and will be further oxidized to NO_3^-. Further investigations are needed to reveal the reaction mechanisms. After reaction, –Cl and PO_4^{3-} (chelated state) in the dye molecules are transformed into Cl^- and $H_2PO_4^-$. The SO_4^{2-} production rates in dye No.3 and No.4 are relatively low, approximately 78% and 81%, respectively. The $-SO_3H$ in other dyes is transformed to SO_4^{2-} completely.

It is concluded that the derivative organic molecules contain hardly any molecules and other groups consisting of S, Cl and P. The pH in degradation solution is lower than if all the NO_2^-, NO_3^-, SO_4^{2-}, $H_2PO_4^-$ and Cl^- were present as acids. It indicates that there are organic acids in the degradation solution.

3.7. Changes to Dye Molecules

Except the dye No.9 belongs to the lush dyes, the chemical structure of the other kinds of dyes are azo and anthraquinone. In this section, Acid Light Yellow G and Acid Blue were selected to represent azo and anthraquinone dyes. For these two kinds of dye, the changes in molecular structure and groups before and after reaction were analyzed by using the infrared spectrum.

3.7.1. Acid Light Yellow G—Azo Dye

Figure 16(1) shows the infrared spectra of Acid Light Yellow G solution before and after ozonation. The upper curve is the infrared spectrum of degraded solution (at 40th min), and the lower curve is the infrared spectrum of the original solution. The azo group is an atoms group containing electrons and is the most important functional group in the dye molecule. It forms a 8-naphtol-3,6-bissulfonate conjugation chromophore group with a benzene ring and a naphthalene ring, and has a characteristic frequency between 1630 and 1575 cm^{-1}. As shown in Figure 16(2), in the infrared spectrum before ozonation, the azo group has a characteristic absorbance peak. After ozonation, this peak is diminished. The azo group wasn't degraded completely in 40 min.

Alcohol generally exists in hydrogen-bonded polymeride in pure solid or liquor, free state in nonpolar diluted solution and in both states in concentrated solutions or mixtures. Strenching vibration absorbance band of hydroxyl of alcohol in free state is near 3300 cm^{-1}

and 3640 cm^{-1}. It has strong characteristics and does not change with different molecular structures. Distortion vibration band of hydroxyl of alcohol is near 1420～1250 cm^{-1} which has weak strength and large width. As shown in Figure16(2), the absorbance peak of free hydroxyl between 3664 cm^{-1} and 3400 cm^{-1} decreased markedly after reaction, which is related to the decrease in pH. The hydroxyl in alcohol has an absorbance band between 1420 and 1250 cm^{-1} and disappeared after degradation. This demonstrates that the hydroxyl in the alcohol has been completely degraded.

The infrared spectra of aromatic compounds have five important absorbance bands: ① 3100 to 3000 cm^{-1} is strenching vibration absorbance band of hydrocarbon bond; ② 2000 to 1600 cm^{-1} is a wide-frequency area for aromatic compounds, where each substitution can produce different absorbance peaks. So these weak absorbance bands are not enough to identify compounds. ③ 1650 to 1450 cm^{-1} is strenching vibration absorbance band for carbon-carbon bonds. Benzene can be identified if there are four peaks in this band. ④ 1250 to 1000 cm^{-1} is the in-plane bending of hydrocarbon bonds; ⑤ 900 to 650 cm^{-1} is the band for synchronous phase out-of-plane bending of hydrocarbon bonds. These vibration frequencies are related to the number of hydrogen atoms linked to rings [37]. As shown in the infrared spectra of solutions before ozonation, there are peaks in more than five absorbance bands which suggest that dye molecules contain complex aromatic rings. After ozonation, the infrared spectra show there is only the peak for benzene between 1650 and 1450 cm^{-1}. The 810 to 875 cm^{-1} band is the peak for the naphthalene ring, and the disappearance of it demonstrates that a complex aromatic compound has been degraded to simpler aromatic compounds.

3.7.2. Acid Blue R- Anthraquinone Type

Figure 16(2) shows the infrared spectrum before and after degradation of Acid Blue R. The curve above is the infrared spectrum after degradation of Acid Blue R (namely the sample solution 40 min after reaction). The curve below shows the infrared spectrum before degradation (original solution).

In anthraquinone dyes, $\overset{\overset{O}{\|}}{\diagup\;\diagdown}$ (circularity) is the most important characteristic group, whose characteristic absorption peaks are in the region of 1700－1800 cm^{-1}; the characteristic absorption peaks of $-\overset{\overset{O}{\|}}{C}-$ (non circularity) are in the regions of 1650－1870 cm^{-1} and 1405－1475 cm^{-1}.

It can be deduced from the infrared spectrum before degradation that two absorption peaks appear, between1716 and 1818 cm^{-1}; the absorption peaks are in the regions of 1869—1648 cm^{-1} and 1450—1405 cm^{-1}. It suggests that $\overset{\overset{O}{\|}}{\diagup\;\diagdown}$ (circularity) has been broken and transformed to $-\overset{\overset{O}{\|}}{C}-$ (non circularity) after the reaction with ozone.

As with the Acid Light Yellow dye, the Acid Blue R molecule contains the –OH group. As show in Figure 16(2), the absorption peak of free –OH is in the region of 3664 cm^{-1} to 3400 cm^{-1} before and after degradation. Before degradation, the alcohol －OH has absorption band from 1420 to 1250 cm^{-1}. After degradation, there is no absorption band in this wave band. This shows that the alcohol –OH has been totally decomposed. Besides, an absorption peak in the region of 670 cm^{-1} in the infrared spectrum of Acid Blue R, is the carbon–carbon double absorption peak.

As similar to the analysis of the degradation of Acid Light Yellow G, the infrared spectrum of Acid Blue R before degradation has the five characteristic absorption regions of benzene. The original dye molecules have complicated benzene structures. It can be deduced from the infrared spectrum after degradation that there are four characteristic absorption peaks in the regions of 1650 to 1400 cm^{-1}. This indicates that the complicated benzene compounds were decomposed to simple benzene compounds.

The absorption peak of —C—N< (amidocyanogen) is in the region of 2800 to 2700 cm^{-1}. In the infrared spectrum before degradation, absorption peaks appear in the region of 2813 to 2707 cm^{-1}. However, this absorption peak does not appear in the infrared spectrum after degradation. The characteristic peak of the nitrogen–carbon double bond appears in the region of 1600 to 1500 cm^{-1}. Those phenomena illustrate that —C—N< (amidocyanogen) was oxidized to a nitrogen–carbon double bond.

Through the above analysis, it can be concluded that after reaction, complicated molecule of Acid Blue R is decomposed to simple organic compounds, such as benzene, —C(=O)—, —N=O, and so on. Because both ketone and ester have absorption peaks in the same regions (1450 to 1400 cm^{-1}), it is uncertain whether the material in the degradation solutions is ketone or ester, which is the oxidization product of ketone.

The infrared spectra for the other nine dyes are shown in Figure 16(3) – (11).

Comparison of the infrared spectra curves of the other nine dyes showed that the molecular groups in each dyes experienced major changes. From the analysis of the infrared spectra, it can be concluded that there are some common features:

(1) There are absorption peaks in the region of 3664 cm^{-1} to 3400 cm^{-1}, which is the free –OH group. This absorption peak decreases significantly after reaction, which is probably due to the decrease in pH.
(2) Absorption peaks appear in the region of 2360 to 2300 cm^{-1}, which is the absorption peak of CO_2.
(3) In the infrared spectra before degradation, the five characteristic absorption regions of benzene have the absorption peaks of aromatic hydrocarbon compounds. These absorption peaks decreased markedly or disappeared. It indicates that the complex benzene compounds are decomposed to simple benzene compounds.
(4) There are absorption peaks in the region of 1869 to 1648 cm^{-1} and 1450 to 1405 cm^{-1}, which is the absorption peak of —C(=O)—. Because both ketone and ester have absorption peaks in the same region (1450 to 1400 cm^{-1}), it is uncertain whether the material in the degradation solutions is ketone or ester.

(5) The absorption peak in the region of 670 cm^{-1} in the infrared spectrum of Acid Blue R is carbon–carbon double absorption peak.

Each kind of dye produces different products after degradation. One common trend exists in all these dyes: complex compounds are decomposed to simpler organic compounds or inorganic ions.

The dye molecules experienced marked changes following oxidization by ozone. Through the analysis of the UV-via visible spectra, it can be concluded that ozone oxidizes and degrades not only certain functional groups, but the whole dye molecule.

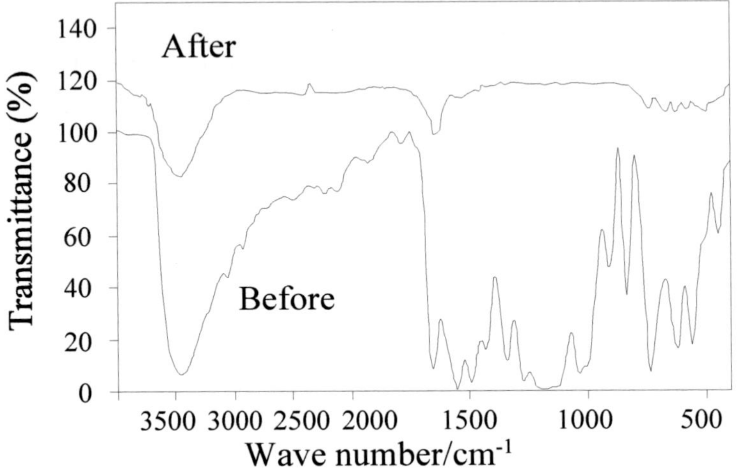

Figure 16 (1). Acid Light Yellow G.

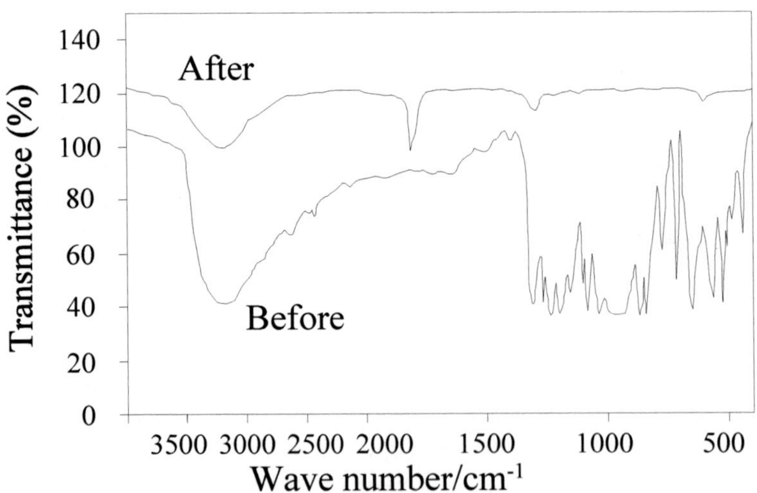

Figure 16(2). Acid Blue R.

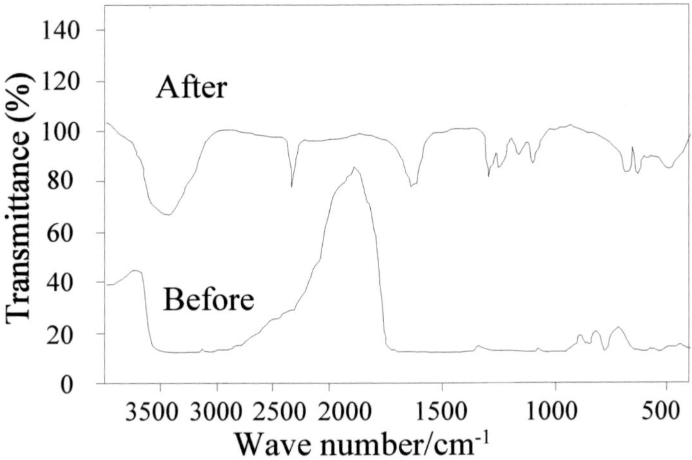

Figure 16 (3). Reactive Light Yellow X-7G.

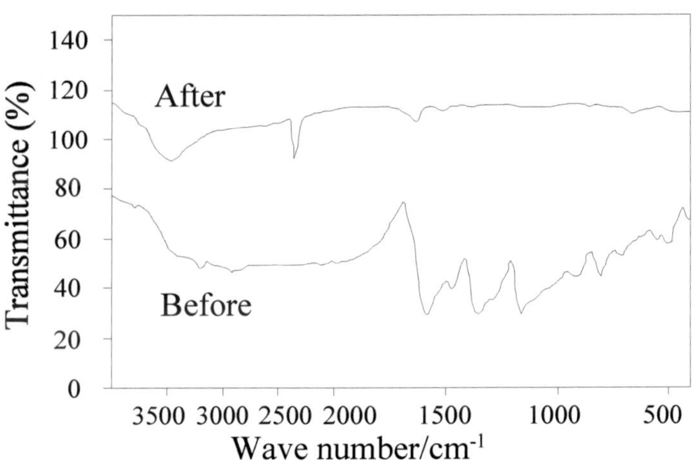

Figure 16 (4). Reactive Brilliant Blue X-BR.

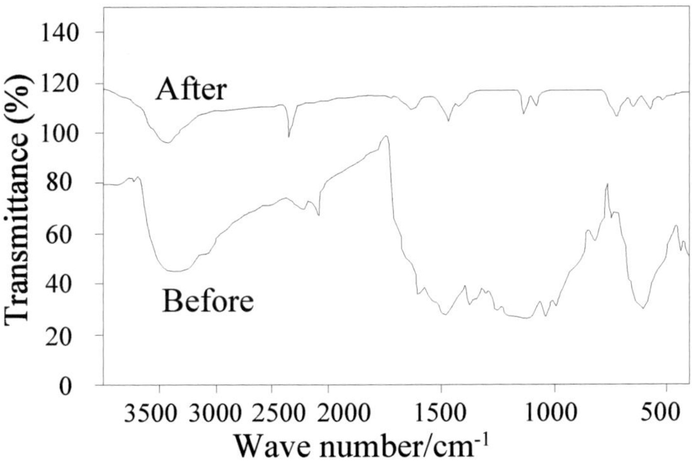

Figure 16(5). Direct Scarlet 4BS.

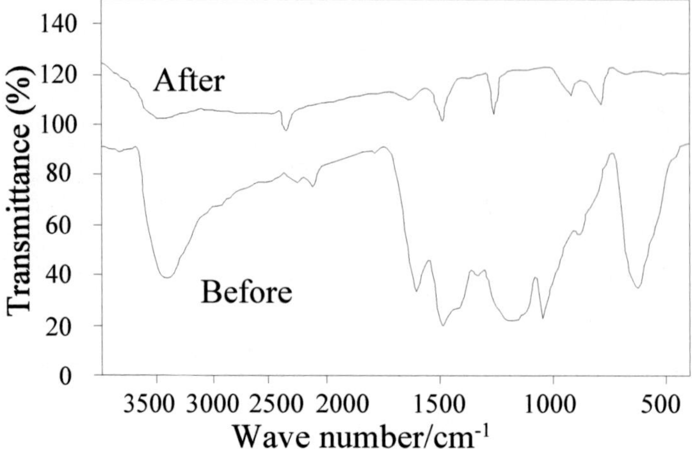

Figure 16 (6). Direct Fast Blue B2RL.

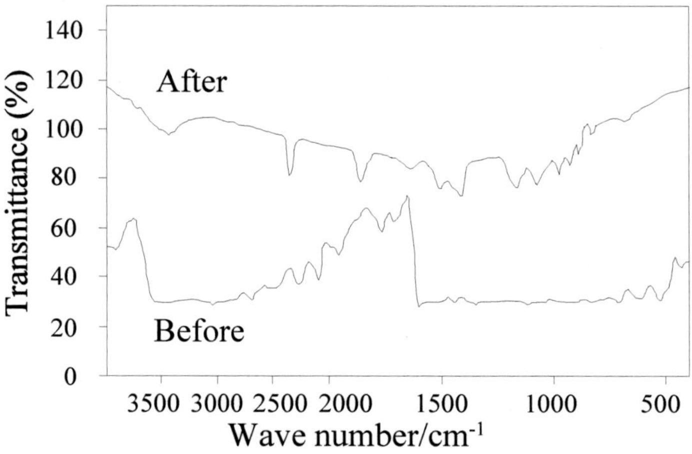

Figure 16(7). Cationic Red GTL.

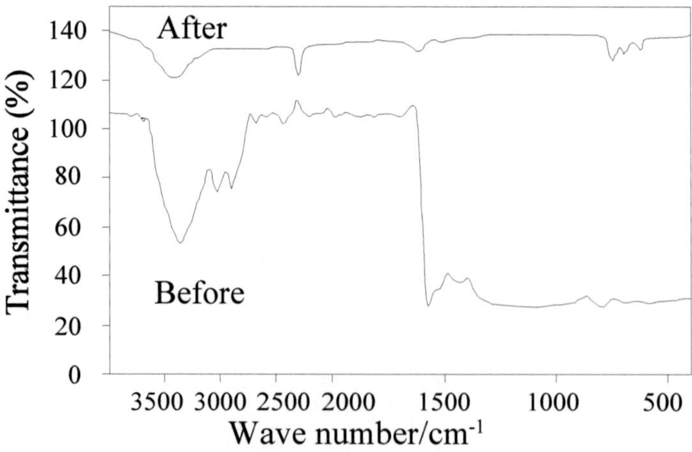

Figure 16(8). Cationic Blue FGL.

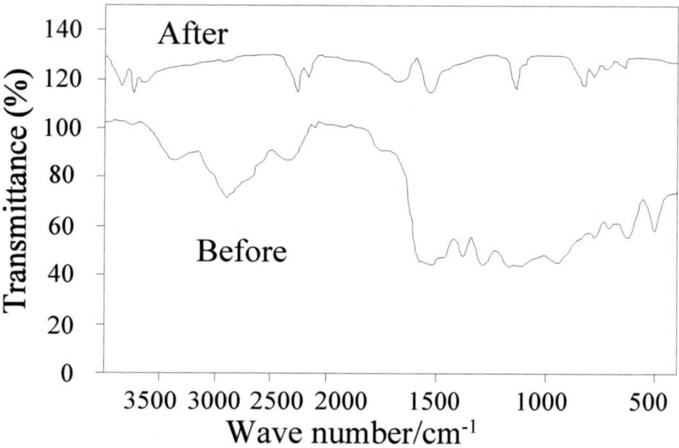

Figure 16(9). Cationic Pink FG.

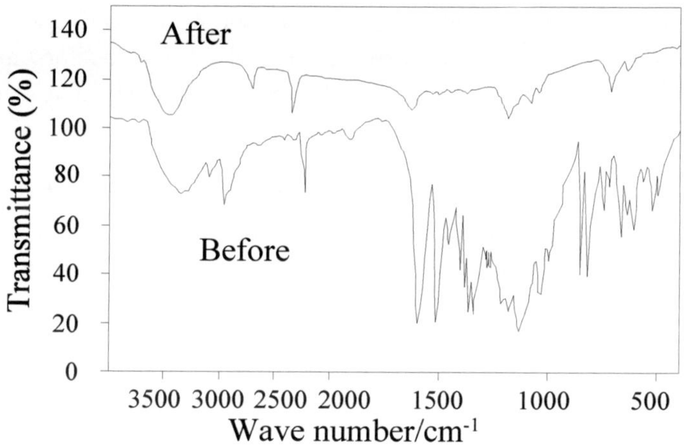

Figure 16(10). Dispersed Orange G.

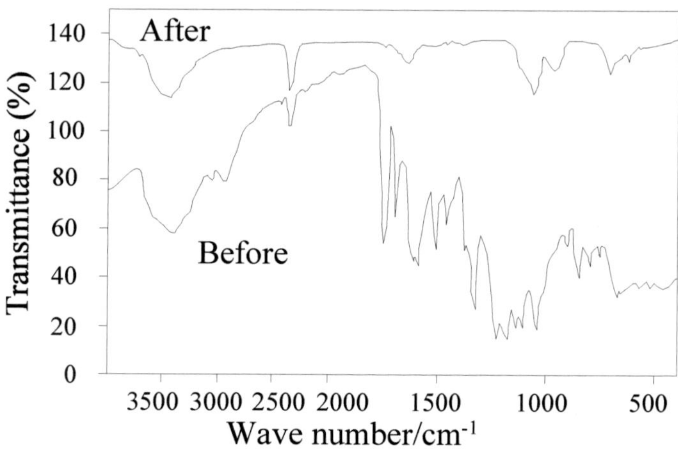

Figure 16(11). Dispersed Red 3B.

Figure 16. Infrared spectrum of original and degraded solution of eleven kinds of dyes.

Through the above analysis, it can be concluded that after reaction, the complex molecule of Acid Blue R is decomposed to simple organic compounds, such as benzene, carbonyl, nitroso group and so on. Although the degradation compounds of other dyes are different, the degradation processes are similar.

3.8. TOC

In order to investigate the reaction mechanisms between the dye and O_3 further, the TOC concentrations of the sample solutions were determined and the variations during the reaction process were analyzed. The relationships among decolorization efficiency, pH, and TOC were studied.

As shown in Figure17, the TOC concentration decreased with increasing reaction time. The TOC in the ultimate solutions was reduced below 83% of the original value after 40min.The relative decrease in TOC concentration in Direct Fast Blue B2RL is the largest to 55%. The changes are consistent with the decolorization efficiency and pH. However, the ratio of the decrease in TOC concentration is less than the decolorization efficiency. It indicates that O_3 can remove the color of the dyes efficaciously, but cannot reduce TOC effectively.

Table 6. △TOC/C value，△TOC and △pH in 0－20min and 20－40min

No.	0－20 min △TOC (mg l^{-1})	20-40 min △TOC (mg l^{-1})	△TOC*/C Atomicity	pH value#	0-20 min △pH	20-40 min △pH
1	10.14	8.09	1.14	5.98	1.78	0.30
2	15.11	4.15	1.38	6.59	2.67	0.45
3	9.27	0.75	0.59	3.99	0.34	0.20
4	14.08	16.67	1.34	5.50	1.75	0.11
5	8.97	3.87	0.37	5.82	1.78	0.16
6	16.81	0.97	0.44	6.95	2.37	0.37
7	10.67	3.33	0.70	5.51	1.67	0.21
8	9.17	8.24	0.79	5.07	1.28	0.18
10	12.29	5.66	0.81	6.18	2.08	0.33
11	12.98	11.41	1.22	7.00	2.51	0.34

* △TOC is 0－40 min △TOC

pH value is the pH value of the original solution.

Besides, Figure 16 and Table 6 both show that except for dispersed dyes, the variation of TOC is rather large whereas the change in decolorization efficiency is small, which indicates that besides chromophores, the other parts of dye molecules also participate in the reaction, generating CO_2 and/or volatile organic compounds [10,38], which cause a decrease in TOC concentration. CO_2 is partly in the form of carbonic acid and the rest is released from solution. Comparing the data in Table 6, the reduction in pH is larger when the pH value of the original solution is high, which can possibly be explained by the increased concentration of carbonic acid. It can be seen that the relative reduction of TOC concentration of azo dyes is less than that of anthraquinone dyes. The reasons are: (1) The chromophore of azo dye is a diazo bond, but the anthraquinone dye is composed of C atoms that can enhance the reaction probability between C and ozone. This will increase the probability of organic carbon being transformed to inorganic carbon or volatile molecules. (2) The decrease of TOC concentration is related to the C atomicity in the dye molecule. The greater the C atomicity, the greater the △TOC value, and an even greater value of △TOC/C atomicity due to ozonation. As shown in Table 4, the values of △TOC/C atomicity for all azo dyes are less than those for anthraquinone dyes, so the reduction rates for TOC of azo dyes are less than those of anthraquinone dyes.

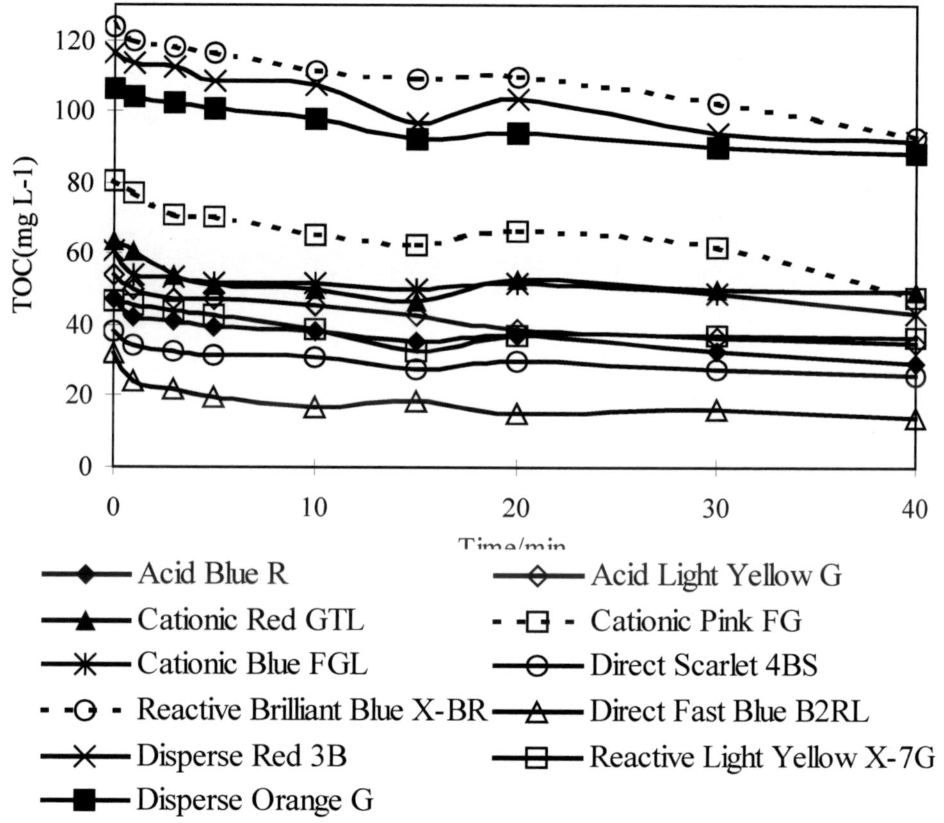

Figure 17. TOC concentrations of samples vs. time.

CO_2 and/or volatile organic compounds are produced, causing a decrease in TOC concentration, and the decrease exceeded 17% after 40 min. The TOC removed from azo dyes is less than from anthraquinone dyes.

3.9. Analysis of Organic Derivatives

In order to analyze the organic materials in dye solutions at 40 min, GC-MS was used to measure the degradation solutions and to compare with a blank. The GC-MS chromatograms are showed in Figure 18(1) – (12).

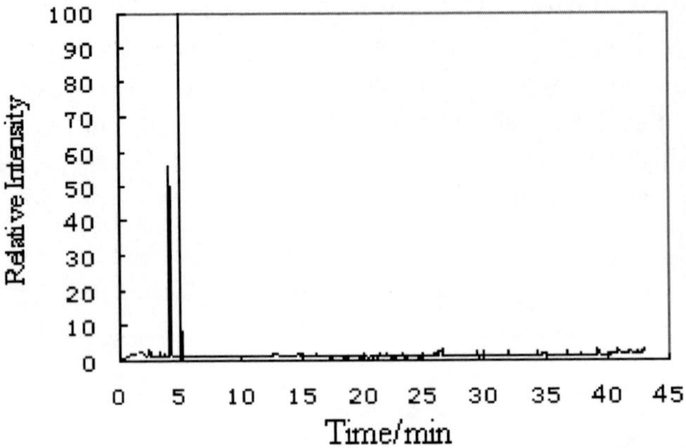

Figure 18(1). Bank

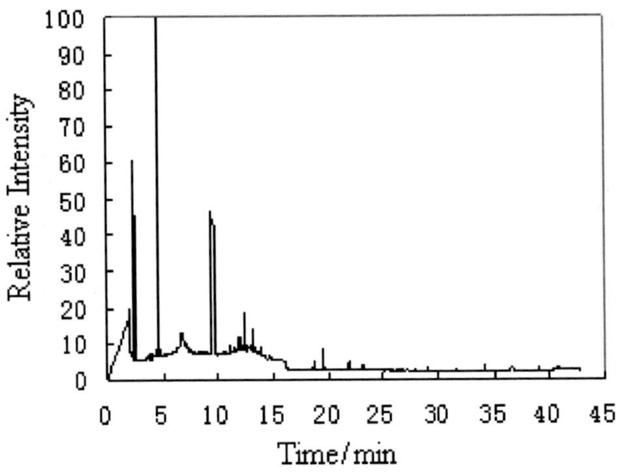

Figure 18(2). Acid Light Yellow G.

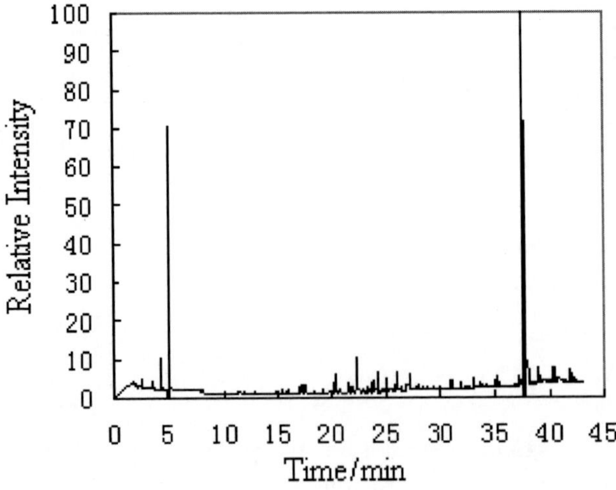

Figure 18(3). Acid Blue R.

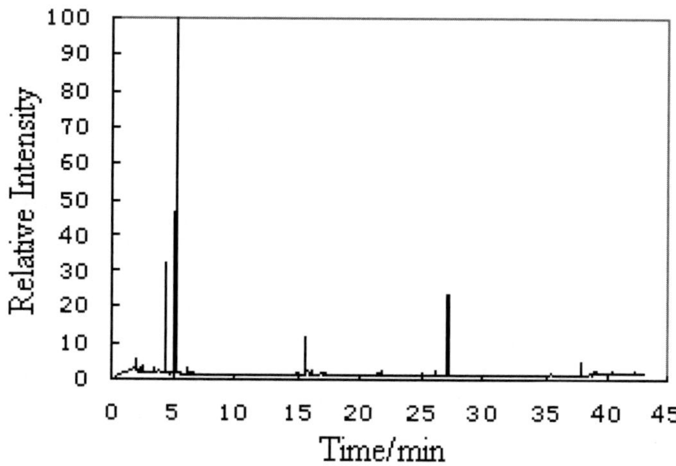

Figure 18(4). Reactive Light Yellow X-7G.

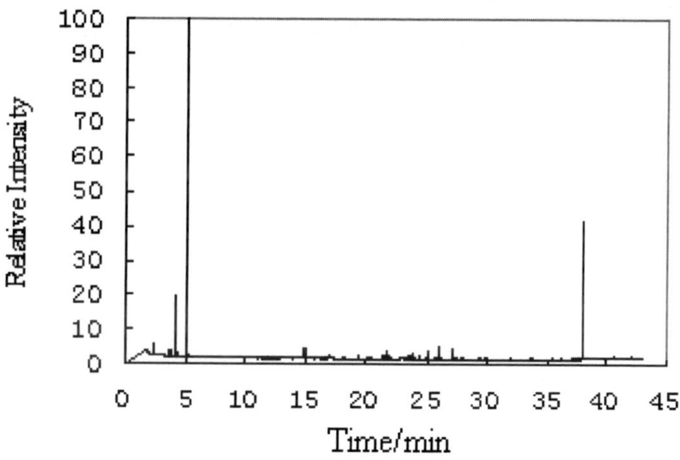

Figure 18(5). Reactive Brilliant Blue X-BR.

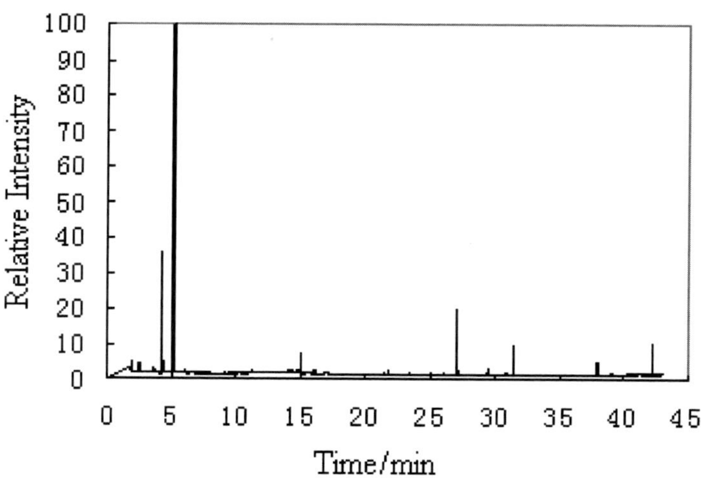

Figure 18(6). Direct Scarlet 4BS.

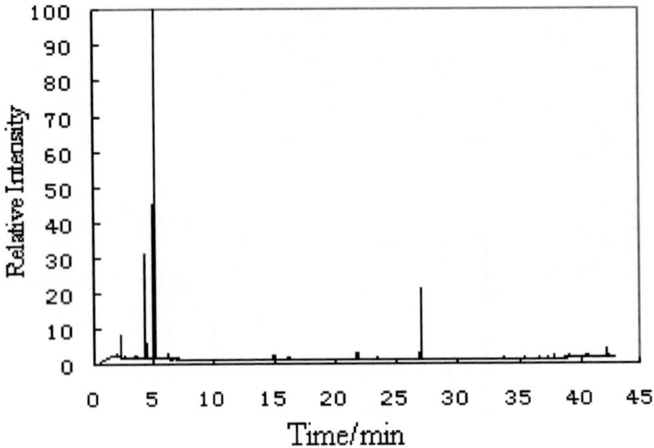

Figure 18(7). Direct Fast Blue B2RL.

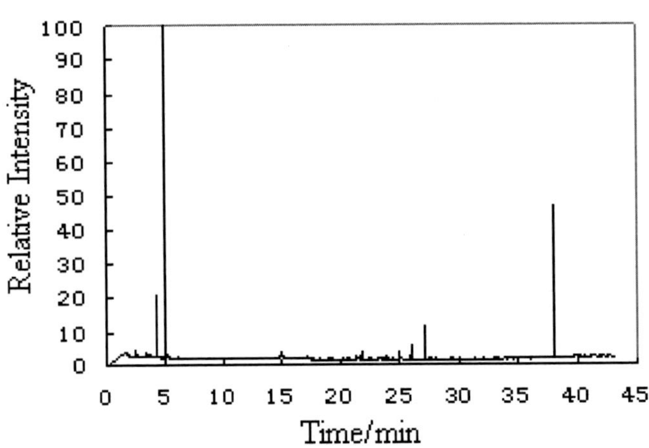

Figure 18(8). Cationic Red GTL.

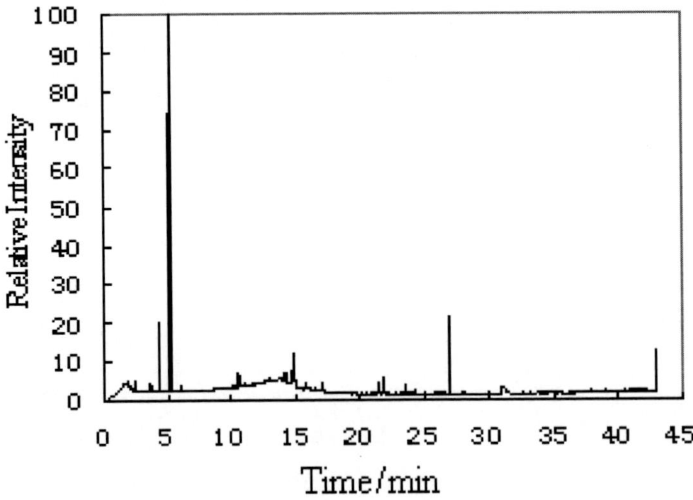

Figure 18(9). Cationic Blue FGL.

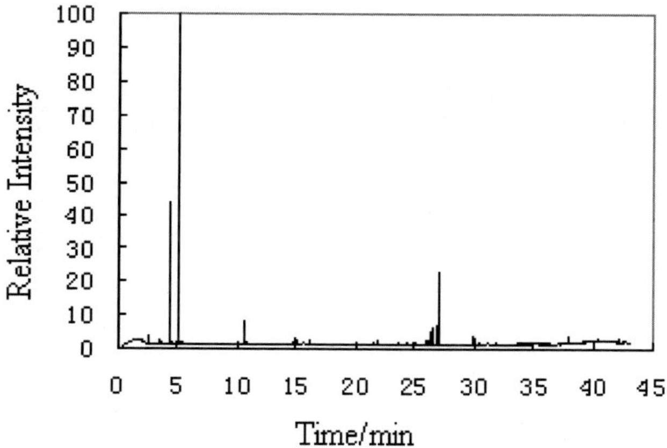

Figure 18(10). Cationic Pink FG.

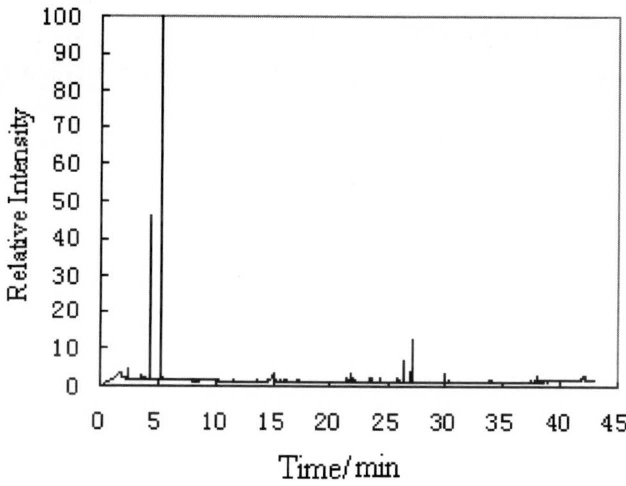

Figure 18(11). Dispersed Orange G.

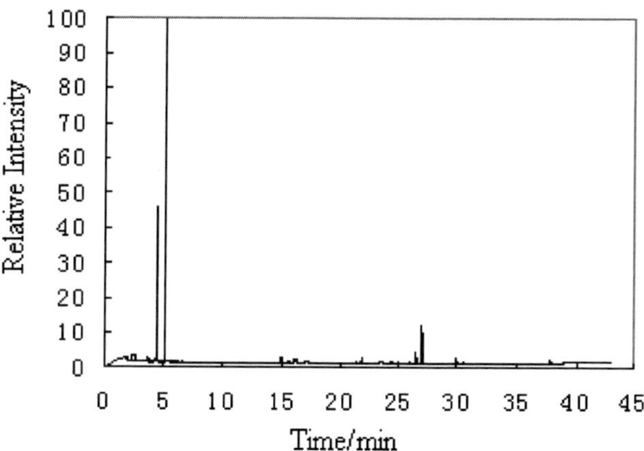

Figure 18(12). Dispersed Red 3B.

Figure 18(1) –(12)The GC-MS spectrum of 11 kinds of dyes and blank.

Comparing the GC-MS chromatograms for the eleven dyes and the blank solution, the degradation solutions of all dyes contain organic derivatives, listed in Table 7(1). Dye No. 2, 5, and 6 all produced organic acids, which validates the analysis that there are organic acids in the degradation solution. It also verifies the analysis of the infrared spectra. In addition to the organic material shown in Table 7(1) for all dyes, each degradation solution has unique organic materials. Table 7(2) lists the substances in the degradation solutions of each.

For example, the degradation solutions of Acid Light Yellow G (No.1) contains benzene ring; 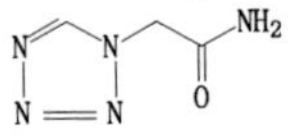 ; it also exists in Acid Light Yellow G (No.1) and Acid Blue R (No.2) degradation solutions. The Degradation solution of Direct Scarlet 4BS (No.5) has 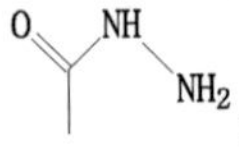 ; Cationic Blue FGL (No.8) produces 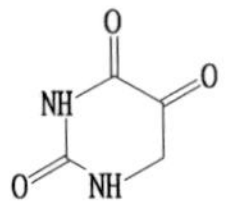 after degradation; 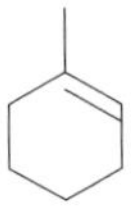 exists in the other ten kinds of dyes except the Reactive Light Yellow X-7G (No.3). There are many other substances in the degradation solutions for eleven kinds of dyes. The properties (toxicity, for instance) of the derivatives need further investigation.

Through the analysis of the derivatives in the eleven kinds of dyes, it appears that dyes with larger molecular mass are decomposed into other substances with comparatively large molecular mass, which is larger than those of the degradation products from dyes with smaller molecules.

Table 7(1). Major organic derivatives identified by GC-MS

No.	Molecular formula	Molecular mass	No.	Molecular formula	Molecula r mass
1	 2H-2, 4a-MethaNonaphthalene, 1, 3, 4, 5, 6, 7-hexahudro-1, 1, 5,5-tetramethyl-,[2S]	204	2	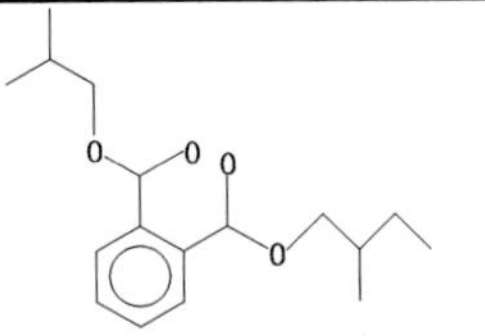 1,2-Benzendicarboxylic acid, bis[2-methylpropyl] ester	278
3	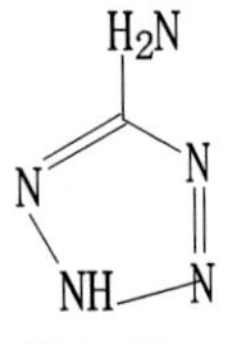 1H—Tetrazol-5-amine	85	4	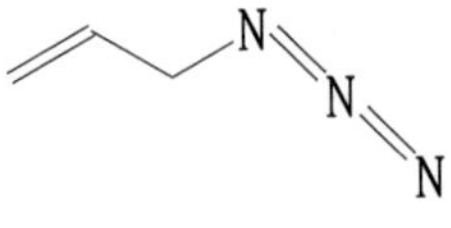 1-Propene,3-azido	83

Table 7(1). Continued

No.	Molecular formula	Molecular mass	No.	Molecular formula	Molecula r mass
5	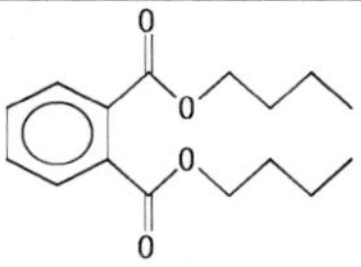 Phthalic acid, dibutyl ester	278	6	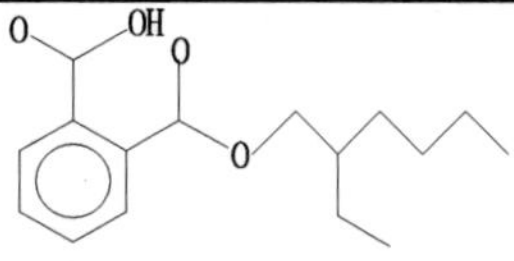 1,2-Benzendicarboxylic acid, moNo[2-ethylhexyl] ester	278
7	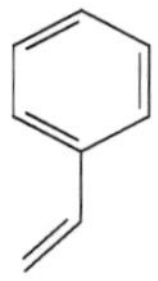 Styrene	104	8	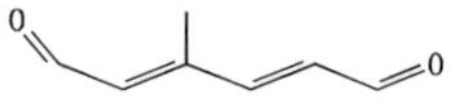 3-methyl-2, 4-diene-hexandial	124

Table 7(2). Organic derivatives by GC-MS

No Organic molecular formula	1	2	3	4	5	6	7	8	9	10	11
	+	—	—	—	—	—	—	—	—	—	—
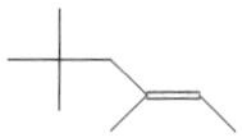	+	—	—	—	—	—	—	—	—	—	—
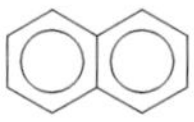	+	—	—	—	+	—	—	+	—	—	—
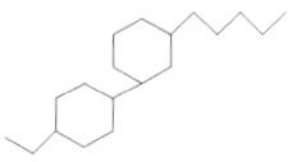	+	—	—	—	—	—	—	—	—	—	—
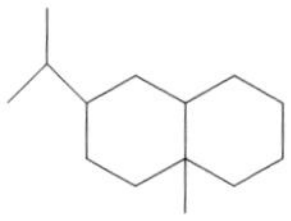	+	—	—	—	—	—	—	—	—	—	—
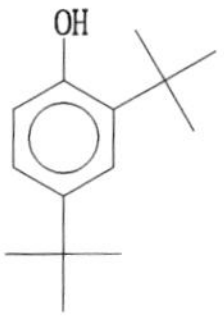	+	—	+	+	+	+	+	—	+	—	+
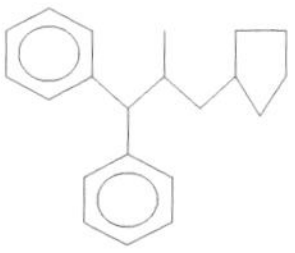	+	—	+	+	+	+	—	+	—	+	+
	+	—	+	+	—	+		+	—	—	—

Table 7(2). Continued

No / Organic molecular formula	1	2	3	4	5	6	7	8	9	10	11
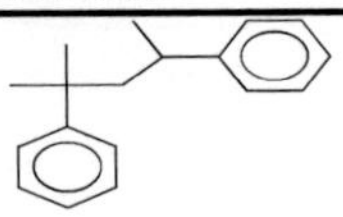	+	−	−	+	+	+	+	+	+	+	+
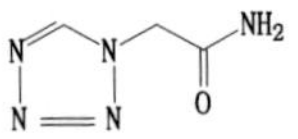	+	+	−	−	−	−	−	−	−	−	−
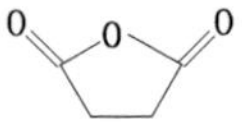	+	+	−	+	+	−	+	+	−	+	+
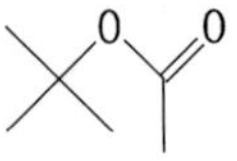	−	+	−	−	−	−	−	+	−	−	−
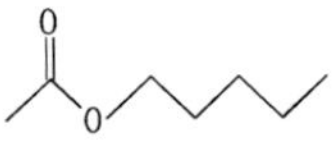	−	+	−	−	−	−	+	−	−	−	−
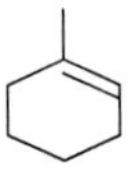	−	+	−	+	+	+	+	+	+	+	+
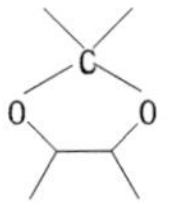	−	−	+	+	+	+	+	+	−	+	+
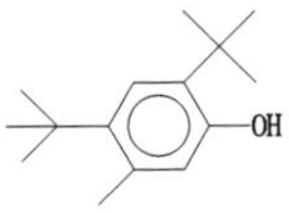	−	−	−	−	−	−	+	−	−	−	−
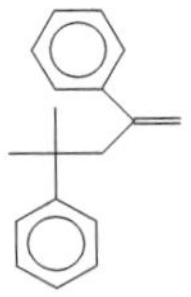	−	−	−	−	−	−	+	−	+	+	+
	−	−	−	−	−	−	−	+	−	−	−
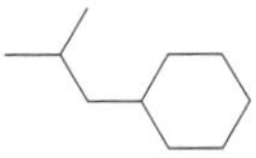	−	−	−	−	−	−	−	+	−	−	−
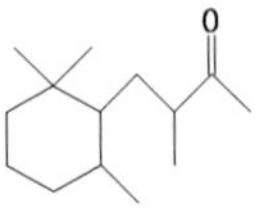	−	−	−	−	+	−	−	+	−	−	−
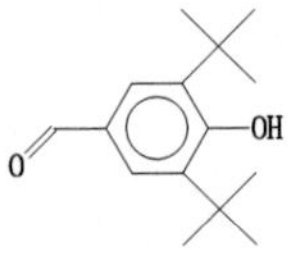	−	−	−	−	−	−	−	+	−	−	+
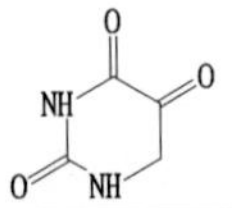	−	−	−	−	−	−	−	+	−	−	−

Table 7(2). Continued

No Organic molecular formula	1	2	3	4	5	6	7	8	9	10	11
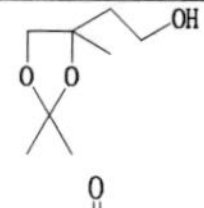	—	—	—	—	—	—	—	—	+	—	—
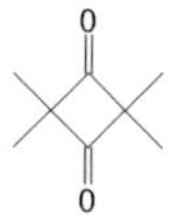	—	—	—	—	—	—	—	—	+	—	—
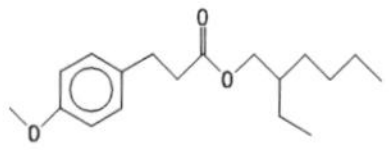	—	—	—	+	—	+	—	—	+	+	+
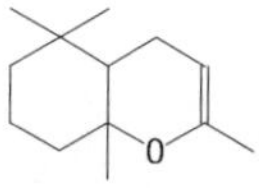	—	—	—	—	—	—	—	—	+	—	—
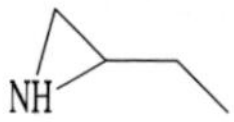	—	—	—	—	—	—	—	—	+	—	—
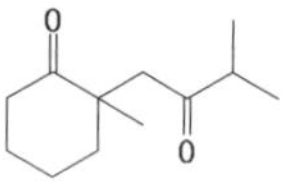	—	—	—	—	—	+	—	—	—	—	+
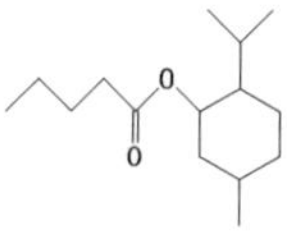	—	—	—	—	—	+	—	—	—	—	—
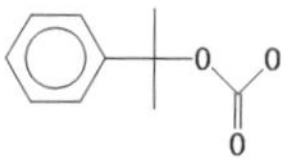	—	—	—	—	+	—	—	—	—	—	—
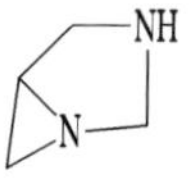	—	—	—	—	—	+	—	—	—	—	—
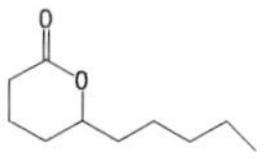	—	—	—	+	+	—	—	—	—	—	—
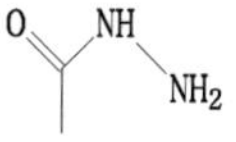	—	—	—	—	+	—	—	—	—	—	—
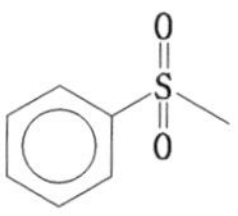	—	—	—	+	—	—	—	—	—	—	—
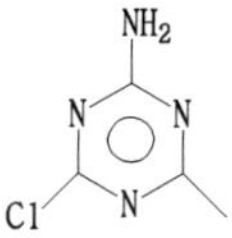	—	—	—	+	—	—	—	—	—	—	—
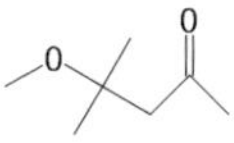	—	—	+	—	—	—	—	—	—	—	—
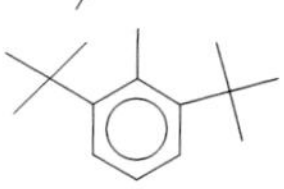	—	—	—	—	—	—	—	—	—	+	—

Table 7(2). Continued

No / Organic molecular formula	1	2	3	4	5	6	7	8	9	10	11
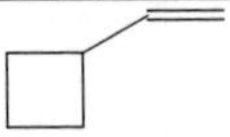	—	—	—	—	—	—	—	—	—	—	+
	—	—	—	—	—	—	—	—	—	—	+
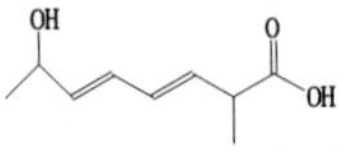 OH O OH	—	—	—	—	—	—	—	—	—	—	+

+ Indicates the organic materials exist in the dye — Indicates the organic materials don't exist in the dye

The degradation solutions contain organic materials. Dyes with larger molecular mass are decomposed into other substances with comparatively large molecular mass, which is higher than that of the degradation products from dyes with smaller molecules.

4. Conclusions

1. Ozone treatment is very effective in the decolorization of soluble dyes. The decolorization efficiencies of all dyes exceeded 79% in 20 min. The sequence of the decolorization reaction rate is: Reactive Brilliant Blue X-BR > Acid Blue R > Reactive Light Yellow X-7G > Acid Light Yellow G > Direct Fast Blue B2RL > Direct Scarlet 4BS > Cationic Pink FG > Cationic Blue FGL > Cationic Red GTL > Dispersed Orange G > Dispersed Red 3B.The sequence of the decolorization rate is: Reactive > Acid > Direct > Cationic > Dispersed. In the same type of soluble dyes, the decolorization rates were quicker for dyes with smaller molecular mass. Some acidic materials are produced, reducing the pH with increasing reaction time, and the final pH values are less than 4.5, and the change in pH is consistent with the decolorization efficiency.
2. The dye molecule structures changes during oxidization. The ozone oxidizes and degrades the whole dye molecule, not just some certain functional groups. The experimental dyes experienced a violet-shift or a red-shift during the reaction process. It shows that the electron cloud density of the color base is altered. Although different kinds of dyes have different degradation products, the same phenomenon existing in all dyes: the complex materials are decomposed to simple organic substances or inorganic ions.
3. The UV-vis visible spectrum analysis showed that due to the strong oxidizability of ozone, absorbance of the 11 dyes decreased not just in a certain wave band, but in the whole ultraviolet-visible spectrum. The dye molecules experienced a violet-shift or red- shift in the degradation process. It can be concluded that the density of the electron cloud of color base and molecular structure have changed. The oxidization of the eleven different kinds of dyes by ozone follows first order kinetics.

4. The electrical conductivity increases with increasing reaction time. This indicates that the dye molecules were decomposed partially to ions during the reaction with ozone. The trend in electrical conductivity was consistent with the decolorization efficiency and pH value. It suggests that the mineralization of solutions and acidic materials increases along with the ozonation of the dye molecules.
5. The complex dye molecules are degraded to simpler organic compounds;—Cl and PO_4^{3-} (chelated state)in the dye molecules are transformed into Cl^-, and PO_4^{3-} (in the form of dissociated $H_2PO_4^-$);nitrogen is partially degraded to NH_4^-, NO_3^- and NO_2^- is absent. Almost all of the —SO_3H is converted into SO_4^{2-}.
6. These imply that the deductions organic derivative contains almost no groups with S, Cl, P. The release of these ions is related to the dye molecules' structures, the bonding places of the groups involving Nitrogen atoms and the number of Nitrogen groups. Since the H^+ concentration in the degradation solution is much more than for the inorganic acids which are produced by combining H^+ with NO_2^-, NO_3^-, SO_4^{2-}, $H_2PO_4^-$ and Cl^-, it is deduced that organic acids are produced in the degradation. The results also show that there are organic acids in the degradation solutions and that the dyes with larger molecular mass are decomposed into other compounds with comparatively large molecular mass.
7. CO_2 and volatile organic compounds are produced, causing a decrease in TOC concentration, and the decrease exceeded 16% in 40 min. Some of the CO_2 produced exists in the form of carbonic acid and the rest escapes from solution. The change in TOC is consistent with the pH change and decolorization efficiency. The TOC removed from azo dyes is less than from anthraquinone dyes, which may be related with the structure and C atomicity of the dyes.
8. Each of the 11 types of dye produces different kinds of products after degradation on the basis of the analysis of infrared spectrums. One common trend exists in all kinds of dyes: the complex compounds were decomposed to organic compounds or inorganic ions.
9. Organic acidic materials and molecules with large benzene or other ring structures were produced in the degradation solutions.

References

[1] Gurnham CF, Ed. Industrial Waste Control. Academic Press, New York 1965.

[2] Lin SH, Lin CM. Treatment of textile waster effluents by ozonation and chemical coagulation. *Water Res.* 1993; 27: 1743–1748.

[3] Naumczyk J, Szpyrkowicz L, Grandi ZF. Electrochemical treatment of textile wastewater. *Water Sci. Technol.* 1996; 34: 17—24.

[4] Szpyrkowicz L, Zilio GF, Canepa P. Performance of a full-scale treatment plant for textile dyeing wastewater. *Toxicol. Environ. Chem.* 1996; 56: 23—34.

[5] Perkowski J, Kos L, Ledaowicz S. Application of ozone in textile wastewater treatment. *Ozone Sci Eng*. 1996; 18: 73—85.

[6] Rana KS, Raizada S. Acute toxicity of tannery and textile dye effluents on common teleost,Lobeo rohita: histological alteration in liver. *J. Environ. Biol.* 1999; 20: 33—6.

[7] Szpyrkowicz L, Juzzolino C, Kaul SN. A compative study on oxidation of disperse dyes by electrochemical process, ozone, hypochlorite and fenton reagent. *Water Res.* 2001; 35: 2129—2136.

[8] Vaghela SS, Jethva AD, Mehta BB, Dave SP, Adimurthy S, Ramachandraiah G. Laboratory studies of electrochemical treatment of industrial azo dye effluent. *Environ.Sci.Technol.* 2005; 39: 2848—2855.

[9] Chu W, Ma CW. Reaction kinetics of UV-decolourization for dye materials. *Chemo.* 1998; 37: 961–974.

[10] Zhao W, Shi H, Wang D. Ozonation of cationic red X-GRL in aqueous solution: degradation and mechanism. *Chemo.* 2004; 57: 1189—1199.

[11] Churchley JH. Ozone for the dye waste colour removal: four years operation at Leak STW. *Inte. Ozone Assoc.* 1998; 20: 111—120.

[12] Feigelson L., Muszkat L, Bir L, Muszkat KA. Dye photoenhancement of TiO_2-photocatalyzed degradation of organic pollutants: the organobromine herbicide bromacil. *Water Sci. Technol.* 2000; 42: 275—279.

[13] Lorimer JP, Mason TJ, Plattes M, Phull SS. Dye effluent decolourization using ultrasonically assisted electro-oxidation. *Ultrason. Sonochem.* 2000; 7: 237—242.

[14] Pereira MFR, Soares SF, Orfao JJM, Figueiredo JJ. Adsorption of dyes on activated carbons: Influence of surface chemical groups. *Carbon* 2003; 41: 811—821.

[15] Brown NW, Roberts EPL, Garforth AA, Dryfe RAW. Treatment of dyehouse effluents with a carbon-based adsorbent using anodic oxidation regeneration. *Water Sci. Technol.* 2004; 49: 219—225.

[16] Netpradit S, Thiravetyan P, Towprayoon S. Adsorption of three azo reactive dyes by metal hydroxide sludge: Effect of temperature, pH, and electrolytes. *J. Colloid Interface Sci.* 2004; 70: 55—261.

[17] Kuo WG. Decolorizing dye wastewater with Fenton's reagent. *Water Res.* 26; 881.

[18] Lin SH, Chen ML. Treatment of textile wastewater by chemical methods for reuse. *Water Res.*1997; 31: 868-876.

[19] Carriere J, Jones P, Broadbent AD. Effect of a dyeing aid on the oxidation reaction of color from an insoluble and a soluble dye in a simulated effluent. Proceedings of the 11th Ozone World Congress, 1993; 1: S10-98-S10-107.

[20] Carriere J, Jones P, Broadbent AD. Decolorization of textile dye solutions. *Ozone Sci Eng.*, 1993; 15: 189—200.

[21] Sarasa J, Roche MP, Ormad MP, Gimeno E, Puig A, Ovelleiro JL. Treatment of a wastewater resulting from dyes manufacturing with ozone and chemical coagulation. *Water Res.* 1998; 32: 2721—2727.

[22] Muthukumara M, Sargunamani D, Selvakumar N, Rao JV. Optimisation of ozone treatment for colour and COD removal of acid dye effluent using central composite design experiment. *Dyes and Pigments* 2004; 63: 127—134.

[23] Hoigne Á J. The chemistry of ozone in water. Process Technologies for Water Treatment, ed. S. Stucki. Plenum Publishing Corporation 1988.

[24] [24] Dore Á M. Les criteries de choix d'un oxydant. *Journal Françair d'Hydrologie* 1990; 21: 9—30.

[25] Muthukumara M, Sargunamani D, Selvakumar N. Statistical analysis of the effect of aromatic, azo and sulphonic acid groups on decolouration of acid dye effluents using advanced oxidation processes. *Dyes and Pigments* 2005; 65: 151—158.

[26] Snider EH, Porter JJ. Ozone treatment of dye waste. *J. Water Pollut. Contr. Fed.* 1974; 46: 886–894.

[27] Peyton GR, Glaze WH. The mechanism of photolytic ozonation. *Abstr. Papers Am. Chem. Soc.* 1985; 189: 5.

[28] Glaze WH, Kang JW, Chapin DH. The chemistry of water treatment processes involving ozone, hydrogen peroxide, and ultraviolet radiation. *Ozone Sci. Eng.* 1987; 9: 335–352.

[29] Gould JP, Groff KA. The kinetics of ozonolysis of synthetic *dyes. Ozone Sci. Eng.* 1987; 9: 153–157.

[30] She HY, Huang CR. Degradation of Commercial Azo Dyes in Water Using Ozonation and UV Enhanced Ozonation Processes. *Chemo.* 1995; 31: 3813—3825.

[31] Tezcanli-Güyer G, Ince NH. Individual and combined effects of ultrasound, ozone and UV irradiation: a case study with textile dyes. *Ultrasonics* 2004; 42: 603—609.

[32] Hou MF, Zhu ZH, Wang RZ. Dye Chemistry. *Chemical Industry Press* 1994.

[33] López-López A, Pic JS and Debellefontaine H. Ozonation of azo dye in a semi-batch reactor: A determination of the molecular and radical contributions. Chemosphere, 2007; 67: 712-717

[34] Herrman JM, Jansen F, Van RA. Water treatment by heterogeneous photocatalysis in environment. Catalysis, Imperial College Press, Catalytic Sci. 1999 Series, Vol.1, London, 1999 (Chapter 9), 171-194.

[35] He ZQ, Song S, Xia M. Mineralization of C.I. Reactive Yellow 84 in aqueous solution by sonolytic ozonation. Chemosphere, 2007; 69: 191-199

[36] Shi HX, Zhao WR, Wang DH. Ozonation mechanism of azo dye. Journal of Zhejiang University (Engineering Science), 2003; 37: 734-738

[37] Yao HW. Spectrometric identification of organic compounds. Science Press (in chinese)1988: 431,(a translated version of Spectrometric Identification of Organic Compounds by Robert M. Silverstein, G. Clayton Bassler, T.C. Morill, and R.M. Silverstein,ISBN 7-03-000389-6).

[38] Koch M, Yediler A, Lienert D, Insel G, Kettrp A. Ozonation of hydrolyzed azo dye reactive yellow 84(CI). *Chemo.* 2002; 46: 109-113.

[39] Glaze WH, Kang JW, Chapin DH. The chemistry of water treatment processes involving ozone, hydrogen peroxide, and ultraviolet radiation. *Ozone Sci. Eng.* 1987; 9: 335–352.

In: Ozone Depletion, Chemistry and Impacts
Editor: Sem H. Bakker, pp. 97-115 ISBN: 978-1-60692-007-7

Chapter 4

OZONE IN THE DEGRADATION OF PHENOLS AND XENOBIOTICS

Maurizio D'Auria, Lucia Emanuele and Rocco Racioppi
Dipartimento di Chimica, Università della Basilicata, Via N. Sauro 85,
85100 Potenza, Italy

Abstract

The irradiation of lignin from pine from steam explosion process in the presence of oxygen, in conditions described for the formation of superoxide ion, for different irradiation time was followed isolating the lignin and determining the average molecular weight. The experiments showed that, until eight hours irradiation, M_n decreases, while M_w and M_z increases. After eight hours irradiation an inverse behaviour was observed, with an increase of M_n and a decrease of M_w and M_z. These results are in agreement with an initial polymerization process followed by a photoinduced degradation. Ozonization was carried out in acetonitrile – methanol solution. The reaction showed a zero order kinetics. After 50 min. the average molecular weight of lignin is the half. The reaction mixture was analyzed by using GC-MS. Oxalic acid was determined. The treatment of diluted olive oil mill waste water with Fenton's reagent reduces COD. The reaction followed a zero order kinetics. The reaction needs to use large amounts of reagents to have an appreciable reduction of COD. Treatment of olive oil mill waste water with ozone reduces COD. The reaction followed a first order kinetics. The uv spectrum of olive oil mill waste water after treatment with ozone did not show absorptions. Different degradation methods have been applied to assess the suitability of advanced oxidation process (AOPs) to promote mineralization of imazethapyr, (*RS*)-5-ethyl-2-(4-isopropyl-4-methyl-5-oxo-2-imidazolin-2-yl)nicotinic acid, a widely used imidazolinone class herbicide, the persistence of which has been demonstrated in surface and ground waters destined to human uses. Independently of the oxidation process assessed, the decomposition of imazethapyr always followed a pseudo-first order kinetic. The direct UV-irradiation of the herbicide as its oxidation with O_3 and H_2O_2/UV-oxidation were sufficiently slow to permit the identification of intermediate products, the formation pathway of which has been proposed. O_3/UV, O_3/UV+TiO_2, TiO_2/UV, and TiO_2/UV+H_2O_2 treatments were characterized by a faster degradation and rapid formation of a lot of little molecules, which were quickly destroyed.

Introduction

Ozone is one of the most important oxidant in the nature. Ozone depletion can represent a potential risk for the environment and for the life on the earth. However, the high oxidation potential of the ozone can represent a potentially useful instrument in our hands. In this chapter we want to report three experiences we performed on the possible use of ozone. In particular, we described the effect of the presence of ozone in lignin degradation, in comparison with the effect of other oxidants. The degradation of lignin could solve the problem of the possible use of this by-product, allowing to obtain important fine chemicals or antioxidants. The degradation of phenol compounds was the object of the second set of experiments we performed. We used ozone in order to reduce the environmental effects of the the olive oil mill wastewater to be spread out on agricultural soils: also in this case, the effect of ozone was compared with the results obtained by using another oxidant. Finally, we described the use of ozone in the degradation of an imidazolinone class herbicide. All these experiments prefigure a possible use of ozone to help us to repair environmental problems.

Wood represents a complex polymeric composition in which lignin and cellulose are the main components. Lignin is a three-dimensional phenylpropanoid polymer mainly linked by ether bonds between monomeric phenylpropane units most of which are not readily hydrolysable. Lignin is constituent of the cell wall of various cell types of plants, e.g. wood fibers, vessels, and tracheids. It constitutes 20-30% of the weight of wood. Lignin encrusts as an amorphic mass the cellulose fibers, which gives the lignified cell wall high mechanical strength and increases the resistance to microbial degradation. Lignin shows a relevant heterogeneity. The heterogeneity of lignin is caused by variations in the polymer composition, size, cross-linking and functional groups. Differences exist in molecular composition and linkage type between the phenylpropane monomers (*p*-hydroxyphenyl, guaiacyl, and syringyl units, derived from respectively coumaryl, coniferyl, and sinapyl alcohol precursors). Lignin composition will be different not only between plants of different genetic origin, but also between different tissues of an individual plant. In softwood lignin, the structural elements are predominantly derives, for more than 95%, from coniferyl alcohol. In hardwoods (and dicotyl crops like flax and hemp) various ratios of coniferyl/sinapyl have been observed, whereas in lignin derived from cereals straws and grasses the presence of coumaryl alcohol is typical.

Lignin is a by-product of pulping and bleaching processes and represents an important environmental problem. Most lignin extracted from lignocellulosic materials for cellulose and paper production in modern pulp mills is burned to generate energy and recover chemicals. The lack of other value added applications are mainly caused by the heterogeneity, odour and colour problems of lignin-based products. The common technologies in pulp industries to extract lignin have been focused on the optimum cellulose yield and maximum deactivation and degradation of lignin.

Steam explosion is a technology useful for the treatment of every lignocellulosic material [1]. In the steam explosion saturated vapor at high pressure is used to rapidly warm the biomass in a digester. The biomass is maintained at the desired temperature for a short time: during this period the hemicellulose is hydrolized and dissolved. At the end of this period, the pressure rapidly decreases to atmospheric one in order to stop the chemical reaction. The explosive decompression gives rise a loss of water from the cells and the cleavage of cellular structures. The time during which the biomass is maintained in the reactor and the temperature defines the severity of the treatment. The severity parameter R_0 is defined as

$$R_0 = t \times \exp\left[\frac{(T-100)}{14.74}\right]$$

where t is the time in minutes and T the temperature in °C. In this paper we describe the characterization of lignin from pine and from corn stalk obtained by using steam explosion.

Recently we showed that lignin degradation could be obtained using photochemically generated singlet oxygen. In particular, we studied the behaviour of singlet oxygen towards lignin model compounds, showing that oxidation reactions occur on the phenoxy part of the molecule [2,3]. The treatment of the non-phenolic β-O-4 aryl ether derivatives gave products deriving from a formal α-C-O cleavage. When the phenoxy part of the molecule showed a lower reactivity towards singlet oxygen, the oxidation of the phenol moiety could occur. The photochemical behavior of these model compounds was rationalized [4]. The same type of results was obtained by other authors working on model compounds [5,6]. Furthermore, we found that the irradiation with visible light in the presence of both a sensitizer and oxygen can degrade isolated lignin from steam explosion and lignin in pulp from steam explosion [7-10]. Singlet oxygen was produced through the reaction of hydrogen peroxide with sodium hypochlorite [11]. The irradiation of lignin in the presence of singlet oxygen gave a residue whose analysis showed the presence of vanillin and sinapyl alcohol [12]. On the contrary, the treatment for different irradiation times gave sinapyl alcohol, 4-hydroxy-3,5-dimethoxybenzaldehyde, and 2,4-dioctylphenol [13]. Four hours irradiation of steam exploded lignin from straw in the presence of singlet oxygen gave in low yield a lactone [14]. A sample of lignin from pine showed the presence of six compounds: *trans*-sinapyl alcohol, 4-hydroxy-3,5-dimethoxybenzaldehyde, 4-hydroxy-3,5-dimethoxyphenylacetone, 4-hydroxy-3-methoxybenzaldehyde, *cis*-sinapyl alcohol, and sinapyl aldehydes [15].

Here we show the possible use of superoxide anion in the photochemical degradation of lignin in the pulp. In this study lignin degradation has been performed by using ozone. Ozonolysis has been employed to prove the aromatic nature of lignin, to determine the nature of the substituents in the aromatic rings, and to study the three-dimensional structure of side-chains in the phenolpropane structural units [16]. The first stage of the reaction of ozone with the aromatic fragments of lignin involves its electrophilic addition with formation of *o*-quinones or muconic acid derivatives [17-19]. Ozonolysis of lignin-containing technical pulp aimed at its bleaching results in the oxidative destruction of the residual lignin, which is accompanied by oxidation and destruction of polysaccharides and cellulose [20]. Ozonation of organosolv lignin from corn stalk showed that some interesting fine chemicals can be obtained after the reaction: the authors showed the presence of glycolic, oxalic, malic, glyceric, *p*-hydroxybenzoic and vanillic acids; furthermore, they identify the presence of *p*-hydroxybenzaldehyde, vanillin, syringaldehyde, and hydroquinone [21,22]. We want to verify the capability of fine chemicals recovery using lignin from steam explosion.

The foodstuff processing industry based on olive oil extraction and table olive oil preparation constitutes an important income source for many regions of the Mediterranean countries. In the first case, the oily juice of olives is extracted from fruit through simple milling or, more recently, by means of centrifugal force. However, despite of increasing concern with regard to environmental matters, during the process relatively important amounts of wastewater are generated. The high contamination level of the wastewater generated results in serious trouble at the time for proper disposal and management. Five million tons of olive oil mill wastewater are produced every years (a quarter in Italy). Anyway, the high concentration of polyphenols, inhibitors of microbial growth, prevents the direct sludge removal. Italian regulation allows the olive oil mill wastewater to be spread out on agricultural soils without contaminating the stratum but, after a short period, the soil becomes unfertile.

Olive oil mill wastewater has been characterized several years ago [23]. COD was found to be into the range 28.9-318.2 g L^{-1} O_2, TOC was in range 12.0-127.5 g L^{-1} C, and BOD was in the range 15-134.8 g L^{-1} O_2. For a sample with COD 101.7 g L^{-1} O_2 the authors found that the total phenol content was 12.7 g L^{-1}.

Several approaches have been tested in order to achieve the reduction of both COD and phenolic content in olive oil mill wastewater: biological methods based on aerobic and anaerobic microorganisms [24-29], enzymatic polymerization of phenols [30], integrated chemical, physical, and biological approaches [31-33], electrocoagulation [34], and several oxidation procedures have been tested: thus, electrochemical oxidation [35], singlet oxygen [36], Fenton and photo-Fenton reagent [37-39], supercritical water oxidation [40], and ozone [41] have been object of extensive work in this field.

Recently, we reported that Fenton's reagent was able to degrade lignin [42]. Fenton's reagent is a mixture of hydrogen peroxide and ferrous sulfate. The reaction between these two components gave HO· as described in the following equation [43-46]:

$$Fe^{2+} + H_2O_2 \rightarrow Fe^{3+} + OH^- + HO\cdot$$

The reagent furnishes a source of HO· that can react with the aromatic part of lignin to give oxidation products. Lignin is a polyphenolic polymer whose components are structurally related to those present in olive oil mill wastewater [47]. We want to compare the efficiency of the treatments of olive oil mill wastewater with Fenton's reagent and ozone. We show that the treatment with ozone is more efficient and more feasible in real plants.

The occurrence of not easily degradable pesticides in the hydrosphere, due to intensive agricultural activities, is of particular concern for surface and groundwater quality. In this respect, it is worthwhile to strengthen existing technologies for purification of human using water, with the aim of enhance the degradation methods for recalcitrant organic compounds [48]. Among the different technologies proposed, are those based on the use of dissolved ozone. Due to its high oxidation potential, ozone has been widely used during the past few years. However, the refractory character of some pollutants or the formation of by-products, which are not further decomposed, often require a more thorough oxidation [49]. A promising way to perform the mineralization of these type of substances is the application of advanced oxidation processes (AOPs), which are characterized by the ''in situ'' production of radicals under mild experimental conditions. More recently, catalytic ozone based systems have been developed in order to enhance the radical production. In this way, it has been proposed the use of metal oxides for the enhancement of the ozonation in heterogeneous processes [48,49].

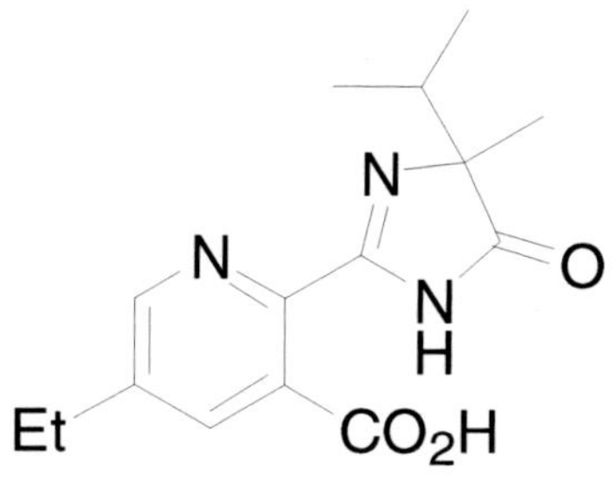

1

Figure 1. Imazethapyr.

Different degradation processes are applied to assess the suitability of AOPs to promote mineralization of imazethapyr, a widely used imidazolinone class herbicide the persistence of which has been demonstrated in surface and ground waters destined to human uses [50-52].

Imazethapyr, (*RS*)-5-ethyl-2-(4-isopropyl-4-methyl-5-oxo-2-imidazolin-2-yl)nicotinic acid (compound **1** in Figure 1), is used to control broadleaf weeds and annual grasses in soybean and peanut.

It is applied preplant incorporated, preemergence, at cracking, and postemergence. This compound controls weeds by reducing the levels of three branched-chain aliphatic amino acids, isoleucine, leucine and valine, through the inhibition of aceto-hydroxyacid synthase, an enzyme common to the biosynthetic pathway for these amino acids. This inhibition causes a disruption in protein synthesis which, in turn, leads to an interference in DNA synthesis and cell growth. Imazethapyr is a weak organic acid (pKa_1, 2.1; pKa_2,3.9), having a water solubility of 1420 mg L^{-1} (pH 7, 25 °C); vapour pressure < 0.013 mPa (60 °C); K_{ow} of 11 at pH 5, 31 at pH 7, and 16 at pH 9. The herbicide molecule is stable and incompatible with strong oxidizing agents. The primary dissipation mechanism of imazethapyr in the environment is through microbial degradation. Imazethapyr has a halflife of 53 - 122 days in aerobic field soil. Imazethapyr loss from hydrolysis is minimal, with virtually none observed at pH 5 or 7 and only minimal degradation occurring at pH 9 ($t_{1/2}$, 9.6 months at 25 °C). Under anaerobic conditions, no significant imazethapyr degradation occurred during a 2-month period when incubated in a variety of soils and sediment. Soil residues of imazethapyr can injure succeeding crops [50].

Ozone and Lignin

GPC analysis of lignin from pine showed a molecular weight distribution reported in Figure 2 with M_w = 5121 and M_z = 2702. The lignin sample showed in the UV spectrum absorptions at λ = 275 nm (D = 26.40 l g^{-1} cm^{-1}) and at λ = 340 nm (D = 11.80 l g^{-1} cm^{-1}). The ^{13}C NMR pectrum showed peaks at δ 148 (C-4 in etherified guaiacyl units), 135 (C-1 in non etherified guaiacyl), 131 (C-2/C-6 in benzoate units), 115 (C-5 in etherified and non etherified guaiacyl units), 107-103 (C-2/C-6 in syringyl units), and 56 ppm (methoxy groups). The ^1H NMR spectrum showed peaks at δ 1.25, 2.0, 2.3, 3.7, 4.2, 4.6, 6 and 6.5-7-1 ppm.

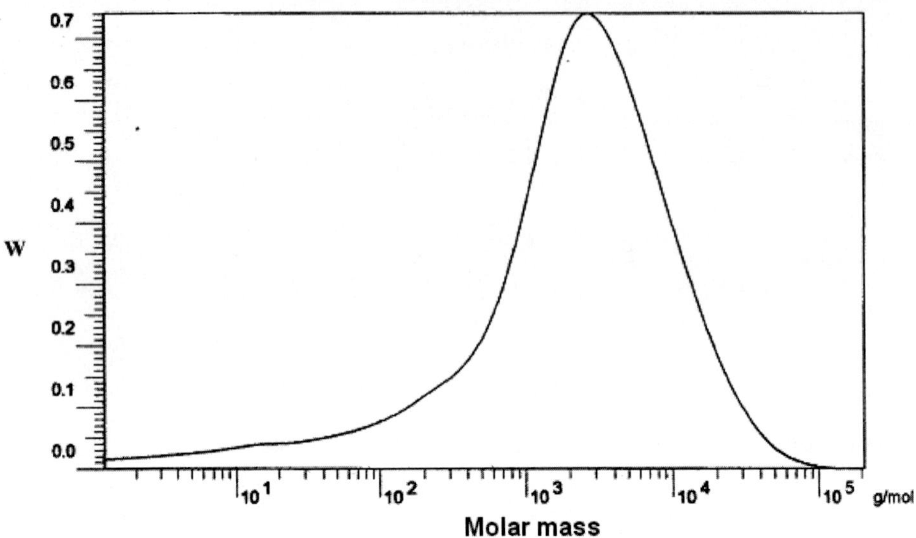

Figure 2. Molecular weight distribution in lignin from pine from steam explosion.

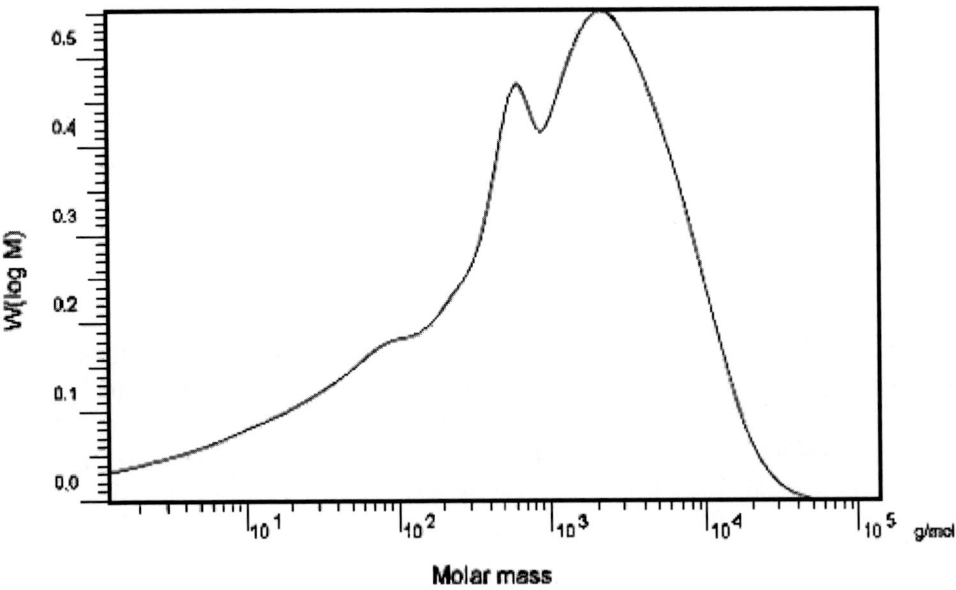

Figure 3. GPC of lignin from corn stalk.

The lignin from corn stalk showed a distribution of the molecular weight as reported in Figure 3. The lignin sample has $M_w = 2640$ and $M_z = 93994$. It is noteworthy that acetylation of the sample was not requested in order to solubilize this lignin in THF. The UV spectrum was recorded in 2-methoxy-1-propanol (4×10^{-2} g l^{-1}). Lignin showed absorptions at $\lambda = 231$ nm (A = 0.552) and 280 nm (A = 0.369). The FTIR spectrum of lignin is reported in Figure 4. We can note the C=O stretching at 1692 and 1651 cm^{-1} while the vibrations attributable to the aromatic rings are at 1604 and 1514 cm^{-1}.

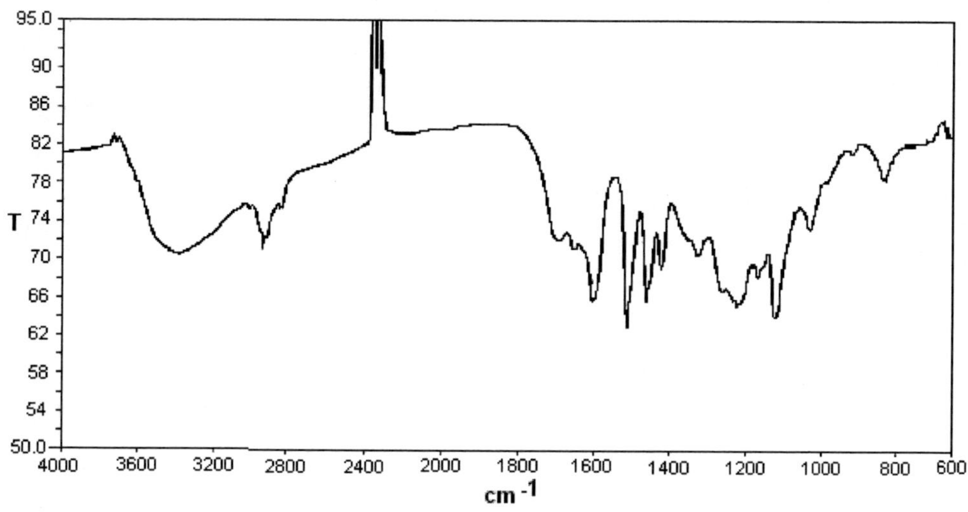

Figure 4. FTIR of lignin from corn stalk.

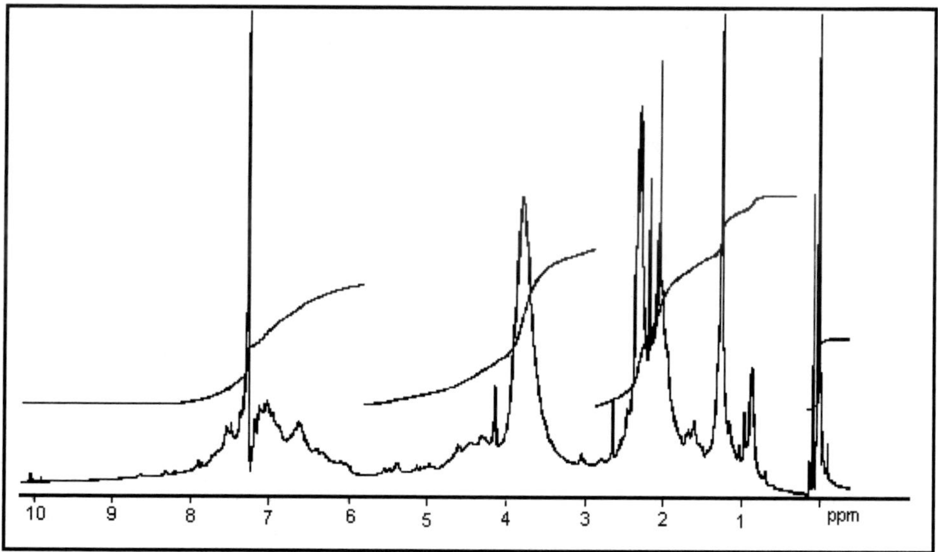

Figure 5. 1H NMR of lignin from corn stalk.

^{1}H NMR spectrum of lignin is reported in Figure 5. We can see signals at ca. 7.4 ppm attributable to vinyl protons on the carbon atoms adjacent to aromatic rings in cinnamaldehyde units, a signal at δ 6.9 ppm that identifies guaiacyl units, and a signal at δ 6.6 ppm typical for syringyl units. From these data, considering the integration of the signals, we have almost 1:1 ratio between guaiacyl and syringyl units. Finally, the characterization of the steam-exploded lignin was completed by using ^{13}C NMR spectroscopy (Figures 6 and 7).

We observed signals at δ 171 (C=O in aliphatic esters), 162 (C-4 in aromatic acids), 148 (C-4 in etherified guaiacyl units), 145 (C-4 in guaiacyl β-O-4 non etherified units), 138 (C-1 in syringyl β-O-4 units), 128 (C-α and C-β in ArCH=CH-CH_2OH), 126 (C-β in cinnamaldehyde units), 122 (C-1 in *p*-hydrophenyl benzoate units), 120 (C-6 in guaiacyl etherified and non etherified units), 115 (C-5 in etherified and non etherified guaiacyl units), 114 (C-3/C-5 in *p*-hydroxyphenyl benzoate units), 111 (C-2 in guaiacyl units, C-2 guaiacyl guaiacyl stilbene units), 104 (C-2/C-6 in syringyl units), and 55.5 ppm (Methoxy groups). We can observe the presence of both guaiacyl and syringyl structures with a prevalence of guaiacyl units. Furthermore, ^{13}C NMR spectrum is in agreement with the presence of cinnamic units in the lignin.

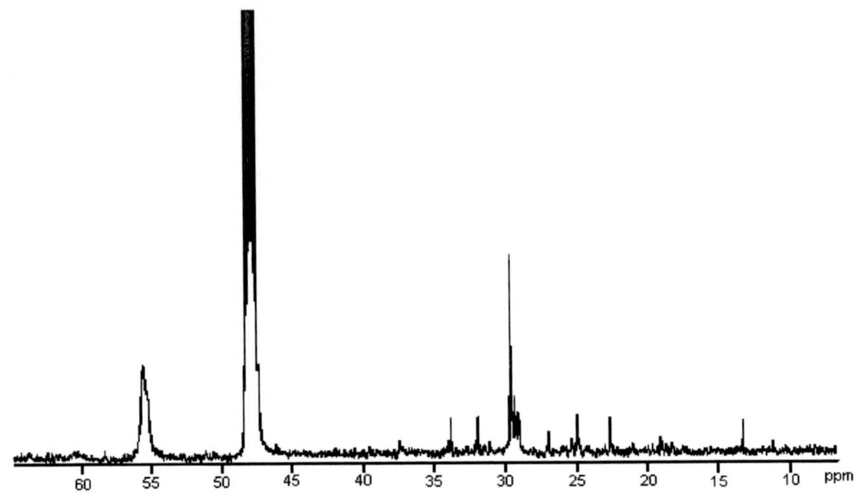

Figure 6. 13C NMR of lignin from corn stalk (high field).

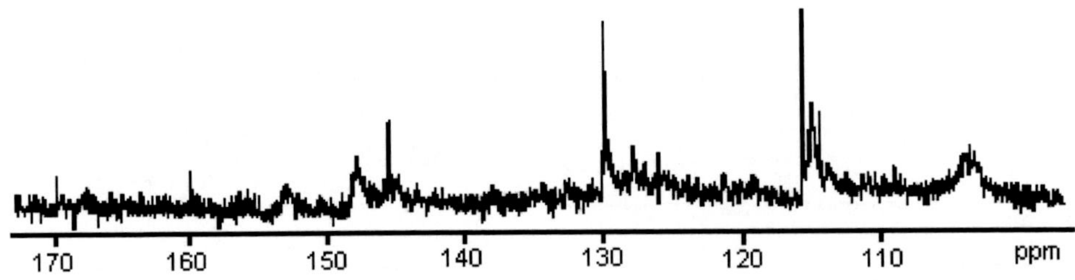

Figure 7. 13C NMR of lignin from corn stalk (low field).

In literature there are several reports on the use of reagents where the formation of hydroxy radical and superoxide anion were supposed to be involved [5,53-57]. Recently Kiwi reported that the irradiation of Orange II in aqueous solution produces several oxidative species in function of the pH [58]. At pH > 4.8 the prevalent oxidant was superoxide anion. We tested the capability of this oxidant to induce degradation on a steam exploded lignin.

We used as starting material the alkaline extracts obtained from the exploded sample after treatment with hot water and treatment with 1.5% NaOH at 90 °C for 15 minutes. This sample was irradiated for different times in the presence of oxygen. After this treatment acid precipitation of lignin was carried out. The irradiation with oxygen for different times gave the results reported in Table 1. We can observe the evolution of M_n in the time (Figure 8).

Table 1. Average molecular weight and treatment of lignin from pine with oxygen in the presence of light

	Irradiation time [h]			
	0	**4**	**8**	**13**
M_n	109	102	99	105
M_w	5121	5648	6115	5043
M_z	18233	19487	244583	17276
M_p	2702	3678	3485	3443

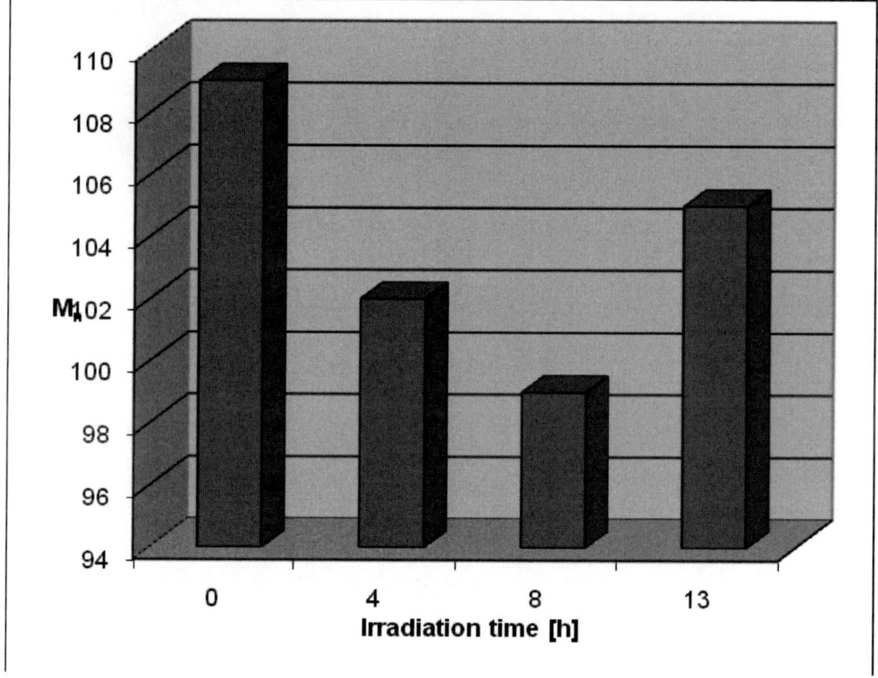

Figure 8. Mn values in function of the irradiation time in the photoirradiation of lignin from pine in the presence of oxygen.

M_n diminishes until 8 h irradiation time. After this period we observed an increment of this parameter. The reduction of M_n could be explained assuming the presence of a degradation process. However, this type of explanation is not able to explain the increment observed after 13 h irradiation time. A more consistent hypothesis could be: a. the reduction of M_n could be explained assuming the presence of a photoinduced polymerization process that uses molecules with low molecular weight to form molecules with higher molecular weight. It is important to note that M_n is more sensible to the presence of low molecular weight components than M_w or M_z. b. The following increment of M_n could be explained assuming the presence of a degradation process that yields an increment of low molecular weight components deriving from the degradation of the polymeric lignin.

This behavior is confirmed by the evolution of M_w and M_z (Figures 9 and 10). In fact, in this case we observed an inverse behavior. Until 8 h irradiation time the molecular weight increases in agreement with a polymerization process with the formation of materials with high molecular weight; after this time a decrease of M_w and M_z was observed: this result is in agreement with the presence of a degradation process.

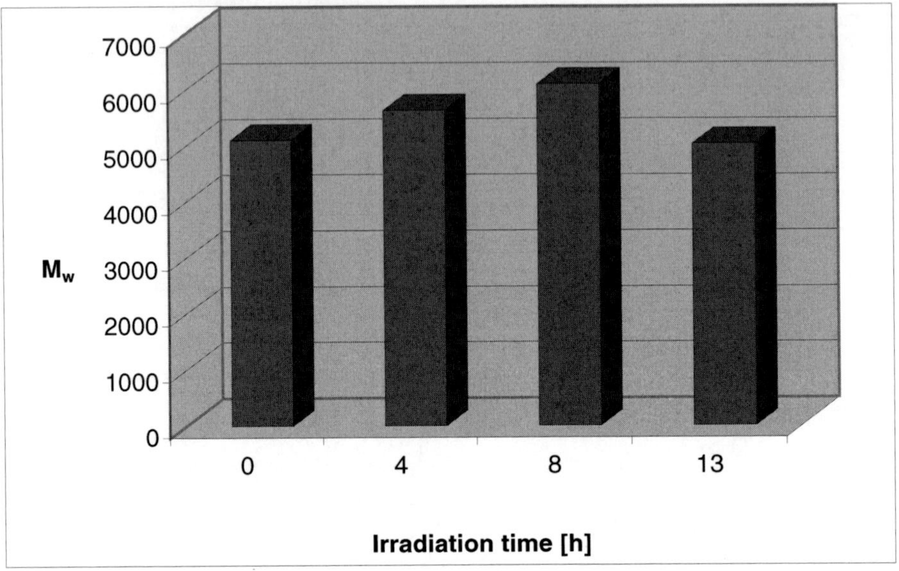

Figure 9. Mw values in function of the irradiation time in the photoirradiation of lignin from pine in the presence of oxygen.

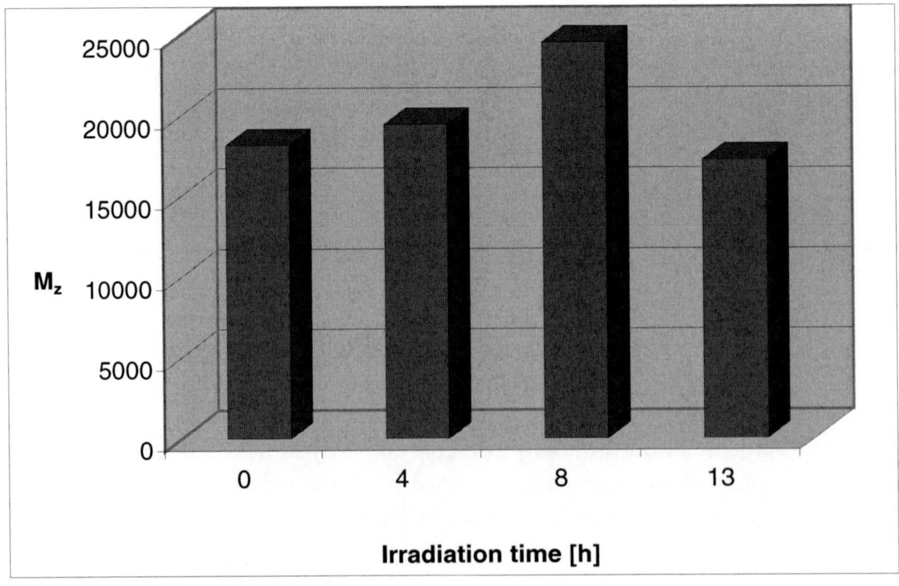

Figure 10. Mz values in function of the irradiation time in the photoirradiation of lignin from pine in the presence of oxygen.

On the basis of these results, we have seen that in the reaction conditions proposed by Kiwi we did not observe an important degradation process. On the contrary, we observed an initial polymerization process, induced by the formation of radicals in lignin components. We have also to note that we have no indications about the formation of superoxide anion as the oxidant species able to induce the degradation process. Nevertheless, superoxide anion could be the responsible species of lignin degradation observed in the final step of our experiments.

We attempted to degrade lignin from corn stalk by using ozone. We selected experimental conditions in order to have the minimum flux obtainable of the oxidant. In our purpose we selected this reaction condition in order to prevent overoxidation of the products allowing us to isolate the oxidation products. We examined the reaction mixtures after different time of treatment following the evolution of M_w. In these experimental conditions the reaction showed a zero order kinetics (Figure 11). This result implies that the reaction kinetics depends on the availability of ozone. The reaction showed a very high efficiency. In fact, after 50 minutes the average molecular weight of lignin was the half.

We analyzed the reaction mixture after ozonolysis by using GC-MS. It is interesting to note that we recovered only the formation of oxalic acid. We did not find any of the compounds found by Quesada after ozonation of organosolv extracts of corn stalk [21]. We think that these different results can be explained considering the acid treatment of the biomass in the steam explosion process. We need the acidic treatment because lignin could not be collected by using neutral conditions.

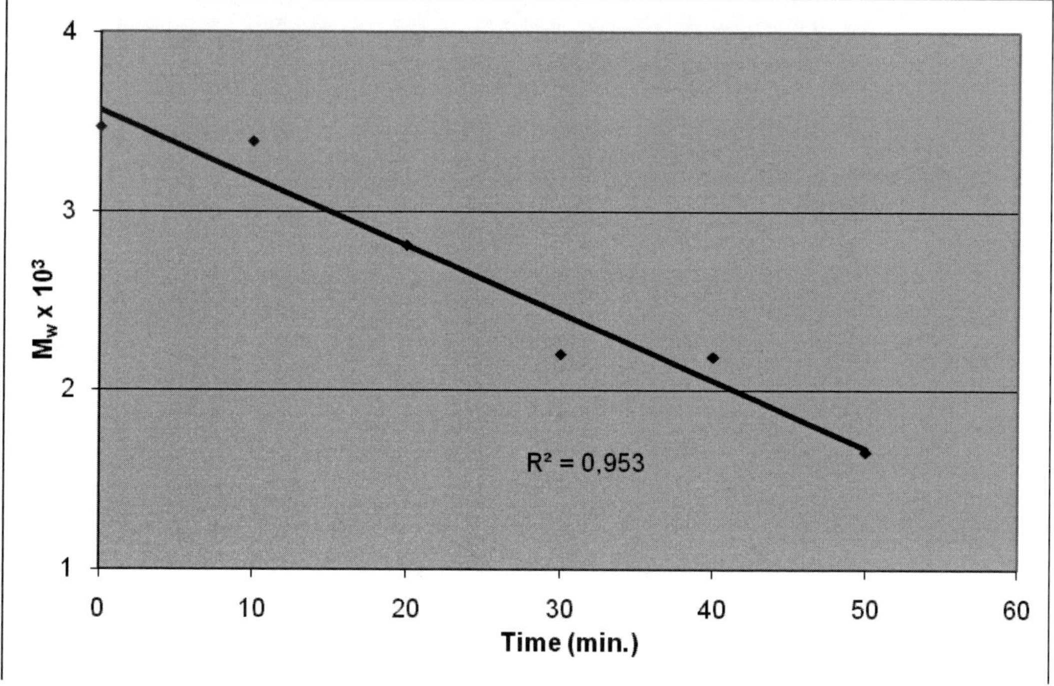

Figure 11. Kinetics of lignin degradation by ozone.

Probably, this treatment induces a gross destructuration of the lignin making it more susceptible towards the ozone oxidation than other treatments. We found that steam explosion could be a very useful method in order to achieve the complete degradation of lignin in pulp by using ozone before cellulose oxidation.

We described that ozonolysis of the steam exploded lignin did not allow us the recovery of fine chemicals with the exception of oxalic acid. On the basis of this result the steam explosion is a method able to afford the highest destructuration of the lignin of corn stalk, making the lignin easily oxidable by ozone.

Ozone and Olive Oil Mill Wastewater

We used a sample of olive oil mill wastewater deriving from a mill in Corleto Perticara (Potenza, Southern Italy). The sample showed a COD value of 130 g L^{-1} O_2. The treatment of this sample with different amounts of Fenton's reagent gave the results depicted in Figure 12 and Table 2.

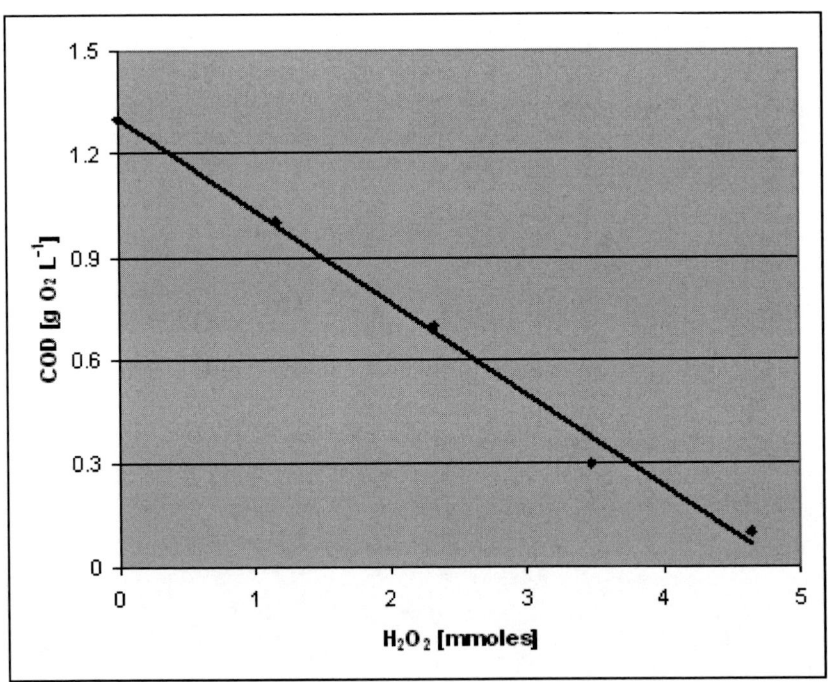

Figure 12. COD of olive oil mill waste water after treatment with Fenton's reagent.

Table 2. COD of olive oil mill wastewater after treatment with Fenton's reagent

Moles of Fenton's reagent [x 10^3]	COD [g O_2 L^{-1}]
0	1.30.
1.16	1.00
2.32	0.70
3.48	0.30
4.64	0.10

We can see that the Fenton's reagent is able to degrade the organic matter present in olive oil mill wastewater. Clearly, the phenolic components in the wastewater were oxidized in the presence of this reagent. This result is in agreement with our previous reported results on the oxidation of phenolic lignin.

The reaction doesn't seem to be quite efficient: the introduction of 4.64 mmoles of Fenton's reagent is able to reduce COD from 1.30 g L^{-1} O_2 to 0.10 g L^{-1} O_2. Assuming that phenols were vanillin, one of the main component in olive oil mill wastewater [47], we started from 0.17 mmoles of phenols to arrive to 0.013 mmoles of phenols: we have a reduction of phenol content of 0.157 mmoles with an efficiency of 3.38%. Furthermore, in the range we have examined, the reaction is linear as clearly depicted in Figure 12 ($r = 0.9928$).

The described method is quite efficient but with low utility in the treatment of a real wastewater. To reduce COD from 130000 mg O_2 L^{-1} to 13000 mg O_2 L^{-1} in 50 L of waste, and this reduction is not sufficient to avoid water pollution if the residue is poured into a river, we need to treat the wastewater with 1100 L of 35% hydrogen peroxide, 650 L of $FeSO_4$ solution with the production of ca. 225 kg of $Fe_2(SO_4)_3$, to be destroyed. Then, the treatment of 50 L olive oil wastewater requires the use of other 1750 L of water and the production of a large quantities of a by-product.

This treatment showed a very large cost. The above reported treatment of 50 L of waste water needs € 35.328 considering only the cost of the reagent.

The treatment of the sample of olive oil mill waste water with ozone gave the results reported in Figure 13 and Table 3.

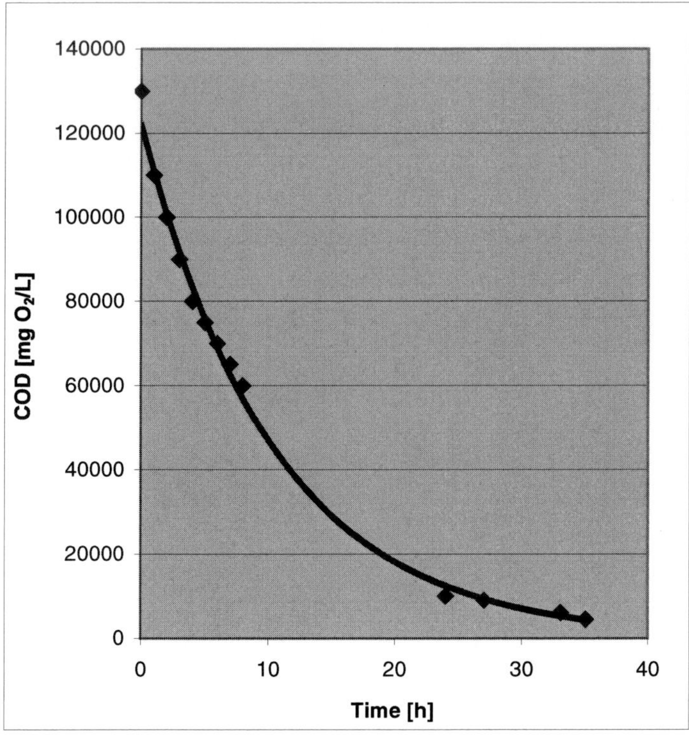

Figure 13. COD of olive oil mill waste water after treatment with ozone.

Table 3. COD of olive oil mill wastewater after treatment with ozone

Time [h]	COD [g O_2 L^{-1}]
0	130
1	110
2	100
3	90
4	80
5	75
6	70
7	65
8	60
24	10
27	9
33	6.1
35	4.5

Ozone is able to reduce the COD efficiently. After 35 h we observed a reduction of 76% of the value of COD. Bondioli *et al.* found that the reduction of COD followed a complex kinetics with two curves with different slopes [41]. On the contrary, we found that the reaction followed a first order kinetics with a slope of 9.51×10^{-2} h^{-1} ($r = 0.9955$).

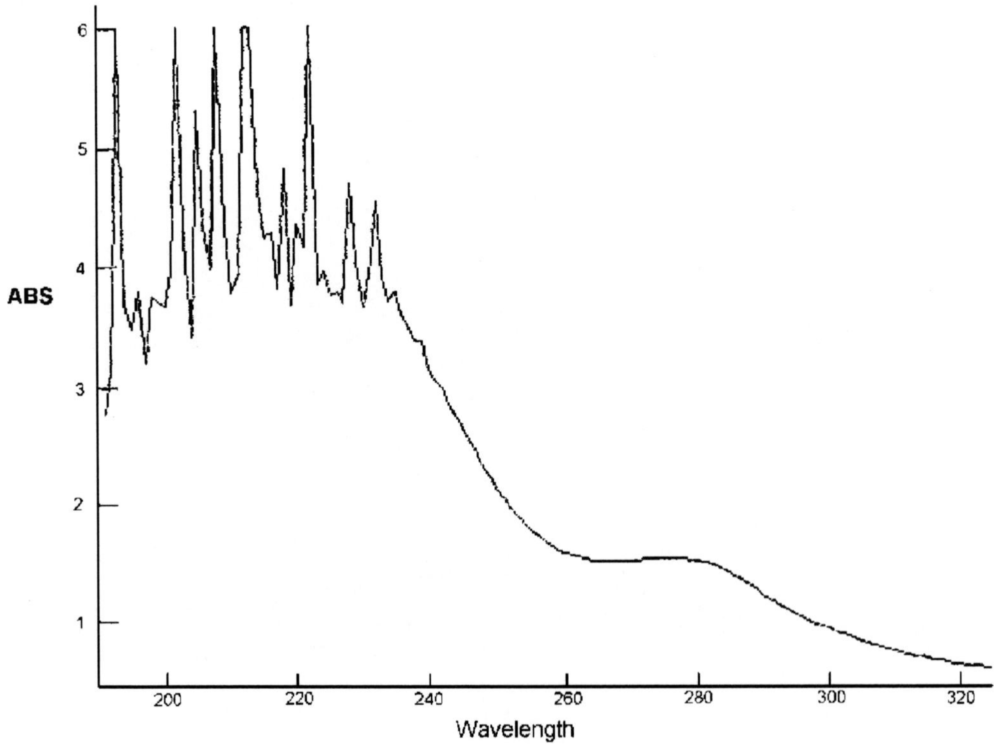

Figure 14. UV spectrum of olive oil mill waste water.

Evidences about the observed degradation were obtained by comparison of the uv spectrum before and after the treatment. The uv spectrum of olive oil mill waste water showed absorptions with λ_{max} at 202, 205, 208, 212, 218, 222, 228, 232, 240, and 280 nm (Figure 14). The uv spectrum of the reaction mixture after the treatment with ozone (35 h) did not show absorptions. All the chromophores present in olive oil mill waste water have been destroyed.

In this case the reduction of the COD of a sample (50 L) of olive oil mill waste water from 130 g L^{-1} to 13 g L^{-1} can be performed, by using an available apparatus able to give 17000 g O_3 h^{-1}, with six hours treatment of the waste. The cost of this treatment, by using air, is connected with the energy consumption. The plant described above shows a consumption of 283 kW h^{-1} with a cost for six hour treatment, assuming a price of € 0.13 for consumed kW h^{-1}, of € 221.

In conclusion, we have showed that both Fenton's reagent and ozone can induce an extensive degradation of olive oil mill waste water. On the basis of our results ozone seems to be more susceptible to be used in a real plant.

Ozone and Herbicides

In Table 4 kinetic parameters of oxidation processes tested have been listed. Independently of the oxidation process assessed, the decomposition of imazethapyr always followed a pseudo-first order kinetic. Kinetic constants ('k' values, in table 4) were calculated from the slope of the linear equation obtained plotting the natural logarithm of the degraded imazethapyr concentration versus the testing time [59].

Table 4. Kinetic parameters of imazethapyr degradation: n, reaction order; $t_{1/2}$, halflife, k, kinetic constant, r^2, determination coefficient. Values were obtained on the basis of three replicate experiments

Assay	Oxidation process	n	$t_{1/2}$ (h)	k (h^{-1})	r^2	k ratios	
1*	Ozone	1	13.4	0.0516	0.9990	k_1/k_4	2.0
2	Ozone/UV	1	1.35	0.5117	0.9992	k_2/k_1	9.9
3	Ozone/UV+TiO_2	1	0.17	3.9784	0.9958	k_3/k_1	77
4*	UV	1	27.1	0.0256	0.9989		
5*	H_2O_2/UV	1	19.1	0.0364	0.9998	k_5/k_4	1.4
6	TiO_2/UV	1	0.39	1.7742	0.9964	k_6/k_4	69
7	TiO_2/UV+H_2O_2	1	0.32	2.1524	0.9990	k_7/k_4	84

* Oxidation process with the formation of intermediate compounds depicted in figure 15: ozone, compounds 3, 4, 5 and 7 (major), 2 and 6 (minor); UV, compounds 2 and 6; H2O2/UV, compounds 2, 3 and 5.

The direct UV photolysis of the herbicide (assay 4 in table 1), as well as its oxidation with ozone in absence of UV-irradiation (assay 1), and photo-oxidation in the presence of hydrogen peroxide (assay 5) were sufficiently slow to permit the identification of intermediate products. All the other cases were characterized by a faster degradation and rapid formation of a lot of little molecules, which were quickly destroyed. During the direct photolysis (UV-irradiation test) compounds 2 and 6 (Figure 15) were found as the main by-products. The number of identified intermediates was the foremost in the case of ozonation in dark conditions, being major products 3, 4, 5, and 7, and minor 2 and 6. Throughout the H_2O_2/UV process, compounds 2, 3 and 5 were identified. The degradation pathways illustrated in figure 15 have been hypothesized on the basis of mass spectrum (single ion monitoring) and the formation curve of each intermediate (not shown).

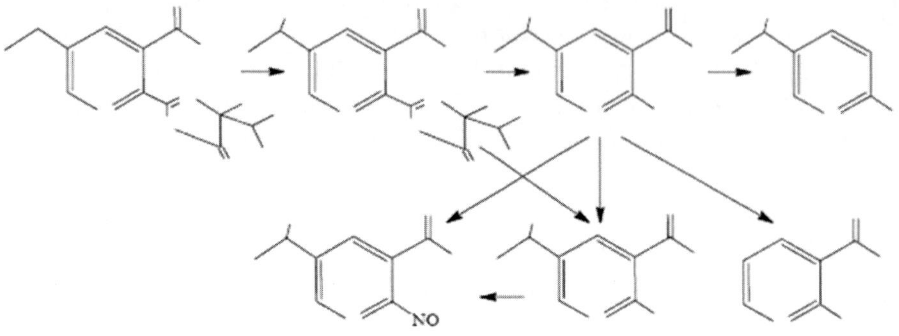

Figure 15. Chemical structure and degradation pathways proposed for imazethapyr (compound 1). Intermediate compounds were identified in assays 1, 4 and 5 (Table 4). Probable ozonation pathway: 1 → 2 (short life) → 3 → 4, 5, 6 (short life), 7; 2 → 6 → 5. Probable UV pathway: 1 → 2 → 6. Probable H2O2/UV pathway: 1 → 2 → 3 → 5.

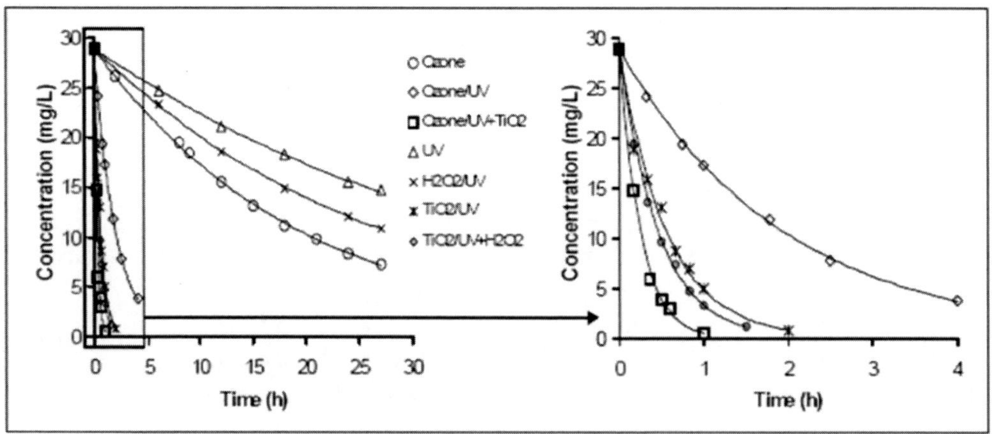

Figure 16. Imazethapyr remaining under different oxidation processes.

Figure 16 illustrates the curve of imazethapyr disappearance as a function of time. On the basis of plotted data, it is possible to group together three curves relative to faster processes (assays 3, 6 and 7), and other three curves referred to slower degradation rates (assays 1, 4 and 5); meanwhile, it is also achievable a medium velocity process (assay 2). In agreement with results shown by Beltrán et al. [51] for phenols' degradation, the most effective means for a rapid destruction of the herbicide seems to be the ozonation in the presence of UV radiations and reinforced with the catalyst TiO_2 (assay 3 in table 4), the specific velocity of which was manifold times (77) faster than the basic ozonation in the dark.

As suggested by Ishiki et al. [52], the TiO_2-photocatalysis (assay 6), and in particular the reinforcement of this process with the hydrogen peroxide (assay 7), revealed also a good capability in the rapid demolition of the herbicide molecules, 69 and 84 times faster than the direct photolysis, respectively. Besides, using these last two oxidation processes as well as the photocatalysis in the presence of ozone, the formation of indecomposable intermediates was avoided.

Comparing our results with literature data, we found a fairly good agreement among calculated kinetic constants ('k' values, in table 4) concerning direct (assay 4) and enhanced (assays 5,6 and 7) UV-photolyses, and those referred by Ishiki et al. [52] containing either the same dispersed quantity of TiO_2 (1.5 g L^{-1}) or an equal concentration of H_2O_2 (5 m*M*), but performed on a more diluted imazethapyr solution in water (0.05 m*M*). Starting from a 'k' value of 0.0378 h^{-1} for direct UV-photolysis, they obtained a fourfold rise adding hydrogen peroxide (*x* 1.4, our result), more than a fiftyfold augment by using titan dioxide (*x* 69, our result), but only *x* 54 increasing factor when both the boosting agents where adjoined (*x* 84, our result). Hydrogen peroxide retained on the TiO_2 surface may indeed swallow the OH radical formed on the surface of the catalyst hydrogen peroxide, but since it is a good electron acceptor, we stressed that interaction with the conduction band electrons may effectively occur and generate the hydroxyl radical necessary for the rapid mineralization of the organic substance [60,61].

According to Avila et al. [50] and Elazzouzi et al. [62], in natural conditions imazethapyr has the potential to be degraded by direct photolysis because it absorbs light above 290 nm and undergoes aqueous photolysis with halflives ranging from 43 h (k = 0.016 h^{-1}) to 57 h (k = 0.012 h^{-1}). Instead, when stronger condition were applied either adding TiO_2 or simply enhancing the hydroxyl radicals concentration the photolysis was quickly improved. Literature findings together with our result let us consider that imazethapyr, as well as other recalcitrant organic compounds, is able to rapidly dissipate absorbed energy from UV radiations and extinguish its excited states, perhaps through a wildly rotation and stretching of the imidazolinone ring (one of the first derivative formed was the compound deprived of this moiety). On the other hand, this molecule is mostly sensible to the presence into the reaction environment of radical species formed by photocatalysis or photocatalysis enhanced by hydrogen peroxide.

In conclusion, the application of advanced oxidation processes promises to be a profitable expedient for the rapid degradation and detoxification of persistent herbicides as imazethapyr. Most of all, AOPs are useful tools to avoid the development of by-products, which in milder oxidation conditions could arise in several structures and could be more toxic than the parent molecule. More effective were oxidation methods which used titan dioxide photocatalysis enhanced either by ozonation or hydrogen peroxide.

References

[1] Zimbardi, F.; Ricci, E.; Braccio, G. *Appl. Biochem. Biotech.* 2002, 98, 89.

[2] Crestini, C.; D'Auria, M. *J. Photochem. Photobiol., A: Chem.*, 1996, 101, 69.

[3] Crestini, C.; D'Auria, M. *Tetrahedron*, 1997, 53, 7877.

[4] D'Auria, M.; Ferri, R *J. Photochem. Photobiol., A: Chem.*, 2003, 157, 1.

[5] Ruggiero, R.; Machado, A. E. H.; Castellan, A.; Grelier, S. *J. Photochem. Photobiol., A: Chem.* 1997, 110, 91.

[6] Machado, A, E. H;, Gomes, A. J.; Campos, C. M. F.; Terrones, M. G. H.; Perez, D. S.; Ruggiero, R.; Castellan, A. *J. Photochem. Photobiol., A: Chem.* 1997, 110, 99.

[7] Bonini, C.; D'Auria, M.; D'Alessio, L.; Mauriello, G.; Tofani, D.; Viggiano, D.; Zimbardi, F. *J. Photochem. Photobiol., A: Chem.* 1998, 113, 119.

[8] Bonini, C.; D'Auria, M.; Mauriello, G.; Viggiano, D.; Zimbardi, F. *J. Photochem. Photobiol., A: Chem.* 1998, 118, 107.

[9] Lucia, L. A.; Hwang, K. O. *The Spectrum* 2001, 14 *(2)*, 8.

[10] Bonini, C.; Carbone, A.; D'Auria, M. *Photochem. Photobiol. Sci.* 2002, 1, 407.

[11] Bentivenga, G.; Bonini, C.; D'Auria, M.; De Bona, A.; Mauriello, G. *Chemosphere* 1999, 39, 2409.

[12] Bentivenga, G.; Bonini, C.; D'Auria, M.; De Bona, A. *J. Photochem. Photobiol., A: Chem.* 1999, 128, 139.

[13] Bentivenga, G.; Bonini, C.; D'Auria, M.; De Bona, A.; Mauriello, G. *J. Photochem. Photobiol., A: Chem.* 2000, 135, 203.

[14] D'Auria, M.; Bonini, C.; Emanuele, L.; Ferri, R. *J. Photochem. Photobiol., A: Chem.* 2002, 147, 153.

[15] Bonini, C.; D'Auria, M.; Ferri, R.. *Photochem. Photobiol. Sci.* 2002, 1, 570.

[16] Demin, V. A.; Shereshovets, V. V.; Monakov, J. B. *Russ. Chem. Rev.* 1999, 68, 937.

[17] Ragnar, M.; Eriksson, T.; Reitberger, T. *Holzforschung* 1999, 53, 292.

[18] Ragnar, M.; Eriksson, T.; Reitberger, T.; Brandt, P. *Holzforschung* 1999, 53, 423.

[19] Olkkonen, C.; Tylli, H.; Forsskåhl, I.; Fuhrmann, A.; Hausalo, T.; Tamminen, T.; Hortling, B.; Janson. J. *Holzfoschung* 2000, 54, 397.

[20] Roncero, M. B.; Colon, J. F.; Vidal, T. *Inv. Téc. Papel* 2001, 147, 175.

[21] Quesada, J.; Rubio, M.; Gomez, D. *J. Agric. Food Chem.* 1998, 46, 692.

[22] Bertaud, F.; Croué, J. P.; Legube, B. *Ozone Sci. Eng.* 2001, 23, 139.

[23] Balice, V.; Carrieri, C.; Cera, O. *Riv. Ital. Sostanze Grasse*, 1990, 67, 9.

[24] Boari, G.; Mancini, I. M.; Trulli, E. *Water Sci. Technol.*, 1993, 28, 27.

[25] Tsonis, S. P.; Grigoropoulos, S. G. *Water Sci. Technol.*, 1993, 28, 35.

[26] Ranalli, A.; De Mattia, G. *Riv. Ital. Sostanze Grasse*, 1996, 73, 61.

[27] Yesilada, O.; Sik, S.; Sam, M. *World J. Microbiol. Biotechnol.*, 1998, 14, 37.

[28] Zenjari, B.; Hafidi, M.; El Hadrami, I.; Bailly., J. –R.; Nejmeddine, A. *Agrochimica*, 1999, 43, 277.

[29] Blanquez, P.; Caminal, G.; Sarra, M.; Vicent, M. T.; Gabarell, X. *Biotech. Progr.*, 2002, 18, 660.

[30] Iamarino, G.; Rao, M. A.; Nocerino, G.; Gianfreda, L. *Polyphenols Actualites*, 2003, 23, 20.

[31] Beccari, M.; Majone, M.; Riccardi, C.; Savarese, F.; Torrisi, L. *Water Sci. Technol.*, 1999, 40, 347.

[32] Fiorentino, A.; Gentili, A.; Isidori, M.; Lavorgna, M.; Parrella, A.; Temussi, F. *J. Agric. Food Chem.*, 2004, 52, 5151.

[33] Majone, M.; Beccari, M.; Petrangeli, P. M.; Torrisi, L. *Eur. Pat. Appl.* EP1167305 [*Chem. Abstr.* 2002, 136, 58113].

[34] Inan, H.; Dimoglo, A.; Simsek, H.; Karpuzcu, M. *Sep. Pur. Technol.*, 2004, 36, 23.

[35] Longhi, P.; Vodopivec, B.; Fiori, G. *Ann. Chim.*, 2001, 91, 169.

[36] Cermola, F.; Della Greca, M.; Iesce, M. R.; Montella, S.; Pollio, A.; Temussi, F. *Chemosphere*, 2004, 55, 1035.

[37] Rivas, F. J.; Beltran, F. J.; Gimeno, O.; Frades, J. *J. Agric. Food Chem.*, 2001, 49, 1873.

[38] Canepa, P.; Cauglia, F.; Caviglia, P.; Chelossi, E. *Environ. Sci. Pollut. Res. Intern.*, 2003, 10, 217.

[39] Bressan, M.; Liberatore, L.; D'Alessandro, N.; Tonucci, L.; Belli, C.; Ranalli, G. *J. Agric. Food Chem.*, 2004, 52, 1228.

[40] Rivas, F. J.; Gimeno, O.; Portela, J. R.; De La Ossa, E. M.; Beltran, F. J. *Ind. Engin. Chem. Res.*, 2001, 40, 3670.

[41] Bondioli, P.; Lanzani, A.; Fedeli, E.; Sala, M.; Gerali, G. *Riv. Ital. Sostanze Grasse*, 1992, 69, 487.

[42] Bentivenga, G.; Bonini, C.; D'Auria, M.; De Bona, A. *Biomass Bioenergy,* 2003, 24, 233.

[43] Jefcoate, C. R. E.; Lindsay Smith, J. R.; Norman, R. O. C. *J. Chem. Soc. B*, 1969, 1013.

[44] Brook, M. A.; Castle, L.; Lindsay Smith, J. R.; Higgins, R.; Morris, K. P. *J. Chem. Soc., Perkin Trans.* 2, 1982, 687.

[45] Lai, C. S.; Piette, L. H. *Tetrahedron Lett.*, 1979, 775-778.

[46] Kunai, A.; Hata, S.; Ito, S.; Sasaki, K. *J. Am. Chem. Soc.*, 1986, 108, 6012.

[47] Sayadi, S.; Allouche, N.; Jaoua, M.; Aloui, F. *Process Biochem.,* 2000, 35, 725.

[48] Farré, M. J.; Franch, M. I.; Malato, S.; Ayllón, J. A.; Peral, J.; Doménech, X. *Chemosphere*, 2005, 58, 1127.

[49] Camel, V.; Bermond, A. *Wat. Res.*, 1998, 32, 3208.

[50] Avila, L. A.; Massey, J. H.; Senseman, S. A.; Armbrust, K. L.; Lancaster, S. R.; McCauley, G. N.; Chandler, J. M. *J. Agric. Food Chem.*, 2006, 54, 2635.

[51] Beltrán, F. J.; Rivas, F. J.; Gimeno, O. *J. Chem. Technol. Biotechnol.*, 2005, 80, 973.

[52] Ishiki, R. R.; Ishiki, H. M.; Takashima, K. *Chemosphere*, 2005, 58, 1461.

[53] Vanucci, C.; Fornier de Violet, P.; Bouas-Laurent, H.; Castellan, A. *J. Photochem. Photobiol., A: Chem.*, 1988, 41, 251.

[54] Castellan, A.; Colombo, N.; Cucuphat, C.; Fornier de Violet, P. *Horzforschung*, 1989, 43, 179.

[55] Machado, A. E. H.; Ruggiero, R.; Neumann, M. G. *J. Photochem. Photobiol., A: Chem.* 1994, 81, 107.

[56] Machado, A. E. H.; Ruggiero, R.; Terrones, M. G. H.; Nourmamode, A.; Grelier, S.; Castellan, A. *J. Photochem. Photobiol., A: Chem.* 1996, 94, 253.

[57] Felicio, C. M.; Machado, A. E. H.; Castellan, A.; Nourmamode, A.; Perez, D. S.; Ruggiero, R. *J. Photochem. Photobiol., A: Chem.* 2003, 156, 253.

[58] Bandara, J.; Kiwi, J. *New J. Chem.* 1999, 23, 717.

[59] Scrano, L.; Bufo, S. A.; D'Auria, M.; Emmelin, C. *J. Photochem. Photobiol. A: Chemistry,* 1999, 129, 65.

[60] Wang, Y.; Hong. C. S. *Water. Res.*, 1999, 33, 2031.

[61] Dionysiou, D. D.; Suidan, M. T.; Bekou, E.; Baudin, I.; Laîné, J. M. *Appl. Catal. B: Environ.*, 2000, 26, 153.

[62] Elazzouzi, M.; Mekkaoui, M.; Zaza, S.; El Madani, M.; Zrineh, A.; Chovelon, J. M. *J. Environ. Sci. Health B*, 2002, 37, 445.

In: Ozone Depletion, Chemistry and Impacts
Editor: Sem H. Bakker, pp. 117-133
ISBN: 978-1-60692-007-7

Chapter 5

AIRWAY INFLAMMATION AND HYPER-RESPONSIVENESS INDUCED BY OZONE EXPOSURE

M.D. Yang Xiang[a] and Xiaoqun Qin[b]
Department of Physiology, Xiangya School of Medicine, Central South University, Xiangya Road 110, Changsha, Hunan 410078, P.R.China

Abstract

Acute ozone exposure is known to decrease pulmonary function. We have successfully constructed an animal airway hyper-responsiveness (AHR) model by ozone stress, mimicking the airway obstruction, airway inflammatory response, and increased airway responsiveness observed in human AHR disease. The mechanisms leading to the increased AHR are not clear, but epithelial injury is involved. The epithelium is not merely a passive barrier but can generate a range of mediators that may play a role in the inflammatory and remodeling responses. Damage of the bronchial epithelium associated with leukocyte infiltration and increased airway responsiveness are consistent features of asthma. It is reasonably hypothesized that disruption of these functional processes or defects in airway epithelium integrity may be the initial steps leading to airway hyper-responsiveness. Therefore, we damaged the airway epithelium with ozone stress in cultured BEC and animal model, focusing in particular on the roles of airway epithelium in airway inflammation and hyper-responsiveness.

Years of research in our laboratory showed that, after repeated stimulation of bronchial epithelial cells by ozone, a serious of events are programmed to occur: defect in function (e.g., anti-oxidation or secretion) or structural integrity (e.g., imbalance in adhesive molecules expression) weaken the protective ability of BEC against exogenous factors or antigens, such that BEC are easily stressed, damaged and even denuded; and thus the sensitivity of the epithelium and sensory nerve ends are enhanced aberrantly. Subsequently the inflammatory mediators are released; antigen-presenting activities are increased, recruiting and activating immune or inflammatory responsive cells with enhanced airway inflammation and hyper-responsiveness. Pulmonary peptidergic innervation remodeling and airway remodeling continued to increase the airway resistance. These events may be involved in the pathogenesis of AHR.

[a] E-mail address: xyww4595@163.com.
[b] E-mail address: xiaoqun1988@xysm.net.

Airway hyper-responsiveness (AHR) diseases, such as asthma and chronic obstructive pulmonary disease (COPD), are typically and commonly characterized as an excessive response of the airway to low level of antigens, slight challenge of environment factors, and even endogenous bioactive substances in physiological condition [1]. Currently, AHR is considered as a chronic airway inflammation, which shows a cluster of pathologic changes both in function and architecture of the airways including inflammatory cells infiltration, excessive secretion of mucus, bronchial smooth muscle spasm, and airway wall remodeling. However, the exact mechanism underlying the pathogenesis of AHR or asthma is still unclear. Take asthma for example, the current theories concerning its pathogenesis, apart from the classic allergic theory, include the neurogenic airway inflammation hypothesis and epithelia defect hypothesis.

The bronchial epithelium is an attractive target in which to identify novel mechanisms and new therapeutic strategies as it is critically involved in disease development as the first cell layer of contact with the environment. More and more basic or clinic questions are imputed to bronchial epithelial cells （BEC） .Bronchial epithelial cells are known to play an integral role in the airway defense mechanism via mucociliary system as well as mechanical barriers. Recent studies further indicate that BEC produce and release biologically active compounds including lipid mediators, growth factors, endothelin and a variety of cytokines/chemokines important in the pathogenesis of airway inflammation, mucosal immunity and remodeling.[2,3]. Damage of the bronchial epithelium associated with leukocyte infiltration and increased airway responsiveness are consistent features of asthma[4]. It is reasonably hypothesized that disruption of these functional processes or defects in airway epithelium integrity may be the initial steps leading to airway hyper-responsiveness. The Global Initiative for Asthma (GINA)'s program pointed that disorders in airway epithelium functions may equally important with IgE-dependent responses in asthma pathogenesis. Therefore, we constructed an animal AHR model damaging the airway epithelium with ozone stress, focusing in particular on the roles of airway epithelium in airway hyper-responsiveness.

1. A Novel Animal Model of Airway Hyperresponsiveness Induced by Ozone Exposure

Animal Model of Ozone Exposure

Balb/c mice were obtained from the experimental animal center, Xiangya School of Medicine, Central South University. All mice were housed under specific pathogen-free conditions and had a libitum access to food and water. The mice housed in whole-body exposure chambers were exposed to ozone concentrations of 2 ppm for 30min/day for 0,1,2,4,8 days respectively (n=6/group), while they were awake and breathing spontaneously in the chamber.

Measurement of Airway Responsiveness in vivo

Airway responsiveness was measured using barometric whole-body plethysmography by recording air flow and respiratory pressure curves (Buxco) in response to inhaled methacholine (Sigma). Airway responsiveness was expressed in pulmonary resistance (R_L), which was determined by multiple linear regressions from transpulmonary pressure and airflow, and dynamic lung compliance (Cdyn).

Briefly, mice were placed in a whole-body chamber, and basal readings were obtained and averaged for 3 min. Aerosolized saline, followed by methacholine (0.39-3.12mg/ml), were nebulized for 3 min, and readings were taken and averaged for 3 min after each nebulization.

The results showed that 4 days of ozone exposure caused a small but significant increase in R_L and decrease in Cdyn. Aerosolized methacholine challenge elicited a dose-dependent change in R_L and Cdyn in ozone-exposed mice and the maximum increase in R_L and maximum fall in Cdyn in these mice were significantly greater than that observed in control mice ($P < 0.05$).

The changes of pulmonary resistance by the challenge of methacholine

	Baseline ($cmH_2O/ml/s$)	**Methacholine –challenged ($cmH_2O/ml/s$)**	**%above baseline**
Control	0.128±0.06	0.346±0.09*	169.9±12
Ozone-stressed	0.894±0.19	3.839±0.96*	329.2±0.469#

n=6 *$p<0.05$ *vs* Baseline #$p<0.05$ *vs* Control.

The changes of Cdyn by the challenge of methacholine

	Baseline ($cmH_2O/ml/s$)	**Methacholine –challenged ($cmH_2O/ml/s$)**	**%above baseline**
Control	1.369±0.22	1.047±0.27 *	23.5±1.7
Ozone-stressed	0.979±0.17	0.519±0.09*	46.9±2.3#

n=6 *$p<0.05$ *vs* Baseline #$p<0.05$ *vs* Control.

Bronchoalveolar Lavage (BAL)

BAL was performed immediately after the last measurement of airway responsiveness. The mice were deeply anesthetized with 50 mg/kg of pentobarbital sodium injected intraperitoneally and were sacrificed by bleeding from femoral artery and subjected to bronchoalveolar lavage. The tracheas were cannulated, and the left lungs were ligated. The right lung was lavaged three times with 1.0 ml of sterilized saline solution (37 °C). Volumes of total BAL fluid were 2±1 ml per animal. Then the recovered fluid was immediately centrifuged for 5 min at 1500×g, 4 °C. The pellet was washed with Tris-buffered ammonium chloride solution (pH 7.2) to lyse the contaminated red blood cells, then centrifuged again. The pellet was re-suspended with saline solution and the total number of cells was counted under light microscope. The viability of cells was confirmed more than 80% by typan blue exclusion technique. Smears were stained with Giemsa to classify different cell types. Total protein concentration in BALF of each animal were determined by the Lowry Method.

The results showed that mice exposed to ozone showed a significant increase in total protein and total cell numbers in BALF compared with the control group, especially the number of macrophages, neutrophils and eosinophils increase obviously.

Preparation of Lung Tissues and Morphological Analysis

After fixation in 4% paraformaldehyde, the left lung tissues were embedded in paraffin and 4-mm thick transverse sections were mounted and stained with hematoxylin and eosin (Experimental Pathology Laboratory, Research Triangle Park, NC). Pathology evaluations were made under light microscopy.

A typical inflammatory pathology changes were observed in pulmonary tissue slides, including neutrophils and eosinophils infiltration, mucus exudation and bronchial epithelial cells (BEC) shedding.

Our results indicate that this ozone exposed animal model mimics the airway obstruction, airway inflammatory response, and increased airway responsiveness observed in human AHR disease, which indicate the success of AHR formation. Compared with the traditional allergic animal model, our model focus on the central role of airway epithelium in airway immune response, airway inflammation, and airway remodeling.

2. Association with Ozone Stressed Epithelial Cells and AHR

Decrease of Antioxidant Activity

The antioxidant activity of bronchial epithelial cells undertakes the first front line of defense or self cytoprotection, which is essential for keeping airway response normally. Most of injury factors cause damages through a common down stream: inducing excessive production of reactive oxygen substances (ROS) in cells. Bronchial epithelial cells are constantly exposed to gases with higher levels of oxygen and insulted by exogenous factors; therefore, an effective antioxidant system should have been evolutionally developed in the cells. Using gently trypsin digestion adding brush, our laboratory established a method for primary culture of rabbit bronchial epithelial cells[5]. On rabbit ciliated bronchial epithelial cells, glutathione synthesis and peroxidase, catalase, and bcl-2 gene expression consist predominant and a nearly perfected antioxidant system, which delighted the injury effects of ozone stress and can be upregulated by heat stress, vasoactive intestinal peptide, epidermal growth factor, and adhesive molecules integrin binding to extracellular matrix [6-13]. In experimental models, oxidants can produce many of the typical features of asthma. They induce bronchoconstriction and increase airway responsiveness to several agonists[14, 15]. On the AHR model, nearly all features in AHR including aberrantly increasing in airway resistance, mucus secretion, protein leaking out to airway lumen and alveolar space, bronchial epithelial cells denude, edema in parenchyma, enosiphils and neutrophils infiltration, and even airway wall thickening were observed. The susceptibility of the airway to oxidative injury will depend in part on the ability to upregulate protective ROS scavenging systems. Numerous studies have shown that cells protect themselves from oxidative stress through the expression of various antioxidant enzymes such as glutathione peroxidase, catalase, superoxide dismutase and Cu/Zn-SOD that detoxify ROS to less toxic species[16, 17]. Extracellular glutathione peroxidase is synthesized by airway epithelium and alveolar macrophages, secreted into the surface epithelial lining fluid, and functions as a first-line defense against inhaled ROS[18]. An imbalance between oxidants and antioxidants is considered to play a role in the pathogenesis of chronic obstructive pulmonary disease (COPD)[19, 20] and asthma [21, 22].

Imbalance of Adhesion Molecule Expression

The bronchial epithelium is a critical site during the pathology of asthma and other airway hyper-responsive disease[23]. During this process, the structure integrity of the bronchial epithelium is usually destroyed and the shedding of bronchial epithelium cell happen often[23, 24]. If any architecture or functional defect exists in bronchial epithelium, the cells hardly anchor to the basement membrane firmly and easy to be denuded, thus, the risk or susceptibility of asthma will increase aberrantly. After damage, activated proliferation, migration, spreading, differentiation activities of bronchial epithelial cells are needed to repair the impaired epithelium and restore normal structure, cells proportion and functions[25]. Ciliated cells are the terminally differentiated columnar cells, which are thought to originate from basal or secretory cells. Beneath the layer of columnar cells, basal cells, via hemidesmosomal complex, anchor to the basement membrane and act as local stem cells. Basal cells differentiate into the secretory or ciliated cell when the epithelial desquamates during some airway disease[26, 27].

More and more evidence define that adhesion molecules play important roles for epithelium function. There are four kinds of adhesion molecules express on airway: integrins, cadherins, selectins and intercellular adhesion molecules-1 (ICAM-1) .

Integrins, a large family of heterodimeric transmembrane glycoproteins, is regarded as the special kind of receptor which mediates cell's adhesion to extracellular matrix[28]. Integrin molecule is consisted of two subunits, α and β. Up to date, in human, totally 18 kind of α-subunits and 8 kind of β-subunits have been identified, which compose 24 kind of integrin heterodimers, and 8 of them found to be expressed on the bronchial epithelial cells: α2β1, α3β1, α6β4, α9β1, α5β1, αvβ6, αvβ8, αvβ6[29]. For example, integrin α6β4 mediates the adhesion of basal cell to basement membrane through the formation of hemidesmosome, and α2β1, α3β1 can also play a role in the homologization adhesion between two closed epithelial cells[30]. A mice with integrin β4 gene knock-out showed cell cycle and adhesion defect [31]. Recent studies indicated that signaling pathways activated by integrins adhesion participate in cell activation and regulation of cell's proliferation, differentiation, spreading, cell cycle, and even cell death[28, 30, 32]. The FAK and Fyn/Src are the two known integrin dependent pathways[28, 32].

There are evidences to indicate that integrity of the tight junctions between epithelial cells could be influenced by E-cadherin, a transmembrane glycoprotein for homophilic connection between columnar and bronchial basal cells[33]. The cytoplasmic domain of E-cadherin associates with three proteins termed α-, β-, and γ-catenin that link E-cadherin to the actin cytoskeleton [34]. In cooperation with cytoskeletal structures, E-cadherin is thought to regulate cell differentiation and morphogenesis. When inflammatory disorders in the epithelia occur, expression of E-cadherin decreased and its localization changed[35].

ICAM and selectins are another two kinds of adhesion molecule associated with migration of inflammatory cells to airway epithelium[36]. Selectins and their ligands are involved in the rolling and tethering of leukocytes on the vascular wall[26]. ICAM-1 binds as ligand for β2 integrins expressed on leukocytes and regulate the infiltration of leukocytes.

Recent years, ICAM-1 has been paid more attention especially. Studies on nasal turbinate showed that ICAM-1 is broadly distributed within the epithelium, on ciliated and basal cells and in the submucousa on the endothelial and infalammatory cells[[37, 38]. It can be stimulated by many inflammatory cytokines, such as IL-1 or TNF-α. The expression of ICAM-1 is greatly increased in stressed bronchial epithelial cells [39] and in inflamed tissues, implicating the pathogenesis of disease states, such as asthma[37].

The balance in these four kind adhesive molecules expression on bronchial epithelial cells should be significant for keeping airway homeostasis. If the spectrum of these four kinds of adhesive molecules expression changes abnormally, the susceptibility of AHR increases consequently. The speculation needs to be confirmed directly, although an alteration in the spectrum of adhesive molecules expression was observed both in leukocytes from asthma patients and bronchial epithelial cells from AHR animal model by our group recently.

Enhanced Airway Inflammation

Bronchial epithelial cells can participate in local chemical signal molecules networks and regulate inflammatory airway events by synthesizing and secreting various chemical signal molecules that communicate in a paracrine manner with various cells. When activated, bronchial epithelial cells produce a variety of cytokines, including interleukins[32], chemokines[40, 41], growth factors[42-45], interferon (IFN)[46, 47], colony-stimulating factor[48], TNF[49] six major categories.

In previous study, we found that, for initiation of airway inflammation, BEC stimulated by ozone can release various proinflammatory cytokines such as IL-1, IL-5, IL-8[50], and increase the expression of ICAM-1 through NF-κB dependent pathway[39], then the inflammatory signals can be transmitted from BEC to eosinophils, neutrophils, lymphocytes, mast cells and monocytes[39, 51-53]. BEC appear capable of generating a series of chemoattractants and chemoattractant cytokines that may act over a greater distance to direct inflammatory cell movement and activation in tissue[54]. Synchronously, a scheme of peripheral blood white cells into inflamed airways was developed that revolved around the regulated expression of distinct families of cell communication molecules, notably the selectins, the integrins, and the cell adhesion molecule (CAM)[55]. In the case of epithelial-leukocyte interaction, this information has enabled the development of a stepwise molecular scheme for leukocyte influx into airway lumen from the circulation.

When stimulated by allergen, virus or physical injury, airway epithelium produces various chemotactic factors and cytokines to start immune response including recruitment of inflammatory cells and repair[39, 50-53].Compared to normal individuals, the bronchial epithelium of patient with asthma can create more IL-1, IL-6, IL-8, IL-11, GM-CSF and TNF-α[56], the expression of RANTES,TARC,eotaxin,MCP-1, MCP-4 are also increased[57], suggesting that epithelium play a critical role in the infiltration of multiple inflammatory cell such as eosinophils, neutrophil and T lymphocyte to the site of inflammation. Airway remodeling maybe ascribe to the imbalance of factors promoting the proliferation of epithelium or mesenchymal cell. Injured bronchial epithelial cells induced the transdifferentiation of sub-epithelial fibroblasts into myofibroblasts, which may be mediated by ET-1 and TGF-β through MAPK pathways such as p38 MAPK and ERK 1/2[58]. Fibronectin (FN) is an important epithelia-derived chemotactic factor which is upregulated by

TGF-β in injury, accelerating the migration of fibroblast and epithelial cells to the site of wound.

Mucus Secretion Control and Transport in AHR

A common pathological feature of airway hyper-responsiveness diseases such as asthma and COPD is mucus hypersecretion. The mucus layer that coats the airway epithelium provides a protective barrier against pathogenic and noxious agents and participates in the mucosal response to inflammation and infection. Airway mucus is composed of water, ions, lung secretions, serum protein transudates, and mucin glycoproteins (mucins). Mucins are the major components of mucus and the macromolecules that impart rheologic properties to airway mucus. MUC5AC is the predominant mucin gene expressed in healthy airways and is increased in asthmatic and COPD patients[59]. Airway mucus is overproduced in the upper and/or lower respiratory tracts during acute challenges and in chronic conditions, thereby contributing to airways obstruction with mucus[60]. Mucus obstruction is the culmination of several complex processes including mucin (MUC) gene regulation, mucin secretion and goblet cell hyperplasia[61]. The mucosa of the airway surface epithelium is lined with a thin layer of fluid called the airway surface liquid (ASL). The composition of the ASL affects its physiological functions, the most important of which are removal of inhaled particles and antimicrobial activity. Acidic ASL has been implicated in airway diseases such as asthma and cystic fibrosis (CF)[62]. In the process of mucus secretion, aquaporin (AQPs) water channels play a major physiological role in airway submucosal gland function through facilitating water transport across cell membrane. In the airways, AQP3 and AQP4 facilitate water transport. AQP5 deletion in submucosal glands in upper airways reduced fluid secretion with over two fold increased- protein content[63, 64].

Active Cl^- transport across the epithelial cells, followed by Na^+ and water movement through the paracellular pathway, are important in the formation of low-viscosity mucus, thereby maintaining a conductive and aseptic environment in the lung [65]. In airways, calcium-activated chloride channels (CaCCs) contribute to fluid secretion. Expression of CaCC1 would be increased in the airways of patients with asthma, which would be particularly associated with mucus overproduction [64]. Cystic fibrosis transmembrane regulator (CFTR) and epithelial sodium channel (ENaC) are the principal rate-limiting steps for Cl^- secretion and Na^+ absorption by bronchial ciliated epithelial[66]. Recently, our group observed that ozone stress could down-regulate the expression of CFTR and decrease CFTR chloride current in HBEC, the signal mechanism of which refered to cascade events in cells including early oxidative stress signal transmission molecules, and subsequently transcription modulator Nrf2 and STAT1

Efficient mucociliary transport is accomplished by the coordinated beating of the cilia which are immersed persistently in the periciliary liquid layer and contact intermittently with the overlying mucus layer[67]. The efficacy of the mucociliary apparatus depend upon morphologic integrity of the cilia structure, mucus components, functional efficacy of synchronism or magnitude of ciliary activity, periciliary fluid depth, and mucus rheological properties. In contrast, in chronic pathologic conditions such as chronic bronchitis (CB), cystic fibrosis and asthma, permanent changes in ciliary structure and function, mucus hypersecretion and/or rheological changes, result in mucus retention[68].

Antigen Presentation Action by BEC in AHR

Several cell types which have the capacity to present antigen are called as professional antigen presentation cells, including dendritic cells, B cells, and macrophages[69]. Additional, various epithelial cells, in certain circumstance, can express MHC class II molecules and have been reported to be also capable of antigen presentation[70], being called as non-professional antigen presentation cells. Bronchial epithelial cells were postulated to have the potential of antigen presentation, because they have been reported to express MHC class I and class II molecules, and to induce T-cell proliferation[71]. Study on antigen presenting function of bronchial epithelial cells and the control mechanism in this process is of great significance for broadening the understanding of physiological function of bronchial epithelial cells and elucidating the mechanism of airway inflammation.

HLA-DR antigens on ciliated bronchial epithelial cells were detected by immunofluorescence, cytoplasmic dot blot analysis, and Northern blot hybridizations[72]. Heat shock protein 70 (HSP70) are recognized to have a role in chaperoning antigenic peptides and in facilitating class II peptide assembly. Upregulations of HSP70 with HLA-DR were observed both on professional and non-professional antigen presenting cells. In asthmatics, high level of HSP70 and HLA-DR expression was detected in BAL and bronchial epithelium. On the other hand, the expression of HLA-DR in bronchial epithelial cells from COPD patients, asthmatic patients, chronic bronchitis patients were significantly higher than those from healthy volunteers, which brings light that upregulation of HLA-DR in epithelial cells correlates with lymphocytes activation[73].

Optimal activation of T cells requires both costimulation and T cell receptor engagement. Antigen presentation in the absence of costimulation may lead to T cell allergy. Costimulatory interactions between the B7 family ligands expressed on antigen-presenting cells and their receptors on T cells play critical roles in the growth, differentiation, and death of T cells. CD4+ T cell activation, proliferation, and cytokine production require two distinct signals from the antigen-presenting cells. The first signal is triggered by interaction of the antigen-specific TCR with the MHC class II-peptide complex, and the second signal is derived from costimulatory molecules. The most widely studied costimulatory complex are CD28, ICOS ,and CTLA-4 express on T cells and their ligands B7-1(CD80), B7-2(CD86),and B7-H2(ICOS ligands) expressed an several types of the antigen-presenting cells, including monocyte/macrophages, dendritic cells, activated B cells, keratinocytes and some activated T cells. CD80 and CD86 are the best-characterized costimulatory molecules[74]. When considering bronchial epithelial cells as potential antigen-presenting cells, it is crucial to clarify appropriate costimulatory molecules expressed on bronchial epithelial cells. Papi investigated class I and class II antigens and B7-1 and B7-2 costimulatory molecule expression in human A549 pulmonary epithelial cells and primary bronchial epithelial cells (HBEC) at baseline and after rhinovirus infection. The data show that respiratory epithelium expresses full antigen-presentation machinery and that rhinovirus infection up-regulates this expression[75]. These findings would support the hypothesis that bronchial epithelial cells may play a role as antigen-presenting cells during the development of airway and inflammatory immune responses.

Salik investigated accessory cell function, antigen trafficking, and uptake of immune complexes in isolated nasal epithelial cells and airway epithelial cells, as well as in the two respiratory epithelial cell lines A549 and BEAS-2B[76]. The kinetics of Ag uptake by the respiratory epithelial cells were dramatically increased by IFN-γ and GM-CSF[71]. These studies showed that Ag followed a class II endocytic pathway inside the airway epithelial cells.

Our group recently confirmed that an immortalized human bronchial epithelial cell line expresses both HLA-DR and CD80, and observed an obvious process of antigen uptake by this cells, which indicates that bronchial epithelial cells are capable of antigen presenting function. We also observed that this antigen presenting function of BEC can be affected by ozone stress and some neuropeptides such as vasoactive intestinal peptide (VIP) and calcitonin gene-related peptides (CGRP). However, there is no experiment evidence up to now to account for a direct relation between the normality of BEC antigen presentation and the susceptibility of AHR or asthma. We hypothesize reasonably that aberrantly hypersensitivity of BEC antigen presentation or targeted subtype of T cells disorder may be involved in allergic asthma.

The Wound Repair in Airway Epithelium and Airway Remodeling

Various stimulants, such as mechanic injury, environmental pollutants, infection agents, can cause a damage of airway epithelium and initiate airway inflammatory response, including the clearance of inflammatory cells and the repair of the damaged area[77]. When epithelium damaged, in general, epithelial cells migrate to the denuded area and start proliferation in order to replace injured cells and establish squamous or mucous cell metaplasia; but some over produced and surplus cells should be discarded; at last the proportions of cell types are restored once the cell numbers have been reduced to normal[26]. It is such a sophisticated and complicated process that many factors will join in the regulation mechanism.

Integrins can participate in the all needed process during repair. In response to epithelial injury, there are dramatic changes either in spatial distribution or in expression level of epithelial integrins. Expression of α2β1, α3β1, α6β4, α5β1, αvβ6, αvβ6 were observed to be dramatically upregulated along the wound edge [26, 29]. A specific subset of integrins (including α1β1, α5β1 and αvβ3) can induce or enhance cell proliferation through interaction of the integrinα-subunit cytoplasmic domain with caveolin-1, a membrane protein that plays a role in organizing membrane microdomains. However, integrins can serve both positive and negative regulation of proliferation[30, 78]. Several studies have found that overexpression of fibronectin receptor α5β1 results in inhibition of proliferation [28]. It somewhat because of the unligated integrin and still need more information. In adhesion cells, simple expression of unligated integrin is sufficient to initiate cell death[28]. Cell type, composition of the surrounding matrix, tissue origin, state of differentiation or state of neoplastic transformation will determine which set of the integrins are critical for transducing these survival signals to the interior of the cell. For airway epithelium, α1β1, α2β1, α3β1, α5β1, α6β1, α6β4, αvβ3 all can exert impacts on cell survival, which one is more important is still unclear[77, 79]. If the integrins molecules express or function defectively, the rate of airway epithelium wound repair decelerated, the airway response becomes hypersensitivity, on which we are focusing.

Airway remodeling maybe ascribe to the imbalance of factors promoting the proliferation of epithelium or mesenchymal cell. Injured bronchial epithelial cells induced the transdifferentiation of sub-epithelial fibroblasts into myofibroblasts, which may be mediated by ET-1 and TGFβ through MAPK pathways such as p38 MAPK and ERK 1/2[58].

Fibronectin (FN) is an important epithelia-derived chemotactic factor which is upregulated by TGF-β in injury, accelerating the migration of fibroblast and epithelial cells to the site of wound. Although being regarded as fixed cells of connective tissue origin, lung fibroblasts and myofibroblast retain the capacity for growth, proliferation and pluripotent differentiation. They may be precursors for various cell types including smooth muscle cells[80]. The myofibroblast may potentially contribute to the regulation of bronchial inflammation via the release of cytokines and to tissue remodeling by its release of ECM components such as elastin, fibronectin, and laminin. Restoration of normal tissue structure after injury often coincides with the loss of the myofibroblast phenotype and the reappearance of normal-appearing fibroblasts[81]. After epithelial perturbation, basic fibroblast growth factor, platelet-derived growth factor, IGF-1, transforming growth factor-beta2, and endothelin-1 are released from BEC[82], which are important for regulating the response of bronchial myofibroblasts to cell damage and repair in airway remodeling. At the same time, mesenchymal fibroblasts have a growth stimulatory effect on surrounding epithelial cells[83]. We also observed that the proliferation activities of BEC and lung fibroblast are inhibited each other under normal state. Stress of BEC induced by ozone can transform the reciprocal inhibition into promoting proliferation. The increased secretion of TGF-β1 and the decreased synthesis of PGE2 from zone-stressed BEC are responsible for the promoted proliferation of lung fibroblast. The experiment may be able to explain the mechanism of airway wall remodeling during AHR.

Pulmonary Peptidergic Innervation Remodeling

Pulmonary neuroendocrine cells (PNECs) usually distribute dispersively in airway epithelium, some get together to form neuroepithelial bodies (NEB) which connect with sensory nerve and act as chemoreceptor. It has been proved that several bioactive substance in the form of amines and peptides located in PNECs, such as VIP, calcitonin gene-related peptides (CGRP), substance P (SP), endothelin-1 (ET-1), 5-HT, somatostatin and so on. These substances act as neurotransmitter and local hormone, permitting to postulate that PNECs act as vaso-and bronchio-modulators and pulmonary growth promoters. It has been generally accepted that tachykinins and CGRP released by an axon reflex mechanism participate in a neurogenic inflammation in asthma. CGRP has complex pulmonary effects including vasodilatation and diverse immunomodulatory actions[84]. CGRP has been described earlier as a bronchial constrictor of human airways, but another study demonstrated that replacement of CGRP following its depletion in allergic mice can reverse the changes in airway responsiveness and suggest that CGRP may have potential for the treatment of allergic AHR[85]. So, targeting the effects of CGRP may be of value for future strategies in nerve modulation.

Previous research in our group noticed that some dense-core particles, a kind of neuro-secretory granule, exist in ciliated cells, and its contents are possibly vasoactive intestinal peptide (VIP)[50]. Indeed, using radioimmuno assay, we determined VIP secreted from the primary cultured rabbit BEC. The secretion of VIP was upregulated by EGF. Furthermore, we identified VIP receptor on BEC using radio-ligand binding[86]. The studies in our group indicated that VIP is a significant beneficial substance to airway based on its ability on anti-inflammatory, anti-oxidative damage and epithelial restoration. Through promoting the antioxidant enzyme system and bcl-2 expression, VIP attenuated the oxidant injury in BEC[9]. In addition, VIP decreased the secretion of IL-1,IL-5 and IL-8 from BEC[51], and inhibited BEC adhering to leukocytes and ICAM-1 expression depending NF-κB/I-κB pathway[39, 50], but promoted BEC chemotactic migration during airway epithelium repair[87]. In the condition of injury, the up-regulated expression of VIP suggests that it may participate in the local protective mechanism in the airway microenvironment.

Our laboratory found that when airway stressed or inflammatory, sensory neuropeptide VIP, CGRP, SP and bombesin receptor subtype-3 (BRS-3) are distributed in lung as temporal and spatial regularly[88-90]. VIP had a downregulatory effect to attenuate the inflammation, while SP and CGRP took an upregulatory effect to enhance the inflammation[88, 90]. The imbalance between the two kinds of neuropeptides probably gives the explanation of the occurrence of AHR[88, 90].

Summary

Summarily, airway epithelium should not be considered continuously as the simple mechanic barrier. Based on it's variety of physiologic functions including anti-oxidation, endocrine, mucus secretion and transportation, stress or inflammatory signal transduction, antigen presentation, constitutive adhesion, injury repair, and airway remodeling regulation, etc., bronchial epithelial cells play critical roles in maintenance of airway local microenvironment homeostasis and participate in airway inflammation. Our airway hyper-responsiveness (AHR) animal model constructed by ozone stress made it reasonably to be hypothesized that the defects in airway epithelium integrity or functions disorder are the initial step of airway hyperresponsiveness such as in asthma and chronic obstructive pulmonary disease. The understand to the functions of bronchial epithelial cells will bring a new sight to realize the novel cellular and molecule mechanism of asthma, a classic and thorny clinic question, and approach to find new targets for airway hyperresponsiveness diseases therapy, although there are still a lot of difficulties in laboratory experiment and clinic research. Whatever, an era of airway epithelial cells is coming really in pulmonologist's sights.

References

[1] O'Byrne PM, I.M., Airway hyperresponsiveness. *Chest*, 2003. 123(3 Suppl): p. 411S-416S.

[2] Takizawa H, Bronchial epithelial cells in allergic reactions. *Curr. Drug Targets Inflamm Allergy*, 2005. 4(3): p. 305-311.

[3] Renauld JC, New insights into the role of cytokines in asthma. *J. Clin. Pathol..*, 2001. 54(8): p. 577-89.

[4] Holgate ST, L.P., Davies DE., The bronchial epithelium as a key regulator of airway inflammation and remodelling in asthma. *Clin. Exp. Allergy*, 1999. 29(Suppl 2): p. 90-95.

[5] Qin XQ, S.X., Zhang CQ., Technique of enzyme digestion adding brushing for isolating bronchial epithelial cells.(Chinese, Abstract in English). *Bull. of Hunan Med. Univ.*, 1999. 24(1): p. 74-76.

[6] Qin XQ, X.Y., Guan CX, Zhang CQ, Sun XH, Integrin-ligands binding reaction upregulates the antioxidant activity of rabbit bronchial epithelial cells.(Chinese, Abstract in English). *Acta Physiologica Sinica,* 2001. 53(1): p. 41-44.

[7] Qin XQ, S.X., Luo ZQ, Guan CX, Zhang CQ, Cytoprotective effect of epidermal growth factor on cultured rabbit airway epithelial cells exposed to ozone.(Chinese, Abstract in English). *Acta Physiologica Sinica, 1*996. 48: p. 190-194.

[8] Qin XQ, S.X., Zhang CQ, Zhang JL, The cytoprotective effect of heat stress on bronchial epithelial cells exposed to ozone. Chin J Appl Physilo, 1996. 12: p. 243-246.

[9] Qin XQ, S.X., Luo ZQ. , Vasoactive intestinal peptide and epidermal growth factor upregulates bcl-2 gene expression in bronchial epithelial cells.(Chinese, Abstract in English). *Acta Physiologica Sinica*, 1999. 51: p. 419-424.

[10] Zhang CQ, Q.X., Guan CX, Xiang Y, Sun XH., Resisting injury and protective effect of fibrinectin on bronchial epithelial cells.(Chinese, Abstract in English). *Chin. J. Appl. Pbysiol.*, 2001. 17: p. 383-387.

[11] Xiang Y, Q.X., Guan CX, Zhang CQ, Luo ZQ, Sun XH., Fibronectin upregulates catalase gene expression in rabbit bronchial epithelial cell (Chinese, Abstract in English). *Acta Physiologica Sinica*, 2004. 56(3): p. 365-368.

[12] Liu HJ, W.Y., Qi MM, Qu F, Xiang Y, Tan YR, Zhang CQ, Qin XQ., Protective effect of bombesin receptor subtype-3 on human brochial epithelia l cells against injury.(Chinese, Abstract in English). *J Cent South Univ (Med Sci),* 2006. 31(2): p. 178-183.

[13] Qin XQ, S.X., Zhang CQ, , Cytoprotective effect of vasoactive intestinal peptide on rabbit bronchial epithelial cells damaged by ozone exposure.(Chinese, Abstract in English). *Chin. J. Appl .Physiol.*, 1998. 14: p. 250-253.

[14] Dworski R, Oxidant stress in asthma. *Thorax*, 2000. 55(suppl 2): p. S51-3.

[15] Henricks PA, N.F., Reactive oxygen species as mediators in asthma. *Pulm Pharmacol Ther*, 2001. 14: p. 409-20.

[16] Kameyama S, K.M., Takeyama K, Atsushi N., Air exposure causes oxidative stress in cultured bovine tracheal epithelial cells and produces a change in cellular glutathione systems. *Experimental Lung Research*, 2003. 29: p. 567 - 583.

[17] Michiels C, R.M., Toussaint O, Remacle J., Importance of Se-glutathione peroxidase, catalase, and Cu/Zn-SOD for cell survival against oxidative stress. *Free Radic. Biol. Med.*, 1994. 17(3): p. 235-48.

[18] Comhair SA, T.M., Erzurum SC., Differential induction of extracellular glutathione peroxidase and nitric oxide synthase 2 in airways of healthy individuals exposed to 100% O(2) or cigarette smoke. *Am. J. Respir. Cell Mol. Biol.*, 2000. 23(3): p. 350-4.

[19] MacNee W, Pulmonary and systemic oxidant/antioxidant imbalance in chronic obstructive pulmonary disease. *Proc Am Thorac Soc*, 2005. 2(1): p. 50-60.

[20] Sandowska AM, L.C., Vints AM, Verbraecken J, VAN ranst D, DE backer WA., Systemic antioxidant defences during acute exacerbation of chronic obstructive pulmonary disease. *Respirology*, 2006. 11: p. 741-747.

[21] Caramori G , P.A., Oxidants and asthma. *Thorax*, 2004. 59: p. 170-173.

[22] Cho YS, L.J., Lee TH, Lee EY, Lee KU, Park JY, Moon HB., alpha-Lipoic acid inhibits airway inflammation and hyperresponsiveness in a mouse model of asthma. *J. Allergy Clin. Immunol..*, 2004. 114(2): p. 429-35.

[23] Chanez P, Severe asthma is an epithelial disease. *Eur Respir J,* 2005. 25: p. 945-946.

[24] Trautmann A, K.K., Akdis M, Muller-Wening D, Akkaya A, Brocker EB, Blaser K, Akdis CA, Apoptosis and Loss of Adhesion of Bronchial Epithelial Cells in Asthma. *Int. Arch. Allergy Immunol.*, 2005(138): p. 142–150.

[25] Tan YR, Q.M., Qin XQ, Xiang Y, Li X, Wang Y, Qu F, Liu HJ, Zhang JS., Wound repair and proliferation of bronchial epithelial cells enhanced by bombesin receptor subtype 3 activation. *Peptides*, 2006. 27: p. 1852-1858.

[26] Coraux C, H.R., Lesimple P, Puchelle E. , In vivo models of human airway epithelium repair and regeneration. *Eur. Respir. Rev.*, 2005. 14: p. 131-136.

[27] Sharhda S, *Cell contacts and airway epithelial damage in asthma*. 2005.

[28] Sheppard D, Functions of Pulmonary Epithelial Integrins: From Development to Disease. *Physiol. Rev.*, 2003. 83(JULY): p. 673-686.

[29] Sheppard D, Airway Epithelial Integrins: Why So Many? *Am J Respi Cell Mol Biol.*, 1998. 19: p. 349-351.

[30] Watt FM, Role of integrin in regulation epidermal adhesion,growth and differentiation *The EMBO Journal*, 2002. 21(16): p. 3919-3926.

[31] Murgia C, B.P., Kim N, Dans M, Petrie HT, Giancotti FG., Cell cycle and adhesion defects in mice carrying a targeted deletion of the integrin b4 cytoplasmic domain. *EMBO J.*, 1998. 17(14): p. 3940-3951.

[32] Grossmann J, Molecular mechanisms of “detachment-induced apoptosis—Anoikis”. *Apoptosis* 2002. 7: p. 247-260.

[33] Roche WR, M.S., Backer J, Holgate ST., Cell adhesion molecule and the bronchial epithelim. *Am. Rev. Respir. Dis* 1993. 148(79-82).

[34] Ozawa M, B.H., Kemler R. , The cytoplasmic domain of the cell adhesion molecule uvomorulin associates with three independent proteins structurally related in different pecies. *EMBO J.*, 1989. 8: p. 1711-1717.

[35] Goto Y, U.Y., Nomura A, Sakamoto T, Ishii Y, Morishima Y, Masuyama K, Sekizawa K. , Dislocation of E-Cadherin in the Airway Epithelium during an Antigen-Induced Asthmatic Response. *Am. J. Respir. Cell Mol. Biol.*, 2000. 23: p. 712-818.

[36] Vignola AM, B.G., Siena L, Melis M, Chiappara G, Gagliardo R, Bousquet J, Merendino AM, ICAM-1 and α3β1 expression by bronchial epithelial cells and their in vitro modulation by infl ammatory and anti-infl ammatory mediators. *Allergy*, 2000. 55: p. 931-939.

[37] Shebani, E., *Ultrastructural studies of the airway epithelium in airway disease*. 2006, UPPSALA.

[38] Shirasaki, H., Watanabe,K., Kanaizumi,E., Sato,J., Konno,N., Narita,S., Himi,T., The effectes of glucocorticosteroids on tumor necrrosis factor induced intercellular adhesion molecule-1 expression in cultured primary human nasal epithelial cells. *Clin. Exp. Allergy*, 2004. 34: p. 945-951.

[39] Tan YR, Q.X., Guan CX, Zhang CQ, Luo ZQ, Sun XH. , Regulatory peptides modulate ICAM-1 gene expression and NF-κB activity in bronchial epithelial cells. *Acta Physiologica Sinica*, 2003. 55: p. 121-127.

[40] Ge N, N.Y., Nakamura Y, Okano Y, Yoneda K, Ogawa H, Sugita A, Yanagawa H, Sone S., Synthesis and secretion of interleukin-15 by freshly isolated human bronchial epithelial cells. *Int Arch Allergy Immunol.*, 2004. 135(3): p. 235-42.

[41] Lilly CM, T.H., Oguma T, Israel E, Sonna LA., Effects of allergen challenge on airway epithelial cell gene expression. *Am J Respir Crit Care Med.*, 2005. 171((6)): p. 579-86.

[42] Hemelaers L, L.R., Eotaxin: an important chemokine in asthma. *Rev .Med. Liege*., 2006. 61(4): p. 223-6.

[43] Meyer-Hoffert U, L.-M.D., Bartels J, Montes-Vizuet AR, Schroder JM, Teran LM., Th2- and to a lesser extent Th1-type cytokines upregulate the production of both CXC (IL-8 and gro-alpha) and CC (RANTES, eotaxin, eotaxin-2, MCP-3 and MCP-4) chemokines in human airway epithelial cells. *Int Arch Allergy Immunol*., 2003. 131(4): p. 264-71.

[44] Montes-Vizuet R, V.-M.A., Valencia-Maqueda E, Negrete-Garcia MC, Velasquez JR, Teran LM., CC chemokine ligand 1 is released into the airways of atopic asthmatics. *Eur Respir J*., 2006. 28(1): p. 59-67.

[45] Reibman J, H.Y., Chen LC, Bleck B, Gordon T., Airway epithelial cells release MIP-3alpha/CCL20 in response to cytokines and ambient particulate matter. *Am. J. Respir. Cell. Mol. Biol.*, 2003. 28(6): p. 648-54.

[46] Booth BW, A.K., Bonner JC, Tournier F, Martin LD., Interleukin-13 induces proliferation of human airway epithelial cells in vitro via a mechanism mediated by transforming growth factor-alpha. *Am J Respir Cell Mol. Biol.*, 2001. 25(6): p. 739-43.

[47] Freund V, F.N., *Nerve growth factor (NGF) in inflammation and asthma. Rev. Mal. Respir.*, 2004. 21(2): p. 328-42.

[48] Wu T, H.C., Shelhamer JH, Involvement of p38 and p42/44 MAP kinases and protein kinase C in the interferon-gamma and interleukin-1alpha-induced phosphorylation of 85-kDa cytosolic phospholipase A(2) in primary human bronchial epithelial cells. *Cytokine*., 2004. 25(1): p. 11-20.

[49] Krull M, B.P., Wuppermann FN, Klucken AC, Muhling J, Schmeck B, Seybold J, Walter C, Maass M, Rosseau S, Hegemann JH, Suttorp N, Hippenstiel S., Mechanisms of Chlamydophila pneumoniae-mediated GM-CSF release in human bronchial epithelial cells. *Am. J. Respir .Cell Mol. Biol.*, 2006. 34(3): p. 375-82.

[50] Tan YR, Q.X., Guan CX, Zhang CQ, Xiang Y, Ren YH, Influence of regulatory peptides on the secretion of interleukins from bronchial epithelial cells of the rabbit(Chinese, Abstract in English). *Acta Physiologica Sinica*, 2002. 54(2): p. 107-10.

[51] Tan YR, Q.X., Guan CX, Zhang CQ. , Effects of regulatory peptides on adhesion of eosinophil to bronchial epithelial cells.(Chinese, Abstract in English). *Acta Physiologica Sinica*, 2002. 54(1): p. 43-46.

[52] Zhang CQ, T.Y., Qin XQ., *Stimulation of ozone stress on the adhesion of inflammatory cells to bronchial epithelial cells.(Chinese, Abstract in English). . Bull. of Human Med. Univ.*, 2002. 27: p. 192-194.

[53] Zhang JS, T.Y., Xiang Y, Luo ZQ, Qin XQ. , Regulatory peptides modulate adhesion of polymorphonuclear leukocytes to bronchial epithelial cells through regulation of interleukins,ICAM-1 and NF-κB/IκB. . *Acta Biochim. Biophys. Sin.*, 2006. 38(2): p. 119-128.

[54] Tabary O, C.H., Boncoeur E, Chadelat K, Fitting C, Cavaillon JM, Clement A, Jacquot J., Adherence of airway neutrophils and inflammatory response are increased in CF airway epithelial cell-neutrophil interactions. *Am. J. Physiol. Lung Cell Mol. Physiol.*, 2006. 290(3): p. 588-96.

[55] Jagels MA, D.P., Zuraw BL, Hugli TE., Mechanisms and regulation of polymorphonuclear leukocyte and eosinophil adherence to human airway epithelial cells. *Am. J. Respir. Cell. Mol. Biol.*, 1999. 21(3): p. 418-27.

[56] Marrin M, V.E., Hollembor J, etal, Exression of the potent inflammator chemokines GM-CSFIL-6 and IL-8,in bronchial epithelialcells of patients with asthma *Allert Clin. Immunol.*, 1992. 89: p. 1001-1009.

[57] Lamkhioued B, R.P., Abi-Younes S, etal, Incresed exression of eotaxin in bronchoalveolar lavage and airway of asthmatics contributes to the chemotaxis of eosinophils to the sites of inflammation. *Immunol*, 1997. 159: p. 4593-4601.

[58] Chen XU, X.J., Effects of airway epithelium injury on the transdifferentiation of sub-epithelial fibroblastsand its role in the development of airway hyperresponsiveness in asthma. *Chin. J. Tuberc. Respir. Dis.*, 2005. 128(110): p. 698-703.

[59] Mata M, S.B., Buenestado A, Cortijo J, Cerdá M, Morcillo EJ. , Phosphodiesterase 4 inhibition decreases MUC5AC expression induced by epidermal growth factor in human airway epithelial cells *Thorax,* 2005. 60: p. 144-152.

[60] Kaliner MJ, S.H., BorsonB, Nadel J, Patow C, Marom Z., Human respiratory mucus. *Am. Rev. Respir. Dis.*, 1986. 134: p. 612-621.

[61] Rose MC, N.T., Voynow JA Airway Mucus Obstruction: Mucin Glycoproteins, MUC Gene Regulation and Goblet Cell Hyperplasia *Am. J. Respir. Cell Mol. Biol,* 2001. 25(5): p. 533-537.

[62] Fischer H, W.J., Illek B Acid secretion and proton conductance in human airway epithelium *Am. J. Physiol. Cell Physiol.* 2002. 282(4): p. C736-C743.

[63] Borok Z, V.A., Lung Edema Clearance: 20 Years of Progress Invited Review: Role of aquaporin water channels in fluid transport in lung and airways. *J. Appl. Physiol.*, 2002. 93(6): p. 2199-2206.

[64] Ruddy MK, D.J., Pitkanen OM, Rafii B, O'Brodovich HM, Harris HW. , Modulation of aquaporin 4 and the amiloride-inhibitable sodium channel in perinatal rat lung epithelial cells *Am J Physiol Lung Cell Mol Physiol*, 1998. 274(6): p. L1066-L1072.

[65] Ito Y, S.S., Son M, Kondo M, Kume H, Takagi K, Yamaki K, *Bisphenol A Inhibits Cl- Secretion by Inhibition of Basolateral K+ Conductance in Human Airway Epithelial Cells* 2002. 302(Issue 1).

[66] Huang P, G.E., Kultgen P, Barnes P, Milgram S. , Local Regulation of Cystic Fibrosis Transmembrane Regulator and Epithelial Sodium Channel in Airway Epithelium *The Proceedings of the American Thoracic Society* 2004: p. 1:33-37

[67] Sinn PL, S.A., Donovan MD, McCray Jr PB Viscoelastic Gel Formulations Enhance Airway Epithelial Gene Transfer with Viral Vectors A*m. J. Respir. Cell Mol. Biol.*, 2005. 32: p. 404-410.

[68] Del Donno M, B.D., Chetta A, Olivieri D, Lopez-Vidriero MT. , The Effect of Inflammation on Mucociliary Clearance in Asthma-An Overview *Chest*, 2000. 118: p. 1142-1149.

[69] Villadangos GP, P.H., Proteolysis in MHC class II antigen presentation: who is in charge? *Immunity*, 2000. 12(233-239).

[70] kalb TH , C.M., Marom Z,Mayer l., Evidence for accessory cell function by class II MHC antigen-expressing airway epithelial cells. *Am. Respir. Cell Mol. Biol.,* 1991. 4(4): p. 320-9.

[71] Kurosawa S, M.A., Chen L, Wang S , Ni J, Plitt JR, Heller NM, Bochner BS, Schleimer RP., Expression of the costimulatory molecule B7-H2 by humam airway epithelial cells. *Am. Respir. Cell Mol. Biol*, 2003. 28: p. 563-573.

[72] Rossi GA, S.O.B.B., Oddera S,Mattioni T,Corte G,Ravazzoni C,Allegra L, Human ciliated bronchial epithelial cells: expression of the HLA-DR antigens and of the HLA-DR alpha gene, modulation of the HLA-DR antigens by gamma-interferon and antigen-presenting function in the mixed leukocyte reaction. *Am. J. Respir. Cell Mol. Biol,* 1990. Nov3(5): p. 431-9.
[73] Qu Q, Z.M., Guo Z., ICAM-1and HLA-DR expression in bronchial epithelial cells from patients with chronic obstructive pulmonary disease. *Chin. J. Tuberc. Respir. Dis.,* 1998. 7: p. 411-414.
[74] Bugeon L, D.M., Costimulation of T cells. *Am J Crit Care*, 2000. 162: p. S164-S168.
[75] Papi A, S.L., Papadopoulos NG, Teran lM, Holgate ST, Johnston SL., Rhinovirus infection induces major histocompatibility complex class I and costimulatory molecule upregulation on respiratory epithelial cells. *J. Infect. Dis.*, 2000. 181(5): p. 1780-4.
[76] Salik E, T.M., Mohan S, George I, Becker K, Oei E, Kalb T, Sperber K. , Antigen Trafficking and Accessory Cell Function in Respiratory Epithelial Cells. *Am Respir Cell Mol Biol,* 1999. 21(3): p. 365-379.
[77] Tesfaigzi Y, Roles of apoptosis in airway epithelia. *AM j Respir .Cell Mol. Biol.*, 2006. 34: p. 537-547.
[78] Schwartz MA, A.R., Integrins and cell proliferation: regulation of cyclindependent kinases via cytoplasmic signaling pathways. *Journal of Cell Science,* 2001. 114: p. 2553-2560.
[79] O'Sullivan MP, T.J., Holtzman MJ., Apoptosis in the Airways--Another Balancing Act in the Epithelial Program. *Am. J. Respir. Cell Mol. Biol.*, 2005. 29: p. 3-7.
[80] Gizycki MJ, A.E., Rogers AV, O'Byrne PM, Jeffery PK Myofibroblast involvement in the allergen-induced late response in mild atopic asthma. *Am. J. Respir. Cell Mol. Biol,* 1997. 16(6): p. 664-673.
[81] Leslie KO, M.J., Low R., Lung myofibroblasts. *Cell Motil Cytoskeleton.*, 1992. 22(2): p. 92-98.
[82] Zhang S, S.H., Holgate ST, Roche WR., Growth factors secreted by bronchial epithelial cells control myofibroblast proliferation: an in vitro co-culture model of airway remodeling in asthma. *Lab Invest.*, 1999. 79(4): p. 395-405.
[83] Holgate ST, D.D., Lackie PM, Wilson SJ, Puddicombe SM, Lordan JL., Epithelial-mesenchymal interactions in the pathogenesis of asthma. *J Allergy Clin Immunol.,* 2000. 105(2 Pt 1): p. 193-204.
[84] Springer J, G.P., Fischer A, *Calcitonin gene-related peptide as inflammatory mediator. Pulm. Pharmacol. Ther.*, 2003. 16(3): p. 121-130.
[85] Azzeddine D, A.K., Mika JM, *Regulation of airway hyperresponsiveness by Calcitonin Gene-related Peptide in Allergen Sensitized and Challenged Mice. Am. J. Res. Cri. Car. Med,* 2002. 165(1137-1144).
[86] Qin XQ, S.X., Luo ZQ, Guan CX, Zhang CQ, Affection of epidermal growth factor on VIP secretion and VIPR expression in airway epithel ial cells(Chinese, Abstract in English). *Bull of Hunan Med Univ*, 1999. 24(2): p. 99-102.
[87] Guan CX, Z.C., Qin XQ, Luo ZQ, Sun XH, Zhou FW., Effects of vasoactive intestinal peptide on the chemotaxis of bronchial epithelial cells. *Acta Physiologica Sinica*, 2002. 54(2): p. 103-107.
[88] Ren YH, Q.X., Guan CX, Luo ZQ, Zhang CQ, Sun XH., Temporal and spatial distribution of VIP,CGRP and their receptors in the development of airway

hyperresponsiveness in the lung. Acta Physiologica Sinica. *Acta Physiologica Sinica.*, 2004. 56: p. 119-128.

[89] Ren YH, Q.X., Guan CX, Luo ZQ, Zhang CQ, Sun XH, The temporal and spatial distribution of vasoactive intestinal peptide and its receptor in the development of airway hyperresponsiveness.(Chinese,Abstract in English). *Chin. J. Tuberc. Respir. Dis,* 2004. 27: p. 224-228.

[90] Wu H, G.C., Qin XQ, Xiang Y, Qi MM, Luo ZQ, Zhang CQ. , Upregulation of substance P receptor expression by calcitonin gene-related peptide, a possible cooperative action of two neuropeptides involved in airway hyperresponsiveness. *Pulm. Pharmacol. Ther..*, 2006. 20: p. 513-524.

In: Ozone Depletion, Chemistry and Impacts
Editor: Sem H. Bakker, pp. 135-145
ISBN: 978-1-60692-007-7

Chapter 6

OZONE HISTORY AND ECOSYSTEMS: A GOLIATH FROM IMPACTS TO ADVANCE INDUSTRIAL BENEFITS AND INTERESTS, TO ENVIRONMENTAL AND THERAPEUTICAL STRATEGIES

Eugenia Bezirtzoglou and Athanasios Alexopoulos
Democritus University of Thrace, Department of Food Science and Technology, Laboratory of Microbiology, Biotechnology and Hygiene. GR68200, Orestiada, Greece

Introduction

Ozone is known to be formed naturally in the atmosphere, as a colorless gas having a pungent odor and associated with the ability to guard against the sun's harmful ultraviolet radiation. Historically[1], the characteristic odor of ozone was noticed by Van Marum at 1785 in his electrostatic machine during passage of electric sparks. Later at 1801, Cruickshank notice this odor at the anode electrode during water electrolysis. Finally, in 1840 Shonbein named this chemical substance as ozone from the Greek word ozein, which means to smell. Siemens 's invention followed the early ozone method production by a corona discharge which is based on passing an electric discharge through dry oxygen or air.

From the side of chemistry, ozone is characterized as a triatomic allotropic form of oxygen (O_3) having a molecular weight of 48.

This study has focused on the general analysis and evaluation of the industry with the introduction of the ozone generating equipment from the perspective of the evolution of its application and usefulness, as well as its interaction with the environmental and social variables, together with the problems and challenges to be faced contributing in the industry's development. Moreover, ozone application during the last decades brought together great advances in the industry contributing to an effective and sustainable production policy based in a main and capital tool, which is a clean technology with environmental benefits.

Applications of Ozone in Food Industry

Ozone is fast becoming the sanitizer of choice in food industry for a number of very important reasons such as, (a) costly chemical replacement, (b) energy and cost saving and (c) efficient bacteria kill.

These facts induce food industry sector to incorporate the ozone technology in a plethora of applications from the perspective of sustainable food development, particularly after the recognition of ozone as safe (GRAS -Generally Recognized as Safe) on 1995 by FDA for treatment of bottled water [2]. In this context, later on, its application was extended as GRAS to the food processing by experts [3].

Ozone is a strong broad – spectrum antimicrobial agent and its action is guided against bacteria, fungi, viruses, protozoa and fungi.

The many benefits connected to its use, has recently made it possible to gain approval for use in food processing industry in many countries as a way to help rid food of dangerous pathogens such as bacteria, parasites, fungi and viruses.

The main advantage of ozone use in food is that treated food products are also free of residuals or other harmful by-products. As it is an unstable gas, ozone decomposes in a short period of time (aprox. 20 minutes) into simple oxygen, leaving no residues on disinfected foods.

Inactivation of bacteria by ozone, seems to be a complex, not yet elucidated processes. Ozone acts against different cellular substances including proteins lipids, peptidoglycans, enzymes, and virus capsids. Microbial inactivation associated with ozone seems to be dependent on several environmental factors. Treatment temperature affects solubility, stability, and reactivity of ozone. The pH, the Relative Humidity (RH), and additives seems also to contribute to the degree of ozone action.

Additionally, ozone is a strong antimicrobial agent in both gaseous and aqueous phases. The oxionism mechanisms of ozone may involve direct reactions of molecular ozone and free radical – mediated destruction. The main advantage of ozone use, consists of its superiority compared to chlorine for 3 capital reasons; firstly, it is reported to be 1.5 times stronger than chlorine [4] and acting 3000 times faster than chlorine [5], without producing harmful decomposition products [6].

Having the merit of a GRAS status, ozone must be widely adopted and extended in the food industry sector. However, as it is a new technology, the lack of experience in management and supervisory personnel to solve the ozone processing problems may hinder its implementation.

Without a doubt, it seems to be a multiple tool in industry as it combines the disinfectant effect of the industrial equipment, together with the treatment of the material exposed or the food itself. Moreover, it is not harmful for the environment or the material involved in ozone treatment.

On the other hand, lack of commercial specifications have not been developed for its application to specific foods able to be exposed to the ozone effect, due to the absence of governmental and legal guidelines.

Different types of food, such as dairy, meat, poultry, fish, canned foods, beverages and water, are impacted by the safe use of ozone.

Ozone is applied to clean and sanitize the equipment in the above places. Specifically, in the clean in place systems (CIP) ozonated water is applied as a final rinse to neutralize the chemical residue.

It should be pointed out that in food, water, and beverage industry, we have to face to important food pathogens such as, *Listeria monocytogenes, Clostridium botulinum, E.coli 0157:H7, Staphylocoocus aureus, Salmonella sp.* Moreover, the role of the personnel microbial microflora seems to be of importance to the food contamination and quality. Food hygiene seems to be an important concern for the industry as spoiled products are associated with an important industrial economic loss extrapolated to the cost for treatment of foodborne occurring disease.

The bactericidal effects of ozone have been documented on a wide variety of organisms, including Gram positive and Gram negative bacteria, as well as spores and vegetative cells.

Treating fruits or vegetables with ozone have been found to increase shelf-life of the products [7,8], as ozone delays the maturation from 20% to 30%. Additionally, ozone inhibits the ethylene gas produced by many vegetables and thus, protects from scents diffusion between vegetables during their processing and transport.

Refrigerator chambers necessary for food conservation are deodorized and disinfected by ozone. It furnishes an important bactericidal effect upon the developed microflora of the refrigerated product, as the developed humidity inside the refrigerating facility create a favorable environment for bacterial and fungal growth.

Ozone destroys harmful organisms from meat and fishes surfaces and prevents from the flesh rotting in meat.

In cheese production, high relative humidity contribute to its quality. However, in this condition cheese surfaces are rapidly prone to mould formation. Ozone keeps a bipolar role by inhibiting mould presence and deodorizing the environment.

Humidity such as 90% is also necessary to egg storage in order to avoid losses and contractions. As discussed previously, humidity offers an ideal support for bacterial and mould growth. Ozone offers the solution to this case as well protecting eggs' quality aspect and hygiene.

In silos containing cerial or sugar plants, especially when storage conditions are changing by exposition to heating, ozone protects from fermentations due to the present microflora in association with the higher created moisture content in the facility space.

Last, but not least, ozone offers important features in packaging in items of extending food-shelf life [9,10].

Factors Influencing Microbial Inactivation in Food Industry

Solubility of ozone in the water, expressed in terms of mg/l O_3 in water per mg/l of O_3 in the gas phase influenced from various factors, with pressure and temperature being the most important ones. Also, the diffusion method (eg bubbling, turbulence of the gas in the water and contact time) as well as the purity of the water for organics and metals, and in some extent the pH, are additional factors which should always must be considered [2,3,4] in the process.

Once solubilized, ozone is relatively unstable in aqueous solutions. The half-life of ozone in distilled water varies according to the factors mentioned earlier, and perhaps this is the reason for the difference in reported values in literature [5,6], but in general a half time of 20 to 25 min in 20° C are acceptable values (18 to 20 min in our experiments [7]). The stability of ozone depends highly in the water quality and this is important when considering the usefulness of ozonation in food processing as water of drinking quality is mostly used. Under such conditions, the half-life of ozone is sort and its decomposition rate increases. Ozone can react with organic compounds directly or through the radicals formed after its decomposition (OH for example). Highly succeptible to oxidation are the sulfhydryl groups which are present in microbial enzymes as well as glucosamine and peptidoglycans which are base components of microbial cell walls. Moreover, ozone reacts quickly with most of the nucleobases [9].

The antimicrobial efficacy of ozone depends highly in factors associated with its solubility, stability and decomposition. Increased temperatures could lower the solubility but various studies have show that the reaction rates with the substrate are steady [10] or became accelerated by a factor of 2 or 3 [5] as in the case of *Giardia* cysts or *E.coli* in the surface of fruits [11] and thus compensates for the decrease in its stability. The presence of organic compounds and pH values also plays an important role in the antimicrobial action but this is also varies with the microorganism and the substrate. In a review article Khadre et al. [10] refers various studies in which gram-negative or positive bacteria and viruses have been tested for inactivation by ozonated water. In any case, the CxT concept (concentration multiplied by contact time) looks applicable as with chlorine inactivation kinetics, when the limitations in extrapolation are taken in account [13].

Experiments in our laboratory, have verify some of those results but others issues must also studied before concrete conclusions reached. For example, the method of measuring ozone concentration and counting of microbes and also the role of viable but not cultivable (or shocked) microorganisms. Or, that is the base of misbehavior of counts after short time ozonation, as the sudden increase in total plate counts after short time exposures of mixed microbe populations to ozonated aqueous solutions (rebound effect) which has been observed both *in vitro* and *in vivo* studies [14,15,16,17]. All those confounding factors are negatively affecting results and conclusions. Therefore, it is advisable, prior to any application, the employment of indicator organisms and the study of the particular factors influence the sanitization process [17].

Applications of Ozone in Husbandry (Animal Breeding)

In veterinary practice, ozone is used successfully. Ozone enriched water is used in the calving operation to sanitize the bottles and nipples used to feed the young animals and the equipment designated to host or transport animals as illness or mortality can be devastating to the pet body and of important economic loss for the farmer. Furthermore, the zoonotic illness can be passed from wild or domesticated animals to humans. It should be pointed here the importance of the species specifically associated disease which can cross the animal species barrier to dwell in man. It is then conceivable that prevention more than therapy is imposed. Prevention in classic terms include animal vaccination. However, strains variations or mutations incline the vaccine effectiveness. An effective disinfection strategy of animal hosting buildings and parlours by the aid of ozone associated to the vaccination protocols should highly increase animal protection from harmful diseases.

The water quality serving for different purposes in a food industry or animal care units seems to be of capital importance as it can be a source of bacteria .The industry also is interesting in using ozone to decontaminate processing water and decrease its chemical and biological oxygen demand (BOD) [18].

This last applies to the use of ozone in aquacultures which is an innovative technology. Moreover, removal of fine and colloidal solids, removal of dissolved organic compounds and nitrites together with its high disinfection effect in aquaculture water treatment systems reveals the multifocal role of this oxidizing agent. Additionally, ozone could greatly reduce the number of spoilage bacteria in the seafood plant and help maintain low levels of spoilage bacteria over time in air, water and on the processing equipment [17,18].

However, ozone efforts to decontaminate bean sprouts and remove biofilms formation have not been successful [18].

Ozone efficacy can be superior when combined with other chemical substances as hydrogen peroxide or physical treatments as ultraviolet radiations or mechanical action [19].

Ozone can be used in the pulp and paper industry in replacement of the traditionally applied chlorine for the delignification and the pulp bleaching at different stages of the processing. It can be applied alone or in combination with other chemicals substances [19].

It is known that the safety of drinking as well as surface waters are of capital importance for preserving human and animal public health [19].

Recently, the importance of the use of ozone in the microelectronics industry was applied in the cleaning of semiconductor devices with promising results [20].

Ozone has been used by the European food industry as a standard for decades, and as a sanitizer for the public water supplies for over a century [10,14,15,19]. Newer studies ascertain its disinfecting superiority and USA adopted ozone in a wide variety of applications recognizing it as safe (GRAS).

Environmental Applications of Ozone

Ozone is a strong oxidizing agent which oxidizes organic substances, pesticides, herbicides, fungicides, other hazardous substances and destroys pathogenic microorganisms.

Due to its ability to oxidize organic chemical substances and pecticides, ozone have been used to clean up sites containing variety of soil contaminants and limit the pathogenic microflora. Moreover, it is used effectively in the processing of waste waters reach in industrial or clinical bioproducts.

Of the major applications of ozone in environmental is the use of treatment of effluents either municipal [15], industrial [14,20,21,22,23] or hazardous like those arose from hospitals [14,24,25]. In that context, the high reactivity of ozone makes it appropriate for achieving certain objectives when applied alone or with other processes. These objectives aiming to achieve higher quality standards prior to final discharge or to meet standards for effluent recycling as in agriculture [20]. The final goal through the ozone use is the removal of colour [26], the degradation of organic loading and COD reduction [27], the enhanced sludge solubilization [28], the disinfection of the final effluent [29] and also the detoxification or elimination of pharmaceutical compounds [30,31] and antibiotics [32] that hardly eliminable by conventional processes. Particularly, the last category of compounds and their metabolites have developed as ecotoxicologically relevant micropollutants in aquatic environments with increasing concern for humans and the environment. Recent studies have shown also the capabilities of ozone in removing and mineralization of synthetic agrochemicals as the carbofuran (an endocrine disruptor) after short time of oxidation [33]. Ozone technology in sole or as a part of an Advanced Oxidation Process (AOP) has also been used for landfill leachate treatment with promising results concerning the COD reduction and increased biodegrability [34], considering that the recirculation of the treated leachate back to landfill cells must meet specific criteria.

Environmental remediation efforts and particularly soil remediation is another field where ozone usage offers promising results. Traditionally, soils contaminated by organic compounds (Diesel range organics and/or PAHs and agrochemicals) are treated by biological methods and often involves the translocation of mass soil volumes and thus raising the remediation cost. Novel techniques which includes a combination of chemical and biological treatment have been studied and are commercially available. In general, oxidation of pollutant organics by ozone should result in reaction products that have a better solubility in water and thus a better bioavailabity to organic scavengers like *Pseudomonas alcaligenes* PA-10 [35] or *Sphingomonas yanoikuyae* [36]. The ozone in its gas phase, when generated *on site*, has the ability of penetrating the soil surface and oxidizes the organic micropollutants leading to overall removal efficiencies of 78% in experimental conditions [37], although various factors can affect that percentages. In agriculture and hydroponics the ozone offers more handy results as its use can sterilize and oxygenate the soil (or water) with benefits aiming to a better development of the crops, shortening of the harvest and increased yields [38,39].

Nowdays, ozone continues to show increasing importance in the municipal water and waste water arena[40]. Water quality is important to protect public health[17,41], especially in immunocompromised patients where any microorganism could reserve a potential danger. The action of ozone in water is instantaneously and after performing its action it reverts back to oxygen.

Because ozone is a strong oxidizer[17], it can clean most water sources[17]. Ozone showed strongly disinfectant effect upon on the one hand, the classic aerobic fecal bacterial indicators and on the presence of *C.perfringens*, which is reported to be the main bacterial anaerobic indicator in waters[17]. Environmental applications of ozone.

Treating water to make it suitable to drink is another application. Elevated levels of nitrite in drinking water can cause cyanosis, a reduction of the oxygen carrying capacity of the blood.

Routinely, ozone seems to be applied in the treatment technology of bottled water in many countries. However, intercountry law regulations seems to restrict its highly promising use for the purpose of water, wine and beverage market application.

Following the EPA (Environmental Protection Agency) in USA the indoor air pollution is incriminated to be the origin of health associated issues [19]. Indoor pollution is provided by many sources such as, burning, cooking, tobacco, carpets, chemical agents used, central heating, cooling or humidification systems, animals and anthropogenic activities. Ozone has been used for professional cleaning in industries and companies indoor spaces.

During the year 2003, SARS spreading in China imposed the choice of an extremely effective procedure to stop the virus expansive ability. The implementation of ozone was shown to stand this criterion.

It is conceivable, that its use must be extrapolated to all hospitality and travel spaces where persons are exposed in stale air and unpleasure odors, as odor offers a fresh smelling and clean space.

Likewise, its high efficacy as disinfectant ,the impact of safety must be stressed when using ozone technology in regard to human exposure, as high ozone levels could involve a potential health risk associated with coughing, asthma, allergy, irritation of the throat, dermatitis, chest pain and other breathing features [19]. These symptoms last only few hours after ozone removement is performed.

Undesirable odors and scents indoors should be removed and treated successfully by the ozone innovative technology.

Ozone should be applied together with other disinfectants in a combined strategy to prevent nosocomial infections, which are known to be a major intrahospital problem due to the developed resistance of multiresistant bacterial strains in the applied schemes of antibiotics and disinfectants [19]. Resistance in ozone can not be developed as ozone acts by lysing the bacterial cell and killing so then the microorganism.

Ozone was used therapeutically as well to purify the blood (Lender,1870)[1]. Since then, medical ozone is shown to be implemented in a plethora of therapeutical purposes. in varying concentrations following the way of treatment.

Ozone therapy is one of the most powerful and versatile therapies known today, as through its action mechanism, ozone shown to have multi therapeutical features.

These features include inactivation of microorganisms by rupturing of their cellular envelope through peroxidation of the phospholipids and lipoproteins [1,19], enhancement of the blood stream circulation, stimulation of oxygen metabolism, oxidation of the outer lipid layer of malignant cells, activation of the immune system through increasing production of interferon and interleukin in the body and healing cellular capacity.

In dentistry, ozone has been used extensively to *in vitro* experiments for the deactivation of oral microorganisms and dental plaque treatment [42] with remarkable results. A practical application of the ozone should be the possibility to use it for prevention from different oral pathogens. Moreover, ozone showed an important healing effect [43].

Another interesting use of ozone should be its role in disinfection of toothbrushes as it was found that the ozone treatment decreased gradually the microbial charge in brushes [16]. Maximum efficacy of ozone treatment was observed after 30 minutes while exposure for short time periods seems to be inefficient or even promote pathogen regrowth [16].

It is obvious that ozone application in all the above discussed domains remain very promising as an important disinfectant effect is associated to very low ozone levels ,such as less than 0.1ppm. However, much research must be done in order to clarify its mode of action in the different settings, determine the better and most accessible ozone technology and decide about the strategy to be followed for a maximal effectiveness in all domains where the substance is applied.

Last but not least, we must pointed out that there is needed a uniform decision about its use internationally in the different forms and for different purposes anticipated by a flexible and innovative legislation standing on its safe and correct application for preserving public health and disease states.

Measurement of Ozone

Of the widest acceptable methods for ozone measurement in water is the indigo colorimetric method developed [40]. The indigo colorimetric method is sensitive, precise, fast, and more selective for ozone than other methods. During the reaction, ozone binds across the carbon-carbon double bond of a sulphonated indigo dye and decolorizes it. There are two indigo colorimetric methods available from commercial sources as test kits : the spectrophotometric and the visual test. The spectrophotometric procedure has a lowest detection level of 0.005 μg/l, while for the visual procedure the detection limit is 10 mg/L. The indigo method exhibit less interference with various ions than any of the other methods (Langlais et al., 1991[41]). Alternatively, there is the iodometric method, which has been approved by the International Ozone Association [44] and relays on the oxidation of iodide ion by ozone, release of iodine, and measurement with endpoint titration with sodium thiosulfate in the presence of starch as indicator.

Ozone Safety

Ozone is a dangerous gas and should be treated accordingly. Although, health risks are possible in the literature there are no direct references or any case studies referring to accidents from the use of ozone. Despite that, the ozone is generated from photochemical reactions contributing to the air pollution, particularly in cities with heavy traffic. Currently, ozone is considered as one of the three priority pollutans whose health risks are evaluated by an experts committee in WHO [45]. One of their key conclusions is the absence of any threshold limit for exposure and therefore measures should be always taken accordingly. The Occupational Safety and Health Administration (OSHA) has set the maximum permissible exposure to airborne concentration of ozone not in excess of 0.1 mg/l averaged over an eight-hour work shift. The American National Standards Institute/American Society for testing Materials, the American Conference of Government Industrial Hygienists (ACGIH), and also the American Industrial Hygiene Association have adopted this limit and have also set as a limit to a ceiling concentration exposure of 0.3 mg/l for not more than ten (or fifteen) minutes. The FDA [46,47,48,49,50] in a guidance manual, advice the use of instrumentation and equipment to initiate an alarm signal at an ambient level of 0.1 mg/l for personnel awareness and an alarm at 0.3 mg/l for automatic ozone generators shut down and emergency ventilation initiation. Beyond, the above actions and the proper signing and education of the personnel, no other special measures (such as the construction of separate building for ozone equipment) are necessary.

Bibliography

[1] Rubin MB.2001. The history of ozone. The Schonbein period, 1839-1868.*bull.hist.Chem.***26**(1):40-56

[2] Schultz CR, Bellamy WD. 2000. The role of mixing in ozone dissolution systems. *Ozone Sci. Eng.* **22**:329-350.

[3] Kim J-G. 1998. Ozone as an antimicrobial agent in minimally processed foods disinfection processes. *Water Sci. Technol.* **21**(3):179-195.

[4] Hoigné J, Bader H. 1985. Rate constants of reactions of ozone with organic and inorganic compounds in water. III:Inorganic compounds and radicals. *Water Res.* **19**:993-996.

[5] Wynn CS, Kirk BS, McNabney R. 1973. Pilot plant for tertiary treatment of wastewater with ozone. EPA Report R2-73-146, US EPA, *Municipal Environ. Res. La*b., Cincinnati, Ohio.

[6] Wickramanayake GB, Rubin AJ, Sproul OJ. 1984. Inactivation of *Giardia lamblia* cysts with ozone. *Appl. Environ. Microbiol.* **48**:671-672.

[7] Ceciu S,Alexopoulos A,Lazar V,Bezirtzoglou E.Effectiveness of ozone against microorganisms colonizing raw vegetables. XXXI Congress of the Society of Microbial Ecology and Disease (SOMED),May 28-30,2008, Djuronaset, Stockholm, Sweden

[8] Staehelin J, Hoigné J. 1985. Decomposition of ozone in water in the presence of organic solutes acting as promoters and inhibitors of radical chain reactions. *Environ. Sci. Technol.***19**:1206-1213.

[9] Ishizaki K, Shinriki N, Ikehata A, Ueda T. 1981. Degradation of nucleic acids with ozone. I. Degradation of nucleobases, ribonucleosides, and ribonucleoside- 5'-monophosphates. *Chem. Pharm. Bull.* **29**:868-872.

[10] Kinman RN. 1975. Water and wastewater disinfection with ozone: A critical review. *Crit. Rev. Environ. Contr.* **5**:141-152.

[11] Rodgers SL, Cash JN, Siddiq M, Ryser ET. 2004. A comparison of different chemical sanitizers for inactivating Escherichia coli O157:H7 and Listeria monocytogenes in solution and on apples, lettuce, strawberries, and cantaloupe. *J. Food. Prot.* **67**(4):721-31.

[12] Khadre MA, Yousef AE, Kim J-G. 2001. Microbiological Aspects of Ozone Applications in Food: A Review. *J. Food. Sci.* **66**(9):1242-1252.

[13] Gyurek LL, Finch GR, Belosevic M. 1997. Modeling chlorine inactivation requirements of Cryptosporidium parvum oocysts. *ASCE J. Environ. Eng.* **123**(9):865-875.

[14] Troyan, J.J. and S.P. Hansen. 1989. Treatment of microbial contaminants in potable water supplies technologies and costs. Noyes Data Corp., Park Ridge, NJ.

[15] Paraskeva P., Graham NJD., 2002. Ozonation of Municipal wastewater effluents. *Water Environ. Res.* **74**(6):569-581.

[16] Bezirtzoglou E, Cretoiu SM, Moldoveanu M, Alexopoulos A, Lazar V, Nakou M. 2008. A quantitative approach to the effectiveness of ozone against microbiota organisms colonizing toothbrushes. J Dent. [http://dx.doi:10.1016/j.jdent.2008.04.007]

[17] Voidarou C., Tzora A., Skoufos I., Vassos D., Galogiannis G., Alexopoulos A. and Bezirtzoglou E. (2007). Experimental effect of ozone upon some indicator bacteria for preservation of an ecologically protected watery system. *Water, Air, and Soil Pollution.* **181**(1-4); 161-171

[18] Guzel-Seydim ZB,Greene Ak,Seydim AC.2004.Use of ozone in the food industry.*Lebenmittel-Wissenschaft und Technologie*.**37**(4):453-460

[19] Franken L.2005.The application of ozone technology for public health and industry.Food Safety and Security at Kansaw State University.http://fss.k-state.edu

[20] De Smedt F ,De Gendt S,Claes M,Heyns MM,Vankerckhoven H,Vinckier C.2002.The increasing importance of the use of ozone in the microelectronics industry.*Ozone :Science and Engineering*,**24**(5) :379-390

[21] Smilanick JL,Crisosto C,Mlikota F.1999.Postharvest use of ozone on fresh fruit. *Perishables handling quarterly*.**99**:10-14

[22] Kim JG,Yousef AE,Khadre MA.2003.Ozone and its current and future application in the food industry.*Adv.Food Nutr.Res*.**45**:167-218.

[23] Brochier B,Kuligowski C,Voiron S,Petit-Conil M.2006.Overview of the use of ozone in the pulp and paper industry.*Ozone News*.**34**(6):21-28

[24] Gimeno O, Rivas FJ, Beltrán FJ, Carbajo M. 2007. Photocatalytic ozonation of winery wastewaters. *J Agric Food Chem*. **55**(24):9944-9950

[25] Machado EL, Kist LT, Schmidt R, Hoeltz JM, Dalberto D, Alcayaga EL. 2007. Secondary hospital wastewater detoxification and disinfection by advanced oxidation processes. *Environ. Technol.* **28**(10):1135-1143.

[26] Oguz E. Keskinler B. 2008. Removal of colour and COD from synthetic textile wastewaters using O3, PAC, H2O2 and HCO3-. *J. Hazard Mater.* **151**(2-3):753-760.

[27] Campos-Reales-Pineda AE, Orta de Velásquez MT, Rojas-Valencia MN.2008. The use of ozone during advanced primary treatment of wastewater for its reuse in agriculture: an approach to enhance coagulation, disinfection and crop productivities. *Water Sci. Technol.* **57**(6):955-962.

[28] Chu LB, Yan ST, Xing XH, Yu AF, Sun XL, Jurcik B.2008. Enhanced sludge solubilization by microbubble ozonation.*Chemosphere*. **72**(2):205-212.

[29] Gong J, Liu Y, Sun X. 2008. O3 and UV/O3 oxidation of organic constituents of biotreated municipal wastewater. *Water Res*. **42**(4-5):1238-1244.

[30] Sumikura M, Hidaka M, Murakami H, Nobutomo Y, Murakami T. 2007. Ozone micro-bubble disinfection method for wastewater reuse system. *Water Sci Technol.* **56**(5): 53-61.

[31] Baumgarten S, Schröder HF, Charwath C, Lange M, Beier S, Pinnekamp J. 2007. Evaluation of advanced treatment technologies for the elimination of pharmaceutical compounds. *Water Sci .Technol.* **56**(5):1-8.

[32] Yargeau V, Leclair C. 2007. Potential of ozonation for the degradation of antibiotics in wastewater.*Water Sci. Technol.* **55**(12):321-326.

[33] Lau TK, Chu W, Graham N. 2007. Degradation of the endocrine disruptor carbofuran by UV, O3 and O3/UV. *Water Sci. Technol.* **55**(12):275-280.

[34] Tizaoui C, Mansouri L, Bousselmi L.2007. Ozone catalysed with solids as an advanced oxidation process for landfill leachate treatment. *Water Sci Technol*. **55**(12):237-43.

[35] O'Mahony MM, Dobson AD, Barnes JD, Singleton I. 2006. The use of ozone in the remediation of polycyclic aromatic hydrocarbon contaminated soil. *Chemosphere.* **63**(2): 307-314.

[36] Stehr J, Müller T, Svensson K, Kamnerdpetch C, Scheper T. 2001. Basic examinations on chemical pre-oxidation by ozone for enhancing bioremediation of phenanthrene contaminated soils. *Appl. Microbiol. Biotechnol.* **57**(5-6):803-809.

[37] Choi H, Lim HN, Kim J, Hwang TM, Kang JW. 2002. Transport characteristics of gas phase ozone in unsaturated porous media for in-situ chemical oxidation. *J. Contam. Hydrol.* **57**(1-2):81-98.

[38] Orta de Velásquez MT, Rojas-Valencia MN, Reales-Pineda AC. 2006. Evaluation of phytotoxic elements, trace elements and nutrients in a standardized crop plant, irrigated with raw wastewater treated by APT and ozone. *Water Sci. Technol.* 54(11-12):165-173.

[39] Rich T, Knueve E. 2002. A look at ozone in hydroponics applications. *Water Technology Magazine* **25**(9): 20-22.

[40] Bader H, Hoigné J. 1981. Determination of ozone in water by the indigo method.*Water Res.* **15**:449-456.

[41] Langlais B, Reckhow DA, Brink DR (editors). 1991. Ozone in Drinking Water Treatment: Application and Engineering. AWWARF and Lewis Publishers, Boca Raton, FL.

[42] Sammons, R. L., Kaur D. and Neal P. (2004). Bacterial survival and biofilm formation on conventional and antibacterial toothbrushes. *Biofilms,* **1** 123-130

[43] Haffajee, A.D., Smith C., Torresyap G, Thompson M., Guerrero D., Socransky S.S. (2001). Efficacy of manual and powered toothbrushes II. Effect on microbiological parameters. *J. Clin. Periodontol.*, **28**: 947-954.

[44] IOA (International Ozone Association). 1989. Photometric measurement of low ozone concentrations in the gas phase. Standardisation Committee-Europe.

[45] Krzyzanowski M. 2008. WHO Air quality guidelines in Europe. *J. Toxicol. Environ. Health* **71**:47-50.

[46] Enviromental Protection Agency. 1999. Guidance Manual: Alternative Disinfectants and oxidants. *EPA* 815-R-99-014

[47] Bocci V., 2002. Oxygen – Ozone Therapy: A Critical Evaluation, Kluwer Academic Publishers (THE NETHERLANDS), ISBN 1 – 4020 – 0588 –1

[48] Graham DM., 1997. Use of ozone for food processing. Food Technol. 51:72-75.

[49] Food and Drug Administration, 1982, 21 CFR 184.1563: Direct Food Substances Affirmed as Generally Recognized as Safe (Washington, DC: Department of Health and Human Services)

[50] Food and Drug Administration, 2001. United States Food and Drug Administration. Secondary direct food additives permitted in food for human consumption, final rule. *Federal Register* **66**, pp. 33829–33830.

In: Ozone Depletion, Chemistry and Impacts
Editor: Sem H. Bakker, pp. 147-161
ISBN: 978-1-60692-007-7

Chapter 7

CATALYTIC OZONIZATION: A NEW APPROACH TO THE TREATMENT OF WASTEWATER

Luciana Serra Soeira and Renato S. Freire[1]
Instituto de Química – Universidade de São Paulo/USP
CP 26077 – 05513-970 – São Paulo/SP – Brazil

Abstract

This chapter is a review about catalytic ozonation, which is a new way of contaminants removal from wastewater. Despite its current application is mainly limited to laboratory use, the results obtained employing this approach in pollutant degradation showed to be promissory to scale large applications, however further investigation about its efficiency and drawback must be carried out. The aim of this chapter is to give a survey of the application of several homo- and heterogeneous catalysts, with special attention given to the last one. In this context, some metals (for example, Fe, Co, Mn, Zn and Ni) under various forms (salt of reduced metal, solid oxide or deposited metal on support) can be used to enhance the efficiency of ozone towards the removal and/or conversion of different organic compounds in aqueous solution. Moreover, the activity and the parameters that influence the efficiency of these two catalytic systems will be present as a short overview.

Keywords: ozone, catalytic ozonation, advanced oxidative process, metal oxides

Introduction

The ozone has been used for decades as an efficient disinfectant of drinking water, especially in Europe, where your first full scale application dates from 1906, in France (Nice) [1]. Since then, many studies have been developed to improve and increase the applications of ozone for purification of ground and surface water for domestic consumption and for treatment of industrial wastewater.

---

[1] E-mail address: rsfreire@iq.usp.br.

Ozone can react through two different reaction pathways: direct attack of the ozone molecule on the organic compounds (direct reaction); or generation of hydroxyl radicals ($HO^{\cdot}$), due to the decomposition of ozone, which subsequently degrade the organic compounds (indirect reaction). The direct reaction has been shown to achieve a very limited mineralization degree of refractory pollutants, especially in acidic solutions [2,3]. Nowadays, because the indirect reaction is more effective then the first pathway, the use of ozone as an Advanced Oxidation Process (AOP) for the treatment of pollutants is gaining the attention of scientific and technological medium.

In this way, another recent alternative for the enhancement of ozonation is to use a metal catalyst to initiate the decomposition of aqueous ozone. The main advantages of catalytic ozonation include the lower ozone consumption and the improvement of the mineralization degree achieve at the end of the process. It is observed that the accumulative refractory substances generated during the ozonation normally react faster in the presence of a metal catalyst, which leads to better mineralization degrees than the use of ozonation alone [4-6].

In this approach, it is possible to use homogeneous catalytic ozonation, which is based on ozone activation by metal ions present in aqueous solution and heterogeneous catalytic ozonation, which is characterizing by the presence of metal oxides or metal/metal oxides on supports in aqueous medium [3,5].

Homogeneous Catalytic Ozonation

The ozone activation during the homogeneous process is given by transition metals in solution such as Fe(II), Mn(II), Ni(II), Co(II), Cd(II), Cu(II), Ag(I), Cr(III) and Zn(II). The nature of transition metal applied determines reaction rate, selectivity, ozone consumption, degradation/mineralization of the pollutant and reaction mechanism [5].

Hewes and Davinson [6] observed that, during the ozonation of wastewater, the activation of ozone by Fe(II), Mn(II), Ni(II) or Co(II) increases the total organic carbon (TOC) removal efficiency as compared to ozonation alone.

Gracia et al. [7] studied the catalytic activity of Mn(II), Fe(II), Fe(III), Cr(III), Ag(I), Cu(II), Zn(II), Co(II) and Cd(II) in the degradation/mineralization of humic substances in water by ozonation. The reaction of ozone with humic substances is quick, with the production of low-molecular weight oxygenated by-products, which are generally more easily biodegradable, polar and hydrophilic than their precursors. But, the mainly problem is that the ozonation alone do not achieve significant reduction in this organic matter. In addition, for high ozone dosages (4,3 g O_3 / 1 g TOC), simple ozonation provides 33% of TOC removal in water. In this study, the authors reported that the application of catalytic ozonation improved the efficiency of humic substances removal from water, under the same experimental conditions. The best results were observed for Mn(II) and Ag(I) with 62% and 61% of TOC removal, respectively. Besides the increase in TOC removal, the use of catalytic ozonation resulted in a decrease of ozone consumption, since radical hydroxyl is a better oxidant agent.

Another metal ion that is gaining attention for the improvement of ozonation is Ce(III). Matheswaran et al. [3] developed a novel regenerative catalytic system using cerium and ozone in nitric acid medium. The efficiency of this hybrid system comprising of ozone and cerium redox pair (Eq. 1) towards organic mineralization was evaluated taking phenol as the model organic pollutant and compared with Ce(III) catalyzed and no-catalyzed ozonation processes.

$$2Ce(III) + O_3 + 2H^+ \rightarrow 2Ce(IV) + O_2 + H_2O \quad (1)$$

The presence of cerium ions increased the degradation efficiency of phenol compared to no-catalyzed ozonation (54% and 71% of TOC removal, respectively), whereas a synergic effect was observed by the authors between the cerium redox pair (Ce(III) and Ce(IV)) and ozone towards phenol mineralization and a maximum TOC removal.

In general, two mechanisms are proposed for the homogeneous catalytic ozonation. The first one is based on the decomposition of ozone by the metal ion, followed by the generation of radical species; the second one consists in the formation of complexes between the catalyze metal and the organic pollutant, with a final oxidation reaction [8].

Ozone decomposition by metal ions, generally, leads to the formation of radical species. Hill [9,10] found experimental evidences that propose the generation of hydroxyl radicals by the direct interaction between ozone and the metal ions in aqueous medium (Eq. 2).

$$M^{n^+} + O_3 + H_2O \rightarrow MOH^{n+} + O_2 + HO^\bullet \quad (2)$$

Hill observed that the original oxidation state could be reestablish due to the parallels reactions with radical species in the reaction medium, for example, hydroperoxyl radical (Eq. 3).

$$HO_2^\bullet + MOH^{n+} \rightarrow M^{n+} + O_2 + H_2O \quad (3)$$

Many authors [1,11,12] have shown that Fe(II) catalyzes ozone decomposition to generate hydroxyl radicals. In fact, the catalytic O_3/Fe(II) system involves direct reaction of the metal ion with the ozone, leading to the formation of hydroxyl radical, within FeO^{2+} as an intermediate (Eqs. 4-5).

$$Fe(II) + O_3 \rightarrow FeO^{2+} + O_2 \quad (4)$$

$$FeO^{2+} + H_2O \rightarrow Fe(III) + HO^\bullet + HO^- \quad (5)$$

The intermediate FeO^{2+} is able to oxidize Fe(II) to Fe(III) at slower rate, thus limiting the generation of hydroxyl radical for high Fe(II) concentration (Eq. 6).

$$FeO^{2+} + Fe(II) + 2H^+ \rightarrow 2Fe(III) + H_2O \quad (6)$$

Some authors [13-15] studied the enhancement of catalytic ozonation through the employ of ultra-violet light, e.g. Fe(II)/UV/O_3 system. Piera et al. [13] compared the degradation of the herbicide 2,4-dichlorophenoxyacetic (2,4-D) at pH 3.0, testing ozonation catalyzed with Fe(II) and/or UV light. The use of Fe(II)/O_3 and UV/O_3 systems led to a partial removal of

2,4-D, whereas the pollutant was completely mineralized by Fe(II) catalyzed ozonation under UV irradiation. The highest oxidative capability of the system Fe(II)/UV/O_3 is due to the combination of many reactions that leads to the generation of hydroxyl radicals. For UV irradiation at $\lambda < 300$ nm, this radical is produced by photodissociation of H_2O_2, previously formed from ozone photolysis (Eqs. 7-8) and for the photoreduction of Fe(III) formed in equations 5 and 6 to give Fe(II) and an additional hydroxyl radicals via photo-Fenton reaction (Eq. 9).

$$O_3 + H_2O + h\nu \rightarrow H_2O_2 + O_2 \quad (7)$$

$$H_2O_2 + h\nu \rightarrow 2\,HO^\bullet \quad (8)$$

$$Fe(III) + H_2O + h\nu \rightarrow Fe(II) + HO^\bullet + H^+ \quad (9)$$

The regeneration of Fe(II) propagates a catalytic reaction (Eq. 4) and more hydroxyl radical can be formed (Eq. 5). Since H_2O_2 is generated in this system (Eq. 7), more $HO^\cdot$ can also be produced from the classical Fenton reaction (Eq. 10).

$$Fe(II) + H_2O_2 \rightarrow Fe(III) + HO^\bullet + HO^- \quad (10)$$

The resulting Fe(III) can be photoreduced (Eq. 9). In addition, the mineralization process can be accelerated by photolysis of complexes of Fe(III) with short organic diacid intermediates, such as oxalic acid [16].

The second mechanism proposes that the enhancement of ozonation by metal ions occur through the formation of complexes between the metal ions and the organic molecules such as carboxylic acids. Pines and Reckhow [17] studying homogeneous catalytic ozonation of oxalic acid using Co(II) (at pH 6.0) found that this degradation follows a two step reaction. In the first step, a Co(II)-oxalate complex is formed, which is subsequently oxidized by ozone to form Co(III)-oxalate (Eqs. 11-14).

$$Co(II) + C_2O_4^{2-} \leftrightarrow CoC_2O_4 \quad (11)$$

$$CoC_2O_4 + O_3 \rightarrow CoC_2O_4^+ + O_3^- \quad (12)$$

$$CoC_2O_4^+ \rightarrow Co(II) + {}^\bullet C_2O_4^- \quad (13)$$

$${}^\bullet C_2O_4^- + O_2 + O_3 + HO^\bullet \rightarrow CO_2 + O_2^- + O_3^- + HO^- \quad (14)$$

The site of the attack is probable the metal center. The partial donation of electron density from oxalate ion to Co(II) may increase the reactivity of Co(II)-oxalate when compared to free Co(II).

The authors observed that the rate of both oxalate removal and ozone decomposition increases with decreasing pH (6.7 to 5.3). This event is very important to justify the proposed mechanism, because it evidences a distinct reaction pathway to the ozonation processes. The reduction of pH, and consequentially the decreasing of hydroxyl ions in solution that normally initiates auto decomposition of ozone, leads to lower mineralization of organic matter. Besides, the authors verified that the use of a scavenger resulted in no changes in the degradation of oxalic acid. These two factors confirm that the direct reaction of the ozone with the complex Co(II)-oxalate is the responsible for the degradation of organic species [17,18].

Andreozzi et al. [19] found that Mn(II) accelerates the oxidation of oxalic acid under acidic conditions. They proposed a mechanism that Mn(II) catalyzed ozonization is proceeding through complexion between oxalic acid and Mn(III), forming an intermediate product which might be easier oxidized by ozone (Eqs. 15-20).

$$\mathrm{Mn(II)} + \mathrm{O_3} + 2\mathrm{H^+} \rightarrow \mathrm{Mn(IV)} + \mathrm{O_2} + \mathrm{H_2O} \tag{15}$$

$$\mathrm{Mn(IV)} + \mathrm{Mn(II)} \rightarrow 2\mathrm{Mn(III)} \tag{16}$$

$$\mathrm{Mn(III)} + \mathrm{nAO^{2-}} \rightarrow \mathrm{Mn(III)(AO^{2-})_n} \tag{17}$$

$$\mathrm{Mn(III)(AO^{2-})_n} \rightarrow \mathrm{Mn(II)} + \mathrm{AO^{-\bullet}} + (\mathrm{n}-1)\mathrm{AO^{2}} \tag{18}$$

$$\mathrm{AO^{-\bullet}} + \mathrm{O_3} + \mathrm{H^+} \rightarrow 2\mathrm{CO_2} + \mathrm{O_2} + \mathrm{HO^\bullet} \tag{19}$$

$$\mathrm{HO^\bullet} + \mathrm{AO^{2-}} \rightarrow \text{radical chain reactions} \rightarrow \text{end products} \tag{20}$$

It is important to emphasize that several factors such as the pH of the solution, reagents concentration and the metallic ion can influence both the efficiency and the mechanism of homogeneous catalytic ozonation [5].

Besides the promissory results obtained from the application of metallic ions to enhance the ozonation process, the direct use of metal ions in solution cannot be recommended for water treatment, mainly because some of these metals are very toxic to many organisms. Another drawback is that, generally, the metallic ions must be removed from the treated solution before it was discharged into local watercourses. Thus, some studies on heterogeneous catalytic ozonation [20,21] have being developed in order of aggregate the benefices brought from the use of metals without these problems [8].

Heterogeneous Catalytic Ozonation

Heterogeneous catalytic ozonation was defined by Gracia et al. [22] as "a novel type of advanced oxidation which combines ozone with the adsorptive and oxidative properties of solid-phase metal oxide catalysts to achieve at room temperature mineralization of dissolved organics". This technology, dating from the 1970s, has been received much attention in recent years, specially because the presence of the catalyst enhance both ozone dissolution and ozone decomposition, increasing the mineralization degree of many refractory organic pollutants, like saturated aldehydes and carboxylic acids formed in the ozonation of more complex organic compounds such as phenol [23,24], at room temperature [5,22].

The main catalysts proposed for the heterogeneous ozonation are metal oxides (MnO_2, TiO_2, Al_2O_3) and also metals or metal oxides on metal oxide supports (e.g. Cu/Al_2O_3, Cu/TiO_2, Ru/CeO_2, V-O/TiO_2, V-O/silica gel, TiO_2/Al_2O_3 or Fe_2O_3/Al_2O_3). Once again, the activity of the solid catalysts is mainly based on the catalytic decomposition of ozone and the enhanced generation of hydroxyl radicals [5].

Ma and Graham [25] found that MnO_2 formed in situ during the ozonation of atrazine in the presence of small quantities of Mn(II) leads to a significant increase of atrazine oxidation compared to ozontion alone. The principle of this case of heterogeneous catalytic ozonation is quite similar to the homogeneous catalytic ozonation, but the main difference between them is that in the first one the reactions occur both on the catalyst surface and in the aqueous phase. The authors assigned these result to the generation of highly oxidative intermediate species.

Al-Hayek et al. [26] showed that catalytic ozonation of phenol in the presence of Fe(III)/Al_2O_3 leads to an increase of the TOC removal compared to ozonation alone. The authors suggested that both a formation of free radicals and an enhancement of nucleophilic sites of adsorbed molecules were responsible for the increased in the degradation of refractory organic species present in the reaction medium.

Farré et al. [20] studied the application of homogeneous (photo-Fenton/Ozone) and heterogeneous (TiO_2-photocatalysis/Ozone) catalytic ozonation for the degradation of some biorecalcitrant pesticides: alachlor, atrazine, chlorfenvinfos, diuron, isoproturon and pentachlorophenol. The best results of pesticides mineralization were obtained when heterogeneous ozonation was applied. The authors suggested that this process leads to the detoxified of the aqueous pesticide solutions, except in the case of atrazine and alachlor.

The TiO_2-photocatalysis/Ozone process is based on a radical attack in the organic molecule. So, in this case, the mechanism proposed is based in the generation of hydroxyl radicals through two ways: reaction of adsorbed H_2O molecules with photogenerated holes at the irradiated TiO_2 particle (Eqs. 21-22), and by reaction of adsorbed O_3 and photogenerated electrons at TiO_2 particle (Eqs. 23-24) [27].

$$TiO_2 + h\nu \rightarrow e^- + h^+ \tag{21}$$

$$H_2O + h^+ \rightarrow HO^\bullet + H^+ \tag{22}$$

$$O_3 + e^- \rightarrow O_3^{-\bullet} \tag{23}$$

$$O_3^{-\bullet} + H+ \rightarrow HO^\bullet + O_2 \tag{24}$$

The presence of dissolved ozone in the irradiated TiO_2 aqueous suspension increases the hydroxyl radical production and decreases the electron-hole recombination, which gives to the TiO_2-photocatalysis/Ozone process the best performance in the degradation of some pesticides [20].

As observed in the reaction between TiO_2 and ozone, the determining step in the hydroxyl radical production is the interaction of the semiconductor surface and the oxidant species present in the reaction medium. Thus, it can be inferred that the efficiency of the catalytic ozonation process depends, in a great extent, on the catalyst and its surface properties. The pH of the solution has a direct effect in the efficiency of this process because it influences the properties of the surface active sites and ozone decomposition reaction in aqueous solutions.

Another important point to discuss is the catalytic properties of metal oxides, which are determined by acidity and basicity. Although hydroxyl groups are present on the surface of all metal oxides, the amount and the properties of these groups depend on the metal. The hydroxyl groups formed at the surface of the metal oxide behave as Brönsted acid sites. Lewis acids and Lewis bases are sites located on the metal cation and coordinated with unsaturated oxygen, respectively. So, the catalytic centers of metal oxides are both Brönsted and Lewis acid sites.

Some oxides (e. g., silica, alumina, titania and zirconia) can be applied on catalytic ozonation as the metal oxide that will catalyze the reaction or as the support of metals/metal oxides for the same purpose. However, their chemistry is quite different from one oxide to another. As opposed to silica, which is only capable of ion exchange, the other metal oxides mentioned above are capable of both ion and ligant exchange. However, depending on the oxide metal and/or the pH of the solution (in case of amphoteric oxide) these metal oxides can behave as cation or anion exchanges. The ion exchange is possible because alumina, titania and zirconia have Lewis acid sites on their surface, which are responsible for the ligant exchange ability [5,28].

Besides, to understand the catalytic activity of heterogeneous surfaces it is important to know the adsorption properties of the metal oxide, because it is often one of the stages of this process and it has to be highlighted that the great adsorption ability of some metal oxides can cause permanent blockage of the catalysts active surface sites and, consequently, leads to the decrease of their catalytic activity [5].

One adsorptive material which is gaining much attention in heterogeneous catalytic ozonation is activated carbon. This material acts both as an adsorbent and as a catalyst in promoting ozone oxidation. Kaptijn [29] showed that catalytic ozonation on activated carbon has a low ozone requirement and is significantly less affected by dissolved radical scavengers. The author also found that this process is equally effective under both highly acidic conditions (when hydroxyl radicals are not formed) as well as under highly alkaline conditions. This study suggests that not only hydroxyl radicals but also other highly active species such as O-radicals can be generated on the surface of the catalyst.

The surface properties rule the capacity of the activated carbon on decompose ozone into highly oxidative species in the reaction medium. The ozone reduction on the surface of carbon can result in hydroxyl ions and hydrogen peroxide formation. Subsequently, these active species initiate the decomposition of ozone in the aqueous phase into more oxidized species, which lead to the enhancement of degradation/mineralization of organic substances [30].

Furthermore, activated carbon accelerates the transformation of ozone into formation of hydroxyl radicals. These radicals are not bound to the surface of carbon; they are free to react in the bulk of the aqueous phase. Thus, it can be inferred that activated carbon is an initiator of the radical-type chain reaction that transforms ozone into hydroxyl radical in the aqueous phase [31]. The investigations promoted by Beltrán et al. [32] confirmed this hypothesis. The authors affirmed that, simultaneously to the heterogeneous reactions, homogeneous reactions take place in the aqueous medium. They also proposed two possible ways for the ozone decomposition into the surface of activated carbon based on the pH of the reaction medium.

The catalytic ozone decomposition's mechanism in the aqueous phase, proposed by Beltrán et al. [32], is divided in three mayor steps: (a) homogeneous decomposition (Eqs. 25-28); (b) heterogeneous decomposition surface reaction (Eqs. 29-35); and (c) homogeneous propagation and termination reactions (Eqs. 36-39):

a) Homogeneous decomposition (pH > 6.0):

$$O_3 + OH^- \rightarrow HO_2^- + O_2 \tag{25}$$

$$O_3 + HO_2^- \rightarrow HO_2^\bullet + O_3^{-\bullet} \tag{26}$$

$$O_3 + In \rightarrow O_3^{-\bullet} + In^+ \tag{27}$$

$$HO_2^\bullet \leftrightarrow O_2^{-\bullet} + H^+ \tag{28}$$

b) Heterogeneous decomposition surface reaction:

i) $2.0 \leq pH \leq 6.0$

$$O_3 + M \leftrightarrow O_3\text{-}M \tag{29}$$

$$O_3\text{-}M \leftrightarrow O\text{-}M + O_2 \tag{30}$$

$$O_3 + O\text{-}M \leftrightarrow 2O_2 + M \tag{31}$$

ii) $pH > 6.0$

$$OH^- + M \leftrightarrow OH\text{-}M \tag{32}$$

$$O_3 + OH\text{-}M \leftrightarrow {}^\bullet O_3\text{-}M + HO^\bullet \tag{33}$$

$$O_3\text{-}M \leftrightarrow {}^\bullet O\text{-}M + O_2 \tag{34}$$

$$O_3 + {}^\bullet O\text{-}M \leftrightarrow O_2^{-\bullet} + M + O_2 \tag{35}$$

c) Homogeneous propagation and termination reaction:

$$O_3 + O_2^{-} \rightarrow O_3^{-\bullet} + O_2 \tag{36}$$

$$O_3^{-\bullet} + H^+ \rightarrow HO_3^{\bullet} \tag{37}$$

$$HO_3^{\bullet} \rightarrow HO^{\bullet} + O_2 \tag{38}$$

$$HO_3^{\bullet} + S \rightarrow \text{end products} \tag{39}$$

where M is the catalyst surface, In the initiator of ozone decomposition, and S is the scavenger of hydroxyl radicals.

As mentioned before, only reactions represented by Eqs. 29-35 are the steps of the surface chemical reaction involving adsorption, surface reaction and desorption process. This heterogeneous process is followed by the reactions 36-39 that take place in aqueous medium. It is also confirmed that ozone decomposition increases with increasing pH of the reaction medium [31,32].

Activated carbon reacts with ozone in a different way when compared with metal oxides. For the first one, Lewis basic sites are thought to be responsible for ozone decomposition. For the others, the proposed mechanism considers that strong Lewis acid sites of the oxides might be the active sites.

Activated carbon can also be used as a support in heterogeneous catalytic ozonation and, as discussed before, the pH can both affect the ozone decomposition and the surface properties. Ma et al. [33] investigated the effect of pH on MnO_x/granular activated carbon (GAC) heterogeneous catalytic ozonation into the degradation of nitrobenzene. The authors found that in catalyst preparation process the manganese oxide can react with the carbon (Eq. 40).

$$MnO_4^{-} + C + H_2O \rightarrow MnO_x + CO + HO^{-} \tag{40}$$

These hydroxyl ions may be physically adsorbed on the surface of the catalyst, and they are also easily desorbed into the aqueous phase. These adsorbed and desorbed ions contribute to the catalyst activity because it can determines if the metal oxide will behave as cation or anion exchange.

Ma et al. [33] found that low pH condition had benefitted the adsorption of nitrobenzene on the catalyst. Thus, the authors concluded that in MnO_x/GAC catalyzed ozonation, the adsorption of organic pollutants, such as nitrobenzene, on the catalyst is an important step, which may have a direct influence on the catalytic effectiveness of heterogeneous ozonation. They also assumed that the organic compounds may be further decomposed on the surface of the catalyst.

As we can see, the aim of heterogeneous catalytic ozonation is the superficial properties of the catalyst, which is directed related with the pH of the reaction medium and with the nature of the catalyst, specially the synthesis methodology that is responsible for differences in superficial area, porosity, chemical stability, Lewis acid sites, etc. Thus, it can be defined that the adsorptive capacity of the catalysts is very important for the efficiency of this process. Based on this, Kasprzyk-Hordern et al. [5] elucidated three hypotheses for catalytic ozonation in heterogeneous systems:

a) chemisorption of ozone on the catalyst surface leading to the formation of active species, which react with non-chemisorbed organic molecule;
b) chemisorption of organic molecule (associative or dissociative) on the catalytic surface and its further reaction with aqueous ozone;
c) chemisorption of both ozone and organic molecule and the subsequence interaction between chemisorbed species.

Based on these three acceptable hypotheses and in the main catalysts proposed for the heterogeneous ozonation, two reaction's mechanisms were proposed [34,35]. The first one is based on the application of a metal oxide as a catalyst, which performs a dual function: the presence of the solid leads to both an enhancement of ozone dissolution and the initiation of ozone decomposition [34]. For the catalytic reaction takes place, both ozone and organic molecules must be transported to the surface of the catalyst. Just after ozone adsorption, it is converted in O-radicals [36] or hydroxyl radicals [37]. In this case $O_2^{-\bullet}$ transfers an electron to another ozone molecule to form an ozonide anion ($O_3^{-\bullet}$), which is the chain reaction promoter and leads to the generation of hydroxyl radicals. These free radicals initiate a radical-type chain reaction both on the surface of the metal oxide and in the bulk of the aqueous phase. While the oxidation proceeds, the affinity of the oxidation products to the catalyst decreases and the final products desorb from the oxide metal surface [34].

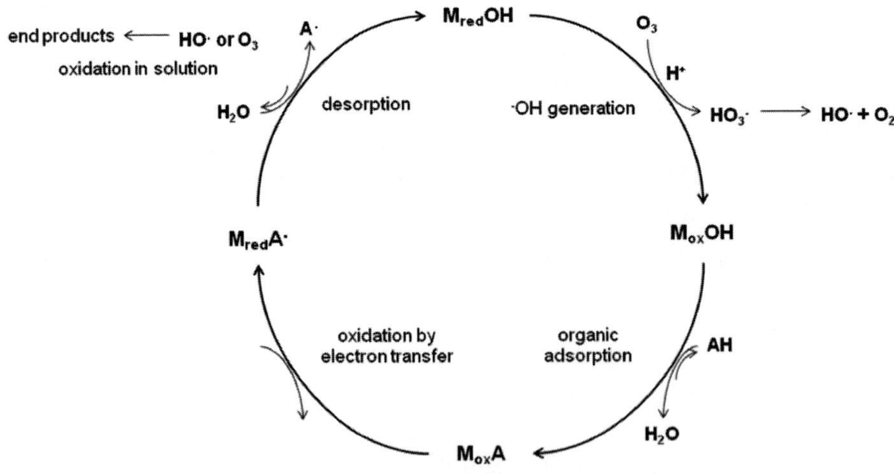

Figure 1. Scheme illustrating the possible mechanism for the degradation of organic compounds by heterogeneous catalytic ozonation in the presence of metals or metal oxides on metal oxide supports (adapted from Ref. [35]).

The second mechanism acceptable for the heterogeneous catalytic ozonation is based on the application of metals or metal oxides on metal oxide supports. The reaction pathway was explained by Legube and Leitner [35]. Firstly, ozone oxidizes the reduced catalysts (M_{red}, Figure 1) leading to the formation of hydroxyl radicals. Organic compounds (AH, e.g. salicylic and succinic acids) adsorb on the surface of the oxidized catalyst, then they are oxidized by an electron-transfer reaction and the reduced catalyst is restored ($M_{red}A^{\cdot}$). The organic radical species ($A^{\cdot}$) can easily desorbed from the catalyst and then oxidized by hydroxyl radical or ozone either in bulk of solution, or more probably, into the thickness of electric double layer.

The catalytic ozonation strongly depends on metal-support interaction, especially because the adsorption and diffusion of organics at the surface of the catalyst are deeply dependent on that; consequently both the kind of metal/oxide metal and the nature of the support are determinant for the efficiency of the process [35,38].

Besides, according to Lin et al. [38], a determining factor for the activity of the catalysts is the activity of the oxygen species present in the reaction medium. The mechanism of ozone decomposition on metals or metal oxides follows some pathways (Table 1).

Table 1. Mechanisms for the catalytic decomposition of ozone in aqueous phase [38].

Case of oxygen not adsorbed on metal surfac	Case of oxygen adsorbed on metal surface
$O_3 \rightarrow O_{3(ads)}$ $O_{3(ads)} \rightarrow O_{(ads)} + O_2$ $O_{(ads)} + O_3 \rightarrow 2O_2$	$O_3 \rightarrow O_{3(ads)}$ $O_{3(ads)} \rightarrow O_{(ads)} + O_2$ $O_{(ads)} + O_3 \rightarrow O_{2(ads)} + O_2$ $O_{2(ads)} \rightarrow O_2$

Then, some abilities of the metal oxides supported or not are determined for the activity of these catalysts in heterogeneous ozonation. Thus, some properties must be taking in consideration to choose a good catalyst, such as the ability of adsorb and desorb the ozone molecules and easily desorb oxygen formed [38]. Based on these facts, some researchers suggested that p-type oxide semiconductors are the most active metal oxides for ozone decomposition [39].

As discussed until now, there are many possibilities of reactions between metal oxides and metals or metal oxides on metal oxide supports. Some of these catalysts have been studied more, such as MnO_2 [2,40-44], TiO_2 [43,45-48] and alumina [21,23,34,47,49-53], and recently others are gaining attention, such as nanosized metal oxides [54-56], natural mineral materials [57-58] and micro and mesoporous materials, specially silicates [59-60]. As commented before some of these metal oxides can be used as a support or as the catalytic itself, such as titania, alumina and micro and mesoporous materials. Some applications of these catalysts are discussed above.

Villaseñor et al. [43] studied the performance of manganese dioxide, titanium dioxide and manganese dioxide supported on titania, as catalysts for phenol photodegradation, whereas ozone was the oxidant agent. The best catalyst was TiO_2 synthesized by sol-gel method in the presence of ozone and UV irradiation. The result was attributed to the high hydroxyl surface population of this catalyst that is stabilizing by the large size of the anatase channel, which can easily accommodate the hydroxyl groups. In the Mn/TiO_2 system, part of the hydroxyl groups was substituted by manganese ions, which may explain the lower activity of this catalyst.

The potential of catalytic ozonation in the presence of alumina for the removal of natural organic matter (NOM) from drinking water was investigated by Kasprzyk-Horden et al. [50]. The application of alumina to the ozonation process doubled the efficiency of NOM removal from water when compared to ozonation alone. Furthermore, the authors observed that catalytic ozonation in the presence of alumina led to twice as low formation of oxalic acid, the main ozonation by-product and much lower biodegradable organic carbon formation when compared with ozonation alone.

Alumina can also be used as support for metal or metal oxides in catalytic ozonation. Konova et al. [51] studied the influence of alumina-supported cobalt oxide with overstoichiometic oxygen (CoO_x/Al_2O_3) for the oxidation of volatile organic compounds (VOC) and carbon monoxide. A significant increase in the catalytic activity and decrease in the reaction temperature is observed using ozone as an oxidant, when compared the activity of the catalyst in presence of oxygen as the oxidizing agent. The authors found two reasons for this result: the high content of active and mobile oxygen obtained during the synthesis on the catalyst surface, and the catalytic active complexes of $O^-[Co^{4+}]$ that are formed during the reaction of ozone decomposition in the catalyst surface and are able to oxidize VOC at room temperature.

Jung and Choi [54] carried out a study in order to evaluate the efficiency of nanosized ZnO as a catalyst for the enhancement of ozonation. The efficiency of the application of this metal oxide was related to the enhancement of ozone's transformation into hydroxyl radicals. This phenomenon was proportional to the amount of nanosized ZnO in the experimental system. The authors concluded that the rate of ozone decomposition on the surface of the metal oxide was strongly dependent on the pH solution. Above pH 6.5, ZnO particles dramatically enlarged to form large clusters, from less than 50 nm at pH 3.0 to 350 nm at pH 7.5. The surface area of the zinc oxide decreases with the increases of pH, leading to an increase of self-aggregation, which diminishes the efficiency of the catalytic ozonation.

Another possibility of enhancement ozonation is the usage of natural minerals as catalysts. The study performed by Dong et al. [58] evaluates the efficiency of magnesia from natural brucite mineral as a catalyst for the increase of the degradation of nitrobenzene and aniline by catalytic ozonation. The authors found that this process accelerates the degradation of the organic molecules present in the reaction medium when compared to the ozonation alone. They inferred that the essential of the catalytic process was the homogeneous catalysis of hydroxyl ions in water, which accelerate the generation of hydroxyl radicals.

Mahmoud [60] investigated the application of Fe_2O_3 and MnO_2 supported on ordered mesoporous silica (OMS) as catalysts for the degradation/mineralization of an azo dye by heterogeneous catalytic ozonation. It was observed that the system OMS/O_3 did not promote a significant increase in the mineralization of the dye, whereas both systems Fe_2O_3/OMS/O_3 and MnO_2/OMS/O_3 led to a strong mineralization enhancement of the organic specie. Thus, the author concluded that the best result obtained from the systems with a metal oxide support on OMS was due to the synergic effect between the catalytic properties of the metal oxides and the high superficial area of the support (740 m^2 g^{-1}).

Conclusion

Based on the studies presented in this chapter, it is possible to conclude that catalytic ozonation is a very efficiency treatment for recalcitrant organic compounds, extending the application horizon of a treatment process known for more than a century. Then, it is possible to affirm that the consensus of some scientists is that catalytic ozonation is one of the most promissory technologies for treatment of water and wastewater. It is also a consensus that this area of knowledge needs to be investigated more in many physic-chemical aspects: pH and temperature of the aqueous phase, presence of radical scavengers in the reaction medium, catalysts properties (such as superficial area, volume of porous, composition and life-time of the solids), the mechanisms that governing the innumerous possibilities of the reaction between ozone and the metal oxides, among others.

Other aspects, such as engineering design and economic studies, must be taken into consideration when the discussion is about a technology that has a potential of being employed in some important areas of human necessities, especially when concerned about drinking water, wastewater and remediation of contaminant aquatic environment.

Besides, it is necessary to remember that there is no miraculous treatment process, but there are many efficient ones and catalytic ozonation is among them. Moreover, catalytic ozonation is also an important, wide and complex area of research that must be explored in detail, in order to be converted in real benefits for our lives and for the maintenance of the natural resources.

References

[1] Jekel, M. In *Ozonation of Water and Waste Water* Gottschalk, C.; Libra, J. A.; Saupe, A.; Eds.; Wiley-VCH: Weinheim, Germany, 2000, Chapter 3.

[2] Wu, C.-H.; Kuo, C.-Y.; Chang, C.-L. *React. Kinet. Catal. Lett.* 2007, 91, 161-168.

[3] Matheswaran, M.; Balaji, S.; Chung, S. J.; Moon, I. S. *Catal. Commun.* 2007, 8, 1497-1501.

[4] Carbajo, M.; Beltrán, F. J.; Medina, F.; Gimena, O.; Rivas, F. J. *Appl. Catal. B* 2006, 67, 177-186.

[5] Kasprzyk-Hordern, B.; Ziółek, M.; Nawrocki, J. *Appl. Catal. B* 2003, 46, 639-669.

[6] Hewes, C. G.; Davison, R. R. *Water AIChE Symposium Series* 1972, 69, 71-80.

[7] Gracia, R.; Aragues, J. L.; Ovelleiro, J. L. *Ozone Sci. Eng.* 1996, 18, 195-208.

[8] Mahmoud, A.; Freire, R. S. *Quim. Nova* 2007, 30, 198-205.

[9] Hill, G. R. *J. Am. Chem. Soc.* 1948, 70, 1306-1307.

[10] Hill, G. R. *J. Am. Chem. Soc.* 1949, 71, 2434-2435.

[11] Conocchioli, T. J.; Hamilton Jr., E. J.; Sutin, N. *J. Am. Chem. Soc.* 1965, 87, 926-927.

[12] Loegager, T.; Holcman, J.; Sehested, K.; Pedersen, T. *Inorg. Chem.* 1992, 31, 3523-3529.

[13] Piera, E.; Calpe, J. C.; Brillas, E.; Domènech, X.; Peral, J. *Appl. Catal. B* 2000, 27, 169-177.

[14] Sauleda, R.; Brillas, E. *Appl. Catal. B* 2001, 29, 135-145.

[15] Canton, C.; Esplugas, S.; Casado, J. *Appl. Catal. B* 2003, 43, 139-149.

[16] Zuo, Y.; Hoigne, J. *Environ. Sci. Technol.* 1992, 26, 1014-1022.

[17] Pines, D. S.; Reckhow, D. A. *Environ. Sci. Technol.* 2002, 36, 4046-4051.
[18] Beltrán, F. J.; Rivas, F. J.; Montero-de-Espinosa, R. *Ind. Eng. Chem. Res.* 2003, 42, 3210-3217.
[19] Andreozzi, R.; Insola, A.; Caprio, V.; D'Amore, M. G. *Water Res.* 1992, 26, 917-921.
[20] Farré, M. J.; Franch, M. I.; Malato, S.; Ayllón, J. A.; Peral, J.; Doménech, X. *Chemosphere* 2005, 58, 1127-1133.
[21] Beltrán, F. J.; Rivas, F. J.; Montero-de-Espinosa, R. *Ind. Eng. Chem. Res.* 2003, 42, 3218-3224.
[22] Gracia, R.; Cortes, S.; Sarasa, J.; Ormad, P.; Ovelleiro, J. L. *Water Res.* 2000, 34, 1525-1532.
[23] Beltrán, F. J.; Rivas, F. J.; Montero-de-Espinosa, R. *J. Chem. Technol. Biotechnol.* 2003, 78, 1225-1233.
[24] Decoret, C.; Royer, J.; Legube, B.; Doré, M. *Environ. Technol. Lett.* 1984, 5, 207-218.
[25] Ma, J.; Graham, N. J. D. *Ozone Sci. Eng.* 1997, 19, 227-240.
[26] Al-Hayek, N.; Legube, B.; Doré, M. *Environ. Technol. Lett.* 1989, 10, 415-426.
[27] Sánchez, L.; Peral, J.; Domènech, X. *Appl. Catal. B* 1998, 19, 59-65.
[28] Nawrocki, J.; Rigney, M. P.; McCormick, A.; Carr, P. W. *J. Chromatogr. A* 1993, 657, 229-282.
[29] Kaptijn, J. P. *Ozone Sci. Eng.* 1997, 19, 297-305.
[30] Rivera-Utrilla, J.; Sánchez-Polo, M. *Appl. Catal. B* 2002, 39, 319-329.
[31] Jans, U.; Hoigné, J. *Ozone Sci. Eng.* 1998, 20, 67-90.
[32] Beltrán, F. J.; Rivas, J.; Alvarez, P.; Montero-de-Espinosa, R. *Ozone Sci. Eng.* 2002, 24, 227-237.
[33] Ma, J.; Sui, M.; Zhang, T.; Guan, C. *Water Res.* 2005, 39, 779-786.
[34] Cooper, C.; Burch, R. *Water Res.* 1999, 33, 3695-3700.
[35] Legube, B.; Leitner, N. K. V. *Catal. Today* 1999, 53, 61-72.
[36] Ni, C. H.; Chen, J. N. *Water Sci. Technol.* 2001, 43, 213-220.
[37] Ma, J.; Graham, N. J. D. *Water Res.* 1999, 33, 785-793.
[38] Lin, J.; Kawai, A.; Nakajima, T. *Appl. Catal. B* 2002, 39, 157-165.
[39] Dhandapani, B.; Oyama, S. T. *Appl. Catal. B* 1997, 11, 129-166.
[40] Andreozzi, R.; Insola, A.; Caprio, V.; Marotta, R.; Tufano, V. *Appl. Catal. A* 1996, 138, 75-81.
[41] Andreozzi, R.; Insola, A.; Caprio, V.; Marotta, R.; Tufano, V. *Ind. Eng. Chem. Res.* 1997, 36, 4774-4778.
[42] Andreozzi, R.; Caprio, V.; Insola, A.; Marotta, R.; Tufano, V. *Water Res.* 1998, 32, 1492-1496.
[43] Villaseñor, J.; Reyes, P.; Pecchi, G. *Catal. Today* 2002, 76, 121-131.
[44] Xi, Y.; Reed, C.; Lee, Y.-K.; Oyama, S. T. *J. Phys. Chem. B* 2005, 109, 17587-17596.
[45] Gracia, R.; Cortés, S.; Sarosa, J.; Ormad, P.; Ovelleiro, J. L. *Water Res.* 2000, 34, 1525-1532.
[46] Beltrán, F. J.; Rivas, J.; Alvarez, P.; Montero-de-Espinosa, R. *Appl. Catal. B* 2002, 39, 221-231.
[47] Beltrán, F. J.; Rivas, J.; Alvarez, P.; Montero-de-Espinosa, R. *Appl. Catal. B* 2004, 47, 101-109.
[48] Domingues, J. R.; Beltrán, F. J.; Rodríguez, O. *Catal. Today* 2005, 101, 389-395.

[49] Karpel, N.; Leitner, V.; Delouane, B.; Legube, B; Luck, F. *Ozone Sci. Eng.* 1999, 21, 261-276.

[50] Kasprzyk-Hordern, B.; Raczyk-Stanislawiak, U.; Swietlik, J.; Nawrocki, J. *Appl. Catal. B* 2006, 62, 345-358.

[51] Konova, P.; Stoyanova, M.; Naydenov, A.; Christoskova, St.; Mehandjiev, D. *Appl. Catal. A* 2006, 298, 109-114.

[52] Stoyanova, M.; Konova, P.; Nikolov, P.; Naydenov, A.; Christoskova, St.; Mehandjiev, D. *Chem. Eng. J.* 2006, 122, 41-46.

[53] Alvarez, P. M.; Beltrán, F. J.; Pocostales, J. P.; Masa, F. J. *Appl. Catal. B* 2007, 72, 322-330.

[54] Jung, H.; Choi, H. *Appl. Catal. B* 2006, 66, 288-294.

[55] Jung, H.; Park, H.; Kim, J.; Lee, J.-H.; Hur, H.-G.; Myung, N. V.; Choi, H. *Environ. Sci. Technol.* 2007, 41, 4741-4747.

[56] Yang, Y.; Ma, J.; Qin, Q.; Zhai, X. *J. Mol. Catal. A* 2007, 267, 41-48.

[57] Dong, Y.; He, K.; Zhao, B.; Yin, V.; Yin, L.; Zhang, A. *Catal. Commun.* 2007, 8, 1599-1603.

[58] Dong, Y.; He, K.; Yin, L.; Zhang, A. *Catal. Lett.* 2007, 119, 222-227.

[59] Cooper, C.; Burch, R. *Water Res.* 1999, 33, 3689-3694.

[60] Mahmoud, A.; *Ms. Thesis*, Universidade de São Paulo, Brazil, 2006.

In: Ozone Depletion, Chemistry and Impacts
Editor: Sem H. Bakker, pp. 163-186
ISBN 978-1-60692-007-7

Chapter 8

BIVARIATE STOCHASTIC VOLATILITY MODELS APPLIED TO MEXICO CITY OZONE POLLUTION DATA

***Jorge A. Achcar* and Henrique C. Zozolotto*†**
Faculdade de Medicina de Ribeirão Preto
Universidade de São Paulo
Ribeirão Preto – SP, Brasil
***Eliane R. Rodrigues*‡**
Instituto de Matemáticas – UNAM
Area de la Investigación Científica
Circuito Exterior, Ciudad Universitaria
México, D.F. 04510, México

Abstract

In this paper, we consider recently introduced bivariate stochastic volatility models commonly used to analyse financial time series, to study problems related to air pollution data. Such models are used here to estimate the volatility of weekly averaged ozone measurements taking into account two different sets of data provided by the monitoring network of Mexico City. A Bayesian analysis is developed using Markov Chain Monte Carlo (MCMC) methods to simulate samples from the joint posterior distributions and perform the estimates of interest.

1. Introduction

Air pollution is one of the main problems in several large cities throughout the world. Among the many existing pollutants, ozone is one that most affect these cities. In many of them, including Mexico City, environmental authorities have implemented measures aiming to reduce the level of pollutants in general. Such measures are very important because

*E-mail address: jorge@icmc.usp.br;
†E-mail address: henrique_ceretta@yahoo.com.br
‡E-mail address: eliane@math.unam.mx

when ozone concentration stays above a given threshold for a certain period of time, individuals exposed to the pollutant may experience serious health problems (see for example Wilson *et al.*, 1980; Loomis *et al.*, 1996; Bell *et al.*, 2004; Bell *et al.*, 2005; Bell *et al.*, 2007). In particular, it is well known that for ozone levels above 0.11 parts per million (0.11ppm) a very sensitive part of the population (such as the newborn and the elderly) experience a deterioration in their health. Therefore, being able to understand the long term behaviour of such pollutant is of great importance for policy makers. If there is a trend in the measurements that is decreasing with time and it is such that the variations are not so large, or otherwise if it is increasing or with large variations, then environmental authorities could have an idea if in the long run the preventive measures taken by them are producing a desirable result or not.

Several methods have been used in order to predict the violation of an air quality standard. Among them we may refer to Roberts (1979a, 1979b), Horowitz (1980) and Smith (1989) when the interest resides in applying extreme values theory to perform predictions. However, other techniques may also be used to study this type of problems. As examples we may quote, multivariate analysis (Guardani *et al.*, 2003), neural networks (Guardani *et al.*, 1999), Poisson models (Javits, 1980; Raftery, 1989; Leadbetter, 1991; Achcar *et al.*, 2007) and Markov chain models (Larsen *et al.*, 1990; Austin and Tran, 1999; Álvarez *et al.*, 2005).

Another possibility is to consider time series modeling the daily or the weekly averaged pollution data (see for example, Loomis *et al.*, 1996; Lanfredi and Macchiato, 1997; Zolghadri and Henry, 2004; Pan and Chen, 2007).

In this paper we use stochastic volatility (SV) models (see for instance Ghysels *et al.*, 1996; Kim *et al.*, 1998; Meyer and Yu, 2000) to study the behaviour of weekly averaged measurements of ozone provided by the monitoring network of Mexico City. The type of volatility models considered here have been extensively used to analyse financial time series (see for example Danielsson, 1994; Yu, 2002), as a powerful alternative for the usual existing ARCH (autoregressive conditional heteroscedastic) models introduced by Engle (1982) and the generalised autoregressive conditional heteroscedastic (GARCH) models introduced by Bollerslev (1986). The SV-type models have many advantages when they are used to analyse time series. The main reason being that they consider two processes to model the series: a process modelling the observations and another modelling the latent volatility.

A Bayesian inference approach using Markov Chain Monte Carlo (MCMC) methods (see for instance Gelfand and Smith, 1990; Smith and Roberts, 1993) has been performed to analyse the SV models considered here. The use of MCMC methods is necessary because we can have great difficulties when using standard classical inference approach. Some of the problems present are high dimensionality, likelihood function with no closed form and also a possible high computational cost.

This paper is organised as follows. In Section 2 the volatility models used here are introduced. Section 3 presents the Bayesian formulation of the problem. In Section 4 we establish the criterion for selecting the best model that fits the data used. An application to the data provided by the Mexico City monitoring network is given in Section 5. Finally, in Section 6 we conclude.

2. Bivariate Stochastic Volatility Models (BSV)

Different classes of multivariate stochastic volatility models are introduced in the literature (see for example Yu and Meyer, 2005). In the present paper we consider six bivariate models. In order to describe those models, we start by considering $N \geq 1$ a fixed integer number that will represent the size of the data set used in the analysis. Let $\mathbf{Z}(t) = (Z_1(t), Z_2(t))'$, $t = 1, 2, \ldots, N$ be the series recording the results of the same event performed in two different locations at the same time. (In here, for v a vector or a matrix we use v' to indicate the transpose of v.) In our case $\mathbf{Z}(t)$, $t = 1, 2, \ldots, N$ will represent the weekly averaged ozone measurements taken in two different regions of the Metropolitan Area of Mexico City. Consider a vector of latent variables $\mathbf{h}(t) = (h_1(t), h_2(t))$, $t = 1, 2, \ldots, N$ where $h_i(t)$ are defined by the following auto-regressive models AR(1)

$$\left\{ \begin{array}{lcl} h_1(1) & = & \mu_1 + \eta_1(1) \\ h_2(1) & = & \mu_2 + \eta_2(1) \end{array} \right. \tag{1}$$

and for $t = 2, 3, \ldots, N$

$$\left\{ \begin{array}{lcl} h_1(t) & = & \mu_1 + \phi_{11}\left[h_1(t-1) - \mu_1\right] + \eta_1(t) \\ h_2(t) & = & \mu_2 + \phi_{22}\left[h_2(t-1) - \mu_2\right] + \eta_2(t) \end{array} \right. \tag{2}$$

where $0 < \phi_{11}, \phi_{22} < 1$ and $\boldsymbol{\eta}(t) = (\eta_1(t), \eta_2(t))$ has a bivariate Normal distribution with mean vector $\mathbf{0} = (0, 0)'$ and variance-covariance matrix given by the 2×2 diagonal matrix $\text{diag}\left(\sigma^2_{\eta_1}, \sigma^2_{\eta_2}\right)$.

Consider $\mathbf{H}(t)$, $t = 1, 2, \ldots, N$ a 2×2 diagonal matrix with diagonal $e^{h_1(t)/2}$ and $e^{h_2(t)/2}$, i.e., $\mathbf{H}(t) = \text{diag}\left(e^{h_1(t)/2}, e^{h_2(t)/2}\right)$. Let $\mathbf{Y}(t) = (Y_1(t), Y_2(t))'$, $t = 1, 2, \ldots, N$ be modelled by

$$\mathbf{Y}(t) = \mathbf{H}(t)\boldsymbol{\epsilon}(t)$$

where $\boldsymbol{\epsilon}(t) = (\epsilon_1(t), \epsilon_2(t))'$ is the vector of error components having a bivariate Normal distribution with mean vector $\mathbf{0}$ and variance-covariance matrix Σ_ϵ given by

$$\Sigma_\epsilon = \begin{pmatrix} 1 & \rho_\epsilon \\ \rho_\epsilon & 1 \end{pmatrix}$$

with $\rho_\epsilon \geq 0$. Hence, $\mathbf{Y}(t)$, $t = 1, 2, \ldots, N$ is such that $Y_1(t) = e^{h_1(t)/2}\epsilon_1(t)$ and $Y_2(t) = e^{h_2(t)/2}\epsilon_2(t)$. (Usually, $Y_1(t)$ and $Y_2(t)$, $t = 1, 2 \ldots, N$ are the logarithms of the returns of $Z_1(t)$ and $Z_2(t)$ centred around their averages.)

Remark. By definition, we observe that $\text{E}\left[\mathbf{Y}(t)\right] = \mathbf{0}$ and the variance-covariance matrix for $\mathbf{Y}(t)$ is given by,

$$\begin{array}{lcl} \Sigma_Y & = & \text{Var}\left(\mathbf{Y}(t)\right) = \mathbf{H}'(t)\Sigma_\epsilon\mathbf{H}(t) \\ & = & \begin{pmatrix} e^{h_1(t)} & \rho_\epsilon e^{h_1(t)/2}e^{h_2(t)/2} \\ \rho_\epsilon e^{h_1(t)/2}e^{h_2(t)/2} & e^{h_2(t)} \end{pmatrix}, \end{array} \tag{3}$$

for $t = 1, 2, \ldots, N$. Furthermore, $\mathbf{Y}(t) = (Y_1(t), Y_2(t))'$, $t = 1, 2, \ldots, N$ has a bivariate Normal distribution with density

$$f(Y_1(t), Y_2(t) \mid h_1(t), h_2(t)) = \frac{1}{2\pi\sqrt{(1-\rho_\epsilon^2)\, e^{h_1(t)+h_2(t)}}} \exp\left\{-\frac{1}{2(1-\rho_\epsilon^2)}\left[\frac{Y_1^2(t)}{e^{h_1(t)}} + \frac{Y_2^2(t)}{e^{h_2(t)}} - \frac{2\rho_\epsilon Y_1(t)Y_2(t)}{e^{h_1(t)/2}e^{h_2(t)/2}}\right]\right\}$$

Also observe from (1), (2) and the definition of $\boldsymbol{\eta}(t)$, that the latent variables $\mathbf{h}(t) = (h_1(t), h_2(t))$ have Normal distributions. In fact the density function of $h_i(1)$ is a Normal distribution $\mathrm{N}(\mu_i, \sigma_{\eta i}^2)$ and given $h_i(t-1)$ the variable $h_i(t)$ has density function $\mathrm{N}(\mu_i + \phi_{ii}\left[h_i(t-1) - \mu_i\right], \sigma_{\eta i}^2)$ for $i = 1, 2$ and $t = 2, \ldots, N$.

The different models considered in this paper are described as follows

2.1. Model I

In this model we assume that the error coordinates $\epsilon_i(t)$, $i = 1, 2$ are independent, i.e., $\rho_\epsilon = 0$. Hence, $\mathbf{Y}(t)$, $t = 1, 2, \ldots, N$ will have Normal distribution with mean vector $\mathbf{0}$ and with variance-covariance matrix a diagonal matrix given by $\Sigma_{\epsilon,0} = \mathrm{diag}(e^{h_1(t)}, e^{h_2(t)})$.

2.2. Model II

In this version of the model the covariance ρ_ϵ between $\epsilon_1(t)$ and $\epsilon_2(t)$ is considered to be an unknown quantity that ought to be estimated. Hence, $\mathbf{Y}(t)$, $t = 1, 2, \ldots, N$ will have Normal distribution with mean vector $\mathbf{0}$ and with variance-covariance matrix given by (3).

2.3. Model III

In a third version of the model we keep the assumption of Model II, except for the way we define the latent variable $h_2(t)$, $t = 1, 2, \ldots, N$. In this version, we assume the presence of the Granger causality when modelling $h_2(t)$, i.e., the latent variable $h_2(t)$ is now given by

$$h_2(t) = \mu_2 + \phi_{21}\left[h_1(t-1) - \mu_1\right] + \phi_{22}\left[h_2(t-1) - \mu_2\right] + \eta_2(t) \tag{4}$$

where $t = 2, 3, \ldots, N$ and $0 < \phi_{21}, \phi_{22} < 1$.

Remark. Note that from (4), that if $\phi_{21} \neq 0$, then the volatility of the second return is considered with the Granger causality induced by the volatility of the first return.

2.4. Model IV

In this version of the model we take the assumptions of Model II except that we consider the additional hypothesis on the correlation between $\epsilon_1(t)$ and $\epsilon_2(t)$. In this way, we assume that $\boldsymbol{\epsilon}(t) = (\epsilon_1(t), \epsilon_2(t))'$ has a bivariate Normal distribution with mean vector $\mathbf{0}$ and variance-covariance matrix $\Sigma_\epsilon(t)$ a 2×2 matrix given, for $t = 1, 2, \ldots, N$, by

$$\Sigma_\epsilon(t) = \begin{pmatrix} 1 & \rho_\epsilon(t) \\ \rho_\epsilon(t) & 1 \end{pmatrix},$$

where $\rho_\epsilon(t) = \left(e^{q(t)} - 1\right) / \left(e^{q(t)} + 1\right)$, $q(1) = \psi_0 + \sigma_\rho v(t)$ and for $t = 1, 2, \ldots, N$, we have $q(t) = \psi_0 + \psi_1 \left[q(t-1) - \psi_0\right] + \sigma_\rho v(t)$, with $v(t), t = 1, 2, \ldots, N$, independent and identically distributed quantities with common distribution a Normal distribution $\mathrm{N}(0, 1)$ (see for example Yu and Meyer, 2005). We further assume that $q(1)$ has a Normal distribution $\mathrm{N}\left(\psi_0, \sigma_\rho^2\right)$ and for $t = 2, 3, \ldots, N$, we also assume that given $q(t-1)$, the quantity $q(t)$ has a Normal distribution $\mathrm{N}(\psi_0 + \psi_1 \left[q(t-1) - \psi_0\right], \sigma_\rho^2)$.

Remark. In this stochastic volatility model with dynamic correlation between the error components we see that the correlation also change as well as the volatility. Also observe that we need to have $-1 < \rho_\epsilon(t) < 1$ to have a well defined variance-covariance matrix $\Sigma_\epsilon(t)$.

2.5. Model V

Consider the same hypothesis of Model IV with the exception of the hypothesis made for $h_2(t)$. In the case of this latent variable we consider the assumption used in Model III. Therefore, we consider the presence of the Granger causality for the latent variable $h_2(t)$, i.e., $h_2(t)$ is given by (4).

2.6. Model VI

In this model we consider a setting similar to that of Model I. The difference being that now the error coordinates $\epsilon_i(t)$, $i = 1, 2$, have a bivariate Student distribution with ν degrees of freedom, $t = 1, 2, \ldots, N$. Hence, we assume that $\boldsymbol{\epsilon}(t)$ has a bivariate Student distribution with mean vector $\mathbf{0}$ and variance-covariance matrix $[\nu/(\nu-2)]\Sigma_{\epsilon,0}$, for $\nu > 2$. Therefore, the vector $\boldsymbol{\epsilon}(t) = (\epsilon_1(t), \epsilon_2(t))'$ has density function a $t\left(\mathbf{0}, \Sigma_{\epsilon,0}, \nu\right)$, where ν is the of degree of freedom, i.e., its density function is given by,

$$f\left(\boldsymbol{\epsilon}(t)\right) = \frac{\Gamma\left[\left(\nu+2\right)/2\right]}{\nu\pi\Gamma\left[\nu/2\right]} \left|\Sigma_{\epsilon,0}\right|^{-1/2} \left(1 + \frac{1}{\nu}\boldsymbol{\epsilon}'(t)\Sigma_{\epsilon,0}^{-1}\boldsymbol{\epsilon}(t)\right)^{-(\nu+2)/2} \tag{5}$$

where $\Gamma\left(x\right)$ denotes a Gamma function.

Remark. Using this heavy tail Student distribution, we could have the presence of extra Kurtosis for the return distributions. It is also interesting to observe that many other modifications for the BSV models could be considered.

3. Bayesian Analysis

In this section we are going to describe the Bayesian formulation of the models splitting them into three categories. Those in the so-called Class I, are the models that have the error vector $\boldsymbol{\epsilon}(t) = (\epsilon_1(t), \epsilon_2(t))'$ Normally distributed with mean vector $\mathbf{0}$ and a constant correlation for the error components model, i.e., the correlation either is zero or is a constant not depending on t. The models falling into this class are Models I, II and III. In Class II, we include models where $\boldsymbol{\epsilon}(t)$ is Normally distributed with mean vector $\mathbf{0}$, but the correlation between $\epsilon_1(t)$ and $\epsilon_2(t)$ also depends on time, i.e., Models IV and V. Finally, in Class III we have Model VI.

Bayesian inference will be performed through a sample taken from the posterior distribution of the parameters of the models. Therefore, we need to specify in each case which prior distributions we are considering and what likelihood functions are used. In all cases we assume prior independence among the parameters of the models.

3.1. Class I

If Model I is considered then the vector of parameters to be estimated is $\boldsymbol{\theta}_I = (\phi_{11}, \phi_{22}, \sigma^2_{\eta 1}, \sigma^2_{\eta 2}, \mu_1, \mu_2)$. We assume that ϕ_{ii}, $\sigma^2_{\eta i}$ and μ_i have as prior distributions a Beta, an Inverse Gamma and a Normal distributions, respectively, $i = 1, 2$, i.e., ϕ_{ii}, $\sigma^2_{\eta i}$ and μ_i have, respectively, prior distributions $\mathrm{Beta}(a_{ii}, b_{ii})$, $\mathrm{IG}(c_i, d_i)$ and $\mathrm{N}(e_i, f_i)$, where the hyperparameters a_{ii}, b_{ii}, c_i, d_i, e_i and f_i are known, $i = 1, 2$. (In here, we are considering the $\mathrm{Beta}(a, b)$ and $\mathrm{IG}(c, d)$ as the Beta and the Inverse Gamma distributions with means $a/(a + b)$ and $d/(c - 1)$ and variances $ab/[(a + b)^2(a + b + 1)]$ and $d^2/[(c - 1)^2(c - 2)]$, $c > 2$, respectively.)

When Model II is considered the vector of parameters is $\boldsymbol{\theta}_{II} = (\boldsymbol{\theta}_I, \rho_\epsilon)$, where we assume that ρ_ϵ has as prior distribution an uniform distribution U(-1, 1) and use the same priors distributions for $\boldsymbol{\theta}_I$ and in Model I, with possibly different values for the hyperparameters.

In Models I and II, the joint density functions of the latent variables $\mathbf{h}(t) = (h_1(t), h_2(t))$ given the vector of parameters $\boldsymbol{\theta}$ are given by

$$\begin{cases} g\left(\mathbf{h}(1)\,|\,\boldsymbol{\theta}\right) \propto \prod_{l=1}^{2} \left(\sigma^2_{\eta_l}\right)^{-1/2} \exp\left[-\frac{1}{2\sigma^2_{\eta_l}} \left(h_l(1) - \mu_l\right)^2\right], \; t = 1 \\ g\left(\mathbf{h}(t)\mid \mathbf{h}(t-1), \boldsymbol{\theta}\right) \propto \prod_{l=1}^{2} \prod_{t=2}^{N} \left(\sigma^2_{\eta_l}\right)^{-1/2} \exp\left\{-\frac{1}{2\sigma^2_{\eta_l}} \left[h_l(t) - \mu_l \right.\right. \\ \qquad \left.\left. -\phi_{ll}\left(h_l(t-1) - \mu_l\right)\right]^2\right\}, \; t = 2, 3, \ldots, N. \end{cases} \tag{6}$$

Let $\boldsymbol{\theta} = \boldsymbol{\theta}_{II}$ and take $\varphi = (\boldsymbol{\theta}, \mathbf{h})$, with $\mathbf{h} = (\mathbf{h}(1), \mathbf{h}(2), \ldots, \mathbf{h}(N))$. Hence, we have that the joint likelihood function of $\boldsymbol{\theta}$ and $\mathbf{h}$ for Model II is given, for $\mathbf{Y} = (\mathbf{Y}(1), \mathbf{Y}(2), \ldots, \mathbf{Y}(N))$, by

$$\begin{aligned} L(\varphi\,|\,\mathbf{Y}) &\propto \prod_{t=1}^{N} p\left(\mathbf{Y}(t)\mid \mathbf{h}(t), \boldsymbol{\theta}\right) \\ &\propto \left(1 - \rho_\epsilon^2\right)^{-N/2} \exp\left\{-\frac{1}{2}\left[\sum_{t=1}^{N} h_1(t) + \sum_{t=1}^{N} h_2(t)\right]\right\} \\ &\quad \exp\left\{-\frac{1}{2\left(1 - \rho_\epsilon^2\right)}\left[\sum_{t=1}^{N} Y_1^2(t) e^{-h_1(t)}\right.\right. \\ &\quad \left.\left. + \sum_{t=1}^{N} Y_2^2(t) e^{-h_2(t)} - 2\rho_\epsilon \sum_{t=1}^{N} Y_1(t) Y_2(t) e^{-h_1(t)/2} e^{-h_2(t)/2}\right]\right\} \end{aligned} \tag{7}$$

In the case of $\boldsymbol{\theta} = \boldsymbol{\theta}_I$ we just set $\rho_\epsilon = 0$ in (7).

Therefore, for $\varphi = (\boldsymbol{\theta}_I, \mathbf{h})$ or $\varphi = (\boldsymbol{\theta}_{II}, \mathbf{h})$ we have that the joint posterior distribution of the vector of parameters and $\mathbf{h}$ is given by

$$\pi(\varphi \mid \mathbf{Y}) \propto \pi(\boldsymbol{\theta})\, g(\mathbf{h}(1) \mid \boldsymbol{\theta}) \left(\prod_{t=2}^{N} g(\mathbf{h}(t) \mid \mathbf{h}(t-1), \boldsymbol{\theta}) \right) L(\varphi \mid \mathbf{Y}) \tag{8}$$

where $\pi(\boldsymbol{\theta})$ is the prior distribution of the vector of parameters with $\boldsymbol{\theta} = \boldsymbol{\theta}_I, \boldsymbol{\theta}_{II}$, $L(\varphi \mid \mathbf{Y})$ is the likelihood function of the model given by (7), and $g(\mathbf{h}(1) \mid \boldsymbol{\theta})$, $g(\mathbf{h}(t) \mid \mathbf{h}(t-1), \boldsymbol{\theta})$, $t = 2, 3, \ldots, N$ are given by the set of recursive functions (6).

When we assume the presence of the Granger causality for $h_2(t)$ and constant correlation for the error components, i.e., Model III, the vector of parameters is $\boldsymbol{\theta}_{III} = (\boldsymbol{\theta}_{II}, \phi_{21})$. We take the same prior distributions for $\boldsymbol{\theta}_{II}$ with possibly different hyperparameters. Additionally, for the variable ϕ_{21} we take as its prior distribution a Beta distribution $\text{Beta}(a_{21}, b_{21})$ with known hyperparameters a_{21} and b_{21}.

The likelihood function of Model III is given by (7). The joint posterior distribution of $\varphi = (\boldsymbol{\theta}, \mathbf{h})$ also has the same expression as in Models I and II, i.e., the expression is given by (8) but now taking $\boldsymbol{\theta} = \boldsymbol{\theta}_{III}$ and replacing the density function $g(\mathbf{h}(t) \mid \mathbf{h}(t-1), \boldsymbol{\theta})$, $t = 2, 3, \ldots, N$ by

$$\begin{aligned} & g(\mathbf{h}(t) \mid \mathbf{h}(t-1), \boldsymbol{\theta}) \propto \\ & \quad \left(\prod_{t=2}^{N} \left(\sigma_{\eta_1}^2\right)^{-1/2} \exp\left\{ -\frac{1}{2\sigma_{\eta_1}^2} \left[h_1(t) - \mu_1 - \phi_{11}\left(h_1(t-1) - \mu_1\right)\right]^2 \right\} \right) \\ & \quad \left(\prod_{t=2}^{N} \left(\sigma_{\eta_2}^2\right)^{-1/2} \exp\left\{ -\frac{1}{2\sigma_{\eta_2}^2} \left[h_2(t) - \mu_2 - \phi_{21}\left(h_1(t-1) - \mu_1\right) \right.\right.\right. \\ & \qquad \left.\left.\left. - \phi_{22}\left(h_2(t-1) - \mu_2\right)\right]^2 \right\} \right). \end{aligned} \tag{9}$$

3.2. Class II

The models in this class have a stochastic volatility model with dynamic correlation for the error components. When Model IV is considered we have that the vector of parameters is $\boldsymbol{\theta}_{IV} = (\boldsymbol{\theta}_I, \psi_0, \psi_1, \sigma_\rho^2)$. The prior distribution for the components of $\boldsymbol{\theta}_I$ are taken to be the same as in Class I. The parameters ψ_0, ψ_1 and σ_ρ^2 have as prior distributions a Normal distribution $\text{N}(0, f_3)$, a Beta distribution $\text{Beta}(f, g)$ and a Inverse Gamma distribution $\text{IG}(c_3, d_3)$, where the hyperparameters c_3, d_3, f_3, f, g are considered to be known.

Hence, the joint likelihood function of this class of models is given, for $\boldsymbol{\theta} = \boldsymbol{\theta}_{IV}$, by,

$$\begin{aligned} L(\boldsymbol{\theta}, \mathbf{h} \,|\, \mathbf{Y}) \;\propto\; & \prod_{t=1}^{N} p\left(\mathbf{Y}(t) \mid \mathbf{h}(t), \boldsymbol{\theta}\right) \propto \left\{ \prod_{t=1}^{N} \left(1 - \rho_\epsilon^2(t)\right)^{-1/2} \right\} \\ & \exp\left\{ -\frac{1}{2} \left[\sum_{t=1}^{N} h_1(t) + \sum_{t=1}^{N} h_2(t) \right] \right\} \\ & \exp\left\{ -\frac{1}{2} \left[\sum_{t=1}^{N} \frac{Y_1^2(t) e^{-h_1(t)}}{1 - \rho_\epsilon^2(t)} + \sum_{t=1}^{N} \frac{Y_2^2(t) e^{-h_2(t)}}{1 - \rho_\epsilon^2(t)} - \right.\right. \\ & \left.\left. -2 \sum_{t=1}^{N} \frac{\rho_\epsilon(t) Y_1(t) Y_2(t) e^{-h_1(t)/2} e^{-h_2(t)/2}}{1 - \rho_\epsilon^2(t)} \right] \right\}. \end{aligned} \tag{10}$$

The joint posterior distribution of $\varphi = (\boldsymbol{\theta}, \mathbf{h}, \mathbf{q})$, where $\boldsymbol{\theta} = \boldsymbol{\theta}_{IV}$ and $\mathbf{q} = (q(1), q(2), \ldots, q(N))$, is given by

$$\begin{aligned} \pi\left(\varphi \mid \mathbf{Y}\right) \;\propto\; & \pi\left(\boldsymbol{\theta}\right) g\left(\mathbf{h}(1) \,|\, \boldsymbol{\theta}\right) \left(\prod_{t=2}^{N} g\left(\mathbf{h}(t) \mid \mathbf{h}(t-1), \boldsymbol{\theta}\right) \right) \\ & f\left(\mathbf{q}(1) \,|\, \boldsymbol{\theta}\right) \left(\prod_{t=2}^{N} f\left(q(t) \mid q(t-1), \boldsymbol{\theta}_{IV}\right) \right) L(\varphi \,|\, \mathbf{Y}) \end{aligned} \tag{11}$$

where $\pi\left(\boldsymbol{\theta}\right)$ is the joint prior distribution for $\boldsymbol{\theta}$; $g\left(\mathbf{h}(1) \,|\boldsymbol{\theta}\right)$ and $g(\mathbf{h}(t) \mid \mathbf{h}(t-1),$ $\boldsymbol{\theta})$ are defined by (6), $f\left(q(1) \,|\, \boldsymbol{\theta}\right)$ and $f\left(q(t) \mid q(t-1), \boldsymbol{\theta}\right)$ are the Normal density functions of $q(1)$ and the conditional Normal density function of $q(t)$ given $q(t-1)$, and $L(\varphi \,|\, \mathbf{Y})$ is the likelihood function defined in (10).

When considering a stochastic volatility model with dynamic correlation for the error components and the presence of the Granger causality for $h_2(t)$, $t = 2, 3, \ldots, N$, i.e., Model V, the vector of parameters here is $\boldsymbol{\theta}_V = (\boldsymbol{\theta}_{IV}, \phi_{21})$. We take the same prior distribution for $\boldsymbol{\theta}_{IV}$ except for the parameters ψ_0 and ψ_1 which now will have as prior distributions, Gamma distributions with appropriate hyperparameters. We still take a Beta(a_{21}, b_{21}) prior distribution for ϕ_{21}.

The likelihood of the model is also given by (10). The posterior distribution is similar to (11), the difference is that $g(\mathbf{h}(t) \,|\, \mathbf{h}(t-1), \boldsymbol{\theta})$, $t = 2, 3, \ldots, N$ are given by (9) with $\boldsymbol{\theta}$ replaced by $\boldsymbol{\theta}_V$.

3.3. Class III

In this class of models we assume that the error components have a bivariate Student distribution with degree of freedom $\nu > 2$. The latent variables $\mathbf{h}(t) = (h_1(t), h_2(t))$, $t = 1, 2, \ldots, N$ are defined as in Model I.

The likelihood function could be obtained directly from (5) or representing the bivariate Student distribution as a mixture of a bivariate Normal distribution with a Gamma distribution (see for example Bernardo and Smith, 1995). Following the latter approach, first we have that the distribution for $\epsilon(t)$ is given by,

$$f_\epsilon(\epsilon(t)) = \int f_{Normal}(\epsilon(t) \mid \boldsymbol{\mu}, \Sigma_Y z_t)\, f_{Gamma}(z_t \mid \alpha, \beta)\, dz_t \tag{12}$$

where $f_{Normal}(\epsilon(t) \mid \boldsymbol{\mu}, \Sigma)$ denotes bivariate Normal density with mean vector $\boldsymbol{\mu}$ and variance-covariance matrix Σ and $f_{Gamma}(z \mid \alpha, \beta)$ is a Gamma density with mean α/β and variance α/β^2.

Therefore, take $z_t = 1/\sqrt{w_t}$, where w_t has as distribution a Gamma(λ_w, λ_w) distribution and is such that $\mathrm{E}(w_t) = 1$ and $\mathrm{Var}(w_t) = \lambda_w^{-1}$. Hence, the conditional distribution of $\mathbf{Y}(t)$ given Σ_Y and w_t is a bivariate Normal distribution with mean vector $\mathbf{0}$ and variance-covariance matrix $\frac{1}{\sqrt{w_t}}\Sigma_Y$, where Σ_Y is given by (3).

The vector of parameters for Model VI is $\boldsymbol{\theta}_{VI} = (\boldsymbol{\theta}_{II}, \lambda_w)$ and, for $\mathbf{w} = (w_1, w_2, \ldots, w_N)$, the likelihood function of the model is given, for $\varphi = (\boldsymbol{\theta}_{IV}$, $\mathbf{h}$, $\mathbf{w})$, by

$$\begin{aligned} L(\varphi, \mid \mathbf{Y}) \;\propto\; & \prod_{t=1}^{N} p(\mathbf{Y}(t) \mid \mathbf{h}(t), w_t) \propto \left(1-\rho_\epsilon^2\right)^{-N/2} \left\{\prod_{t=1}^{N} w_t^{1/2}\right\} \\ & \exp\left\{-\frac{1}{2}\left[\sum_{t=1}^{N} h_1(t) + \sum_{t=1}^{N} h_2(t)\right]\right\} \\ & \exp\left\{-\frac{1}{2\left(1-\rho_\epsilon^2\right)}\left[\sum_{t=1}^{N} Y_1^2(t) w_t^{1/2} e^{-h_1(t)} + \right.\right. \\ & \quad + \sum_{t=1}^{N} Y_2^2(t) w_t^{1/2} e^{-h_2(t)} \\ & \quad \left.\left. -2\rho_\epsilon \sum_{t=1}^{N} Y_1(t) Y_2(t) w_t^{1/2} e^{-h_1(t)/2} e^{-h_2(t)/2}\right]\right\}. \end{aligned} \tag{13}$$

The prior distributions of $\boldsymbol{\theta}_{II}$ are the same as used in Class I models (with possibly different hyperparameters) and we take a Gamma prior distribution Gamma(f_λ, g_λ) for λ_w. We also have that f_λ and g_λ are known hyperparameters.

Hence, the joint posterior distribution of $\varphi = (\boldsymbol{\theta}_{VI}, \mathbf{h}, \mathbf{w})$, is given by

$$\begin{aligned} \pi(\varphi \mid \mathbf{Y}) \;=\; & \pi(\boldsymbol{\theta}_{VI})\, g(\mathbf{h}(1) \mid \boldsymbol{\theta}_{VI}) \left(\prod_{t=2}^{N} g(\mathbf{h}(t) \mid \mathbf{h}(t), \boldsymbol{\theta}_{VI})\right) \\ & \left(\prod_{t=1}^{N} g_w(w_t \mid \boldsymbol{\theta}_{VI})\right) L(\varphi \mid \mathbf{Y}), \end{aligned} \tag{14}$$

where $\pi(\boldsymbol{\theta}_{VI})$ is the joint prior distribution of $\boldsymbol{\theta}_{VI}$, $g(\mathbf{h}(1) \mid \boldsymbol{\theta}_{VI})$ and $g(\mathbf{h}(t) \mid \mathbf{h}(t-1), \boldsymbol{\theta}_{VI})$ are given by (6), and $\prod_{t=1}^{N} g_w(w_t \mid \boldsymbol{\theta}_{VI})$ is the product of Gamma$(\lambda_w$, $\lambda_w)$ densities, that is,

$$\prod_{t=1}^{N} g_w(w_t \mid \boldsymbol{\theta}_{VI}) \propto \prod_{t=1}^{N} w_t^{\lambda_w - 1} e^{-\lambda_w w_t}$$

and $L(\varphi \mid \mathbf{Y})$ is the likelihood function (13).

Samples of the joint posterior distributions are generated using MCMC methods such as the Gibbs sampling algorithm or the Metropolis-Hastings algorithm (see for example Gelfand and Smith, 1990; Smith and Roberts, 1993). A great simplification in the simulation of the samples is obtained using the software WinBugs (Spiegelhalter *et al.*, 2003), where we only need to specify a distribution for the data and the prior distribution of the parameters.

4. Model Selection

In this paper we use the Deviance Information Criterion to select the model that best fit the data. The Deviance Information Criterion (DIC) (see Spiegelhalter *et al.*, 2002) is given by,

$$DIC = \widehat{D} + 2p_D \tag{15}$$

where $\widehat{D}$ is the deviance evaluated at the posterior mean and p_D is the effective number of parameters in the model which is given by $p_D = \overline{D} - \widehat{D}$, where $\overline{D}$ is the posterior mean deviance.

Smaller value of DIC indicates the best BSV models, these values also could be negatives.

5. An Application to the Weekly Ozone Averages in Mexico City

In this section we present a description of the data to which the theory described earlier in this paper is applied. We also present the results obtained.

5.1. Description of the Data

The data was obtained from the monitoring network of the Metropolitan Area of Mexico City (www.sma.df.gob.mx/simat/). The Metropolitan Area is split into five regions or sectors corresponding to the Northeast (NE), Northwest (NW), Centre (CE), Southeast (SE) and Southwest (SW) and the ozone monitoring stations are placed throughout the city (see Álvarez *et al.*, 2005; Achcar *et al.*, 2007).

The primary data used in the analysis corresponds to sixteen years of the daily maximum ozone measurements, ranging from 01 January 1990 to 31 December 2005 (inclusive), of the daily maximum measurements in regions NE, CE and SW. The measurements are obtained minute by minute and the averaged hourly result is reported at each station. The daily maximum measurement for a given region is the maximum over all the maximum averaged values recorded hourly during a 24-hour period by each station placed in the region. The sixteen-year average measurements in regions NE, CE and SW are 0.1, 0.141 and 0.132, respectively, with standard deviations 0.042, 0.055 and 0.048. Regions NE, CE and SW were chosen because the wind direction in the Metropolitan Area of Mexico City is mainly from region NE to SW. Many of the ozone precursors are produced in regions NE and CE. In general, region SW is the one with frequently high levels of ozone.

We perform two analysis in separate. In the first analysis we consider jointly the data from regions NE and CE. In a second instance we consider the joint information provided by regions CE and SW. The data actually used in the analysis were produced by the weekly averaged ozone measurements for regions NE, CE and SW. Hence, for instance, if $X_1(t)$, $t = 1, 2, \ldots, T$, indicate the daily maximum measurements of ozone in region NE, then $Z_1(t)$ is the mean measurement in week t in region NE, i.e., $Z_1(1) = (1/7)\sum_{i=1}^{7} X_1(i)$, $Z_1(2) = (1/7)\sum_{i=8}^{14} X_1(i)$, and so on.

Inference was performed using a sample drawn from the joint posterior distributions after a burn-in period of size 5000. After the burn-in period, we selected 1000 Gibbs samples by keeping every 10th sample in order to have approximately uncorrelated values.

In Figure 1, we have the plots of the weekly ozone averages versus weeks for regions NE, CE and SW.

Observing Figure 1 it is possible to note a decreasing trend for the weekly ozone averages for these three regions of Mexico City, especially after the 400th week (close to the year 1998).

In Figure 2, we have the plots of the log-returns $Y_j(t) = \log(Z_j(t)/Z_j(t-1))$, $t = 1, 2, \ldots, N$, centred on their averages for regions NE, CE and SW. (In here we have $N = 834$.)

5.2. Application of the Models and Results

We are going to analyse each model and pair of data separately.

5.2.1. Model I

The hyperparameters of the prior distributions are given as follows. When either the pair NE and CE or the pair CE and SW is considered, we have that the hyperparameters of the prior distributions of ϕ_{ii} and $\sigma_{\eta i}$ are $a_{ii} = b_{ii} = 1$, $c_i = d_i = 1$, $i = 1, 2$. When the pair NE and CE is considered the hyperparameters of the prior distribution of μ_i are $e_i = 0$, $f_i = 100$, $i = 1, 2$. However, when the pair CE and SW is considered then we have $e_i = 0$, $f_i = 1$, $i = 1, 2$. In Table 1 we have the summary of the quantities estimated when the pair NE-CE is considered as well as when regions CE and SW are taken into account. (Note that in Table 1, and whenever suitable, we report the estimates of $\tau_{\eta i} = 1/\sigma_{\eta i}^2$ and the estimate of $\tau_\rho = 1/\sigma_\rho^2$ instead of reporting the estimates of $\sigma_{\eta i}^2$ and σ_ρ^2, $i = 1, 2$.)

In Figure 3 we have the plots of the estimated square roots of the volatility when regions NE and CE (top two plots) and when regions CE and SW (bottom two plots) are taken into account and when Model I is used.

5.2.2. Model II

In the case of Model II, the hyperparameters of the prior distributions are given as follows. In both cases, i.e., when regions NE and CE are considered and when the pair CE-SW is taken into account, we have that the hyperparameters of the prior distributions of ϕ_{ii} and μ_i are $a_{ii} = b_{ii} = 1$ and $e_i = -3$, $f_i = 100$, $i = 1, 2$. In the case of the parameters $\sigma_{\eta i}$, $i = 1, 2$, we have that if we consider regions NE and CE then $c_1 = 5$, $c_2 = 4$ and $d_1 = d_2 = 1$, $i = 1, 2$. When data from regions CE and SW are used, we have that $c_1 = 3$,

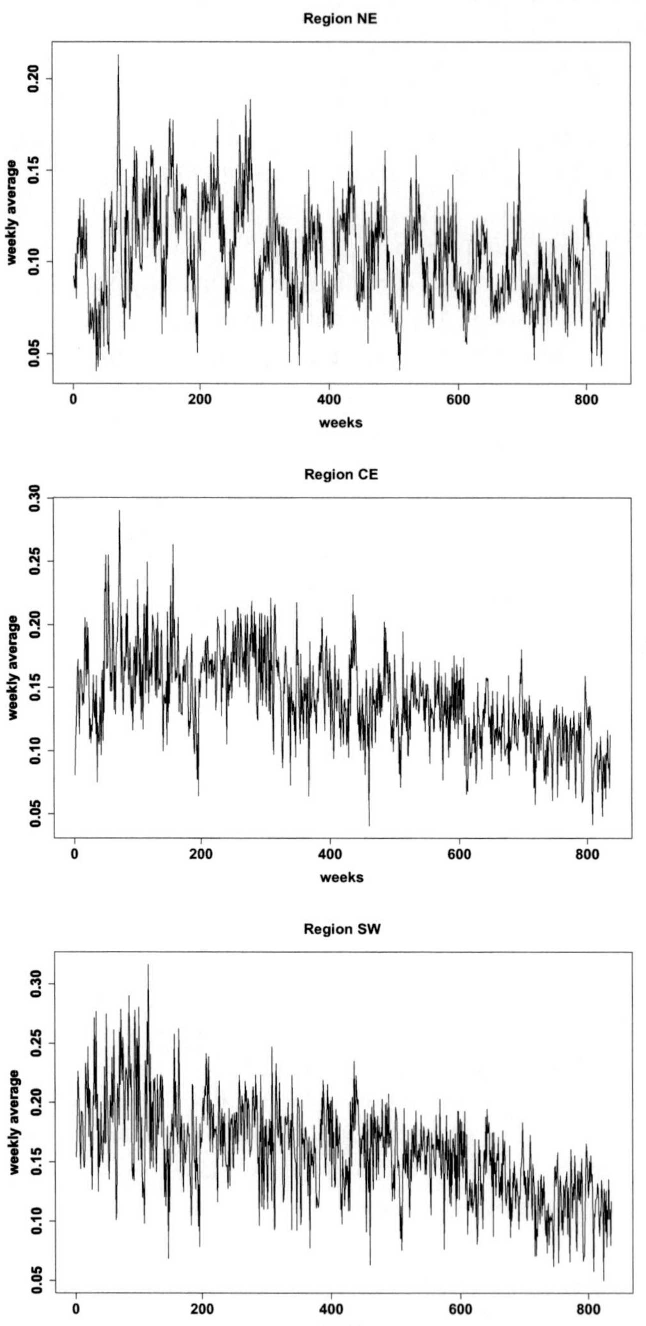

Figure 1. Weekly ozone averages for the regions NE, CE and SW from 01 January 1990 to 31 December 2005.

$c_2 = 4$ and $d_1 = d_2 = 1$. We would like to call attention to the fact that the change in the parameters is because we are using information provided by the results given by Model I. Hence, we are using an empirical Bayes approach to analyse the problem (see Carlin and Louis, 2000). In Table 2 we present the summary of the quantities estimated.

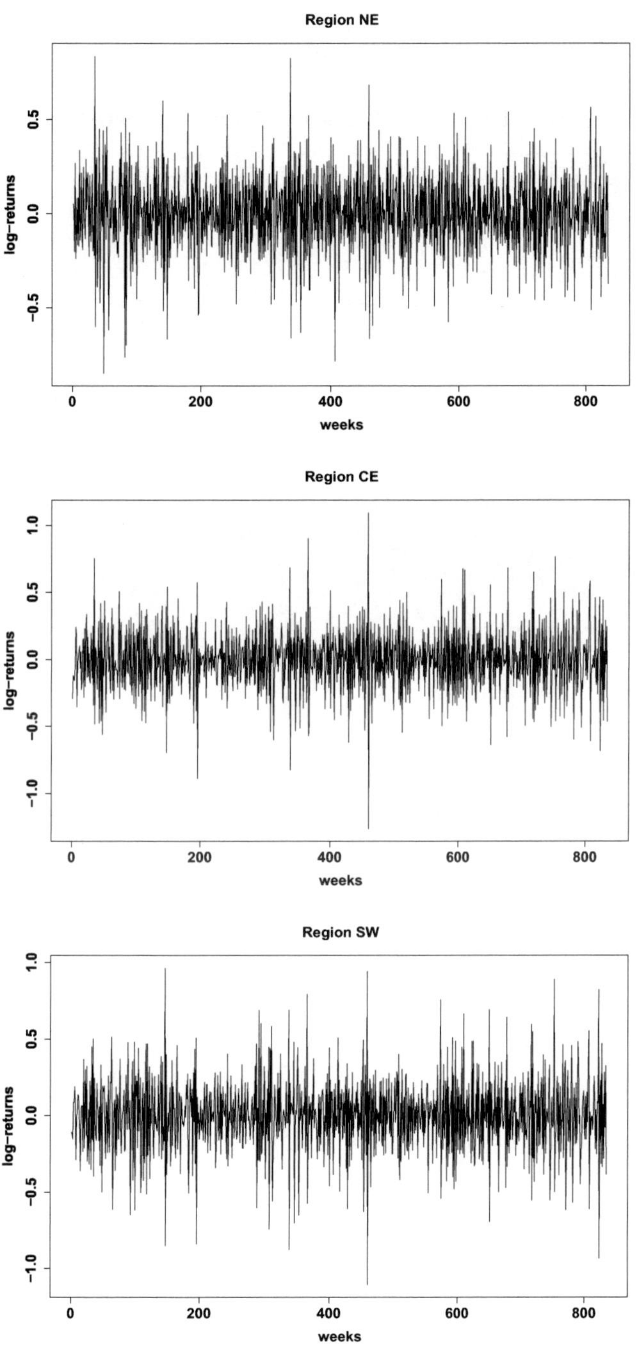

Figure 2. Log-returns centred in their averages for regions NE, CE and SW from 01 January 1990 to 31 December 2005.

Figure 4 shows similar plots to those of Figure 3, but now we use Model II instead of Model I.

Table 1. Posterior estimated mean, standard deviation (SD) and 95% Credible Interval for the quantities of interest when Model I is used and the pairs of regions NE and CE, and CE and SW are considered.

Regions	Parameters	Mean	SD	95% Credible Interval
NE – CE	$\tau_{\eta 1}$	4.827	1.387	(2.85; 8.128)
	$\tau_{\eta 2}$	3.957	1.088	(2.3930; 6.48)
	μ_1	-3.07	0.0633	(-3.19; -2.951)
	μ_2	-2.986	0.0763	(-3.13; -2.838)
	ϕ_{11}	0.4945	0.1491	(0.184; 0.7526)
	ϕ_{22}	0.6334	0.0855	(0.4595; 0.7791)
CE – SW	$\tau_{\eta 1}$	3.291	0.9023	(1.929; 5.214)
	$\tau_{\eta 2}$	4.058	1.185	(2.26; 6.954)
	μ_1	-2.936	0.07911	(-3.091; -2.781)
	μ_2	-2.969	0.07879	(-3.124; -2.82)
	ϕ_{11}	0.6277	0.0865	(0.4359; 0.7712)
	ϕ_{22}	0.6327	0.1012	(0.4128; 0.8007)

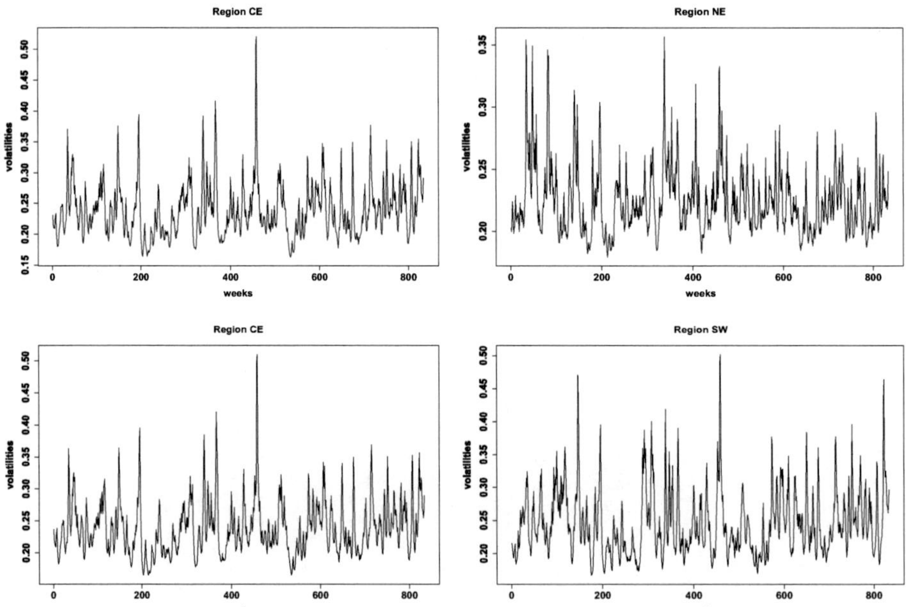

Figure 3. Square roots of the volatility for regions NE, CE and SW when Model I is used and when regions are paired in regions NE and CE (top row) and regions CE and SW (bottom row).

5.2.3. Model III

Consider now Model III. The hyperparameters of the prior distributions of ϕ_{ii}, μ_i, $i = 1, 2$ and ρ_ϵ are the same as taken in Model II for both pair of regions NE and CE and CE and SW. The hyperparameters of the prior distributions of $\sigma^2_{\eta 1}$ and $\sigma^2_{\eta 2}$ are $c_1 = 9$, $c_2 = 7$ and $d_1 = d_2 = 1$ when regions NE and CE are considered and are $c_1 = 8$, $c_2 = 10$ and

Table 2. Posterior estimated mean, standard deviation (SD) and 95% Credible Interval for the quantities of interest when Model II is used and the pairs of regions NE and CE, and CE and SW are considered.

Regions	Parameters	Mean	SD	95% Credible Interval
NE – CE	$\tau_{\eta 1}$	9.426	2.389	(5.3840; 14.65)
	$\tau_{\eta 2}$	7.436	1.86	(4.6590; 11.68)
	μ_1	-3.022	0.07	(-3.163; -2.888)
	μ_2	-2.916	0.0744	(-3.074; -2.772)
	ϕ_{11}	0.7307	0.0818	(0.535; 0.858)
	ϕ_{22}	0.6985	0.0807	(0.5263; 0.8335)
	ρ_ϵ	0.7546	0.0165	(0.7209; 0.7849)
CE – SW	$\tau_{\eta 1}$	7.96	1.49	(5.247; 10.99)
	$\tau_{\eta 2}$	9.628	2.116	(6.425; 14.38)
	μ_1	-2.853	0.06426	(-2.981; -2.731)
	μ_2	-2.892	0.0649	(-3.023; -2.766)
	ϕ_{11}	0.6924	0.07087	(0.5383; 0.8049)
	ϕ_{22}	0.7207	0.07001	(0.5629; 0.8338)
	ρ_ϵ	0.8825	0.008724	(0.865; 0.8989)

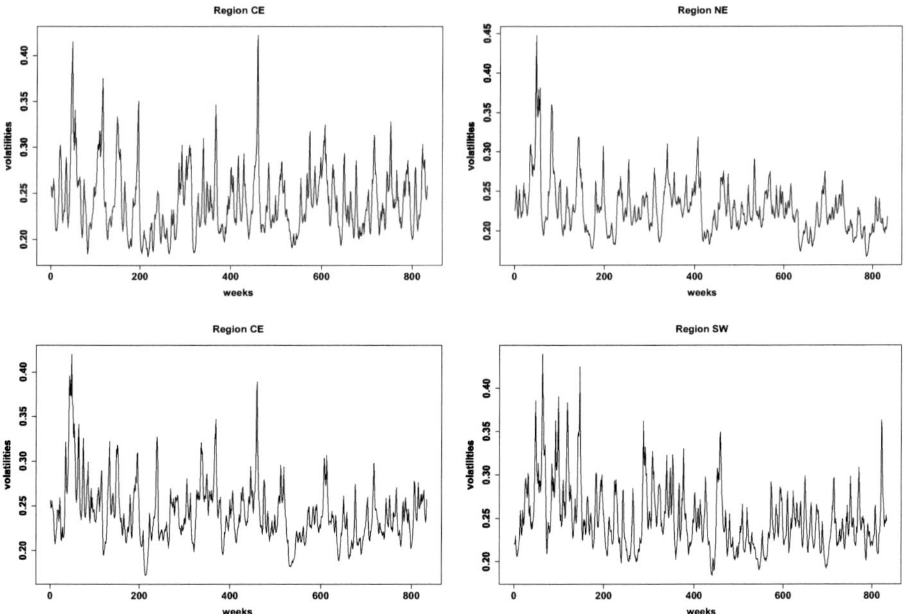

Figure 4. Square roots of the volatility for regions NE, CE and SW when Model II is used and when regions are paired in regions NE and CE (top row) and regions CE and SW (bottom row).

$d_1 = d_2 = 1$ when regions CE and SW are taken into account. The Beta prior distribution of quantity ϕ_{21} has hyperparameters $a_{21} = b_{21} = 1$ when either regions NE and CE or regions CE and SW are considered. Table 3 gives the summary of the estimated quantities.

In Figure 5, we have the plots of the estimated square roots of the volatility when Model

Table 3. Posterior estimated mean, standard deviation (SD) and 95% Credible Interval for the quantities of interest when Model III is used and the pairs of regions NE and CE, and CE and SW are considered.

Regions	Parameters	Mean	SD	95% Credible Interval
NE – CE	$\tau_{\eta 1}$	12.97	2.875	(8.1970; 19.5)
	$\tau_{\eta 2}$	9.233	2.308	(5.645; 14.32)
	μ_1	-3.04	0.0826	(-3.205; -2.882)
	μ_2	-2.917	0.0734	(-3.061; -2.777)
	ϕ_{11}	0.8346	0.0405	(0.7379; 0.9028)
	ϕ_{22}	0.4009	0.1997	(0.0408; 0.7466)
	ϕ_{21}	0.4851	0.2302	(0.1391; 0.9634)
	ρ_ϵ	0.7473	0.0177	(0.7131; 0.781)
CE – SW	$\tau_{\eta 1}$	13.83	2.755	(9.063; 19.34)
	$\tau_{\eta 2}$	13.39	2.694	(8.944; 19.41)
	μ_1	-2.838	0.09852	(-3.038; -2.639)
	μ_2	-2.903	0.08818	(-3.077; -2.733)
	ϕ_{11}	0.8848	0.02978	(0.823; 0.9338)
	ϕ_{22}	0.3309	0.1318	(0.06118; 0.5704)
	ϕ_{21}	0.5628	0.1278	(0.3077; 0.827)
	ρ_ϵ	0.8808	0.008337	(0.8627; 0.78962)

III is considered. The top two plots correspond to the case where regions NE and CE are jointly considered and the two plots at the bottom correspond to the case where regions CE and SW are used.

5.2.4. Model IV

When Model IV is used the hyperparameters of the prior distributions of $\sigma^2_{\eta i}$, $i = 1, 2$ are $c_1 = 13$, $c_2 = 9$ and $d_1 = d_2 = 1$ when regions NE and CE are considered and are $c_1 = 14$, $c_2 = 13$ and $d_1 = d_2 = 1$ when data from regions CE and SW are used. The hyperparameters of the prior distributions of the quantities ϕ_{ii}, $i = 1, 2$ are as in Model II and III in both sets of data, i.e., the one produced by regions NE and CE and the one produced by regions CE and SW. We also have that in both cases the hyperparameters of the prior distributions of μ_i are $e_i = -3$ and $d_i = 10$, $i = 1, 2$. The parameter ψ_0 has a Normal $\mathrm{N}(0, 10)$ prior distribution and ψ_1 has a $\mathrm{Beta}(20, 1)$ prior distribution and the hyperparameters of the prior distribution of σ^2_ρ is $c_3 = d_3 = 1$ when either regions NE and CE or regions CE and SW are taken into account. In Table 4, the estimates of the quantities of interest is presented.

Figure 6 shows similar plots to the ones given by Figures 3, 4 and 5, but now we have used Model IV.

5.2.5. Model V

In the case of Model V, i.e., when the Granger causality is present in $h_2(t)$, $t = 1, 2 \ldots, N$, we have that the hyperparameters of the prior distributions of the quantities ϕ_{ii}, μ_i, $i = 1, 2$ and σ^2_ρ are the same as in Model IV and the hyperparameters of the prior distribution of

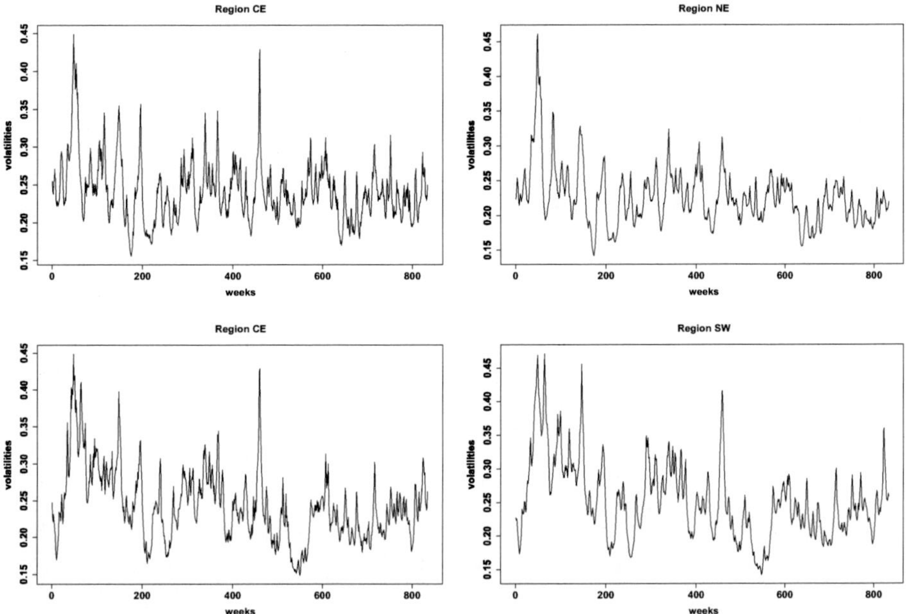

Figure 5. Square roots of the volatility for regions NE, CE and SW when Model III is used and when regions are paired in regions NE and CE (top row) and regions CE and SW (bottom row).

Table 4. Posterior estimated mean, standard deviation (SD) and 95% Credible Interval for the quantities of interest when Model IV is used and the pairs of regions NE and CE, and CE and SW are considered.

Regions	Parameters	Mean	SD	95% Credible Interval
NE – CE	τ_ρ	4.126	1.196	(2.3870; 7.19)
	$\tau_{\eta 1}$	17.05	3.605	(10.46; 24.39)
	$\tau_{\eta 2}$	11.61	2.891	(6.645; 18.94)
	μ_1	-3.017	0.0591	(-3.132; -2.898)
	μ_2	-2.891	0.0662	(-3.017; -2.763)
	ϕ_{11}	0.7238	0.1254	(0.3807; 0.8692)
	ϕ_{22}	0.7188	0.0782	(0.5527; 0.85)
	ψ_1	0.8492	0.0424	(0.7584; 0.9197)
	ψ_0	2.1280	0.1481	(1.838; 2.427)
CE – SW	τ_ρ	6.787	2.047	(3.161; 11.05)
	$\tau_{\eta 1}$	15.41	2.974	(10.66; 22.13)
	$\tau_{\eta 2}$	16.43	3.486	(10.66; 24.15)
	μ_1	-2.792	0.04915	(-2.888; -2.693)
	μ_2	-2.844	0.05712	(-2.961; -2.74)
	ϕ_{11}	0.5875	0.1198	(0.3019; 0.7707)
	ϕ_{22}	0.5518	0.1328	(0.25; 0.772)
	ψ_1	0.9281	0.02514	(0.8721; 0.9684)
	ψ_0	2.989	0.2412	(2.488; 3.423)

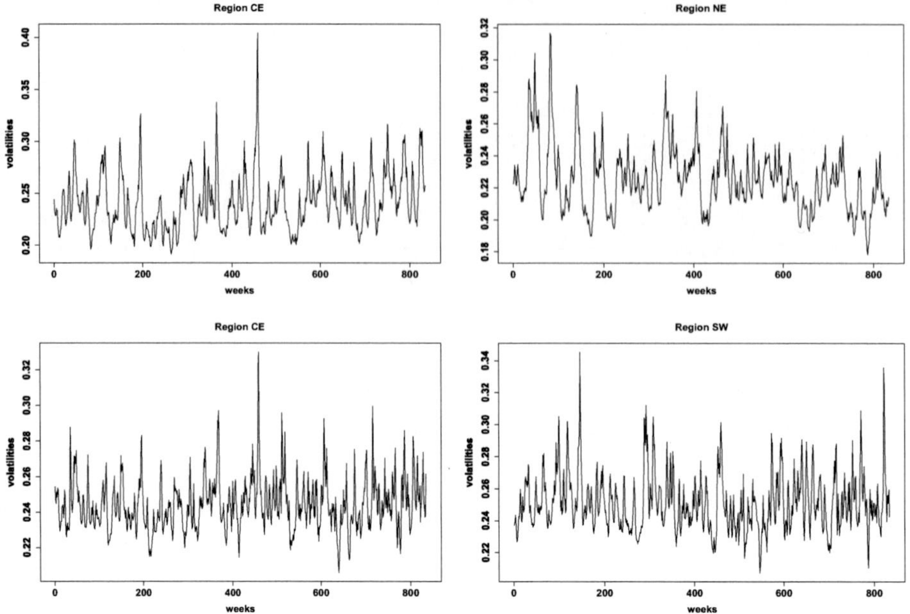

Figure 6. Square roots of the volatility for regions NE, CE and SW when Model IV is used and when regions are paired in regions NE and CE (top row) and regions CE and SW (bottom row).

ϕ_{21} are the same as in Model III for both sets of data, i.e., the one produced by regions NE and CE and the one produced by regions CE and SW. The hyperparameters of the prior distributions of $\sigma^2_{\eta 1}$ and $\sigma^2_{\eta 2}$ are $c_1 = 17$, $c_2 = 12$ and $d_1 = d_2 = 1$ in the case of regions NE and CE, and are $c_1 = 15$, $c_2 = 16$ and $d_1 = d_2 = 1$ in the case of regions CE and SW, ψ_0 has a Gamma prior distribution whose hyperparameters are (2,1) and (3,1) when regions NE and CE and regions CE and SW are considered, respectively. The quantity ψ_1 also has a Gamma prior distribution with hyperparameters (1,1) when either regions NE and CE or regions CE and SW are considered. (In here we are considering a Gamma(α, β) distribution with mean α/β and variance α/β^2.) Table 5 gives the summary of the quantities of interest.

The plots of the estimated volatility produced by Model V and using both data set (i.e., the one considering measurements from regions NE and CE and the one considering those from regions CE and SW) are given in Figure 7.

5.2.6. Model VI

When Model VI is taken into account, we have that the quantities ϕ_{ii}, μ_i, $i = 1, 2$ and ρ_ϵ have prior distributions with the same hyperparameters as those in Model IV in both cases, i.e., when regions NE and CE are considered and when region CE and SW are taken into account. The hyperparameters of the prior distributions of $\sigma^2_{\eta 1}$ and $\sigma^2_{\eta 2}$ are $c_1 = 20$, $c_2 = 13$ and $d_1 = d_2 = 1$ when regions NE and CE are considered and are $c_1 = 17$, $c_2 = 19$ and $d_1 = d_2 = 1$ when we use data from regions CE and SW. The hyperparameters of the prior distribution of quantity λ_w are $f_\lambda = g_\lambda = 1$ when regions NE and CE are considered and is $f_\lambda = 1$ and $g_\lambda = 2$ when the data from regions CE and SW is used. In Table 6 we present

Table 5. Posterior estimated mean, standard deviation (SD) and 95% Credible Interval for the quantities of interest when Model V is used and the pairs of regions NE and CE, and CE and SW are considered.

Regions	Parameters	Mean	SD	95% Credible Interval
NE – CE	τ_ρ	3.026	1.018	(1.592; 5.96)
	$\tau_{\eta 1}$	19.99	4.338	(12.26; 30.15)
	$\tau_{\eta 2}$	13.1	2.618	(8.77; 19.48)
	μ_1	-3.041	0.0722	(-3.177; -2.902)
	μ_2	-2.899	0.0654	(-3.032; -2.773)
	ϕ_{11}	0.8006	0.1086	(0.44; 0.906)
	ϕ_{22}	0.6795	0.1082	(0.4001; 0.8329)
	ϕ_{21}	0.2122	0.1471	(0.023; 0.6022)
	ψ_1	0.7811	0.0605	(0.6516; 0.8925)
	ψ_0	2.107	0.127	(1.861; 2.3640)
CE – SW	τ_ρ	6.241	2.303	(2.323; 11.18)
	$\tau_{\eta 1}$	16.72	3.248	(10.78; 24.01)
	$\tau_{\eta 2}$	18.96	3.744	(12.710; 27.17)
	μ_1	-2.835	0.07064	(-2.974; -2.693)
	μ_2	-2.883	0.06774	(-3.015; -2.738)
	ϕ_{11}	0.8215	0.0503	(0.7096; 0.9102)
	ϕ_{22}	0.4195	0.1152	(0.1955; 0.6503)
	ϕ_{21}	0.434	0.1126	(0.2492; 0.6855)
	ψ_1	0.9124	0.03554	(0.8235; 0.9644)
	ψ_0	2.953	0.205	(2.541; 3.342)

the summary of the quantities of interest.

Table 6. Posterior estimated mean, standard deviation (SD) and 95% Credible Interval for the quantities of interest when Model VI is used and the pairs of regions NE and CE, and CE and SW are considered.

Regions	Parameters	Mean	SD	95% Credible Interval
NE – CE	λ_w	6.028	1.329	(3.918; 9.111)
	$\tau_{\eta 1}$	23.13	4.632	(14.53; 32.98)
	$\tau_{\eta 2}$	13.71	2.954	(8.616; 19.93)
	μ_1	-3.155	0.0726	(-3.283; -3.012)
	μ_2	-3.028	0.0688	(-3.167; -2.895)
	ϕ_{11}	0.7818	0.0848	(0.5817; 0.9136)
	ϕ_{22}	0.7097	0.1062	(0.3987; 0.8477)
	ρ_ϵ	0.7482	0.0169	(0.7143; 0.7816)
CE – SW	λ_w	4.385	0.8142	(3.104; 6.257
	$\tau_{\eta 1}$	19.41	3.721	(13.08; 27.61)
	$\tau_{\eta 2}$	21.45	4.869	(13.54; 33.2)
	μ_1	-2.993	0.07343	(-3.13; -2.855)
	μ_2	-3.035	0.06504	(-3.166; -2.912)
	ϕ_{11}	0.6176	0.1358	(0.2897; 0.8083)
	ϕ_{22}	0.5509	0.2054	(0.06438; 0.8096)
	ρ_ϵ	0.878	0.009271	(0.8582; 0.895)

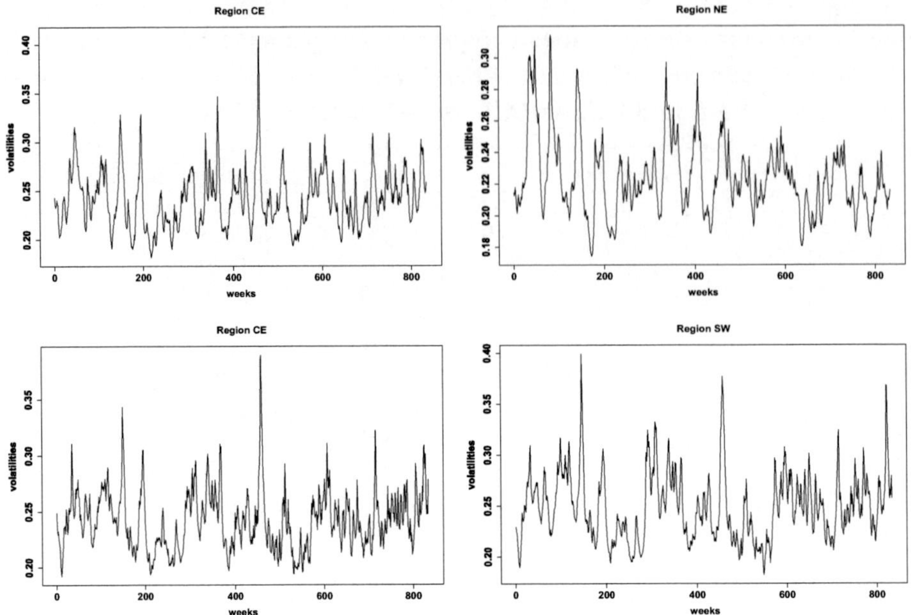

Figure 7. Square roots of the volatility for regions NE, CE and SW when Model V is used and when regions are paired in regions NE and CE (top row) and regions CE and SW (bottom row).

In Figure 8, we have similar plots to those in Figures 3, 4, 5, 6 and 7, but now using Model VI.

5.2.7. Model Selection

The selection of the best model to fit the data was performed using the Deviance Information Criterion. Table 7 presents the estimated values of DIC for each model each set of data.

Table 7. Deviance Information Criterion for Models I, II, III, IV, V and VI when the pairs of regions NE and CE, and CE and SW are considered.

	Deviance Information Criterion					
Regions	Model I	Model II	Model III	Model IV	Model V	Model VI
NE – CE	- 128.708	- 745.865	- 753.381	- 877.872	- 882.2	- 740.968
CE – SW	4.382	- 1090.84	- 1118.85	- 1251.66	- 1268.99	- 1080.25

6. Conclusion

Observe that Model V (a stochastic volatility model with dynamic correlation and the presence of the Granger causality for $h_2(t)$, $t = 2, \ldots, N$ is the best model to explain the behaviour of the ozone pollution data of Mexico City when regions NE and CE are jointly

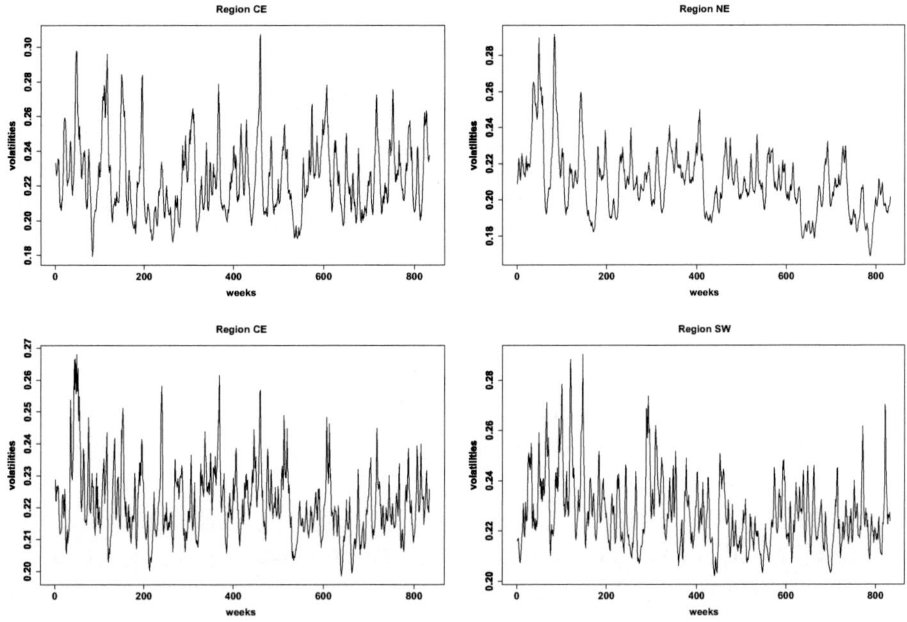

Figure 8. Square roots of the volatility for regions NE, CE and SW when Model VI is used and when regions are paired in regions NE and CE (top row) and regions CE and SW (bottom row).

considered, since it produces the smallest value of DIC given by - 882.200 and it is also the model producing the best fit when regions CE and SW are jointly considered having the smallest DIC value of -1268.99.

It is important to observe from Figures 3, 4, 5, 6, 7 and 8, that for both regions NE and CE assuming all six models, there is smaller and more stable volatility for the ozone weekly measurements after the 400th week (close to the year 1998). Similar conclusion can be reached when jointly considering regions CE and SW.

Note that even though the volatility decreases in all cases, we have a more stable plot, with rare exceptions, after week 600, specially when considering Model V. This week corresponds roughly to the end of year 2001 and beginning of year 2002. It is interesting to call attention to the fact that in the year 2000 we have the culmination of a series of measures that had been taken by the environmental authorities of Mexico since 1990 aiming to reduce the levels of ozone in big cities in the country and in particular in Mexico City. The series of measures may be roughly described as follows. In 1990, cars circulating in the Metropolitan Area of Mexico City would have to undergo periodic inspection of their mechanical condition. Additionally, restrictions were imposed on the circulation of car according to the ending number of their license plates. In 1997, further restrictions were implemented in terms of allowing clean cars to circulate freely. Additionally, in 1999, manufacturers were encouraged to produce cars with cleaner and modern technology. The production of such cars was made compulsory in 2001.

Therefore, we may see that some measures taken by the environmental authorities have produced a positive effect towards the aim of decreasing and stabilising the level of ozone measurements in Mexico City and in modifying its long term behaviour.

Finally, we would like to point out that besides their advantages when compared to the usual ARCH and GARCH models (mentioned earlier in this paper), stochastic volatility models also have an advantage when compared to non-homogeneous Poisson models (see Achcar *et al.*, 2007). The advantage is that besides indicating on average, the number of peaks that occur in a given period of time, stochastic volatility models also permit to analyse if the variability of the measurements stays stable. That is important since environmental authorities are interested not only in decreasing pollution levels but also in keeping the measurements stabilised in a low level.

Acknowledgements

The authors thank Guadalupe Tzintzun for providing the data from the monitoring network and also for providing information on the environmental issues. J.A.A. was partially funded by a CNPq grant number 300235/2005-4. H.C.Z. thanks to FAEPA from Hospital das Clínicas de Ribeirão Preto, FMRP-USP for financial support. E.R.R. thanks the Department of Statistics at the University of Oxford, where part of this work was developed, for all the support received during her stay at the Department. E.R.R was partially funded by DGAPA-UNAM grant number 968SFA/2007.

References

Achcar, J. A.; Fernández-Bremauntz, A. A.; Rodrigues, E. R.; Tzintzun, G. (2007). Estimating the number of ozone peaks in Mexico City using a non-homogeneous Poisson model. *Environmetrics* (http://www.interscience.wiley. com/10.1002/env.890).

Álvarez, L. J.; Fernández-Bremauntz, A. A.; Rodrigues, E. R.; Tzintzun, G. (2005). Maximum a posteriori estimation of the daily ozone peaks in Mexico City. *Journal of Agricultural, Biological, and Environmental Statistics*, **10**, 276–290.

Austin, J.; Tran, H. (1999). A characterization of the weekday-weekend behavior of ambient ozone concentrations in California. In: *Air Pollution* **VII**, 645–661. UK: WIT Press.

Bell, M. L.; McDermontt, A.; Zeger, S. L.; Samet, J. M.; Dominici, F. (2004). Ozone and short-term mortality in 95 US urban communities, 1987-2000. *Journal of the American Medical Society*, **292**, 2372–2378.

Bell, M. L.; Peng, R.; Dominici, F. (2005). The exposure-response curve for ozone and risk of mortality and the adequacy of current ozone regulations. *Environmental Health Perspectives*, **114**, 532-536.

Bell, M. L.; Goldberg, R.; Rogrefe, C.; Kinney, P. L.; Knowlton, K.; Lynn, B.; Rosenthal, J.; Rosenzweig, C.; Patz, J. A. (2007). Climate change, ambient ozone, and health in 50 US cities. *Climate Change*, **82**, 61-76.

Bernardo, J. M.; Smith, A. F. M. (1995). *Bayesian Theory*. New York: John Wiley and Sons.

Bollerslev, T. (1986). Generalized autoregressive Conditional heterocedasticity. *Journal of Econometrics*, **31**, 307–327.

Carlin, B. P.; Louis, T. A. (2000). *Bayes and Empirical Bayes methods for data analysis*. Second Edition. Boca Raton: Chapman and Hall Press.

Danielsson, J. (1994). Stochastic volatility in asset prices: estimation with simulated maximum likelihood. *Journal of Econometrics*, **64**, 375–400.

Engle, R. F. (1982). Autoregressive conditional heteroscedasticity with estimates of the variance of United Kingdon inflation. *Econometrica*, **50**, 987–1007.

Gelfand, A. E.; Smith, A. F. M. (1990). Sampling-based approaches to calculating marginal densities. *Journal of the American Statistical Association*, **85**, 398–409.

Ghysels, E.; Harvey, A. C.; Renault, E. (1996). A stochastic volatility. In: Rao, C.R.; Maddala, G.S., editors. *Statistical Models in Finance*. Amsterdam: North-Holland.

Guardani, R.; Nascimento, C. A. O.; Guardani, M. L. G.; Martins, M. H. R. B.; Romano, J. (1999). Study of atmospheric ozone formation by means of a neural network based model. *J. Air and Waste Management Assoc.*, **49**, 316–323.

Guardani, R.; Aguiar, J. L.; Nascimento, C. A. O.; Lacava, C. I. V.; Yanagi, Y. (2003). Ground-level ozone mapping in large urban areas using multivariate analysis: application to the São Paulo Metropolitan Area. *J. Air and Waste Management Assoc.*, **53**, 553–559.

Horowitz, J. (1980). Extreme values from a nonstationary stochastic process: an application to air quality analysis. *Technometrics*, **22**, 469–482.

Javits, J. S. (1980). Statistical interdependencies in the ozone national ambiente air quality standard. *J. Air Poll. Control Assoc.*, **30**, 58–59.

Kim, S.; Shepard, N.; Chib, S. (1998). Stochastic volatility: likelihood inference and comparison with ARCH models. *Review of Economic Studies*, **65**, 361–393.

Lanfredi, M.; Macchiato, M. (1997). Searching for low dimensionality in air pollution time series. *Europhysics Lett.*, **40**, 589-594.

Larsen, L. C.; Bradley, R. A.; Honcoop, G. L. (1990). A new method of characterizing the variability of air quality-related indicators. In: *Air and Waste Management Association's International Specialty Conference of Tropospheric Ozone and the Environment*. California. USA.

Leadbetter, M. R. (1991). On a basis for "peak over threshold" modeling. *Statistics and Probability Letters*, **12**, 357–362.

Loomis, D. P.; Borja-Arbuto, V. H.; Bangdiwala, S. I.; Shy, C. M. (1996). Ozone exposure and daily mortality in Mexico City: a time series analysis. *Health Effects Institute Research Report*, **75**, 1–46.

Meyer, R.; Yu, J. (2000). BUGS for a Bayesian analysis of stochastic volatility models. *Econometrics Journal*, **3,** 198–215.

Pan, J.-N.; Chen, S.-T. (2007). Monitoring long-memory air quality data using ARFIMA model. *Environmetrics* (http://www.interscience.wiley.com/10.1002 /env.882).

Raftery, A. E. (1989). Are ozone exceedance rate decreasing?, Comment of the paper "*Extreme value analysis of environmental time series: an application to trend detection in ground-level ozone*" by R. L. Smith. *Statistical Sciences*, **4,** 378–381.

Roberts, E. M. (1979a). Review of statistics extreme values with applications to air quality data. Part I. Review. *Journal of the Air Pollution Control Association*, **29**, 632–637.

Roberts, E. M. (1979b). Review of statistics extreme values with applications to air quality data. Part II. Applications. *Journal of the Air Pollution Control Association*, **29**, 733–740.

Smith, R. L. (1989). Extreme value analysis of environmental time series: an application to trend detection in ground-level ozone. *Statistical Sciences*, **4**, 367–393.

Smith, A. F. M.; Roberts, G. O. (1993). Bayesian Computation via the Gibbs Samples and related Markov Chain Monte Carlo Methods (with discussion). *Journal of the Royal Statistical Society Series B*, **55**, 3–23.

Spiegelhalter, D. J.; Best, N. G.; Carlin, B. P.; Van der Linde, A. (2002). Bayesian measures of model complexity and fit. *Journal of the Royal Statistical Society Series B*, **64**,583–639.

Spiegelhalter, D. J.; Thomas, A.; Best, N. G.; Lund, D. (2003). *Winbugs user manual.* Cambridge. United Kingdon: MRC Biostatistics Unit.

Wilson, R.; Colone, S. D.; Spengler, J. D.; Wilson, D. G. (1980). *Health effects in fossil fuel burning: assessment and mitigation.* Cambridge, USA:Ballenger.

Yu, J., (2002). Forecasting volatility in the New Zeland stock market. *Applied Financial Economics*, **12**,193–202.

Yu, J.; Meyer, R. (2005). Multivariate Stochastic Volatility models: Bayesian estimation and model comparison. To appear in *Econometrics Review.*

Zolghadri, A.; Henry, D. (2004). Minmax statistical models for air pollution time series. Application to ozone time series data measured in Bordeaux. *Environmental Monitoring and Assessment*, **98**, 275-294.

In: Ozone Depletion, Chemistry and Impacts
Editor: Sem H. Bakker, pp. 187-194
ISBN: 978-1-60692-007-7

Chapter 9

RATE CONSTANTS OF THE GAS-PHASE REACTION OF OZONE WITH ORGANOSULFIDES AT ROOM TEMPERATURE

Maofa Ge, Lin Du and Kun Wang
Beijing National Laboratory for Molecular Sciences (BNLMS),
Institute of Chemistry, Chinese Academy of Sciences,
Beijing 100080, China

Abstract

The atmospheric sulfur cycle has been the subject of intensive investigation for several decades because of the need to assess the contribution of anthropogenically produced sulfur to such problems as acid rain, visibility reduction, and climate modification. The atmospheric chemistry of sulfur-containing compounds is directly relevant to the formation of sulfur aerosol in marine air. Reduced organic sulfur compounds have been estimated to account for approximately 25% of the total global gaseous sulfur budget. Besides the predominant CH_3SCH_3 (dimethyl sulfide, DMS), other reduced sulfur compounds should also be estimated, such as $C_2H_5SCH_3$ (ethylmethyl sulfide, EMS), n-$C_3H_7SCH_3$ (n-propylmethyl sulfide, PMS), and $C_2H_5SC_2H_5$ (diethyl sulfide, DES), of which there are only a few kinetics investigations. Based on our previous work of DMS and DES, we have measured the rate constants of the gas-phase reactions of ozone with EMS and PMS at room temperature in our self-made smog chamber. Experiments were conducted under supposedly pseudo-first-order decay conditions, keeping $[\text{sulfide}]_0 > 50[O_3]_0$, but having different combinations of $[\text{sulfide}]_0$ and $[O_3]_0$. Cyclohexane was added into the reactor to eliminate the effect of OH radicals. The rate constants of the gas-phase reactions of ozone with EMS and PMS were determined to be $(1.12\pm0.18)\times10^{-19}$ and $(1.24\pm0.15)\times10^{-19}$ cm^3 $molecule^{-1}$ s^{-1}, respectively. Our results will enrich the kinetics data of atmospheric chemistry, and provide some useful information for evaluating the loss processes of reduced organic sulfur compounds.

Introduction

Sulfur compounds play an important role in both the tropospheric and stratospheric budget of atmospheric gases [1,2]. The atmospheric chemistry of sulfur-containing compounds is directly relevant to the formation of sulfur aerosol in marine air [3,4]. Very recently, Barnes et al. [5] reviewed dimethyl sulfide (DMS), dimethyl sulfoxide (DMSO) and their oxidation in the atmosphere, which emphasized the importance of sulfur-containing compounds. Reduced organic sulfur compounds have been estimated to account for approximately 25% of the total global gaseous sulfur budget [3]. The species play an important role in the atmospheric sulfur cycle and it is important to understand their fate in the troposphere. The reduced organic sulfur compounds are mainly emitted from the sea surface, although anthropogenic emissions might also be of importance [6,7].

Among the reduced sulfur compounds released into the atmosphere, natural emissions of COS, H_2S and DMS play a major role, but other species including ethylmethyl sulfide ($C_2H_5SCH_3$, EMS), *n*-propylmethyl sulfide (*n*-$C_3H_7SCH_3$, PMS), should also be considered. EMS was present in the samples of coffee aroma isolated from dry ground coffee as detected with direct mass spectrometric analysis [8]. PMS was found in the volatile organic compounds from garden waste with GC/MS system [9]. The homogeneous oxidation of these trace gases into sulfur dioxide is expected to involve ozone as oxidant as well as other reactive molecules and radicals present in the atmosphere. Very little, however, is known about the kinetics of these oxidation processes. As for the atmospheric chemistry of EMS, its reactions with OH [10], OD [11], NO_3 [12], and Cl [13] have been reported. No other available kinetics data of EMS can be found in the literature. As for that of PMS, there is totally no kinetics work reported. These rate constants are badly needed for evaluating the atmospheric effects of EMS and PMS [14].

In the present study, we have determined the rate constants of ozone reactions with EMS and PMS under nearly real atmospheric conditions in our self-made apparatus of smog chamber. The atmospheric applications of these constants are also discussed.

Experimental

Absolute rate constants for the ozone reactions with EMS and PMS were determined at room temperature by monitoring the O_3 decay rates in the presence of known concentrations of the reactant sulfide. The ozone concentrations were monitored as a function of time. Under these conditions, the reactions removing O_3 are:

$$O_3 + \text{wall} \rightarrow \text{loss of } O_3 \quad (1)$$

$$O_3 + \text{sulfide} \rightarrow \text{products} \quad (2)$$

and hence,

$$-d[O_3]/dt = (k_1 + k_2[\text{sulfide}])[O_3] \quad \text{I}$$

The reactant concentrations were always greatly in excess of the initial O_3 concentrations (i.e., $[\text{sulfide}]/[O_3]_0 > 50$) and equation I may then be rearranged to yield

$$-d\ln[O_3]/dt = k_1 + k_2[\text{sulfide}] \quad \text{II}$$

From the ozone decay rates, *-dln[O₃]/dt*, measured at various sulfide concentrations and with a knowledge of the background O_3 decay rate, k_1, the rate constants k_2 were obtained.

Experiments were carried out in a 70 L FEP Teflon reaction chamber. At the two ends of the reactor, there is an inlet and an outlet made of Teflon, respectively, which are used for introducing reactants and sampling. The reactor and the analytical instruments are linked via Teflon tubes. Ozone and a known volume of the liquid sulfide were injected into the chamber with injectors. The concentrations of sulfides and cyclohexane in the entire chamber were calculated from the amount of organics introduced and the total volume of the reactor. It was observed that for both gaseous and liquid organics, these calculated concentrations agreed to within better than ±10% with the concentrations quantitatively measured by gas chromatography [15].

For a typical experiment, the experimental procedure is described as follows. At first, the reactor was purged with N_2 (99.999%) until there was no detectable ozone and VOCs. Residual gas was pumped out, and a 70 L volume of N_2 was introduced into the reactor. Ozone in O_2 was introduced into the reactor by an injector. The accurate concentration of ozone was measured by the ozone analyzer thereafter. Cyclohexane was injected into the reactor, which was used to eliminate the effect of OH radicals generated in the subsequent reactions. The reactor was shaken fiercely to make it mixed fully. When it became gas completely, a certain amount of sulfide was injected into the reactor, and the reactor was shaken again. At the same time, time was recorded. During the experiment, ozone concentrations were measured every 30 min. All experiments were conducted at (301 ± 1) K and lasted for 3.5 ~4.5 h.

When O_3 reacts with sulfide, OH radicals will probably generate as a secondary radical [16]. Because OH radicals are very active, its presence will result in certain error of the rate constant for the reaction of sulfide and O_3. In order to avoid the impacts of OH radicals, high concentrations of cyclohexane were added into the reactor to react with OH radicals. Cyclohexane cannot react with O_3, so the results will be more accurate. Furthermore, during our experiments, the sulfide was in large excess of ozone, so the reactions between sulfide and secondary radicals might consume a relatively small amount of sulfide. The concentration of the sulfide changed less than 1% after four hours reaction, which was determined both at the beginning and ending of experiments by gas chromatography. Therefore, the concentration of the sulfide can still be considered as a constant.

Ozone in O_2 was produced by an ozone generator based on the silent electric discharge technique. The purity of the O_2 used was 99.995%. Ethylmethyl sulfide ($C_2H_5SCH_3$, EMS) and *n*-propylmethyl sulfide (n-$C_3H_7SCH_3$, PMS) were purchased from Alfa Aesar and used without further purification. C_6H_{12} (cyclohexane) in a purity of 99.5% was from Beijing Beihua Fine Chemicals Company. The instrument used for analyzing ozone concentrations is ozone analyzer (Model 49C, Thermo Electron Corporation). A gas chromatograph (Agilent 6820) equipped with a flame ionization detector (FID) was used for the analysis of the sulfides.

Results and Discussion

The ozone losses were measured with and without cyclohexane in our previous work [17,18], which showed that 8.50×10^{14} molecule cm^{-3} of cyclohexane is enough for scavenging OH radicals in the experiments between DMS and DES with ozone. In this work, the concentration of the cyclohexane used was 2.13×10^{15} molecule cm^{-3}, which was 2.5 times larger than that of the previous work. Therefore, we can assure the OH radicals were completely eliminated and they would have no effects on the determination of the rate constants between ozone and sulfides. In the present work, the loss rate of ozone was measured in a control experiment in the presence of 2.13×10^{15} molecule cm^{-3} of cyclohexane in the bath gas of pure N_2. The wall decay loss profile is shown in Figure 1. As clearly seen, the ozone concentration-time profile is exponential.

As for the mechanism of the reactions between ozone and sulfides, the reaction of DMS and ozone is initiated by the primary attack of ozone on the sulfide [16], which involves C-S bond scission, as evidenced by the SO_2 chemiluminescence [19,20]. The major products of the reactions are H_2CO, SO_2, H_2O and CO. OH radicals, SO radicals and H atoms are important intermediates in the DMS-O_3 system [16]. The reactions between ozone and other sulfides should be similar. The results under different initial concentrations of ozone and sulfides in the bath gas of N_2 are listed in Table 1. The initial concentrations of EMS were in a range of 4.75×10^{14} molecule cm^{-3} -14.25×10^{14} molecule cm^{-3}, and those of PMS change from 4.05×10^{14} molecule cm^{-3} to 9.71×10^{14} molecule cm^{-3}. Plots of $\ln([O_3]_0/[O_3])$ vs. time were constructed, where $[O_3]_0$ is the initial ozone concentration and $[O_3]$ is the ozone concentration at time t. The plots of the EMS-ozone reactions are shown in Figure 2.

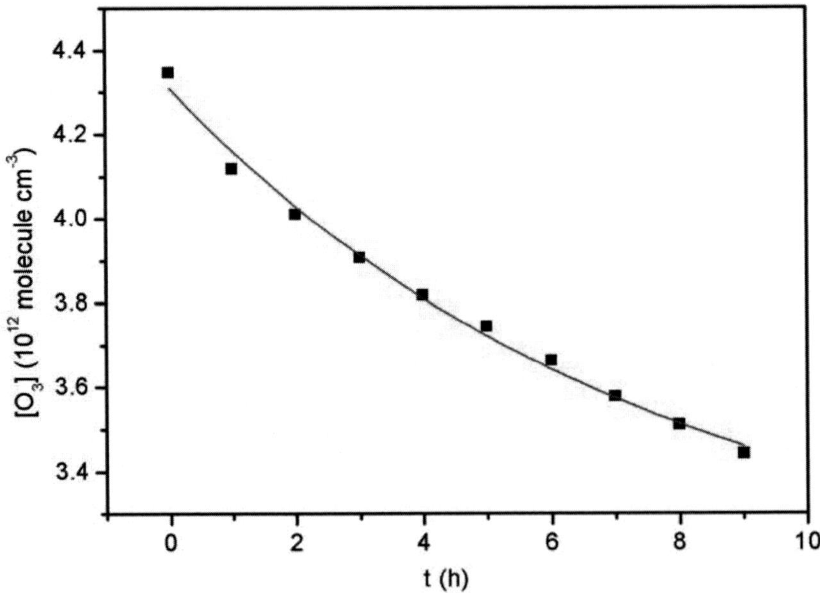

Figure 1. Plot of ozone wall decay in the presence of 2.13×10^{15} molecule cm^{-3} of cyclohexane in the bath gas of N_2.

Linear regression of the data (unit-weighted least squares) yielded high correlation coefficients, near-zero intercepts and slopes, i.e. pseudo first-order rate constants, which were used to calculate the second-order reaction rate constants that are listed in Table 1. A good linearity in Figure 2 implies that equation I is correctly used for the sulfide-ozone reaction. Even if there are some secondary reactions, the influence of these reactions on the rate constant is very small. The data of the PMS-O_3 reactions have been dealt with the same methods, and the results are also listed in Table 1.

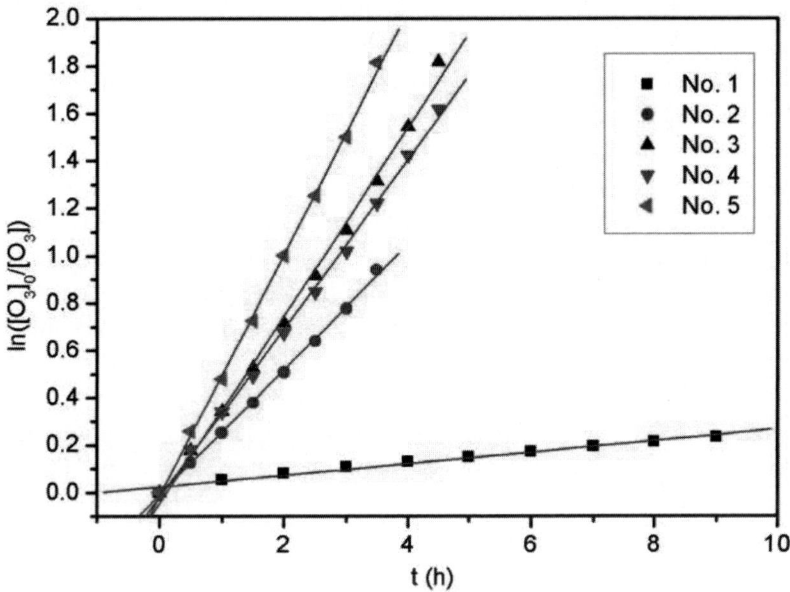

Figure 2. Plots of $\ln([O_3]_0/[O_3])$ against time for ozone decay (No. 1) and EMS reaction experiments (No. 2~5).

Table 1. Experimental result under different initial concentrations of ozone and sulfides

No.	sulfide	$[\text{sulfide}]_0$ ($\times10^{14}$ molecule cm^{-3})	$[O_3]_0$ ($\times10^{12}$ molecule cm^{-3})	$-d\ln[O_3]/dt$ ($\times10^{-5}$ s^{-1})	k ($\times10^{-19}$ cm^3 $molecule^{-1}$ s^{-1})
1	Wall decay	0	4.35	0.618	-
2	EMS	4.75	5.51	7.36	1.42
3	EMS	9.50	5.51	11.02	1.10
4	EMS	9.50	6.74	9.90	0.98
5	EMS	14.25	3.84	14.23	0.96
6	PMS	4.05	3.58	6.03	1.34
7	PMS	6.48	5.04	9.85	1.43
8	PMS	8.10	3.95	10.06	1.17
9	PMS	9.71	7.04	10.68	1.04

Based on the experiments, the obtained wall decay rate is 6.18×10^{-6} s^{-1}, which is ca. one order of magnitude lower than the pseudo first-order loss rates of ozone in the ozone-sulfide-cyclohexane experiments. The corresponding half-life time of ozone loss is 31 h. As seen from the values of $-dln[O_3]/dt$ in Table 1, the effect of ozone wall decay cannot be neglected. When the EMS concentration is 4.75×10^{14} molecule cm^{-3}, $-dln[O_3]/dt$ value is 7.36×10^{-5} s^{-1}, in which the ozone wall decay accounts for 8%. Such an effect must be deducted from the experimental results. It can be known from the data in Table 1 that *-dln[O3]/dt* values increase linearly with the increasing concentration of EMS or PMS. According to equation □, the rate constants of sulfides and ozone can be calculated from the data of wall decay rate, *-dln[O3]/dt* value, and $[\text{sulfide}]_0$. The results are listed in Table 1, and these rate constants data can be averaged for EMS and PMS, respectively. Therefore, the rate constants for ozone reaction with EMS and PMS under room temperature are $(1.12\pm0.18)\times10^{-19}$ and $(1.24\pm0.15)\times10^{-19}$ cm^3 $molecule^{-1}$ s^{-1}, respectively.

At present, the available experimental information on the reactivity of sulfides with the different atmospheric oxidants is scarce. Thus, it is not possible to make an extended evaluation even on a qualitative basis of the relevance and dependence of the different contributions to the reactivity of sulfides. Clearly more experimental measurements are needed in order to increase the existing kinetic database on these reactions and to fully understand the reactivity changes among different sulfides. The rate constants available and estimated atmospheric lifetimes of similar sulfides can be compared and are listed in Table 2, including the reactions with OH radicals, Cl atoms, NO_3 radicals and ozone.

The lifetime of sulfides with respect to oxidation by OH radicals may be estimated using a 12-hour average concentration of OH radicals of 1.6×10^6 molecule cm^{-3} [21]. Although the global concentration of Cl atoms is likely to be small, significant concentrations are expected to be present in costal areas.

Table 2. Rate constants and estimated atmospheric lifetime of sulfides with OH radicals, Cl atoms, NO_3 radicals and O_3 [a]

	CH_3SCH_3	CH_3SCH_2Cl	$C_2H_5SCH_3$	$C_2H_5SC_2H_5$	*n*-$C_3H_7SCH_3$
OH	1.7×10^{-12} [25]	2.5×10^{-12} [26]	8.5×10^{-12} [10]	1.16×10^{-11} [27]	-
τ_{OH} [b]	4.26 d	2.89 d	0.85 d	0.62 d	-
Cl	3.3×10^{-10} [25]	1.0×10^{-10} [26]	4.88×10^{-10} [13]	5.98×10^{-10} [13]	-
τ_{Cl} [c]	5.08 d	16.8 d	3.44 d	2.81 d	-
NO_3	1.1×10^{-12} [25]	-	-	4.8×10^{-12} [28]	-
τ_{NO3} [d]	0.51 h	-	-	0.12 h	-
O_3	1.04×10^{-19} [17]	-	1.12×10^{-19}	2.77×10^{-19} [18]	1.24×10^{-19}
τ_{O3} [e]	55.6 d	-	51.7 d	20.9 d	46.7 d

[a] All the rate constants are in units of cm^3 $molecule^{-1}$ s^{-1}, and the temperature of these reactions is around 298 K (room temperature).

[b] $\tau_{OH} = 1/(k[OH])$, where k is the rate constant, $[OH] = 1.6\times10^6$ molecule cm^{-3} [21], and d = days.

[c] $\tau_{Cl} = 1/(k[Cl])$, where k is the rate constant, $[Cl] = 6.9\times10^3$ molecule cm^{-3} [23], and d = days.

[d] $\tau_{NO3} = 1/(k[NO_3])$, where k is the rate constant, $[NO_3] = 5\times10^8$ molecule cm^{-3} [24], and h = hours.

[e] $\tau_{O3} = 1/(k[O_3])$, where k is the rate constant, $[O_3] = 2.0\times10^{12}$ molecule cm^{-3} [23], and d = days.

Wingenter et al. [22] predict a Cl atom concentration of 3.3×10^4 molecule cm^{-3} in marine air for the first 5 h after dawn. Using this value for the Cl atom concentration during the morning, an effective 24-hour concentration of Cl atom of $(3.3\times10^4\times5/24) = 6.9\times10^3$ molecule cm^{-3} may be calculated [23]. The natural lifetime of sulfides with respect to OH radicals and Cl atoms in coastal areas would thus be calculated to be several days as shown in

Table 2. The global concentration of NO_3 radicals is estimated to be 5×10^{8} molecule cm^{-3} [24], which is a 12-hour average concentration. The corresponding lifetimes with respect to NO_3 radicals for DMS and PMS are within one hour. Similar calculations, employing a fairly high concentration of O_3 of 2.0×10^{12} molecule cm^{-3} (80 ppb) [23], suggest that the lifetime of sulfides with respect to oxidation by O_3 range from 20.9 days to 55.6 days.

From the data in Table 2, we could draw the conclusion that reaction with OH radicals during the daytime and NO_3 radicals during the nighttime are the dominating ways of chemical loss of sulfides. This study further confirms that under atmospheric conditions, the reaction of sulfides with ozone is too slow to represent an important oxidation pathway of sulfides in the atmosphere.

Summary

With our self-made smog chamber, we determined the rate constants of ozone with EMS and PMS, respectively. These reactions are relatively slow, so the wall decay of ozone cannot be neglected. Under the condition of this work, the rate of ozone wall decay is 6.18×10^{-6} s^{-1}, and the half-life time of ozone is 31 h. In order to eliminate the effect of OH radicals, cyclohexane was added into the reactor. The rate constants for ozone reaction with EMS and PMS derived from different concentrations of reactants under room temperature are determined to be $(1.12\pm0.18)\times10^{-19}$ and $(1.24\pm0.15)\times10^{-19}$ cm^3 $molecule^{-1}$ s^{-1}, respectively. Meanwhile, our results further confirm that the gas phase reaction of sulfides with O_3 is too slow to represent a significant pathway for the oxidation of sulfides in the atmosphere.

Acknowledgements

This project was supported by Knowledge Innovation Program (Grant No. KZCX2-YW-205) of the Chinese Academy of Sciences, the 973 Program (No. 2006CB403701) and 863 Program (No. 2006AA06A301) of Ministry of Science and Technology of China, and the National Natural Science Foundation of China (Contract No. 20577052, 20673123).

References

[1] Nguyen, B. C.; Bonsang, B.; Gaudry, A. *J. Geophys. Res.* 1983, 88, 10903-10914.

[2] Chatfield, R. B.; Crutzen, P. J. *J. Geophys. Res.* 1984, 89, 7111-7132.

[3] Andreae, M. O.; Raemdonck, H. *Science* 1983, 221, 744-747.

[4] Harvey, G. R.; Lang, R. F. *Geophys. Res. Lett.* 1986, 13, 49-51.

[5] Barnes, I.; Hjorth, J.; Mihalopoulos, N. *Chem. Rev.* 2006, 106, 940-975.

[6] Khoroshko, L. O.; Petrova, V. N.; Takhistov, V. V.; Viktorovskii, IV; Lahtiperä, M.; Paasivirta, *J. Env. Sci. Pollut. Res.* 2007, 14, 366–376.

[7] Kima, K.-H.; Jeona, E.-C.; Koob, Y.-S.; Im, M.-S.; Younc, Y.-H. *Atmos. Environ.* 2007, 41, 3829–3840.

[8] Merritt, Jr. C.; Bazinet, M. L.; Sullivan, J. H.; Robertson, D. H. *J. Agr. Food Chem.* 1963, 11, 152-155.

[9] Wilkins, K.; Larsen, K. *Chemosphere* 1996, 32, 2049-2055.
[10] Hynes, A. J.; Wine, P. H.; Semmes, D. H. *J. Phys. Chem.* 1986, 90, 4148-4156.
[11] Stickel, R. E.; Zhao, Z.; Wine, P. H. *Chem. Phys. Lett.* 1993, 212, 312-318.
[12] Jensen, N. R.; Hjorth, J.; Lohse, C.; Skov, H.; Restelli, G. *Int. J. Chem. Kinet.* 1992, 24, 839-850.
[13] Kinnison, D. J.; Mengon, W.; Kerr, J. A. *J. Chem. Soc. Faraday Trans.* 1996, 92, 369-372.
[14] Yin, F. D.; Grosjean, D.; Seinfeld, J. *J. Geophys. Res.* 1986, 91, 14417-14438.
[15] Atkinson, R.; Aschmann, S. M. *Int. J. Chem. Kinet.* 1984, 16, 259-268.
[16] Martinez, R. I.; Herron, J. T. *Int. J. Chem. Kinet.* 1978, 10, 433-452.
[17] Du, L.; Xu, Y.; Ge, M.; Jia, L.; Yao, L.; Wang, W. *Chem. Phys. Lett.* 2007, 436, 36-40.
[18] Du, L.; Xu, Y.; Ge, M.; Jia, L. *Atmos. Environ.* 2007, 41, 7434–7439.
[19] Becker, K. H.; Inocencio, M. A.; Schurath, U. *Int. J. Chem. Kinet.* 1975, S1, 205-220.
[20] Akimoto, H.; Finlayson, B. J.; Pitts, J. N. *Chem. Phys. Lett.* 1971, 12, 199-201.
[21] Prinn, R. G.; Weiss, R. F.; Miller, B. R.; Huang, J.; Alyea, F. N.; Cunnold, D. M.; Fraser, P. J.; Hartley, D. E.; Simmonds, P. G. *Science* 1995, 269, 187–192.
[22] Wingenter, O. W.; Kubo, M. K.; Blake, N. J.; Smith Jr., T. W.; Blake, D. R.; Rowland, F. S. *J. Geophys. Res.* 1996, 101, 4331–4340.
[23] Canosa-Masa, C. E.; Duffya, J. M.; Kingb, M. D.; Thompsona, K. C. *Atmos. Environ.* 2002, 36, 2201–2205.
[24] Shu, Y. H.; Atkinson, R. J. *Geophys. Res.* 1995, 100, 7275–7281.
[25] Atkinson, R.; Baulch, D. L.; Cox, R. A.; Crowley, J. N.; Hampson, R. F.; Hynes, R. G.; Jenkin, M. E.; Rossi, M. J.; Troe, J. *Atmos. Chem. Phys.* 2004, 4, 1461-1738.
[26] Shallcross, D. E.; Vaughan, S.; Trease, D. R.; Canosa-Mas, C. E.; Ghosh, M. V.; Dyke, J. M.; Wayne, R. P. *Atmos. Environ.* 2006, 40, 6899-6904.
[27] Nielsen, O. J.; Sidebottom, H. W.; Nelson, L.; Rattigan, O.; Treacy, J. J.; O'Farrell, D. J. *Int. J. Chem. Kinet.* 1990, 22, 603-612.
[28] Daykin, E. P.; Wine, P. H. *Int. J. Chem. Kinet.* 1990, 22, 1083-1094.

In: Ozone Depletion, Chemistry and Impacts
Editor: Sem H. Bakker, pp. 195-237

ISBN: 978-1-60692-007-7

Chapter 10

A MODEL-BASED WARNING SYSTEM FOR AIR POLLUTION MONITORING (APPLICATION TO GROUND-LEVEL OZONE AND PM10 TIME SERIES MEASURED IN BORDEAUX, FRANCE)

Ali Zolghadri*
Université Bordeaux 1; Laboratoire IMS ;
351 cours de la Libération ; 33405 Talence cedex ; France

Abstract

This chapter describes a model based approach to develop an operational public warning system for air pollution monitoring. The proposed methodology is based on hard and soft computing techniques and combines an adaptive nonlinear state space-based prediction mechanism, a gain scheduling strategy and neural network techniques to develop an integrated operational warning system. The overall method was applied to ground-level ozone and PM_{10} time-series data measured in Bordeaux, France over four years (1998 to 2001). The aim of model building was to develop predictive models in order to provide forecasts of the maximal daily ground-level ozone and the daily mean PM_{10} concentrations. The goals of the forecast are to provide information in order to satisfy needs for public information and to further reduction and prevention of exposure, in cases where pollutant limit values are exceeded over a specified length of time. A key characteristic of such a system is that it is constantly fed with new available data, so its behaviour can be adapted to the short term changes of air pollution. Moreover, the warning system provides additional information regarding the extent of a smog episode. This is an important aspect for planning of counter actions and for assessments of human health hazards and negative environmental effects. Finally, he chapter discuss some inherent shortcomings associated with the commonly used statistical techniques for air pollution modelling, and an alternative solution based on l_∞ optimization techniques is proposed.

Keywords: Air quality monitoring Ozone, PM_{10}, Nonlinear adaptive estimation, Minimax optimization, Gain scheduling, Short-term forecasting.

---

* E-mail address: ali.zolghadri@laps.u-bordeaux1.fr

1. Introduction

1.1. Problem Setting

Elevated concentrations of ground level ozone (O_3) and Particulate Matter (PM) constitute the major concern for air quality in most urban areas around the world.

Ground-level ozone is the primary constituent of photochemical smog. Ozone results from chemical reactions in the atmosphere and is formed when the primary pollutants (ozone precursors) interact under the action of suitable ambient meteorological conditions. These primary pollutants are emitted largely by industry and automobiles and fall into two related groups: various oxides of nitrogen (NOx), and a range of volatile organic components (VOCs) such as evaporatives solvents and other hydrocarbons (Derwent, Davies 1994). Ultraviolet radiation causes the precursors to interact photochemically in a set of reactions that result in the formation of ozone. A part of ozone is simultaneously destroyed via oxidation of NOx. These relations can be expressed conceptually as

$$NO_x + VOCs + UV \rightarrow O_3 \qquad \lambda < 420\,\text{nm}$$

$$O_3 + NO \rightarrow NO_2 + O_2$$

where the roles of catalysts and minor chemical species are ignored. In typical urban areas, at least half of the VOCs and NOx come from cars, buses, trucks, and off-highway mobile sources.

Particulate Matter (PM) is the general term used for a mixture of solid particles and liquid droplets found in the air. There is a considerable variety in the chemical makeup and size of particles. $PM_{2.5}$ refer to particles that are less than or equal to 2.5 micrometers in aerodynamic diameter and PM_{10} are those that are between 2.5 and 10 micrometers in diameter. Both the "fine" $PM_{2.5}$ and "coarse" PM_{10} particles can accumulate in the respiratory system and are associated with numerous health effects.

Epidemiological studies indicate that exposure to an elevated concentration of tropospheric ozone is a potential human health hazard (Frischer et al. 1999, Smith-Dorion et al. 2000) and affects vegetation and ecosystems adversely (Matyssek et al. 1997). Other epidemiological studies have indicated a strong link between increased PM_{10} concentrations and increased mortality and morbidity from all causes (Tittanen et al., 1999; Schwartz 1999; McDonnell et al. 2000; Dockery et al., 1993). An incremental increase of PM_{10} by 10 $\mu g/m^3$ can lead to a significant increase in heart, lung and respiratory diseases.

The institution of public information and of possible counter actions in cases where pollutant limit values are exceeded over a specified length of time, has increased the demand for effective short-term predictive models. During the past two decades, models have been developed to predict future concentrations under a variety of different scenarios. Statistical models based on linear or nonlinear regression analysis have been largely applied to address issues of air pollution. See (Thompson et al., 2001, Schlink et al. 2005) for a survey. The goal is the identification of a functional relationship which would allow the prediction of the concentration of pollutants if the emissions and other meteorological variables are given (see for instance Thompson et al. 2001; Cobourn et al., 2000; Hubbard and Cobourn 1998; Gardner and Dorling, 2000; Bloomfield et al., 1996, Schlink et al. 2000, 2005, 2006).

In this chapter, an attempt is made to provide a general framework for reliable model building to predict one day ahead concentrations for ozone and PM_{10}. A new strategy is proposed which combines an adaptive nonlinear state space-based prediction mechanism, a gain scheduling strategy and neural network techniques to develop an integrated operational warning system. Recursive nonlinear state space estimation techniques can be found frequently in the literature dealing with instrumentation and control technology, but are rather uncommon in the context of air pollution or indeed the atmospheric sciences as a whole. We believe that new techniques for recursive nonlinear estimation, recently developed within the field of control theory have important potential benefits for air pollution time series modelling, as it will be demonstrated in this chapter. A key characteristic of such a system is that it is constantly fed with new available data, so its behaviour can be adapted to the short term changes of air pollution. So, contrary to other techniques reported in the related literature, the proposed system can handle the time-evolving nature of the phenomena and does not need frequent adjustments. Moreover, the warning system provides additional information regarding the extent of a smog episode. In case of ozone pollution, this is an important aspect with regards to the planning of counter actions. In fact, although ozone prediction methods exist and some are in use, they mainly concentrate on maximum ozone concentrations and not the temporal extent of the event that is more meaningful for assessments of human health hazards and negative environmental effects. The chapter discus also some inherent shortcomings associated with the commonly used statistical techniques for air pollution modelling. In fact, in the context of air pollution, much attention has been paid to site-specific model structure determination and model validation. However, there is very little guidance about how to choose a suitable method for parameter estimation. Usually, the problem of parameter fitting is solved using a standard "least squares" criterion. A direct consequence is that the resulting models, by nature, do not make any distinction between low and high levels of the time series, as they model the "average" behaviour by minimizing a quadratic loss function, whereas with regards to air quality standards, it is largely the high levels that are of interest and concern. It is shown that l_∞ filtering techniques provide a solution to this problem.

The results are illustrated using air pollution data (ozone and PM_{10}) measured in Bordeaux (France). The period covered is four years, between 01-01-1998 00:00 and 31-12-2001 24:00.

1.2. Air Quality Standards in Europe

In Europe, the importance of ozone as an air quality index has induced the European Commission (EC) to take legislative measures establishing thresholds and air quality standards. A new Commission Directive 2002/3/EC relating to ozone in ambient air came in force in March 2002 (ETCACC 2002). The population information threshold (180 $\mu g/m^3$) and the population warning (360 $\mu g/m^3$) threshold remained the same in accordance with the old Directive and a new alert threshold of 240 $\mu g/m^3$ as daily maximum 1hour-average value has been defined. At the end of summer 2003, the Member States must have implemented this new Directive in their national legislation and the old directive will be repealed (ETCACC 2002).

For PM_{10}, the first European Daughter Directive under Air Quality Framework Directive 96/62/EC set limits for PM_{10}. The EU Daughter Directive states that by the end of 2004 the daily mean PM_{10} should not exceed 50 $\mu g/m^3$ on more than 35 days per year and the annual mean should not exceed 40 $\mu g/m^3$. By 31 December 2009, the daily mean PM_{10} should not exceed 50 $\mu g/m^3$, on more than 7 days per year, and annual mean should be < 20$\mu g/m^3$ (Fuller et al., 2002).

Agencies charged with the responsibility of air quality management often need to make daily air pollution forecasts. The goals of the forecast are to provide information in order to satisfy needs for public information and to further reduction and prevention of exposure, in cases where pollutant limit values are exceeded over a specified length of time.

1.3. Brief Review of Model Based Approaches for Ozone and PM_{10} Monitoring

The institution of public information and of possible counter actions in cases where ozone limit values are exceeded over a specified length of time, has increased the demand for effective short-term predictive models. Several models have been developed previously to predict the ground level ozone concentrations. Deterministic models incorporate complex photochemical reaction mechanisms to account for the transformation of the chemical species and the chemical algorithms are embedded within atmospheric dispersion and advection models (Millan et al. 1996, Vautard et al. 2001, Zlatev et al. 1993). The implementation of such models is complex, as they require significant computer and staffing commitments as well as many detailed and complex emission data and meteorological inputs for the region of interest. In addition, some input data are not easily acquired by environmental protection agencies or local industries. Thus, such models are not very practical choices in many locations for the fast prediction task.

On the other hand, empirical, black-box and grey-box models have been proposed as fast prediction tools (see for instance Jorquera et al. 1998, Davis and Speckman 1999, Chen et al. 1998, Gardner et al. 1998, 2000a, 2000b, Schlink et al. 2000, 2005, 2006, references in Thompson et al. 2001, see the also references of this chapter). They rely on the statistical analysis of current and previous atmospheric conditions and precursors. Linear Kalman filters were used in (Schlink and Volta 2000) as on-line forecast tools for environmental time series. Generally, high ozone events are predicted poorly by black-box linear models. The basic problem is that the ozone formation process, involving precursor emissions, atmospheric transport and mixing, and a complex system of photochemical reactions, is highly nonlinear and non stationary. To deal with non stationary and changing dynamics, (Young et al. 1991, 1999) developed a fully and unified recursive approach to the modelling and forecasting of non-stationary time series. The recursive formulation allows for self adaptive implementation of state-space forecasting and seasonal adjustment. It is shown in (Young et al. 1991, 1999) that the time-variable parameter models can be characterized by parameter variations which are a mixture of smooth, gradual changes and more rapid fluctuations. So, the modelling approach can handle systems with changing dynamics and mild nonlinear behaviour. See also (Zolghadri 1996) for a discussion on this topic. Nonlinear regression models and neural networks seem also to be well situated for modelling ozone time series since they allow for nonlinear relations between variables (Cobourn et al. 1999, 2000, Hubbard and Cobourn 1998, Elkamel et al. 2001, Kolehmainen et al. 2001, Gardner et al. 2000). A comparison of nonlinear regression and neural network models for surface-level ozone forecasting was made in Louisville, KY, by Cobourn et al. 2000. Several research papers have been published, discussing the role that neural networks can play in predicting photochemical pollution. See for instance Corani 2005 and Dutot el al. 2007.

Unfortunately, PM_{10} concentrations are among the most difficult to predict accurately using deterministic models, due to our incomplete understanding and knowledge of the sources, dispersion and sinks of PM_{10} and the processes, which link them (Zickus et al., 2002, Kukkonen et al., 2001, 2002, 2003). Recently, attempts have been made to model PM_{10} concentrations using empirical, semi-empirical or statistical approaches (Perez and Reyes 2002, Van der Wal and Janssen 2000, Zickus et al., 2002, Kukkonen et al., 2001, Kukkonen et al., 2002). Linear Kalman filters have already been used as on-line forecast tools for environmental time series (see for instance Young et al., 1991; Van der Wal and Janssen 2000, Ordieres et al. 2005, Schlink et al. 2005, Kurun et al. 2005, Nittis et al. 2004, Sivakumar 2005). However, it is well known that if the time series presents a very irregular and erratic character, linear Kalman filtering may differ very little from a persistence model (predicted value for tomorrow ≈ measured value for today). PM_{10} formation process, involving precursor emissions, atmospheric transport and meteorology, is highly nonlinear and non stationary. This nonlinear behaviour can hardly be captured by means of linear adaptive filtering techniques.

To overcome the above mentioned drawbacks, in this chapter a complete and general methodology based on hard and soft computing techniques is developed. Also, some inherent inconveniences of the commonly used statistical techniques for air pollution modelling are underlined and discussed.

2. Air Pollution Measurements and Meteorological Data

2.1. Meteorology and Meteorological Data

Bordeaux has a maritime climate with about 2080 hours of sunshine and 930 mm of rain per year and an average temperature of about 13.7° C. The city is located at a distance of 50 km (≈ 90 m above sea level) from the Atlantic coast and responds to ocean air that makes summers relatively hot and long and winters mild and rainy. There is no mountain surrounding Bordeaux. Under see breeze conditions, the prevailing wind direction is mostly from west and northwest. The ground-level meteorological data used in this research came from the French meteorological agency. It consisted of hourly observations taken by local meteorological agency, located in the proximity of the Bordeaux international airport (at about 5 km from the city). Weather forecasting is based on the use of atmospheric numerical simulation. The job of the local weather bureau is to refine regional weather scenarios. The forecasting specialists, armed with knowledge of the regional climate and of the models' limits, adjust the results of the simulation, and convert them into directly observable weather data. At short range (from a few hours to 1 day in advance), local meteorological indicators show that the various components of the weather are quite accurately forecasted for Bordeaux.

Meteorological data were measured in one hour intervals and transformed into 24 h mean, minimum or maximum values.

In this study, we used the measured meteorological variables and not the forecasted ones as they were not available at the time of this study. So, we need to take into account that, under real world conditions, only measurements taken before the prediction date are available together with the weather forecast for the day we want to predict the pollutants level. The accuracy of weather forecast influences the quality of the prediction. It is obvious that the higher the noise level due to errors in input variables for the pollutant forecast, the lower the final prediction quality. As it will be seen in section 3, this aspect is taken into account by introducing additional noise sources in the dynamic state space model. The covariance matrix of the state and observation noises can be considered as high level design parameters. They allow for controlling the flexibility of the state space model.

2.2. Ozone and Precursors Data

The period covered is four years, between 01-01-1998 00:00 and 31-12-2001 24:00.

The ozone data used for our investigations came from the local air quality monitoring agency (AIRAQ) in Bordeaux. The city of Bordeaux is covered by four fixed and a number of mobile measuring stations for NO, NO_2, SO_2, CO, PM_{10} and O_3. Three stations are located in the proximity of the densely populated areas. The forth measuring station is located in an industrialised area.

The monitoring sites are represented in figure 1.

Figure 1. Fixed monitoring sites (☆) in Bordeaux.

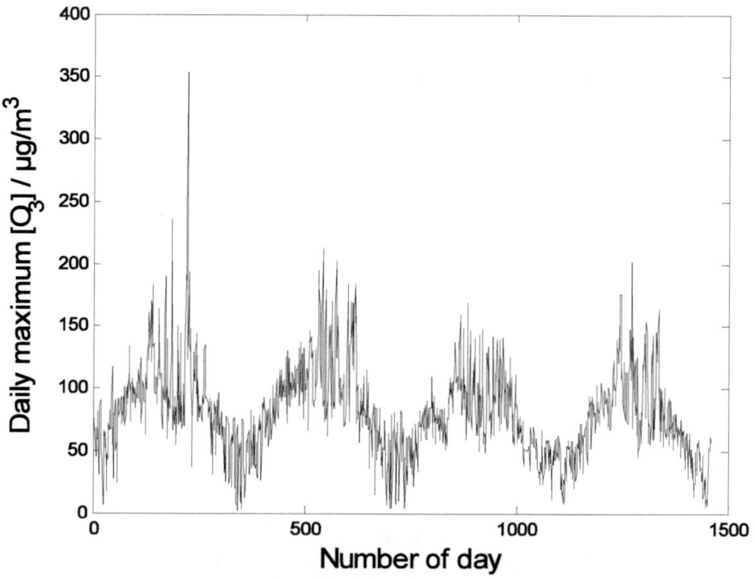

Figure 2. Measured ozone, station Bordeaux Grand Parc, years 1998-2001.

The measuring site "Grand parc" was selected during this study as it represents a typical urban area. This monitoring station is located in the northwest area of Bordeaux, in the proximity of the densely populated areas and is not influenced by any specific pollutant source. Air pollution measurements were recorded every fifteen minutes. The data used contain the average hourly measurements (taken from four readings). NO/NO_2 concentrations are measured using chemiluminescent oxides of nitrogen analysers and ozone concentrations are measured by UV photometric analyser. The 24 maximum hourly ozone concentrations for the four years period is depicted in figure 2 below. Note the erratic character of the time series and its periodic component.

2.3. PM_{10} Data

Concentrations of particulates (PM_{10}) are measured continuously using MP101M EX analysers, manufactured by Environnement S.A™. The principle of operation is based upon low energy beta rays that are absorbed by collision with electrons, whose number is proportional to density. Absorption is thus a function of the mass of the material traversed by the radiation, independently of its physicochemical nature. A zone of the filter ribbon is weighted automatically before and after the formation of the deposit of dust taken from ambient air though a sampling head. Differential measurement thereby overcomes problems of filter heterogeneity in the calculation of the mass of deposited dust. The daily mean PM_{10} concentrations for the three years period (1999-2001) is depicted in figure 3.

The fraction of missing data ranged from 0.2 to 5% for various meteorological and pollutant variables. Because of this low percentage, no estimation of the effect of the replacement technique on the results was judged necessary. Missing data items were filled in using a combination of linear interpolation and Self-Organizing Map. A missing data

MATLAB® toolbox is developed within the European APPETISE project (Junninen *et al.* 2004).

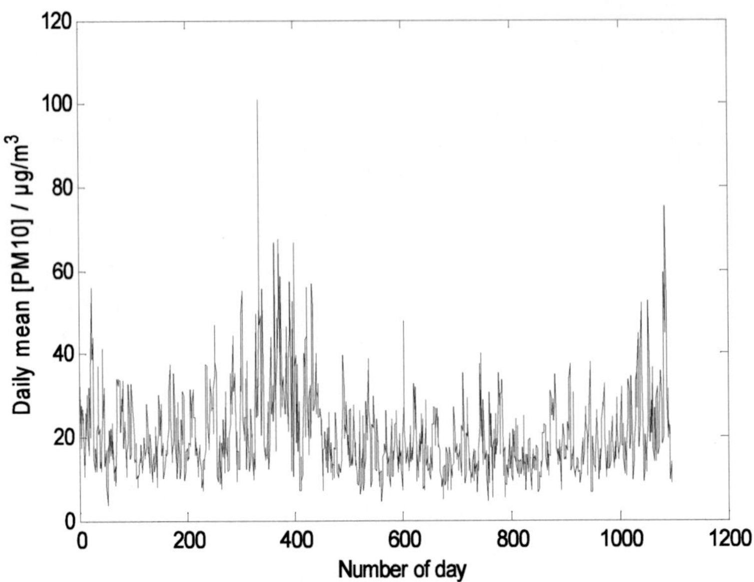

Figure 3. Measured daily mean PM_{10} concentrations, station Bordeaux Grand Parc, years 1999-2001.

A forecast of air pollution for the following 24 hours is prepared by AIRAQ each afternoon, for inclusion in the 18:00 air pollution bulletin. The forecast is currently provided by a team of air quality experts. The decisions are based on pollutant concentration measurements as well as other related information, such as traffic and industrial sources behaviour. A need has emerged recently for automation of the process. A fast short-term prediction process is needed in a day-by-day basis, which would determine the likelihood of a pollution episode (mainly by ozone and PM).

3. Selection of Input Variables

3.1. Input Variables for Ozone Modelling

It is well known that daily peak ozone concentrations are correlated with meteorological parameters such as temperature, wind speed and relative humidity (figure 4).

Based on exploratory analysis using scatterplots and correlation analysis between ozone time series and available ground-level meteorological observations, the selected meteorological variables are surface temperature, wind (speed/direction), relative humidity, pressure, solar intensity and radiation. The choice of the above meteorological variables was also based upon selecting those which were generally available and routinely forecasted; since this would facilitate the results of this work being applied under real-world operational conditions.

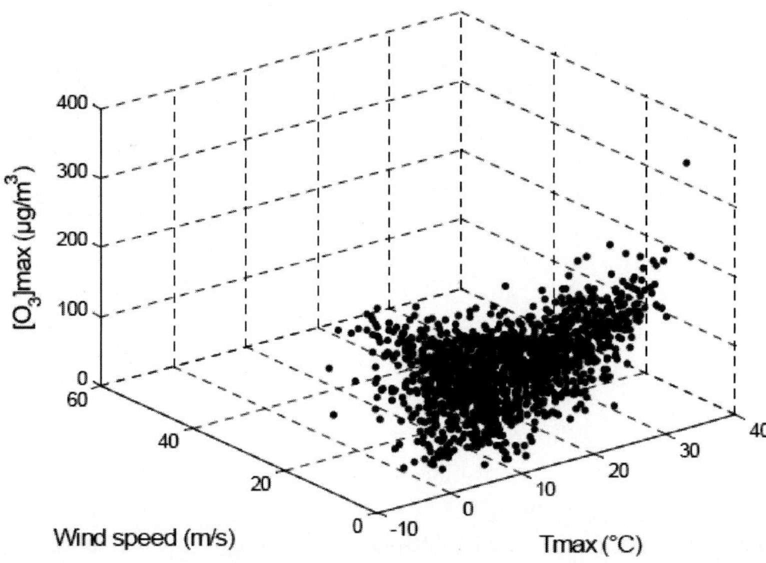

Figure 4. Daily maximum ozone concentrations as a function of measured daily maxima of temperature and mean wind speed, years 1998-2001.

In order to incorporate the variable wind direction as an input to the model and to avoid the discontinuous transition from 0° to 360°, a common method is to calculate the east-west and north-south components of the wind as the cosine and sine of the wind direction multiplied by the wind speed: $u = -WS.\sin(2\pi.WD/360)$ and $w = -WS.\cos(2\pi.WD/360)$. In this investigation, a different approach has been used. To overcome the problem related to the circular nature of the wind, the ozone wind rose (the pattern of wind direction versus concentration) was analysed. As an example, the ozone wind rose for concentrations between 90 µg/m3 and 180 µg/m3 and for different ranges of wind speed (v) is represented in figure 5 below.

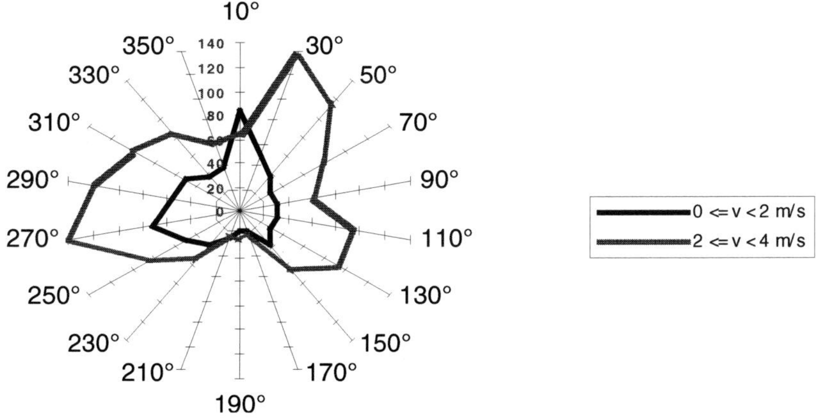

Figure 5. Pattern of wind direction versus ozone concentrations.

It can be seen from the figure that ozone concentrations are significantly lower in the wind sectors (330°-10°) and (170°-230°). Based on the analysis of the ozone wind roses for different ozone levels, it was decided to introduce the variable wind direction in the model as follows: the wind rose is divided into 8 sectors. The wind sector having the lowest average concentration got number 1 and the sector having the highest concentration got the number 8.

Based on correlation results and previous studies (Thompson et al. 2001), the input emission variables that required consideration where NO and NO_2. We used the average concentration for these emissions between 16:00 p.m. and 20:00 p.m of the previous day of interest (vehicular NO_x emissions due to the city activities in the late afternoon). Moreover, the possibility of occurrence of pollution episodes is increased if the previous day's pollution levels where higher than normal. This is attributed to the fact that pollution episodes are "built up" when meteorological conditions favouring high pollutant concentrations occur during successive days. So, the maximum hourly concentration of O_3 of the previous day could indicate the potential of a pollution episode for the next 24 hours.

Finally, a trend function defining the seasonal variation of ozone has also been added as input. This function has been obtained by low pass filtering of the ozone time series. For the years 1998-2001, the resulting trend function is depicted in figure 6 below. The time series was modelled using a standard Kalman filter.

In summary, the selected input variables for ozone prediction are the followings:

i) Surface temperature (maximum measured value of the current day (day k) and maximum forecasted value for the next day (day k+1)).
ii) Wind: speed (forecasted mean value between 13:00 p.m. and 17:00 p.m. of the next day) and direction .
iii) Relative humidity (forecasted mean value between 10:00 a.m. and 14:00 p.m. of the next day).
iv) Pressure (forecasted mean value between 13:00 p.m. and 17:00 p.m. of the next day).

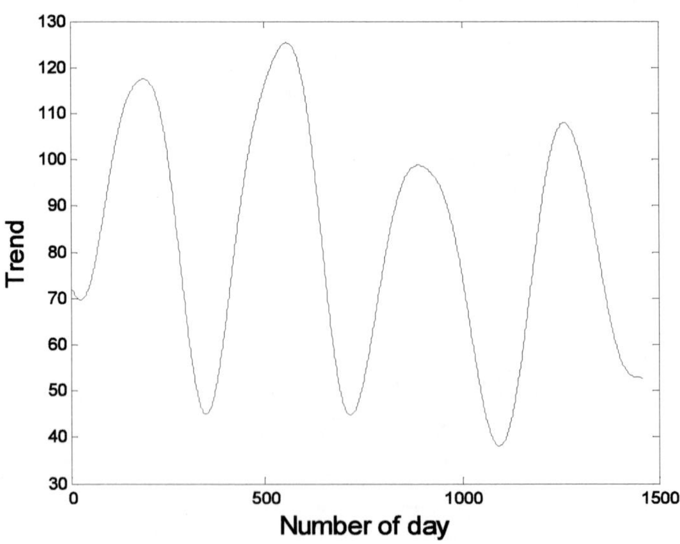

Figure 6. Seasonal variation of ozone.

v) Global radiation (forecasted mean value between 8:00 a.m. and 18:00 p.m. of the next day). All radiation fluxes are expressed as energy per unit area (J/m^2). About 50% of solar (or shortwave) radiation is reflected back into space, while the remaining shortwave radiation is absorbed by the earth's surface and re-radiated as thermal infrared (or long wave) radiation. The total radiation is then indirectly forecasted by Méteo France.

vi) Sunshine duration (forecasted mean value between 8:00 a.m. and 18:00 p.m. of the next day). This parameter is derived from cloud cover forecasts and is expressed as number of minutes of sunshine per hour.

vii) Maximum measured relative amount of $[NO_2]$ and [NO] of the current day: $\mathrm{Max}([\mathrm{NO}_2](t)/[\mathrm{NO}](t))$ where t is the time between 14:00 p.m. and 18:00 p.m. In typical urban areas, at least half of the VOCs (volatile organic compounds) and NO_x come from cars, buses, trucks, and off-highway mobile sources. So, the ratio $[NO_2]/[NO]$ affects the ozone formation/destruction cycle.

viii) A trend function defining the seasonal variation of ozone.

The input variables are resumed in the table 1. In this table, k denotes the current day and $k+1$ the next day for which we want to make a forecast.

3.2. Input Variables for PM_{10}

The daily mean PM_{10} concentrations are also influenced by meteorological conditions, such as wind speed, wind direction and temperature. For instance, light winds associated with high-pressure systems typically increase the chance that pollutants will accumulate in the atmospheric boundary layer.

It was found that the forecasted difference between maximum and minimum temperature presents a relatively high correlation with measured PM_{10} concentrations. A very low correlation was found between PM_{10} time series and relative humidity and solar intensity. So, these variables were not included in the set of input variables.

Table 1. Definition of input variables for ozone

Variable	Unit	Acronym
Temperature (day k+1 – day k)	°C	u_1
Temperature (day k+1)	°C	u_2
Global radiation (day k+1)	J/m^2	u_3
Sunshine duration (day k+1)	min	u_4
Barometric pressure (day k+1)	hp	u_5
Wind speed (day k+1)	m/s	u_6
Relative humidity (day k+1)	%	u_7
Wind direction (day k+1)	-	u_8
Trend (day k+1)	-	u_9
[NO2]/[NO] (day k)	-	u_{10}

The variable "wind direction" has been incorporated as described in section 3.1. Figure 7 shows the PM_{10} wind rose for different ranges of wind speed (v) and concentrations.

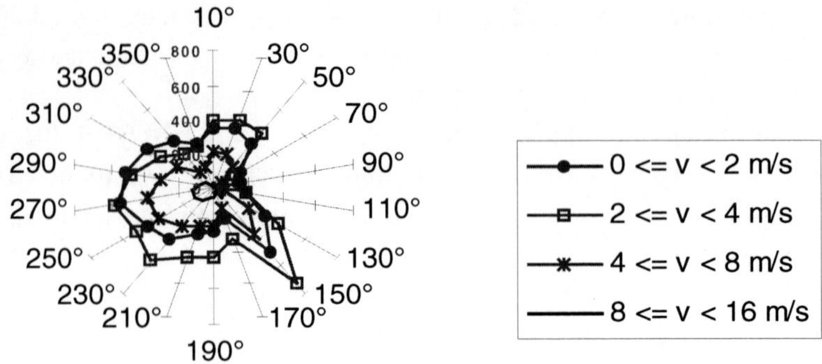

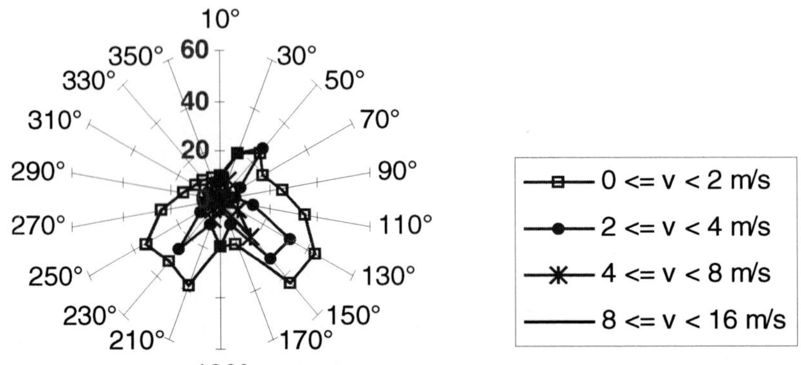

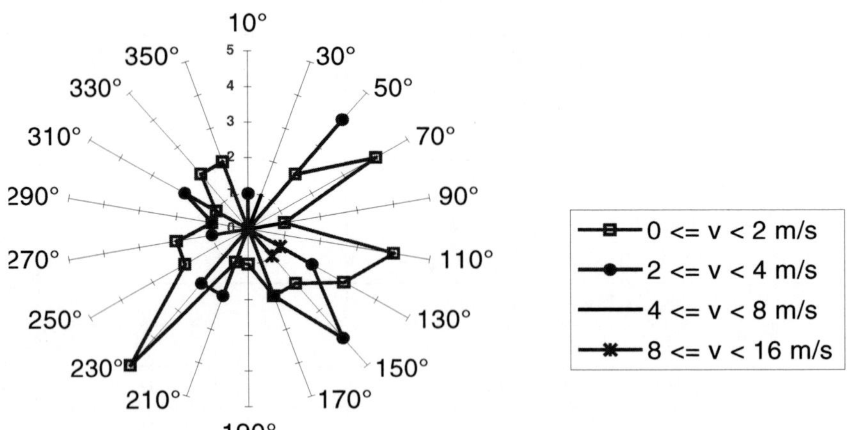

Figure 7. Pattern of wind direction versus PM_{10} concentrations.

If a direction would exist that corresponds with high average PM_{10} concentrations, deviations with respect to this bearing could be selected. However, as it can be seen from the figure 7, no such direction is obvious. Based on this analysis, the variable wind direction was introduced in a different way: the wind rose is divided into 8 sectors and ordered. As for the ozone case, the wind sector having the lowest average concentration got number 1 and the sector having the highest concentration got the number 8. For each day, 24 h mean wind direction was calculated as follows: first, 360° direction was replaced by 0° direction to avoid problems in calculating the mean of directions ranging from 0 to 360°. A sector number was then affected to this mean direction. This approach has been also applied in (Van der Wal and Janssen, 2000). For the day of interest, we also calculated how many hours the wind came from a direction of less than 45° from the mean wind direction. If this was the case during more than 8 h, the sector number was changed to take into account this prevailing wind direction.

The pollutant data have been analysed for relationships between PM_{10} and other pollutants, and close correlations with nitrogen dioxide (NO_2), oxide of nitrogen (NO) and oxide of carbon (CO) were observed. So, vehicle exhaust emissions are among the main local sources of PM_{10} in the studied urban area.

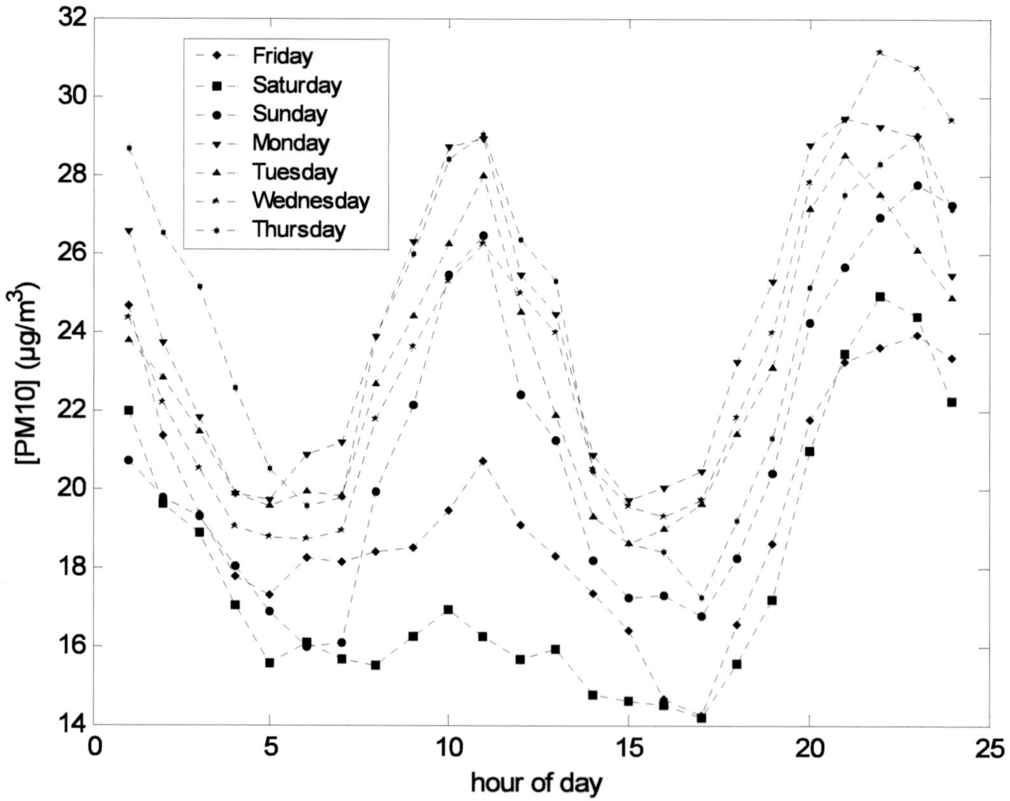

Figure 8. Average daily variation of PM_{10} concentrations; station Bordeaux Grand Parc, years 1999-2001.

Figure 8 shows the averages of one-hour averages of PM_{10} concentrations per hour of the day for each day of the week for years 1999, 2000 and 2001. Note that the hour of day is expressed in UT and there is a time-lag of two hours with respect to the local time. The peak

of PM_{10} concentrations around 11 a.m and 22 p.m (9 a.m and 20 p.m, local time) can be related mainly to morning and evening rush hours. Mean PM_{10} concentrations are higher on week-days than on Sunday and Saturday.

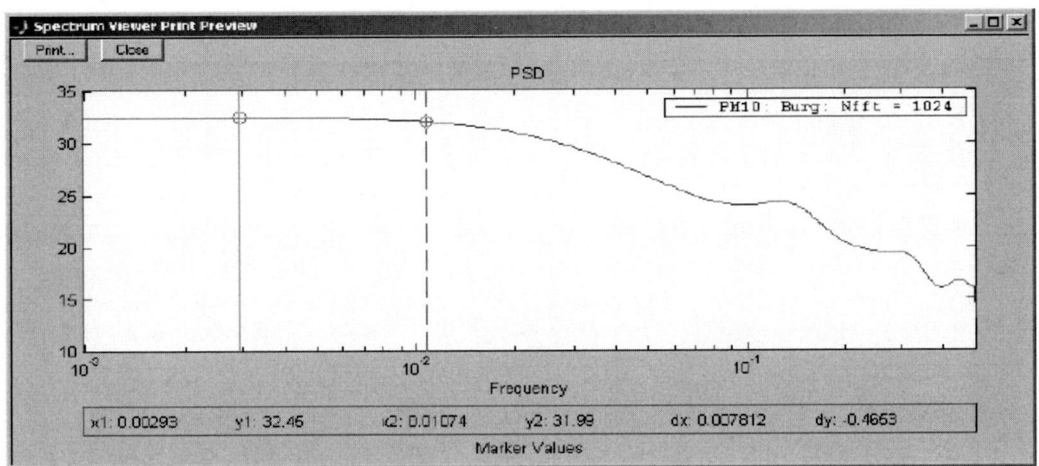

Figure 9. Power spectral density estimate of the daily mean PM_{10} time series.

Surprisingly enough, the highest average concentrations occur on Tuesdays. To take into account this gap between days of the week, we incorporate the "day of the week" as an input variable to the model. Finally, figure 9 represents the power spectral density estimate of the PM_{10} time series using a logarithmic scale and a parametric PSD estimation (Burg method).

The spectral analysis does not reveal any strong long term or annual cycle in the data (frequency at about 0.0027). This fact is probably a characteristic of Bordeaux metropolitan area and cannot be generalized to other urban areas.

Finally, the selected input variables for modelling PM_{10} concentrations are summarized in table 2.

Table 2. Definition of input variables for PM_{10}.

Variable	Unit	Acronym
[NO]mean (day k)	$\mu g/m^3$	u_1
[NO_2]mean (day k)	$\mu g/m^3$	u_2
[CO]mean (day k)	$\mu g/m^3$	u_3
Wind speed (day k+1)	m/s	u_4
Wind direction (day k+1)	-	u_5
Barometric pressure (day k+1)	hp	u_6
Temperature (Tmax-Tmin, (day k+1))	°C	u_7
DoW (Day of Week, (day k+1))	-	u_8

4. Monitoring Strategy

4.1. Methodology

For ozone, he monitoring system is based on three modules:

i) A Nonlinear Adaptive State Space Estimator (NASSE) which aims at prediction of maximum next day ozone level and its uncertainty bounds. The estimated state and parameters are updated every day.
ii) A gain scheduling system which is fed by the output of NASSE and a subset of input variables. This module, called Auxiliary Module (AM), aims at modelling only "threshold exceedance" for extreme ozone concentrations. The thresholds are defined by control authorities (see section 1). The output of AM is the information "exceedance/non exceedance".
iii) A Multi-Layer Perceptron (MLP) neural network which aims at prediction of the smog duration with respect to a given threshold.

For PM_{10}, only the first module is used, because only the daily mean concentrations are of interest. In the following, the overall strategy is developed for the case of ozone.

The general interconnection block diagram is represented in figure 10.

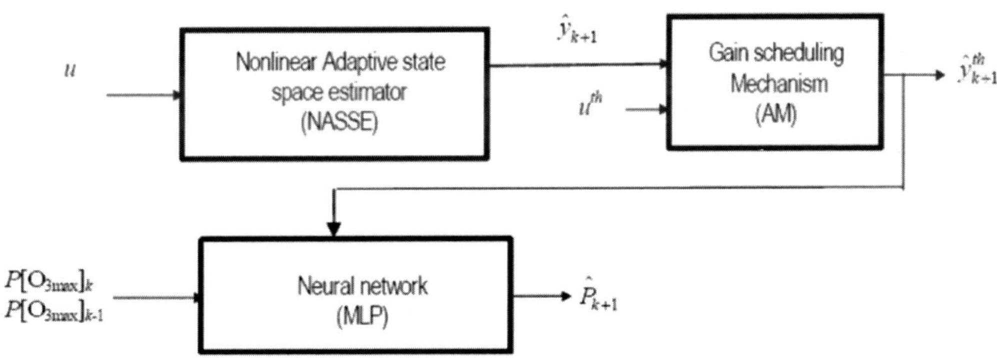

u is a vector which contains all previously described ozone precursors and meteorological variables,
k denotes the discrete time (current day),
$[O_{3max}]_k$ and $[O_{3max}]_{k-1}$ denote the maximum daily ozone concentration for day k and k-1 respectively,
$P[O_{3max}]_k$ and $P[O_{3max}]_{k-1}$ represent respectively the temporal extent of $[O_{3max}]_k$ and $[O_{3max}]_{k-1}$ when they exceed a prescribed threshold,
$\hat{y}_{k+1}$, the output of the nonlinear predictor, is the estimated $[O_{3max}]$ for the next day,
$\hat{y}^{th}_{k+1}$ is the estimated $[O_{3max}]$ with respect to a given threshold,
u^{th} is a subset of u used as input for gain scheduling,
$\hat{P}_{k+1}$, the output of the second layer, is the estimated value of P for the day k+1.

Figure 10. Interconnection scheme of the monitoring system.

4.2. Adaptive Non Linear State-Space Based Prediction

4.2.1. Model Structure for Ozone Forecasting

The observed ozone time series y can be decomposed as

$$y = \hat{y} + e \tag{1}$$

where $\hat{y}$ is the model output and e is the error or residual resulting from the approximation of y by $\hat{y}$ (i.e. the part of the data which is not explained by the model). e represents a general stochastic input that accounts for the combined effects of other factors affecting the original time series, such as noise, other stochastic inputs or model limitations. $\hat{y}$ is the output of a nonlinear dynamic model :

$$\hat{y}_{k+1} = \Phi(y_k, u_k, u_{k+1}, \hat{\theta}_k) \tag{2}$$

where Φ is a nonlinear function, u represents the input variables to the model, k is the discrete time (current day). θ is a vector of free parameters to be computed and $\hat{\theta}$ represents an estimation of θ issued from some optimization procedure.

The exploratory multivariate graphical displays reveal interactions in the relationships between the various meteorological variables in their association with ozone. The graphical displays can motivate various non linear functional forms for the ozone model. In order to optimize the nonlinear model structure, many variable transformations and groupings of terms were evaluated. The suggested multiplicative non linear relation (1) has the following form:

$$\begin{aligned}
\hat{y}(k+1) = {} & \\
& \theta_1(k)y(k) + \theta_2(k)y(k)u_1(k) + \\
& [\theta_3(k) + \theta_4(k)u_2(k+1) + \theta_5(k)u_2(k+1)^2 + \theta_6(k)u_2(k+1)^3 + \\
& \theta_7(k)u_3(k+1) + \theta_8(k)u_4(k+1) + \theta_9(k)u_5(k+1)] \times \\
& \tanh(\theta_{10}(k)/u_6(k+1)) \times \tanh(\theta_{11}(k)/u_7(k+1)) \times \tanh(\theta_{12}(k)/u_8(k+1)) \\
& + \theta_{13}(k)u_9(k+1) + \theta_{14}(k)u_{10}(k+1) + \theta_{15}(k)
\end{aligned} \tag{3}$$

Generally, the problem of minimal model structure in nonlinear dynamic systems is not an easy task and the problem is largely open. Detecting the model structure or determining which terms to include in the model is an important issue in nonlinear system identification. Various approaches have been proposed to address this problem (see for instance Mao and Billings 1997). One of the most efficient and most frequently used model structure detection techniques is the orthogonal algorithm. The advantage of using the orthogonal algorithm is that the contributions of candidate terms are decoupled and consequently the significance of model terms can be measured based on the corresponding error reduction ratios. In this work,

we used an empirical approach to select a suitable model structure as follows. The model structure has been obtained through nonlinear and polynomial regressions of ozone against different input variables. The effect of relative humidity, wind speed and wind direction is modelled by tanh (hyperbolic tangent) functions which are multiplied by polynomial functions of temperature, solar intensity and radiation and pressure. The procedure could be relatively long. Note that a similar approach has been used in (Bloomfield et al. 1996) for defining non linear relationships between ozone concentrations and meteorological variables in the Chicago area. However, with respect to the work reported in (Bloomfield et al. 1996), the structure of the model proposed here is more evolutionary and flexible. In fact, the overall evaluation of the model performance can be done easily by excluding or adding input variables. Finally, strictly speaking, a "general" solution for model structure selection will probably not be found, since each urban area will have to be studied with respect to its meteorology, environment and local sources. The nonlinear model structure is then site-specific.

4.2.2. Model Structure for PM_{10} Forecasting

The observed PM_{10} time series is modelled in a similar way. The suggested multiplicative non linear relation has the following form (see table 2 for the definition of input variables):

$$\begin{aligned}\hat{y}(k+1) &= \theta_1(k)y(k) + \theta_2(k)y(k)u_5(k+1)\\ &+ [\theta_3(k)u_1(k) + \theta_4(k)u_2(k) + \theta_5(k)u_3(k) + \theta_6(k)y(k) + \theta_7(k)] \times \tanh(\theta_8(k)/u_4(k+1))\\ &+ \theta_9(k)u_5(k+1) + \theta_{10}(k)u_6(k+1) + \theta_{11}(k)u_7(k+1) + \theta_{12}(k)u_8(k+1) + \theta_{13}(k)\end{aligned} \tag{4}$$

The effect of wind speed is modelled by a tanh (hyperbolic tangent) function which is multiplied by a linear combination of input pollutants for the day *k* ([CO], [NO], [NO_2] and [PM_{10}]). The effect of the variable “DoW” is modelled using a classification variable. In contrast to other variables which take on continuous values, this classification variable is used to represent categories or discrete levels of information.

The model involves 13 free time varying parameters which will be automatically updated every day. Note again that the nonlinear model structure is site-specific. However, the proposed model structure (4) offers the advantage to be evolutionary and flexible enough, in the sense that the overall evaluation of the model performance can be done easily by excluding or adding input variables.

4.2.3. Extended State Estimation and Adaptation Mechanism

Extended state estimation and adaptation mechanism

The relation (2) can be expressed by the following nonlinear state space representation:

$$x_{k+1} = f(x_k, u_k, v_k) \tag{5}$$

$$y_k = g(x_k, w_k) \tag{6}$$

where x is the augmented state vector of the process (containing $[O_{3max}]$ and θ), v is the measurement noise and w is the process noise.

Note that for ozone we have $\theta \in R^{15}$ (see equation 3) and so $x \in R^{16}$ and for PM_{10} $\theta \in R^{13}$ (see equation 4) and so $x \in R^{14}$.

It is assumed that v and w are both stationary white sequences, Gaussian with covariance matrices

$$Q = E\{v_k v_k^T\} \tag{7}$$

$$R = E\{w_k w_k^T\} \tag{8}$$

The initial estimates of state and covariance matrix are denoted by

$$\bar{x}_0 = E\{x_0\} \tag{9}$$

$$P_0 = E\{(x_0 - \bar{x}_0)(x_0 - \bar{x}_0)^T\} \tag{10}$$

Each parameter θ_i is modelled by a dynamic state equation

$$\theta_i(k+1) = \theta_i(k) + v_i(k) \tag{11}$$

The last state ($[O_{3max}]$) is modelled as in the relations (3) and (4), with additional state disturbances affecting input variables in order to take into account the inaccuracy of weather forecasts. The problem of recursively estimating x can now be formulated as a nonlinear filtering problem. According to a specification of the uncertainties in model (Q) and measurements (R), the filter calculates an optimal forecast of the state and its covariance matrix, whenever observations become available, the forecast and the observations are merged during an analysis step. A classical way to solve the filtering equations is to use the extended Kalman filter (EKF). The EKF formalism uses matrices of derivatives of the process and measurement functions, starting with an initial guess. The state transition and observation equations (6) and (7) are approximated by first-order polynomials

$$x_{k+1} \approx f(\hat{x}_k, u_k) + F_x(k)(x_k - \hat{x}_k) + F_v(k)(v_k) \tag{12}$$

$$y_k \approx g(\bar{x}_k) + G_x(k)(x_k - \bar{x}_k) + G_w(k)(w_k) \tag{13}$$

where

$$F_x(k) = \left.\frac{\partial f(x, u_k)}{\partial x}\right|_{x=\hat{x}_k} \tag{14a}$$

$$F_v(k) = \left.\frac{\partial f(\hat{x}, u_k, v)}{\partial v}\right|_{v=0} \tag{14b}$$

$$G_x(k) = \left.\frac{\partial g(x)}{\partial x}\right|_{x=\bar{x}_k} \tag{14c}$$

$$G_w(k) = \left.\frac{\partial f(\bar{x}_k, w)}{\partial w}\right|_{w=0} \tag{14d}$$

when these approximations are inserted we arrive at (See for instance Ljung and Söderström 1983, Nogaard et al. 2000a):

A priori updates

$$\bar{x}_{k+1} = f(\hat{x}_k, u_k, v_k) \tag{15}$$

$$\bar{y}_k = g(\bar{x}_k, w_k) \tag{16}$$

$$\bar{P}(k+1) = F_x(k)\hat{P}(k)F_x(k)^T + F_v(k)Q(k)F_v(k)^T \tag{17}$$

A posteriori updates

$$K_k = \bar{P}(k)G_x(k)^T\left[G_x(k)\bar{P}(k)G_x(k)^T + G_w(k)R(k)G_w(k)^T\right]^{-1} \tag{18}$$

$$\hat{x}_k = \bar{x}_k + K_k(y_k - \bar{y}_k) \tag{19}$$

$$\hat{P}(k) = \left[I - K_k G_w(k)\right]\bar{P}(k) \tag{20}$$

At each sampling time, the *a priori* updates give then a one step ahead prediction of the state and its covariance. The confidence boundaries can be placed for each estimate based on the estimated covariance matrix $\bar{P}(k+1)$. Note that the estimated covariance matrix $\bar{P}(k+1)$ that comes out of the model estimation procedure has meaning only if (or to the extent that) the prediction errors actually are normally distributed.

In the context of this study, the main practical problems with the above EKF algorithm are the followings. First, convergence to a reasonable estimate may not be obtained if the initial guess is poor or if the disturbances are so large that the linearization is inadequate to describe the dynamics of ozone variation. Second, if the structure of the state model change (for instance if it becomes necessary to add or to exclude some input variables or if the model is to be used for another city), the user has to restart all the analytical matrix derivatives calculations (equations 14). This is a key feature with respect to the practical portability of the resulting software. The prediction software is expected to be used in different sites and so adaptation and reconfiguration tasks are to be performed easily.

The EKF has some well known disadvantages. For instance, it is well known that the parameter estimates can converge far slower than the state estimates, when compared with an algorithm that estimates the parameters separately to the state and then incorporates this in an adaptive state estimation algorithm. Also, convergence to a reasonable estimate may not be obtained if the initial guess is poor or if the disturbances are so large that the linearization is inadequate to describe the dynamics of ozone (or PM_{10}) variation. Recently, state estimators for nonlinear systems are derived which performs much better than the EKF. (Wan and Van der Merwe 2000) proposed the so-called unscented Kalman filter (UKF) for nonlinear state space model estimation. In (Nogaard et al. 2000a), the authors developed a method based on polynomial approximation of the nonlinear approximations obtained with a multi-dimensional extension of Stirling's interpolation formula. It is shown in (Nogaard et al. 2000a) that under certain assumptions, such estimators perform better than estimators based on Taylor approximations. They are fundamentally different from estimators based on Taylor approximations in that the polynomial approximations used take into account the uncertainty on the state estimate. The Taylor approximation underlying conventional filter designs for nonlinear systems, such as the EKF, depends only on the current state estimate and not on its variance. Moreover, the filter implementation is significantly simpler as no derivatives are required. The experimental results presented in the following section have been obtained from the estimation algorithms in (Nogaard et al. 2000a). In the interest of brevity and to avoid duplicating material presented in (Nogaard et al. 2000a), the derived mechanization equations for nonlinear prediction are not presented here. The interested reader can refer to (Nogaard et al. 2000a) for detailed background and proofs.

4.3. Optimization of Hyper Parameters

The matrix Q (equation 7) controls the flexibility of the state space model (equation 5). In this application, noise has been added to all states and parameters. By choosing values for the diagonal elements of the matrix Q, the flexibility of the model is controlled by the amount of noise allowed. Similarly, the matrix R controls the flexibility of the measurement equation (equation 6). Generally Q and R are called "hyper-parameters". There is a lot of literature on the optimization of hyper-parameters in recursive estimation algorithms. See for instance (Young et al. 1991, Ng and Young 1990, Young 1999, Young et al. 1999) and the references herein for a discussion. In (Young et al. 1991), the observation noise is normalised to unity and the optimization is done over the "noise variance ratio" (NVR) matrix. In this study, the optimization of the hyper parameters is done according to the above procedure, adjusted by iteratively testing different values and evaluating the results over a test period.

4.4. Auxiliary Module for Threshold Exceedance (Ozone)

The recursive algorithm for estimation of the extended state, minimizes "average" estimation error. As it will be seen from the experimental results, the time varying non linear model can account for a substantial portion of the daily maximum ozone variation. However, if the threshold exceeding process is just based on the predicted maximum value, under-predicted values could cause low detection rate, even if the average behaviour is quite well captured due to the adaptation algorithm.

As it will be seen in section 6, minimax or l_∞ filtering (see for instance Simon and El-Sherief 1996) would be more appropriate if only good estimation of extreme values is of interest. Minimax filtering minimizes the maximum singular value of the transfer function from the noise to the estimation error. A different approach is to use extreme value theory and to compute a forecast probability of exceedance for a day, given its forecasted meteorology and current-day maximum ozone. See for instance (Cox and Chu, 1996), where a Weibull hazard model is applied.

Here, a simple and useful alternative approach is adopted. To improve the detection performance for high level concentrations, an auxiliary module (AM) based on a gain scheduling mechanism is used. The main idea is to emphasize factors which contribute to extreme ozone values. The input of the AM contains the output of the NASSE a subset of meteorological variables. The AM aims at modelling only "threshold exceedance" for extreme ozone concentrations. The thresholds are defined by control authorities (see section 1). The output of the AM is the information "exceedance/non exceedance (no numerical prediction). Further correlation analyses of those days with high ozone levels, reveal that underestimations more frequently occurred while some meteorological variables were at their extreme values. Here, the appropriate subset of the meteorological variables was found to be maximum temperature, sunshine duration and wind speed. The gain scheduling system is based on an auxiliary linear regression model to emphasize such conditions. The parameters of this model, denoted σ^{th}, depend on the threshold i of interest and are estimated off-line (prior scheduling of gains). The output of the AM is computed as:

$$\hat{y}_{k+1}^{th} = \sigma_0^{th} + \sigma_1^{th}\hat{y}_{k+1} + \sum_{i=2}^{n}\sigma_i^{th}u_i^{th}(k+1) \tag{21}$$

where σ_i^s, i=0…n represent the AM parameters. These parameters have been estimated and fitted using only the data from high ozone days (days when ozone exceed a given threshold. See figure 11 below.

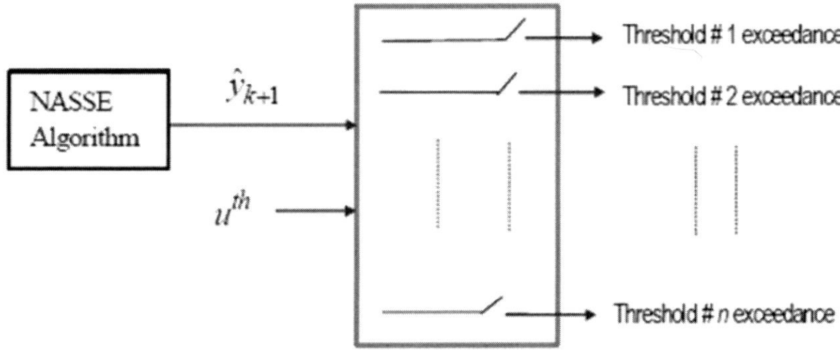

Figure 11. Gain scheduling.

Note that the AM module does not increase the "average" performance (prediction errors) of the NASSE algorithm; it only increases the detection rate with respect to the extreme values.

4.5. Prediction of the Temporal Extent of a Smog Episode (Ozone)

The warning system is expected also to provide information regarding the temporal extent of a smog episode. The "area over threshold" represents the amount by which the regulatory level is exceeded and may by of interest as an indicator of ambient ozone exposure. This information is quantified by the length of time for which the ozone level remains beyond a prescribed alarm threshold. Basically, the proportional hazard approach can be used to model a temporal extent (survival time). Another practically relevant way to model the smog duration is to use a functional estimator. In this paper a multi-layer perceptron (MLP, (Haykin 1999)) is used to model this phenomenon. The main reason for this is its ability to model very complex functional relationships. In this work, the learning algorithm used was Levenberg-Marquardt back-propagation of MATLAB® Neural network toolbox. The activation functions selected for the layers were hyperbolic tangent for the hidden layer and linear for the output layer. One hidden layer has been implemented and the number of neurons in the hidden layer was the optimum found by experimentation. The model structure corresponds to NARX (Nonlinear Auto-Regressive with eXogenous inputs). This model structure was selected because the predictor is always stable (there is a pure algebraic relationship between prediction and past measurements and input. This is particularly important because, as already mentioned, the relations between smog duration and ozone levels/meteorological variables are highly non linear and complex. The input variables are ones represented in figure 12.

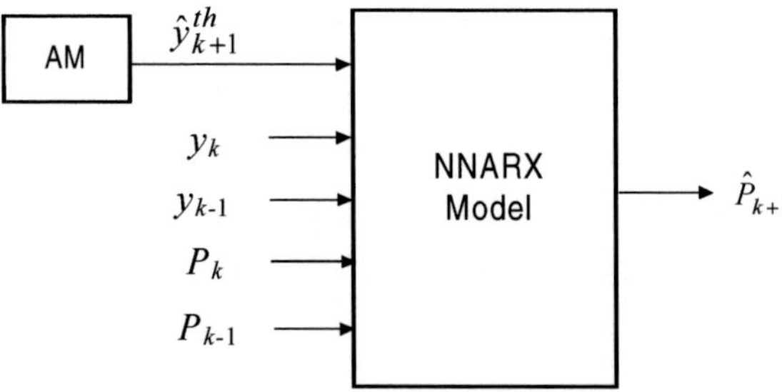

Figure 12. NNARX model structure.

The choice of the lag space, i.e. the number of delayed signals used as regressors was optimized by experimentation. The process of sequential cross-validation is done : the database is split in two parts. The first part is used for model tuning (three years) and the second part (one year) is dedicated to testing and evaluation. In this way, four sets of evaluation results are constructed. The main reason for this process is to improve the statistical reliability.

Input-output data were normalised to avoid overflows due to very large or very small weights. The network architecture was optimised using the so-called Optimal Brain Surgeon strategy (Haykin 1999) to avoid “over fitting”.

5. Experimental Results

5.1. Ozone

A range of model validation statistical indicators were computed to provide a numerical description of the goodness of the predictions. Performance indicators

The selected performance indicators are absolute mean error (AME) and percentage Variance Accounted For (VAF). Moreover, as already mentioned, the capability of predicting concentrations exceeding specified threshold levels is a critical aspect with respect to air quality standards. The detection rate (DR) is defined as the fraction of observed threshold exceedances predicted by the model. DR is a useful indicator of the model performance in predicting high threshold values.

For the years 1998-2001, the performance indicators corresponding for the predicted values from the NSSAE are given in the table 3, where N is the number of data points, O_i the observed data point, P_i the predicted data point and $\overline{O}$ the average of observed data.

The indicators are calculated for the selected measuring site (Bordeaux Grand Parc). The prediction results are quite good with respect to other works reported in the literature (see references). The absolute mean error over the four years period is inferior to 6 ppb (1 ppb = 1.963 μg/m^3 at 25°C). Figure 14 represents the histogram of the one step ahead prediction errors for the four years period. Figure 15 illustrates observed and forecasted ozone level as well as its confidence bands for ten days during July 2001. As expected, due to the Gaussian assumption of the prediction errors sequence, some observed levels do not fall within the confidence bands.

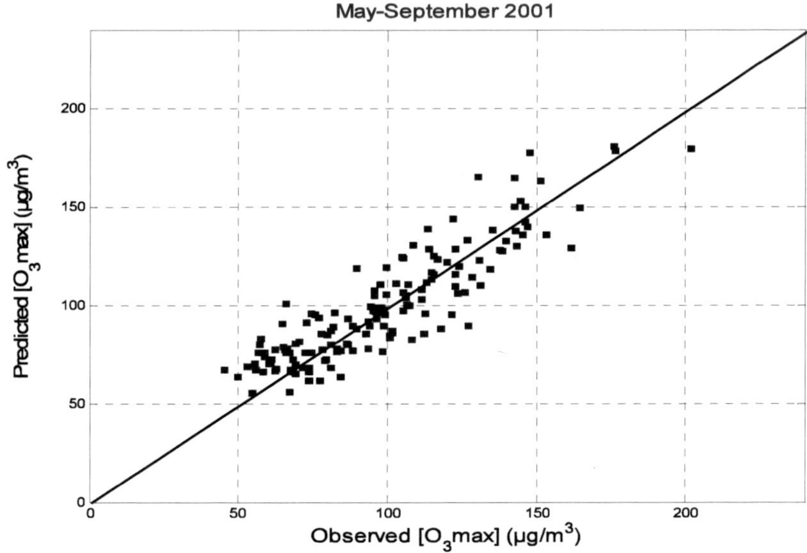

Figure 13. Cross plot of predicted and actual ozone levels for the NASSE.

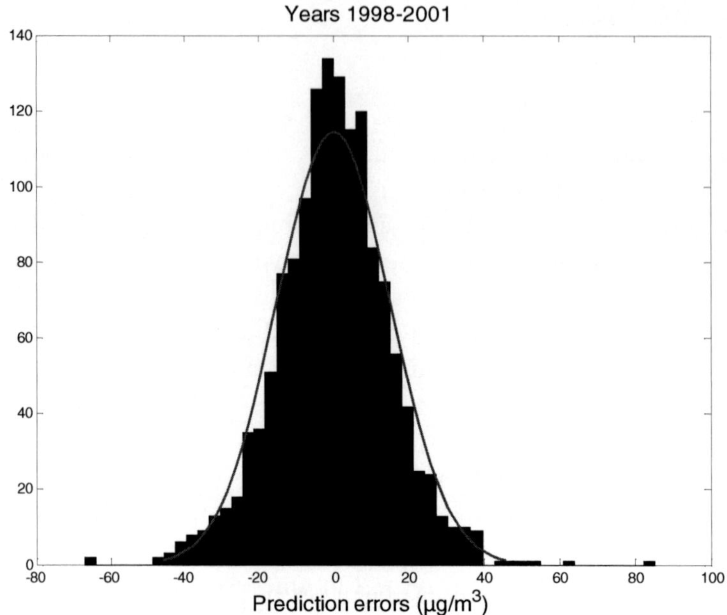

Figure 14. Histogram of the prediction errors (NASSE).

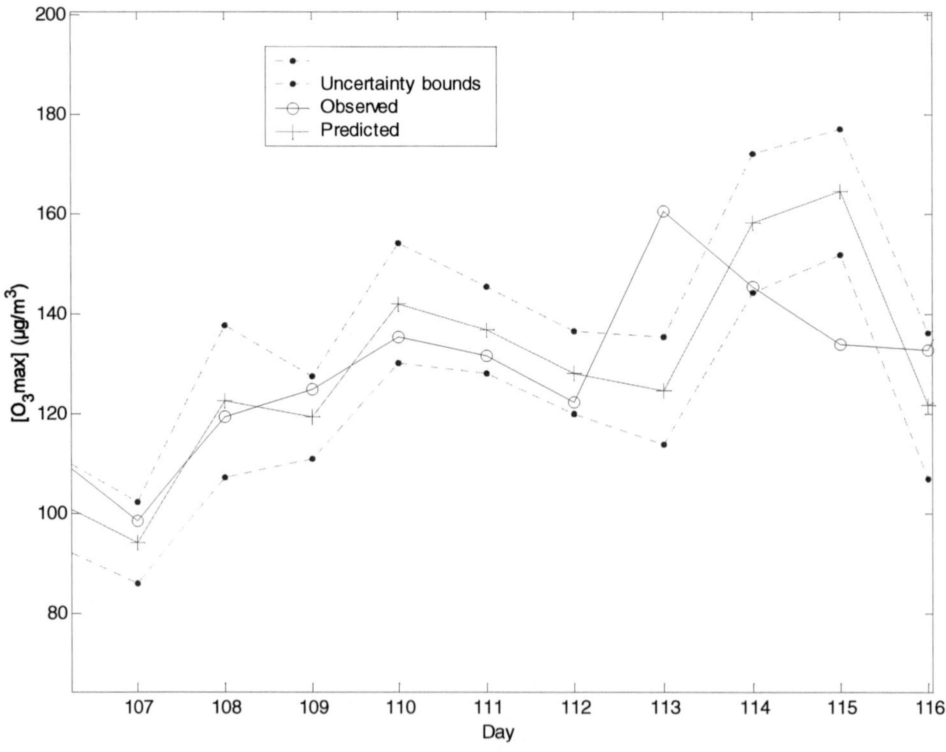

Figure 15. Example of predicted ozone time series and its uncertainty bounds.

Table 3. Performance indicators calculated for the output of the NSSAE

Indicator	value
$AME = \frac{1}{N}\sum_{i=1}^{N}\lvert P_i - O_i \rvert$	11.6 (μg/m³)
$VAF = \left(1 - \frac{\frac{1}{N}\sum_{i=1}^{N}[O_i - P_i]^2}{\frac{1}{N}\sum_{i=1}^{N}[O_i]^2}\right) \times 100$	81%

As it can be seen from figures and table 3, the average behaviour of the phenomena is well captured. The predicted output from the NSSAE, track the day-to-day variation in ozone concentrations well during the ozone season for the four years. However, generally, high level concentrations are underestimated, as it can be seen from the figure 8. To get a deeper insight, detection rates are calculated for some threshold (see table 3). It is true that the detection rate is not the only indicator to consider. To be more precise, air pollution guidelines suggest the following indicators: Let be m the total number of observed exceedances, f the total number of predicted exceedances and a the number of correctly forecasted exceedances. Then, three skill parameters can be defined : rate of detection, false alarms and success index (see appendix).

We observe that the average behaviour of the phenomena is well captured from the NSSAE. However, generally, high level concentrations are underestimated. The reason is that the problem of parameter fitting is solved using an average criterion. A direct consequence is that the NSSAE, by nature, does not make any distinction between low and high levels of the time series. The process aims at modelling the "average" behaviour by minimizing a quadratic loss function, whereas with regards to air quality standards, it is largely the high levels that are of interest and concern. During our study, we noted rarely false alarms with respect to the thresholds. That is why, in this paper, we present just the detection rates. Note that the predicted values just under the threshold are regarded as undetected events, without considering the uncertainty intervals which come with the nominal predicted value.

Table 4. Detection calculated for the NSSAE forecasts

Threshold (μg/m³)	120	130	140	150	160	170	180
Number of observed days exceeding the threshold	188	134	87	53	39	25	15
Number of detected days exceeding the threshold	144	96	58	32	20	13	6
Detection rate ()%	77	72	67	60	51	52	40

For studying the limitations of the nonlinear predictor, those days with high differences between observed and predicted ozone indices were further scrutinized. It was observed that underestimations more frequently occurred on days with high temperature and sunshine duration and low wind speed. Overestimations occurred less frequently than underestimations. As the decision label (polluted/non polluted event) is based on threshold exceedances, the gain scheduling system was designed to improve underestimated events. As already mentioned the parameters of the AM depend on the threshold of interest and are designed based on a linear regression model to emphasize the most important meteorological parameters which produce high level ozone concentrations.

The results are now summarized in the table 5 below:

Table 5. Improved detection rates by AM

Threshold (μg/m^3)	120	130	140	150	160	170	180
Number of observed days exceeding the threshold	188	134	87	53	39	25	15
Number of detected days exceeding the threshold	176	125	79	50	36	21	12
Detection rate (%)	93	93	90	94	92	84	80

The detection rates are significantly improved by using the auxiliary model. This fact can be observed on figure 16. The results are depicted for thresholds 180 μg/m^3.

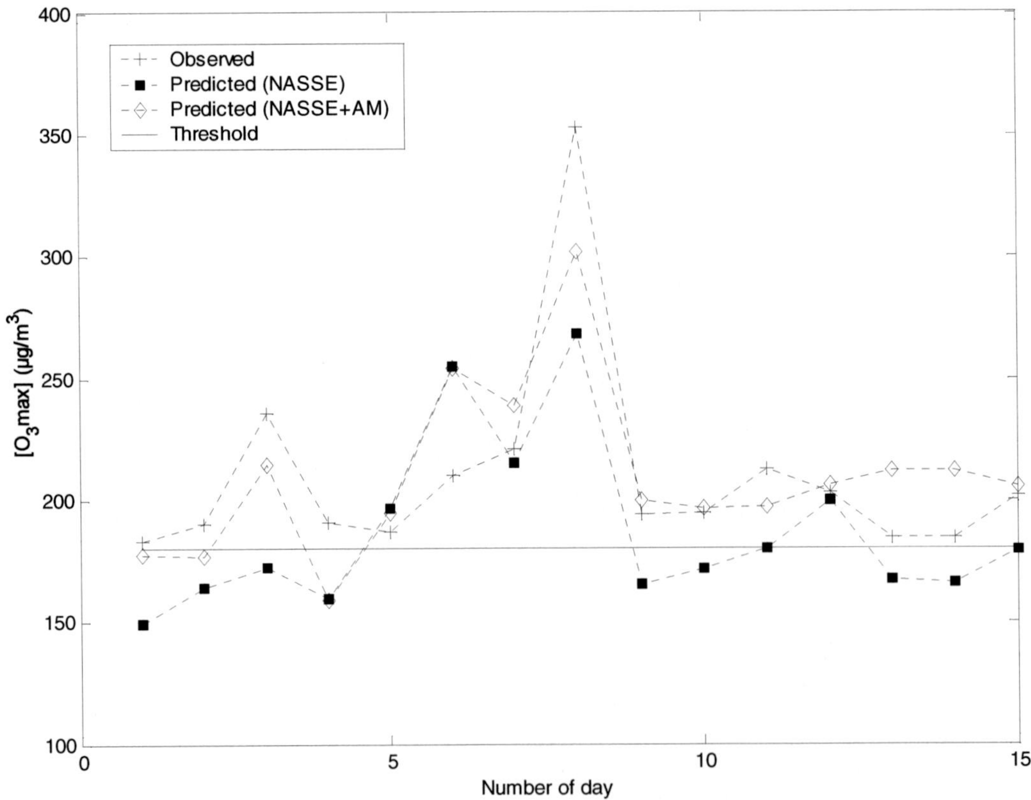

Figure 16. Improvement of the detection rate by gain scheduling. Threshold =180μg/m^3.

Finally, the results of the MLP for temporal extent of a smog episode are examined. This information is quantified by the length of time for which the ozone level remains beyond a prescribed alarm threshold. Despite the strong nonlinear character of the phenomenon, the MLP gives rather good predictions for temporal extents of threshold exceedance (figure 17). Note that the threshold was set to 120 μg/m^3 to have a higher number of detected events than in the case of a 180 μg/m^3 threshold. For the months June, July and August 2001, the absolute mean error is about 3 hours.

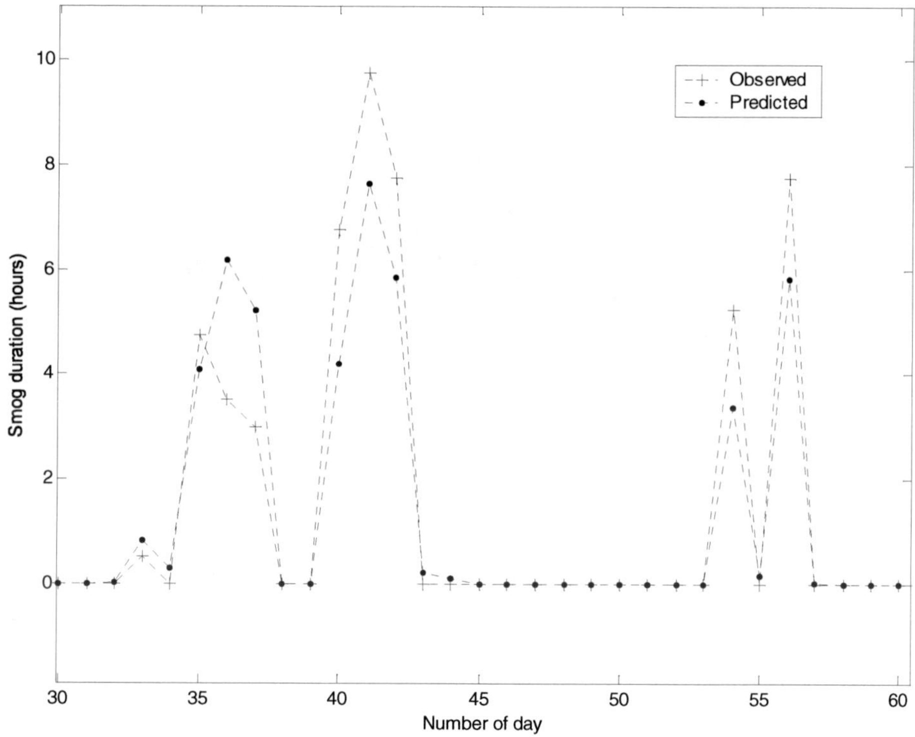

Figure 17. Observed and one day ahead predicted smog duration during July 2001.

5.2. PM_{10}

The initial state x_0 and its covariance P_0 were set up off-line using a restricted training set of data (year 1998). This initialisation stage can be done by means of a nonlinear least squares optimization algorithm, based on the Levenberg-Marquadrt or Gauss-Newton methods.

As already mentioned, air quality standards for PM_{10} are established taking into account the daily mean concentration values (not the maximal values). The choice of the relevant set of performance indicators is then different.

The following statistical indicators were computed to provide a numerical description of the goodness of fit measures of forecast quality: squared correlation coefficient (R^2), index of agreement (IA) and absolute mean error (AME). For the definition of the selected performance indicators, the interested reader can refer to (Gardner and Dorling 2000) and the references therein. For comparison, the results obtained from a persistent model are also presented. A persistent model is simply a prediction model defined as follows: $\hat{y}(k+1) = y(k)$. The corresponding results are given in table 8. The one-day ahead mean PM_{10} predictions for the years 1999, 2000 and 2001 are depicted on figure 18. The corresponding indicators are given in table 7.

Table 7. Statistical indicators for the predicted and measured time series of PM_{10} concentrations

	1999	2000	2001
AME (μg/m^3)	4.33	4.77	4.15
R^2	0.72	0.68	0.70
IA	0.91	0.90	0.91

Table 8. Persistance model

	1999	2000	2001
AME (μg/m^3)	6.3	6.13	5.7
R^2	0.35	0.34	0.37
IA	0.78	0.8	0.81

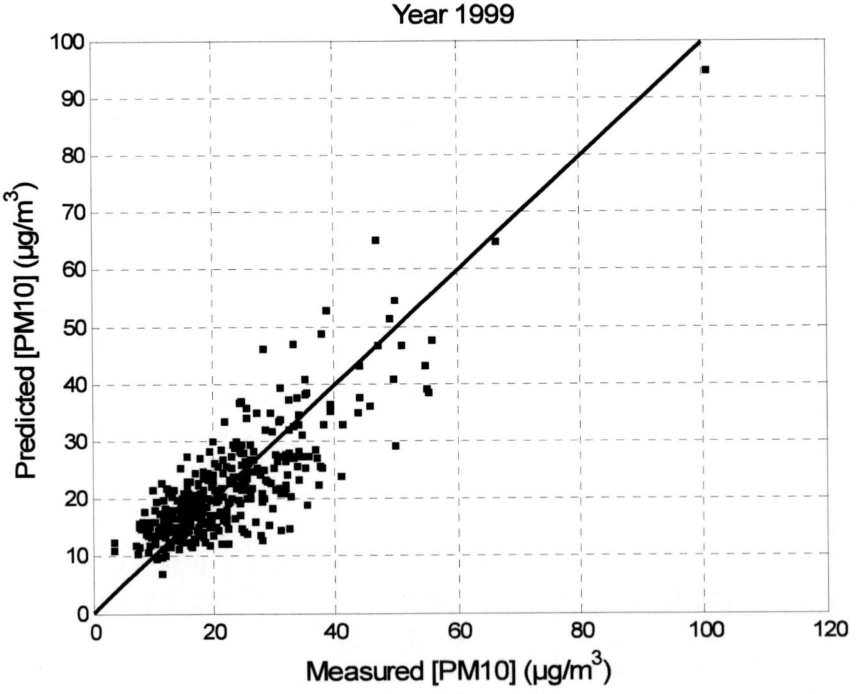

Figure 18. Continued on next page.

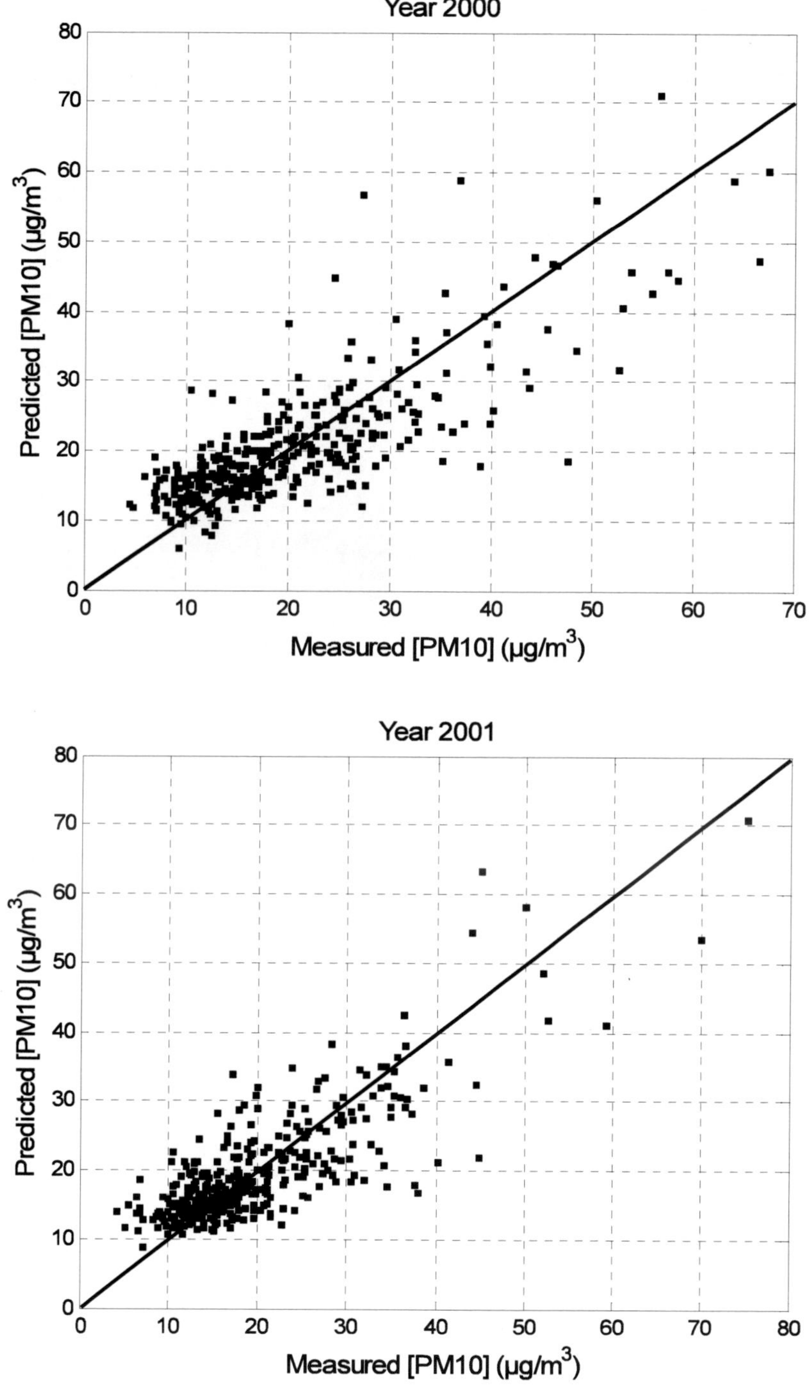

Figure 18. Cross plots of predicted and measured PM_{10} levels (years 1999, 2000, 2001).

The forecast results are good with respect to other works reported in the literature for the prediction of daily mean PM_{10} and the simple persistence model. Figure 19 represents the histogram of the one step ahead prediction errors for the three years period. As an example, figure 20 represents the evolution of the estimated parameters θ_1 (weight for the contribution of the past day PM_{10}) and θ_8 (weight for wind speed) as a function of time. The parameters are updated every day to cope with changing dynamics of PM_{10} concentrations.

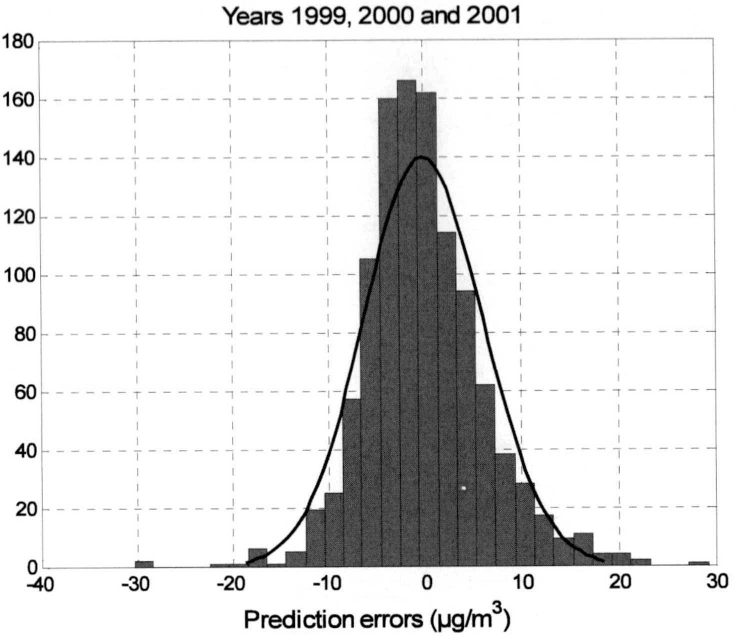

Figure 19. Histogram of the prediction errors.

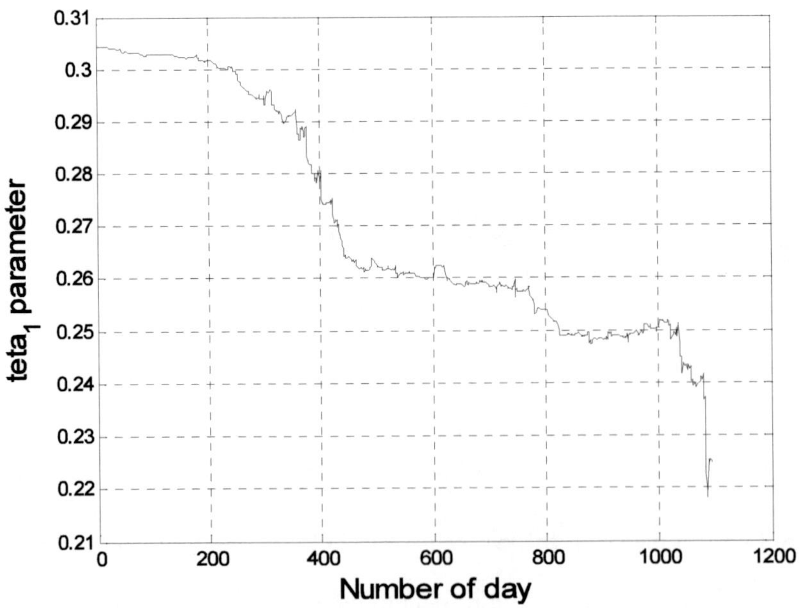

Figure 20. Continued on next page.

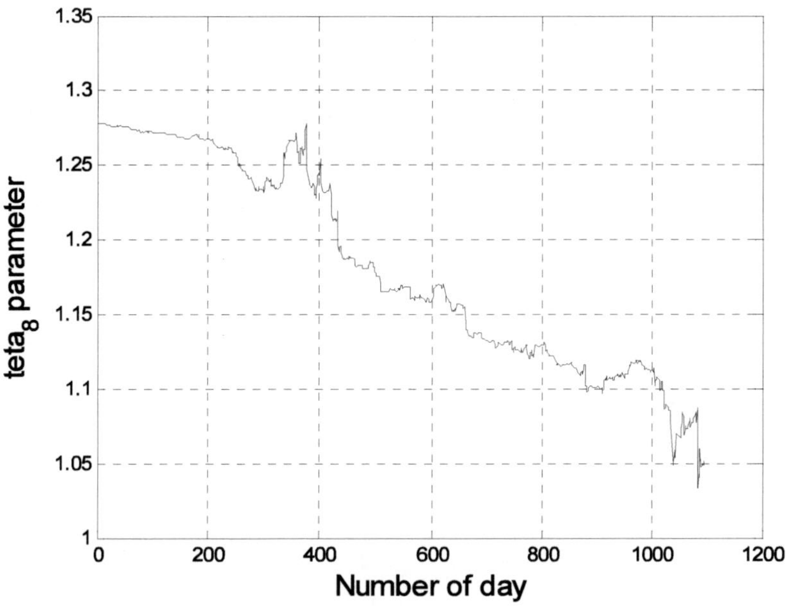

Figure 20. Evolution of estimated parameters θ_1 and θ_8.

5.3. Concluding Remarks

In the previous sections, a methodology is proposed which combines an adaptive nonlinear state space-based prediction mechanism, gain scheduling and neural network techniques to develop an integrated operational warning system for ground-level ozone and PM_{10} monitoring. The goal is to obtain one-day-ahead forecasts, the confidence bands and an estimate of the next day's duration of a smog episode (ozone case).

For ozone monitoring, the first important advantage of the proposed system is that reliable estimated can be obtained, while the implementation of the overall method is simple and the computational effort is rationally low (in contrast to deterministic models, for instance). Another key characteristic of such a system is that its behaviour (model parameters) can be adapted to the short term changes of air pollution time series. So, contrary to rigid and unchanging nonlinear models reported in the literature, the proposed system can handle the time-evolving nature of the phenomena and does not need frequent adjustments. This is an important aspect when considering the testability and certificability of the software implementation. Moreover, the warning system provides additional information regarding the extent of a smog episode that is very important for assessments of human health hazards and negative environmental effects. Finally, while the prototype developed in this work has been fine-tuned for the specific case of Bordeaux, the process can be repeated in new problems and the proposed techniques can be applied in many similar air quality monitoring problems.

PM_{10} concentrations are generally associated with highly erratic and irregular dynamics. That is why the prediction-oriented modelling task is quite complex. The proposed adaptive non linear state space approach appears to be very useful for predicting daily mean concentrations of PM_{10}. The nonlinear model structure has been partly determined by exploring empirical relationships between the measured PM_{10} and other primary pollutants and meteorological variables. Once an appropriate initial state has been set up off-line, the parameters of the model are updated every day (on-line), due to the adaptation mechanism. Mean PM_{10} levels recorded at Bordeaux metropolitan area were not very high. In the study period (years 1999, 2000 and 2001), twenty five daily PM_{10} values exceeded the EU limit value of 50 μg/m^3. The model should be validated in the future for an urban area with higher levels of PM_{10}. If available, traffic emissions (traffic volumes and averages travel speeds) could also be easily incorporated into the model.

Generally speaking, statistical models used in air pollution studies are usually optimized by minimizing an "average" criterion. The resulting models, by nature, fall short of adequately modelling extreme values, as the high values are usually predicted too low. As a consequence, they generally lead to poor results in terms of "threshold exceeding", which is of great concern with respect to air quality standards. In the next section, some inherent shortcomings associated with the commonly used statistical techniques for air pollution modelling are underlined and an alternative appealing solution based on l_∞ (or minimax) optimization techniques is proposed.

6. Modelling Extreme Values for Air Quality Monitoring

6.1. Problem Statement

As already discussed in this chapter, statistical models based on linear or nonlinear regression analysis have been largely applied to address issues of air pollution. See (Milionis and Davies, 1994) and (Thompson et al., 2001) for a survey. The goal is the identification of a functional relationship which would allow the prediction of the concentration of pollutants if the emissions and other meteorological variables are given (Thompson et al. 2001; Shi and Harrison, 1997; Jorquera et al. 1998; Davis and Speckman, 1999; Chen et al., 1998; Schlink and Volta, 2000; Schlink et al., 1997; Cobourn et al., 2000; Hubbard and Cobourn 1998; Gardner and Dorling, 2000; Bloomfield et al., 1996; Young et al., 1991; Ziomas et al., 1995; Van der Wal and Janssen, 2000; Chaloulakou et al., 1999). A statistical model building process involves three main steps: determination of a class of *model structures* or model sets, i.e. families of models with adjustable parameters, *parameter estimation* which amounts to finding the "best" values of these parameters and *model validation*. The goal is to find those numerical values of the free parameters that give the best agreement between the model's predicted output and the measured one. Model validation is achieved by statistical numerical tools and graphical residual analysis to check the ability of the model to reproduce the behaviour of some testing data sets.

In the context of air pollution, much attention has been paid to site-specific model structure determination and model validation. However, there is very little guidance about how to choose a suitable method for parameter estimation. Usually, the problem of parameter fitting is solved using a standard "least squares" criterion. A direct consequence is that the resulting models, by nature, do not make any distinction between low and high levels of the time series, as they model the "average" behaviour by minimizing a quadratic loss function, whereas with regards to air quality standards, it is largely the high levels that are of interest and concern. It is well known that such models fall short of adequately modelling extreme values, as the high values are usually predicted too low (Thompson et al., 2001). Moreover, generally we need to assume a Gaussian distribution in order for the model output to have the desired optimality properties. However, it is well known that most air pollution data (for instance ozone time series) involves a statistics that is non-Gaussian (Thompson et al., 2001).

One solution to overcome the above mentioned difficulties is to use the so called extreme value approach. Extreme value theory is concerned with statistics related to very high (or very low) values in stochastic processes. See (Smith 2003 and Davison and Smith 1990) for a general survey and for applications to environmental time series. In (Smith and Shively 1996) an approach based on extreme value theory was used to compute a forecast probability of exceedance for a day, given its forecasted meteorology and current-day maximum ozone. See also (Cox and Chu 1996), where a Weibull hazard model is applied.

In this section, an attempt is made to underline some inherent shortcomings associated with the commonly used parameter estimation techniques for air pollution statistical modelling and to propose an alternative solution. Here, the parameter estimation stage is based on l_∞ (or minimax) optimization techniques. In linear case, this corresponds to the so-called "l_∞ filtering" techniques which can be found in the literature dealing with modern robust estimation (Shen and Deng, 1997; Burl 1999; Simon and El-Sherief, 1996). It will be shown that that l_∞ filtering techniques have important potential benefits for air pollution time series modelling. They are more appropriate if only good estimation of extreme values is of interest. In addition to the case of linear model structure for which the theory is now well established (Burl 1999), a nonlinear model structure is also tested. Its free parameters are estimated in an l_∞ setting by means of sequential quadratic programming.

Notations Used in this Section

Let $x \in R^N$ be a discrete time series. The l_p norm of x is defined as $\|x\|_p = \left(\sum_{i=0}^{N-1} |x_i|^p\right)^{1/p}$

p=1: The l_1, "absolute value" norm.
p=2: The l_2, "Euclidean", "root energy" or "least squares" norm.
p=∞: The l_∞, "l-infinity" or "minimax" norm: $\|x\|_\infty = \max\limits_{0 \le i \le N} |x_i|$

The above norms can be used for modelling uncertain data. The l_1 norm is well-situated for removing a small set of outliers from a set of data point with otherwise high precision. The l_2 norm is typically used for data sets in which every data point is subject to a small, normally distributed noise. Finally, the l_∞ norm is appropriate if the approximation error should be evenly distributed amongst the data points.

6.2. L_∞ Parameter Estimation

To start, let the observed time series y be modelled as in equation (1) with $\hat{y} = \Phi(u, \hat{\theta})$.

A common choice to measure the misfit between model and observed data is the average quadratic loss of the residual (l_2 norm). In the context of air pollution modelling, the l_2 norm has traditionally been used almost exclusively. However, if it is desirable to put emphasis on modelling extreme values, a more appealing choice is to consider a minimax or worst-case estimation problem, where one seeks to find an estimator minimizing the worst-case error in the face of uncertainty which satisfies a certain stochastic uncertainty constraint. Here, we define the misfit by the l_∞ or amplitude norm of the residual. This approach differs from the traditional l_2 estimation approach in the following two aspects: 1) no statistical assumption on the exogenous noise signals is required; instead the noise signals are only assumed to have finite energy; 2) the estimation criterion for parameter fitting is to minimize the worst possible amplification of the estimation error signal in terms of modelling errors and additive noises. In applications where peak values modelling is of greater interest that the mean performance of estimators, this approach leads to highly robust and more appropriate design. The corresponding optimization problem can be formulated as

$$\arg\min\left\{\inf\|y-\hat{y}\|_\infty\right\} \tag{22}$$

or equivalently

$$\min_{\theta}\max_{\Phi}\|y-\Phi(\theta,u)\|_\infty \tag{23}$$

subject to some user specified nonlinear inequalities or equalities constraints.

In the linear case (linear model structure), the estimation problem can be easily formulated as a linear programming optimization problem. For the nonlinear case, the nonlinear optimization problem can be solved using a Sequential Quadratic Programming (SQP) method (Boggs and Tolle, 1995). SQP methods are generally considered to some of the most efficient algorithms available today. Based on Lagrangian methods, it can be shown that the solution $\hat{\theta}$ of the minimax optimization problem can be obtained by solving, at successive approximation θ^i to $\hat{\theta}$, a sequence of corresponding Quadratic Programming subproblems (QP[i], i=1,2, …) containing linearized constraints (Snyman and Hay, 2002). The solution to subproblem is denoted by s^i and the point at which the next subproblem QP[i+1] is constructed is $\theta^{i+1} = \theta^i + s^i$. If successful, the SQP method yields a sequence θ^0, θ^1, θ^2, … that converges to $\hat{\theta}$. See for instance (Brayton et al., 1979) where modifications are made to the line research and Hessian. Note that the algorithm may give local solutions.

6.3. Experiments

In this paragraph, some experimental results are presented in order to evaluate the predictive ability of both a linear regression model and a nonlinear model, optimized by means of l_∞ techniques. The case study is related to one-day-ahead prediction of ground level ozone concentrations for Bordeaux metropolitan area. See section 2 for a description of ozone and meteorological data and also the input variables.

6.3.1. Model Structures

For this study, two model structures have been selected. The first one corresponds to a linear model structure:

$$y(k)=\theta_0+\sum_{i=1}^{10}\theta_i u_i(k)+\theta_{11}y(k-1)+e(k) \tag{24}$$

u is a vector which contains all previously described ozone precursors and meteorological variables, k denotes the discrete time (current day), y is the $[O_{3max}]$, θ is the vector of unknown parameters and e is a disturbance term. Note that lagged ozone data $y(k\text{-}1)$ (the 1-hour ozone maximum from the previous day) are included as an predictor variable. The noise term represents the part of ozone which cannot be captured this linear model.

As already discussed in the previous sections, the ozone formation process, involving precursors, atmospheric transport and mixing, and a complex system of photochemical reactions, is highly nonlinear. The use of nonlinear techniques is often recommended to deal with the ozone prediction (Schlink et al. 2006). The simple linear model in equation (24) may give poor results, even if the free parameters are optimized in an l_∞ setting. That is why, as in section 3 (see equation 3), a nonlinear model structure is also selected.

6.3.2. Results and Discussion

The model is expected to provide qualitative information as an end-product. As numerical information is available in the step prior to the classification, some usual performance indicators are also evaluated (AME, VAF, d1 and d2, see appendix). d1 and d2 indicate the degree to which predicted deviations differ from observed deviations, considering both sign and magnitude. Moreover, as far as the ability of the model to forecast true episode-days is concerned, the detection rate (number of days with correct alarms) and the false alarms score (number of days with false alarms) are the most critical performance index (De Leeuw 2000). See appendix for the definition of all statistical indicators used for the comparative assessment of models. As already mentioned, for sequential cross-validation, the available dataset has been used as follows: model tuning has been performed using three years of data and the remaining year was used for model evaluation. There was some variability from year to year. The average results for the four testing years (1998-2001) are summarized in the tables 9 and 10. The threshold was set to 130 μg/m^3 to have a higher number of detected events than in the case of a 180 μg/m^3 threshold. For comparison, the mean values obtained from a persistent model are also presented. A persistent model is simply a prediction model defined as follows: $\hat{y}(k+1) = y(k)$. The indicators are calculated for the selected measuring site (Bordeaux Grand Parc).

Table 9. Mean performance indicators and forecast skill (occurrences of $[O_3]$>130 μg/m^3), linear model

Indicator	l_2 parameter optimization	l_∞ parameter optimization	Persistence model
AME (μg/m^3)	11.3	12.9	16.1
VAF (%)	80	76	58
d1 (0→1)	0.76	0.74	0.66
d2 (0→1)	0.93	0.91	0.85
SP (%)	48	75	52
SR (%)	91	78	55
FA (%)	9	22	44
SI (%)	49	73	48

Table 10. Mean performance indicators and forecast skill (occurrences of $[O_3]$>130 $\mu g/m^3$), nonlinear model

Indicator (mean value 1998-2001)	l_2 parameter optimization	l_∞ parameter optimization	Persistence model
AME ($\mu g/m^3$)	10.2	11.4	16.1
VAF (%)	84	81	58
d1 (0→1)	0.79	0.77	0.66
d2 (0→1)	0.95	0.93	0.85
SP (%)	70	79	52
SR (%)	81	74	55
FA (%)	19	24	44
SI (%)	70	87	48

As expected, globally the nonlinear model performs better than the linear one. When the free parameters are optimized by minimizing the l_2 norm of the residuals, the average behaviour of both models is superior to the case where these parameters are estimated in an l_∞ setting (see indicators AME, VAF, d_1 and d_2). However, in the l_∞ case, both models give better performance in terms of "threshold exceeding". See figures 21 and 22. In the l_2 case, both models tend to underestimate high-ozone episodes and overestimate low-ozone events. In this case, the global Success Index (see appendix) is significantly higher. It should be noticed that in the l_∞ case, the false alarm rate is slightly more important with respect to the l_2 case.

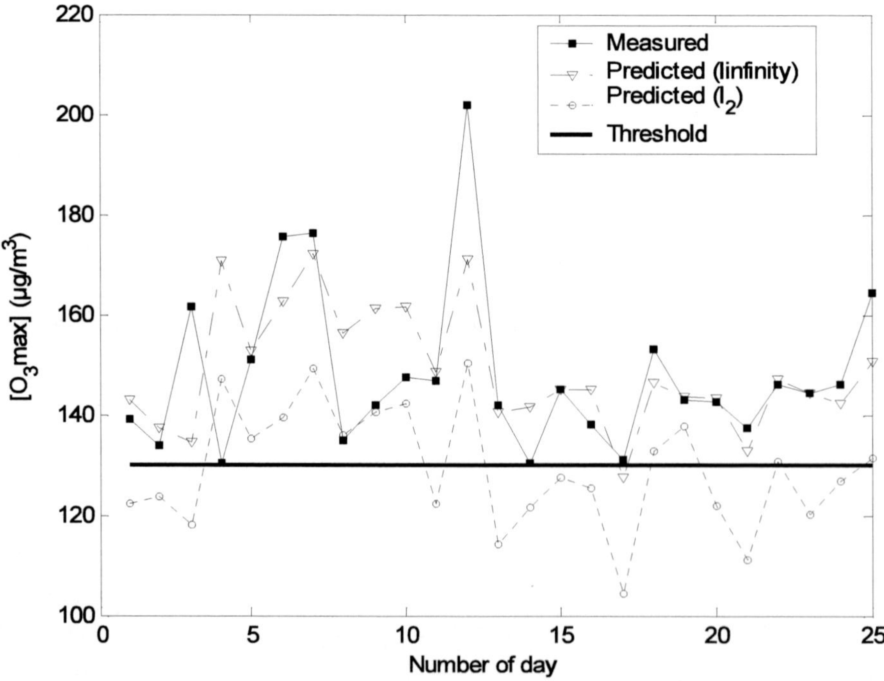

Figure 21. Produced forecasts by the linear model for the days exceeding the threshold (130 $\mu g/m^3$). Testing year 2001.

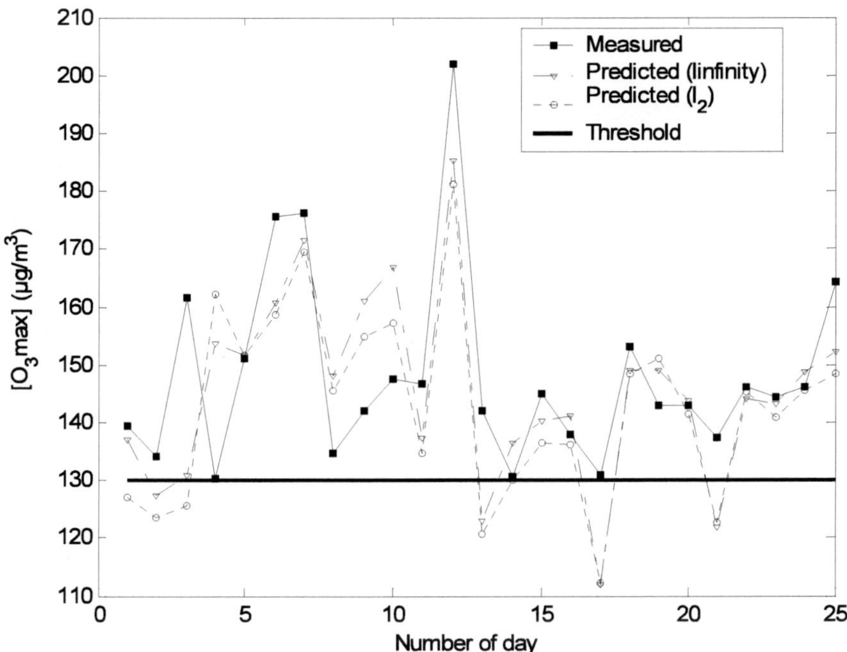

Figure 22. Produced forecasts by the nonlinear model for the days exceeding the threshold (130 μg/m^3). Testing year 2001.

It is clear that, in an operational setting for any specific city, model building could be improved trough experimentation and fine-tuning. However, when the free parameters are fitted by means of l_∞ optimization techniques, the modelling errors would be more evenly distributed amongst the data points, resulting in a more adequate modelling of high events at the expense of slightly less average performance.

6.4. Concluding Remarks

The statistical models commonly used in air pollution studies fall short of adequately modelling extreme values, as the high values are usually predicted too low. As a consequence, they generally lead to poor results in terms of "threshold exceeding", which is of great concern with respect to air quality standards. In this section an alternative and appealing strategy is proposed for estimating the free parameters of these models. The method was applied to ground-level ozone time-series data measured in Bordeaux during the period 1998-2001. The aim of model building was to provide predictions of the maximal daily ground-level ozone concentration, with weather forecasts, measured ozone precursors and lagged ozone data as input to the model. We compared the quality of forecasts produced by both a linear and a nonlinear model, using l_2 and l_∞ parameter optimization. From this case study, it appears that the performance indicators in terms of limit value exceeding are globally better when the parameters are optimized by means of a minimax criterion. Obviously, this result cannot be generalized as the l_∞ model building techniques should be tested and validated using similar air quality monitoring problems and other urban areas with higher levels of ozone. However, the method could provide an useful solution in cases where the detection of high events is judged to be of primary interest.

Appendix

Performance Indicators

Let be N the total number of data points, O_i the observed data point, P_i the predicted data point, $\overline{O}$ the average of observed data. Then:

i) Absolute Mean Error : $AME = \frac{1}{N}\sum_{i=1}^{N}|P_i - O_i|$

ii) Variance Accounted For : $VAF = \left(1 - \frac{\frac{1}{N}\sum_{i=1}^{N}[O_i - P_i]^2}{\frac{1}{N}\sum_{i=1}^{N}[O_i]^2}\right) \times 100$

iii) Index of agreement (d_1 and d_2):

iv) $$d_1 = 1 - \frac{\sum_{i=1}^{N}|P_i - O_i|}{\sum_{i=1}^{N}\left(|P_i - \overline{O}| + |O_i - \overline{O}|\right)} \qquad d_2 = 1 - \frac{\sum_{i=1}^{N}|P_i - O_i|^2}{\sum_{i=1}^{N}\left(|P_i - \overline{O}| + |O_i - \overline{O}|\right)^2}$$

Let be m the total number of observed exceedances, f the total number of predicted exceedances and a the number of correctly forecasted exceedances. Then, three skill parameters can be defined (De Leeuw, 2000):

v) $SP = \left(\frac{a}{m}\right)100\%$ is the fraction of correct forecast smog events (rate of detection, range from 0 to 100 with a best value of 100). The fraction of unexpected events is given by (100-SP)%.

vi) $SR = \left(\frac{a}{f}\right)100\%$ is the fraction of realised forecast smog events (range from 0 to 100 with a best value of 100). The fraction of false alarms is given by FA = (100-SR)%.

Assuming an equal weight to the correct forecasting of smog events and of a non-smog event, the scoring parameters SP and SR can be combined to a success index:

vii) $SI = \left(\frac{a}{m} + \frac{N + a - m - f}{N - m} - 1\right)100\%$ ranging from -100 to 100 with a best value of 100.

Acknowledgements

The financial support by the "Conseil Régional d'Aquitaine" and "Fonds Européen De Développement Régional" is gratefully acknowledged.

The author would like to thank Olivier Petrique who provided meterological and pollution data.

References

[1] Bloomfield, P.J., Royle, A.J., Steinberg, L.J., Yang, Q., 1996. Accounting for meteorological effects in measuring urban ozone levels and trends. *Atmospheric Environment.* **30** (17), 3067-3077.

[2] Boggs, P.T., J.W. Tolle, 1995. Sequential quadratic programming. *Acta Numer.* 1-52.

[3] Burl, J., 1999. *Linear Optimal Control.* (Ed). Addison Wesley.

[4] Brayton, R.K., Directo, S.W., Hachtal, G.D., Vidigal, L., 1979. A new algorithm for statistical circuit design based on quasi-Newton methods and function splitting. *IEEE Transactions on Circuits and Systems,* **26**, 784-794.

[5] Chaloulakou, A., D. Assimacopoulos, T. Lekkas, 1999. Forecasting daily maximum ozone concentrations in the Athens basin. *Environmental Monitoring and Assessment.* **30**, 97-112.

[6] Chen, J.L., Islam, S., Biswas, P., 1998. Nonlinear dynamics of hourly ozone concentrations: nonparametric short term prediction. *Atmospheric Environment.* **32**, 1839-1848.

[7] Cobourn, W.G., Dolcine, L., French, M., Hubbard, M.C., 2000. A comparison of nonlinear regression and neural network models for ground-level ozone forecasting. *Journal of the Air and Waste Management Association.* **50**: 1999-2009.

[8] Corani, G., 2005. Air quality prediction in Milan: feed-forward neural networks, pruned neural networks and lazy learning . *Ecological Modelling,* Vol 185, Issues 2-4, Pages 513-529

[9] Cox, W., Chu, S., 1996. Assessment of international ozone variation in urban areas from a climatological perspective. *Atmospheric Environment.* **30**, 2615-2625.

[10] Davis, J.M., Speckman, P., 1999. A model for predicting maximum and 8h average ozone in Houston. *Atmospheric Environment.* **33**, 2487-2500.

[11] Davis, J.M., B.K. Eder, D. Nychka, Q. Yang, 1998. Modelling the effects of meteorology on ozone in Houston using cluster analysis and generalized additive models. *Atmospheric Environment.* **32**, 2505-2520.

[12] Davison, A.C., Smith, R.L., 1990. Models for exceedances over high thresholds (with discussion). *Journal of the Royal Statistical Society.* B 52, 393-442.

[13] De Leeuw, F., 2000. Criteria for evaluation of smog forecast systems. *Environmental Monitoring and Assessment,* **60**, 1-14.

[14] Derwent, D.G., Davies, T.D., 1994, Modelling the impact of NOx or hydrocarbon control on photochemical ozone in Europe. *Atmospheric Environment.* **28**, 2039-2052.

[15] Derwent D., Van den Hout D., 1995. Computer modelling of ozone formation In Europe. Doc.Ref XI/364/95, European Commission, Brussels.

[16] Dockery, D.W., Pope III, C.A., Xu, X., Spengler, J.D., Ware, J.H., Fay, M.E., Ferris, B.G., Speizer, F.E., 1993. An association between air pollution and mortality in six U.S. cities. *The New England Journal of Medicine.* **329** (24), 1753-1759.

[17] Dutot A.L., Rynkiewicz J., Steiner F.E. and Rude J., 2007. A 24-h forecast of ozone peaks and exceedance levels using neural classifiers and weather predictions. *Environmental Modelling and Software,* In Press, Corrected Proof, Available online 3 November 2006 from ScienceDirect.

[18] Elkamel, A., Abdul-Wahab, S., Bouhamra, W., Alper, E., 2001. Measurement and prediction of ozone levels around a heavily industrialized area: a neural network approach. *Advances in Environmental Research.* **5**, 47-59.

[19] ETCACC (European Topic Centre on Air and Climate Change 2002. Available at the web site http//air-climate.eionet.eu.int/databases/2002o3excess.htm.

[20] Fuller, G.W., Carslaw, D.C., Lodge, H.W., 2002. An empirical approach for the prediction of daily mean PM10 concentrations. *Atmospheric Environment,* **35**, 4433-4442.

[21] Frischer T., Studnicka M., Gartner C., Tauber E., Horak F. Veiter A., Spengler J., Kuhr J., Urbanek R., 1999. Lung function growth and ambient ozone. A three-year population study in school children. *Am. J. Respir. Crit. Care Med.* **160**, 390-396.

[22] Hubbard, M.C., Cobourn, W.G., 1998. Development of a regression model to forecast ground-level ozone concentration in Louisville, KY. *Atmospheric Environment* **32**, 2637-2647.

[23] Gardner, M.W., Dorling, S.R., 1998. Artificial neural networks (the multilayer perceptron). A review of applications in atmospheric sciences. *Atmospheric Environment,* **32**, 2627-2636.

[24] Gardner, M.W., Dorling, S.R., 2000a. Statistical surface ozone models: an improved methodology to account for nonlinear behaviour. *Atmospheric Environment,* **34**, 21-34.

[25] Gardner, M.W., Dorling, S.R., 2000b. Meteorologically adjusted trends in UK daily maximum surface ozone concentrations. *Atmospheric Environment,* **34**, 171-176.

[26] Haykin, S., 1999. *Neural networks: a comprehensive foundation.* Prentice Hall, New Jersey.

[27] Kukkonen, J., Harkonen, J., Karppinen, A., Pohjola, M., Pietarila, H., Koskentalo, T., 2001. A semi-emprical model for urban PM10 concentrations, and its evaluation agains data fron an urban measurement network. *Atmospheric Environment,* **36**, 1431-1441.

[28] Kukkonen, J., Karppinen, Wallenius, L., Ruuskanen, J., Patama, T, Kolehmainen, M., Dorling, S., Foxall, R., Gawley, G., Mandic, D., Chatterton, T., Zickus, M., Greig, A., 2002. Evaluation of neural network, statistical and deterministic models against the measured concentrations of NO_2, PM_{10} and $PM_{2.5}$ in an urban area. *8th Int. Conf. on Harmonisation within Atmospheric Dispersion Modelling for Regulatory Purposes*. 63-67.

[29] Kukkonen J., Partanen L., Karppinen A., Ruuskanen J., Junninen H., Kolehmainen M., Niska H., Dorling S., Chatterton T., Foxall R. and Cawley G.: 2003. Extensive evaluation of neural network models for the prediction of NO2 and PM10 concentrations, compared with a deterministic modelling system and measurements in central Helsinki, *Atmospheric Environment,* **37**, 4539-4550.

[30] Kurun A., Yrekli K. and Evik O. 2005. Performance of two stochastic approaches for forecasting water quality and streamflow data from Yeilrmak River, Turkey. *Environmental Modelling and Software,* Available online 7 January 2005.

[31] Joe, H., Steyn, D.G., Susko, E., 1996. Analysis of trends in tropospheric ozone in the lower Fraster Vally. British Columbia. *Atmospheric Environment.* **30**, 3413-3421.

[32] Jorquera, H., Perez, R., Cipriano, A., Espejo, A., Letelier, M.V., Acuna, G., 1998. Forecasting ozone daily maximum levels at Santiago, Chilli. *Atmospheric Environment,* **32**, 3415-3424.

[33] Junninen H., Niska H., Tuppurainen K., Ruuskanen J., Kolehmainen M.:2004. Methods for imputation of missing values in air quality data sets, *Atmospheric Environment.* **38**, 2895-2907.

[34] Lewis, F.L. 1986 (Ed). *Optimal estimation,* New York: Wiley.

[35] Ljung, L., T. Söderström (1983) (Ed). Theory and Practice of Recursive Identification. *MIT Press*, Cambridge, Mass.

[36] Mao K.Z., S.A. Billings, 1997. Algorithms for nominal model structure detection in nonlinear dynamic system identification. *Int. J. Control,* vol 68, n° 2, 331-330.

[37] Matyssek, R., Havranek, W.M., Wieser, G., Innes, J.L., 1997. *Forest Decline and ozone, a comparison of controlled chamber and field experiments.* Springer, Berlin.

[38] McDonnell, W.F., Nishikawa, N., Peterson, F.F., Chen, L.H., Abbey, D.E., 2000. Relationships of mortality with the fine and coarse fractions of long term ambient PM10 concentrations in nonsmokers. *Journal of Exposure Analysis and Environmental Epidemiology.* **10**, 427-436.

[39] Milionis, A.E., T.D. Davies, 1994. Regression and stochastic models for air pollution – I. Review, comments and suggestions. *Atmospheric Environment.* **28**, 2801-2810.

[40] Millan, M., Salvador, R., Mantilla, E., 1996. Meteorology and photochemical air pollution in southern Europe: experimental results from EC research projects. *Atmospheric Environment.* **30**, 1909-1924.

[41] Ng, C.N., Young, P.C., 1990. Recursive estimation and forecasting of non-stationary time series. *Journal of forecasting.* **9**, 173-204.

[42] Nittis K., Perivoliotis L., Korres G., Tziavos C. and Thanos I., 2005.Operational monitoring and forecasting for marine environmental applications in the Aegean Sea. *Environmental Modelling and Software.* Available online 15 December 2004

[43] Niu, X.F., 1996. Nonlinear additive models for environmental time series with applications to ground-level ozone data analysis. *Journal of the American Statistical Association.* **91**, 1310-1321.

[44] Nogaard M., Ravn, O., Poulsen, N., 2000a. New developments in state estimation for nonlinear systems. *Automatica.* **36**,1627-1638.

[45] Nogaard M., Poulsen, N., Ravn, O., 2000b (Ed). *Neural networks for modelling and control of dynamic systems*. Springer-Verlag London.

[46] Ordieres B., Vergara E.B., Capuz R.S. Salazar R.E. 2005. Neural network prediction model for fine particulate matter ($PM_{2.5}$) on the US–Mexico border in El Paso (Texas) and Ciudad Juárez (Chihuahua). *Environmental Modelling and Software,* Volume 20, Issue 5, May 2005, Pages 547-559.

[47] Perez, P., Reyes, J., 2002. Prediction of maximum of 24-h average of PM10 concentrations 30h in advance in Santiago, Chile. *Atmospheric Environment.* **36**, 4555-4561.

[48] Schlink U., Volta M., 2000. Grey box and component models to forecast ozone episodes: a comparison study. *Environmental modelling and assessment,* Vol 65, pp 313-321.

[49] Schlink, U. et al. 2005. A Rigorous Inter-comparison of Ground-level Ozone Predictions, *Atmospheric Environment,* **37**, 3237-3253.

[50] Schlink, U., Herbarth, O., Richter M., Dorling S., Nunnari G., Cawley G. and Pelikan E., 2006. Statistical models to assess the health effects and to forecast ground-level ozone. *Environmental Modelling and Software,* Vol 21, Issue 4, Pages 547-558.

[51] Schwartz, J., 1996. Air pollution and hospital admissions for respiratory disease. *Epidemiology.* **7**, 20-28.

[52] Shen, W., L. Deng, 1997. Game theory approach to discrete H_∞ filter design. *IEEE Transactions on signal processing,* 1092-1095.

[53] Shi, J.P., R.M. Harrison, 1997. Regression modelling of hourly NO_x and NO_2 concentrations in urbain air in London. *Atmospheric Environment.* **31**, 4081-4094.

[54] Sivakumar B. 2005. Hydrologic modeling and forecasting: role of *thresholds Environmental Modelling and Software*, Volume 20, Issue 5, May 2005, Pages 515-519

[55] Simon, D., El-Sherief H., 1996. Hybrid Kalman/minimax filtering in phase-locked loops. *Atmospheric Environment.* **27A**, 921-943.

[56] Smith-Doiron, M., Stieb, D., Raizenne, M., Brook, J., Dales, R., Leech, J., Cakmak, S., Krewski, D., 2000. Association between ozone and hospitalization for acute respiratory diseases in children less than two years of age. *American Journal of Epidemiology.* **153**, 444-452.

[57] Smith, R.L., 1993. Statistics of extremes, with applications in environment, insurance and finance. http://www.stat.unc.edu/postscript/rs/semstatrls.pdf.

[58] Smith, S., Trevor Stribley, F., Milligan, P., Barratt, B., 2001. Factors influencing measurements of PM10 during 1995-1997 in London. *Atmospheric Environment.* **35**, 4651-4662.

[59] Snyman, J.A., A.M. Hay, 2002. The dynamic-Q optimization method: An alternative to SQP. *Computer and Mathematics with applications.* **44**, 1589-1598.

[60] Thompson, M.L., Reynolds, J., Cox L.H., Guttorp, P., Sampson, P.D., 2001. A review of statistical methods for the meteorological adjustment of tropospheric ozone. *Atmospheric Environment,* **35**, 617-630.

[61] Tittanen, P., Timonen, K.L., Ruuskanen, J., Mirme, A., Pekanen, J., 1999. Fine particulate air pollution, resuspended road dust and respiratory health among symptomatic children. *European Respiratory Journal.* **13**, 266-273.

[62] Tych, W., Pedregal, D. J. , Young, P. C. and Davies, J., 2002. An unobserved component model for multi-rate forecasting of telephone call demand: the design of the forecasting support system, *Int. Journal of*

[63] Van der zee, S.C., Hoek, G., Harssema, H., Brunekreef, B., 1998. Characterization of particulate air pollution in urban and non-urban areas in the Netherlands. *Atmospheric Environment*. **32**, 3717-3729.

[64] Van der Wal, J.T., Janssen, L.H.J.M., 2000. Analysis of spatial and temporal variations of PM10 concentrations in the Netherlands using Kalman filtering. *Atmospheric Environment,* **34**, 3675-3687.

[65] Vautard R., Beekmann, M., Roux, J., Gombert, D., 2001. Validation of a hybrid forecasting system for the ozone concentrations over the Paris area. *Atmospheric Environment,* **35**, 2449-2461.

[66] Wan A., Van der Merwe R., 2000. The Unscented Kalman Filter for Nonlinear Estimation. Proceedings of Symposium 2000 on Adaptive Systems for Signal Processing, Communication and Control (AS-SPCC), IEEE, Lake Louise, Alberta, Canada.

[67] Young, P.C., Ng, C.N., Lane, K., Parker, D., 1991. Recursive forecasting, smoothing and seasonal adjustment of non-stationary environmental data. *Journal of forecasting.* **10**, 57-89.

[68] Young, P.C., 1999. Nonstationary time series analysis and forecasting. *Progress in Environmental Science*, 3-48.

[69] Young, P.C., Pedregal, D.J. Tych, W., 1999. Dynamic Harmonique Regression. *Journal of forecasting,* 369-394.

[70] Young, P. C. 2002. Advances in real-time flood forecasting. *Philosophical Trans. Royal Society, Physical and Engineering Sciences,* **360**, 1433-1450.

[71] Zickus, M., Greig, A.J., Niranjan, M., 2002. Evaluation the variable selection and prediction performance of several machine learning algorithms applied to PM_{10} data set. Available from *http://www.uea.ac.uk/env/appetise*

[72] Ziomas, I.C., D. Melas, C.S. Zerefos, A.F. Bais, 1995. Forecasting peak pollutant levels from meteorological variables. *Atmospheric Environment.* **29**, 3703-3711.

[73] Zolghadri, A., 1996. An algorithm for real-time failure detection in Kalman filters. *IEEE Transactions on Automatic Control,* **41** (10), 1537-1541.

[74] Zolghadri A., M. Monsion, D. Henry, C. Marcionini, O.Petrique (2004). Development of an operational model-based warning system for tropospheric ozone concentrations in Bordeaux, France. *Environmental Modelling and Software,* Elsevier, vol. 19, pp 369-382

[75] Zolghadri A., D. Henry (2004). Minimax statistical models for air pollution time series. Application to ozone time series data measured in Bordeaux. Environmental Monitoring and Assessment, Kluwer Academic Publishers, vol 98, pp 275-294.

[76] Zolghadri A, F. Cazaurang.(2006). Adaptive nonlinear state-space modelling for the prediction of daily mean PM10 concentrations. *Environmental Modelling and Software,* vol 21, pp 885-894.

In: Ozone Depletion, Chemistry and Impacts
Editor: Sem H. Bakker, pp. 239-268

ISBN: 978-1-60692-007-7

Chapter 11

THE QUASI-BIENNIAL OSCILLATIONS IN THE EQUATORIAL STRATOSPHERE: SEASONAL REGULARITIES, DEPENDENCE ON THE SOLAR UV FLUX, AND RELATION TO OZONE DEPLETION IN ANTARCTICA

I. Gabis [*] ***and O. Troshichev*** [**]
Arctic and Antarctic Research Institute, St. Petersburg, Russia

Abstract

The term Quasi-Biennial Oscillation (QBO) was designated to describe the regular reversals of zonal winds in the equatorial stratosphere occurred with periodicity about 28 months. The phase of QBO cycle, East (E) or West (W) is determined by direction of the zonal winds. Although the wind QBO is mainly a tropical circulation feature, its effects are displayed well beyond the equatorial region. Indeed, the year-to-year fluctuations in the extratropical atmosphere are controlled by QBO, which influence is observed in such parameters as temperature, the total ozone distribution, and others. The most important thing is the well known at present QBO property to modulate the solar activity impact on the stratosphere and troposphere.

Today the dissipation of upward propagating equatorial waves usually examined as a theoretical mechanism for QBO generation. So, it is generally accepted that the proximity of the average QBO period to the double annual one being merely statistical result, whereas the actual period varies from 19 to 36 months. As a result of the high variability and elusive relation to the seasonal cycles, the QBO phase duration seemed to be unpredictable.

A new method for the investigation of the QBO cycle evolution is presented in this chapter. The analysis of the height wind profiles makes it possible to reveal new regularities in the equatorial stratosphere wind reversals. The stagnation stage of descending easterlies always starts in solstice, either in December–January or in June–July, and always completes by the equinox being of quantized duration in different QBO-cycles (about 3, 9, or 15 months). So the length of the complete QBO-cycle, if defined as a period between the

[*] E-mail address: gabis@aari.nw.ru
[**] E-mail address: olegtro@aari.nw.ru

inceptions of the successive stagnation stages, depends on their durations and may be equal to 24, 30, or 36 months. Consequently, the full QBO cycle length can be predicted as soon as end of the stagnation stage is found while observing the zonal wind transformation.

In turn the duration of the stagnation stage depends on the intensity of solar UV flux. The short stalling period (3–4 months) is observed under condition of the high level or steady increase of the UV irradiance during the first equinox in course of the partial QBO cycle. If the UV flux is low or decreases during the proper equinox, the easterly winds do not descend from the upper layer to the lower stratosphere, and the stalling period length at about 20-40 hPa increases to 9–10 or 15–16 months. The intensity of the spring ozone depletion within Antarctic polar vortex is thought to be also dependent on the phase of the equatorial QBO cycle. It is shown the relation of the season regularities of the equatorial QBO-circulation to the interannual fluctuations and the intensity of total ozone (TOZ) destruction over Antarctica during "ozone hole" phenomenon.

Key words: equatorial stratosphere, quasi-biennial oscillations, mean zonal winds, height wind profiles, solar UV irradiance, Antarctic ozone hole.

1. Introduction

Investigation of the solar variability effects on the Earth's atmosphere and, hence, on the weather and climate is one of the key tasks at present. The topic is of long history. Considerable advances in study of the solar activity impact on the troposphere and stratosphere have been made after the finding of K. Labitzke [*Labitzke, K., 1987*], that the character of the solar-atmosphere relations depends to a great extent on the quasi-biennial oscillation (QBO) of zonal winds in equatorial stratosphere. The atmospheric parameter (pressure, temperature, ozone and others) showed a clear correlation/anticorrelation with solar activity when the experimental data were grouped according to the phase of the QBO-cycle [*Labitzke, 1987, 2005; Labitzke and van Loon, 1988, 2000; Chanin et al., 1989; Besprozvannaya et al., 1997; van Loon and Labitzke, 1999, 2000; Georgieva et al., 2005*]. The QBO-phases, East (E) or West (W), were defined by the direction of the zonal wind in the tropical stratosphere.

The QBO is a fundamental property of the equatorial stratosphere circulation. Beginning with first regular measurements of the zonal wind speed in tropical stratosphere in 1953 a long series of observation were accumulated. It made it possible to study the QBO-cycle behavior in detail [*Reed, 1961; Veryard and Edbon, 1961; Naujokat, 1986; Baldwin, et al., 2001 and references in*]. The QBO-phenomenon is observed in the tropical band within a latitude range of about ±12 degrees. The QBO-phase change occurs practically simultaneously in the near-equatorial region irrespective of geographical longitude. It means that the phase identification can be made by the wind observations at any equatorial station. At present the measurements at station Singapore (01°22'N, 103°55'E) are used as a rule. Reversal of the zonal wind occurs in the lower and middle stratosphere from about 70 hPa (~18.5 km) to the ~3 hPa (~40 km) level, the maximum of wind speed (~30 m/s for easterly and ~15 m/s for westerly) being observed at 15-20 hPa.

The full cycle of the wind reversal, from easterly to westerly and back to easterly again, has the average period about 28 months [*Naujokat, 1986; Baldwin, et al., 2001*]. Solely easterly or westerly wind in the entire height region from 70 to 10 hPa occurs rarely: over the 50-year period (1953-2003) the westerly (easterly) at all heights is observed in 3% (0.3%) of time. The oppositely directed winds at higher and lower heights present a common phenomenon. Zonal winds blowing in one direction descend in altitude with time being replaced by winds blowing in the opposite direction. As a result, the time of the wind reversal and, accordingly, the beginning and duration of QBO-phases depends on the altitude. This fact gave rise to the problem of a proper choice of the pressure level for determination of the QBO-phase, because it was unclear a priory which QBO level can be responsible for the extratropical effects. To define the QBO-phase during any given month the different studies choose the various levels in range of 20-50 hPa, or some combination of levels [*Lait, et al., 1989*] or representation of the wind QBO by first two EOFs (empirical orthogonal functions) of monthly wind observations [*Wallace, et al., 1993; Baldwin, 1998*]. It turned out that the correlations involving extratropical parameters depend on level chosen to define the QBO-phase. In this connection it would be emphasized that the phase of QBO-cycle determined by wind direction at a certain altitude level corresponds to quite different height structure of equatorial circulation for different moments of the same QBO-phase. Furthermore, the circulation structures prove to be opposite at the beginning and end of the same QBO-phase.

Altitude of the wind reversal (zero line between the winds of opposite direction) is denoted as a shear zone – easterly or westerly shear zone, when the corresponding wind is higher of the zero line altitude. The time of QBO-phase changes and, consequently, the duration of alternating QBO-phases at given height level is evidently related to rate of the shear zones descending. Analysis of observations shows that westerly shear zones descend more evenly and quickly than easterly one. The later process is frequently slowed in the layer of about 30-40 hPa [*Naujokat, 1986; Dunkerton and Delisi, 1985, 1997; Dunkerton, 1990; Kinnersley and Pawson, 1996; Baldwin, et al., 2001*]. The true rate of descending of the easterly and westerly wind regimes during particular QBO period can significantly differs from the mean rate (~1 km/month). Experimental and model investigations revealed the dependence of the shear zone descending rate and the QBO-phase duration on the 11-year solar cycle [*Soukharev and Hood, 2001; McCormack, 2003; Pascoe 2005*]. Hence the process of wind descending may be influenced by the variability of the solar flux. However, this dependency was vanished during the 1990`s [*Pascoe et al., 2005*]. As was shown in [*Gabis and Troshichev, 2006*], the QBO-phase duration is affected by the solar UV irradiance intensity, but this influence is crucial only at the certain stages in evolution of the wind QBO-cycle.

Up to now the upward propagating tropical waves dissipation is usually examined as a mechanism for QBO generation [*Holton and Lindzen, 1972; Baldwin, et al., 2001 and references in*]. The theoretical QBO models presume that the troposphere wave sources are constant in time. So, it is generally accepted that the QBO period is not controlled by the year or half-year cyclicity, and consistency of the average QBO period to the double annual interval is examined as merely statistical result, the actual period being varied from 19 to 36 months. Nevertheless, there was some experimental evidence that the seasonal cycle might influence on the timing of QBO-phase changes. It was suggested [*Dunkerton and Delisi, 1985, 1997; Dunkerton, 1990*] that the QBO is modulated, but not exactly synchronized, by the seasonal cycle. Despite of the comprehensive information about the main basic morphological features of the equatorial wind QBO the forecast of the moment of the QBO-phase change remains to be unresolved and the problem of the QBO cycle duration remained unpredictable.

Meanwhile, the strong seasonal regularities of the QBO-cycle evolution and relation of the QBO-cycle length to the solar UV flux intensity was shown in the recent publications of authors [*Gabis and Troshichev, 2004, 2005, 2006*]. The revealed regularities can be used to predict the full length of the current QBO cycle and hence the onset of the next one. In the present chapter we summarize and update these earlier results. We also demonstrate relation of the revealed season regularities in the equatorial QBO-circulation to the interannual fluctuations and to the descending trend of total ozone (TOZ) over Antarctica during "ozone hole" phenomenon.

2. Data and Method of the Analysis

Our analysis is based on the wind speed measurements performed at the following equatorial stations: Canton Island (02°46'S, 171°43'W) for January 1953 – August 1967, Gan/Maledive Islands (00°41'S, 73°09'E) for September 1967 – December 1975, and Singapore (01°22'N, 103°55'E) after January 1976. Sets of the height profiles of the monthly mean zonal wind component in layer between 70 and 10 hPa were used to reveal the regularities in the wind changes during QBO-cycle evolution.

The Figure 1 presents, as an example, a transformation of the wind height profiles from August 2004 to April 2007. Positive and negative speeds correspond to the westerly and easterly, respectively. The each plot in Figure 1 contains profiles for two successive months, the previous and current months being shown by curves with solid and open circles, respectively. One can see that the westerly wind is observed from 70 (~18.5 km) to almost 10 hPa (~31.2 km) level in August 2004, at 10 hPa the wind speed U=0.0 m/s. The easterly wind was appeared in September 2004 at altitudes 10-20 hPa and then after four months came down to altitude nearly 30 hPa (~23.5 km). So the mean rate of the easterly shear zone descending during this period was ≈1.9 km/month. From December 2004 to March 2005 the altitude of the shear zone remained approximately invariant (in December, January, February and March at 30 hPa U = 1.4, -1.8, 1.8 and -0.9 m/s, respectively). The westerly and easterly winds were observed at altitudes lower and higher ~30 hPa, respectively. In April 2005 the descent of the easterly shear was resumed and the easterly wind was appeared below 30 hPa, firstly at 35 hPa, then at 40 hPa (in May 2005), and so on. In September 2005 the easterly wind was observed practically in the entire altitude range from 70 to 10 hPa (the wind speeds at 10 and 70 hPa was equal 0.1 and 0.0 m/s, respectively). Consequently, the mean rate of the easterly shear zone descending was ≈0.8 km/month during the period from March 2005 to September 2005. The averaged rate of the easterly shear zone descent for the full period from August 2004 to September 2005 was ≈ 1.0 km/month, but for all that the easterly shear zone during three-four months in December 2004 – March 2005 remained at constant altitude near 30 hPa.

In September 2005 westerly shear was observed at 10 hPa (U=0.1 m/s) and then the westerly wind moved down in time. Consequently W-phase was observed in the altitude range 10-12 hPa in October, at 10-15 hPa in November, at 10-20 hPa in December and so on. In period from October 2005 to August 2006 the westerly came down to the lower layer (70 hPa). It is significant that the westerly shear zone demonstrated the permanent descending without stopping in contrast to the easterly shear zone. The mean rate of the westerly shear zone descent for the period from September 2005 to August 2006 amounted to ≈1.1 km/month.

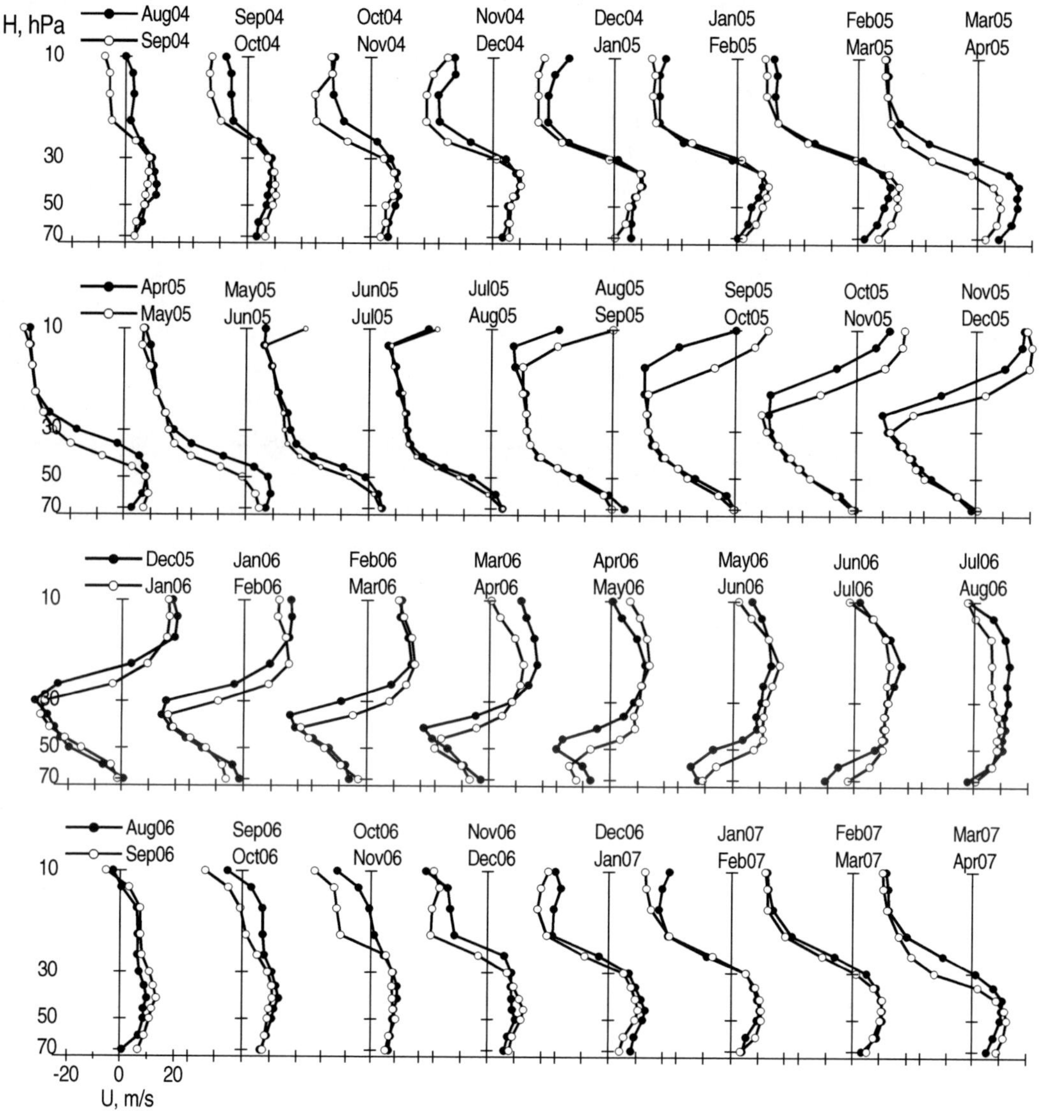

Figure 1. Changes in vertical profiles of the mean zonal wind for the period from August 2004 to April 2007. The each plot contains profiles for two successive months, the current and previous months being marked by open and solid circles, respectively. Positive and negative speeds correspond to the westerly and easterly, respectively.

Commencement of the next easterly shear zone descending occurred in June-July 2006 (the wind speed at 10 hPa level was equal to 1.9 m/s and -2.1 m/s, respectively). Then the easterly wind appeared at the progressively lower levels in the course of event. The easterly shear descending was continued till December 2006. The rate of easterly shear descent during period from June to December 2006 averaged to about ≈1.2 km/month. From December 2006 to March 2007 the changes of height wind profiles near the shear zone altitude were again insignificant, as in previous case, – the differences between wind speed for two adjacent months were below ~5 m/s/month. Hence the altitude of the shear zone remained approximately invariant, but in this case the easterly shear zone halted at slightly higher level than in previous case. The interpolation between measurements at 25 and 30 hPa levels gives the value about 27-29 hPa for the zero wind speed altitude. The wind profiles with easterly in upper layer above the stalling region and westerly lower of the stalling region were observed during three–four months in December 2006 – March 2007. In April 2007 the easterly shear descent was resumed like that in April 2005 .

As Figure 1 shows, the QBO phase, determined by wind direction at the certain level, corresponds to the quite different height structure in the equatorial circulation for different periods of the same QBO-phase, and the circulation structures at the beginning and end of the same QBO-phase occurred to be opposite. For example, if the QBO-phase is defined by the wind direction at 40 hPa level, the E-phase starts in May 2005 during descending easterly and terminates in April 2006 during descending westerly. Hence, the E-phase beginning corresponded to westerly in the layer lower ~40 hPa, and to easterly in layer higher ~40 hPa; however, the height wind structure was opposite for the same E-phase end: easterly (westerly) in the layer lower (higher) 40 hPa.

It will be shown in the next section that the above example for 2004-2007 demonstrates the regularly repeating features of the easterly shear behavior. Irrespective of the time for the easterly shear location at 10 hPa level, the three-four months stage of stagnation is observed during the interval from the solstice (December-January) to the first equinox (March-April), and then the easterly shear resumes the descent to the lower stratosphere.

3. Results and Discussion

3.1. Three Types of the QBO-Cycle Evolution SCENARIOS

Analysis of the height profiles for the monthly mean zonal wind during the 55-years period of measurements (1953–2007) makes it possible to conclude that there are three types of the QBO-cycle evolution scenarios. The main peculiarity, which defines this classification, is duration of the stalling period of the easterly shear zone descent.

3.1.1. 24-Month Scenarios

The ten cases with the stalling period, lasting three-four months, took place within the interval from January 1953 to December 2006. It is very likely that just another stalling period of the same duration already started in the beginning of 2007 (see section 3.4). Figure 2 presents the wind profiles occurred from July-August of preceding year to July-August of the subsequent year during ten events. Like Figure 1, each plot in the horizontal row, except the central one, contains profiles for two successive months, the central plot being contained four profiles for interval from December of the preceding year to March of the subsequent year. The December and March profiles are marked by solid and open circles, respectively, January and February profiles are shown by thin lines. One can see that the nearly full cycle of easterly shear zone descent from ~10 hPa to ~70 hPa is completed during every one year period.

During July-December of the preceding years the E-shear moves down from upper layer to ~20-30 hPa, and during March-August of the subsequent years the E-shear descends to the lower stratosphere. In the middle of the examined periods during three-four months (December-March) a spectacular stagnation stages are observed in majority cases. In seven cases of ten the shear height changes not more 5 hPa for December-March. In 1955-1956, 1959-1960 and 1971-1972 the height during December-March changes by 11, 15 and 6-7 hPa, respectively, however before and after this period the shear descent magnitude was higher. Taking into account this repeated regularity, we can say about the short stalling period in process of the easterly shear zone descent.

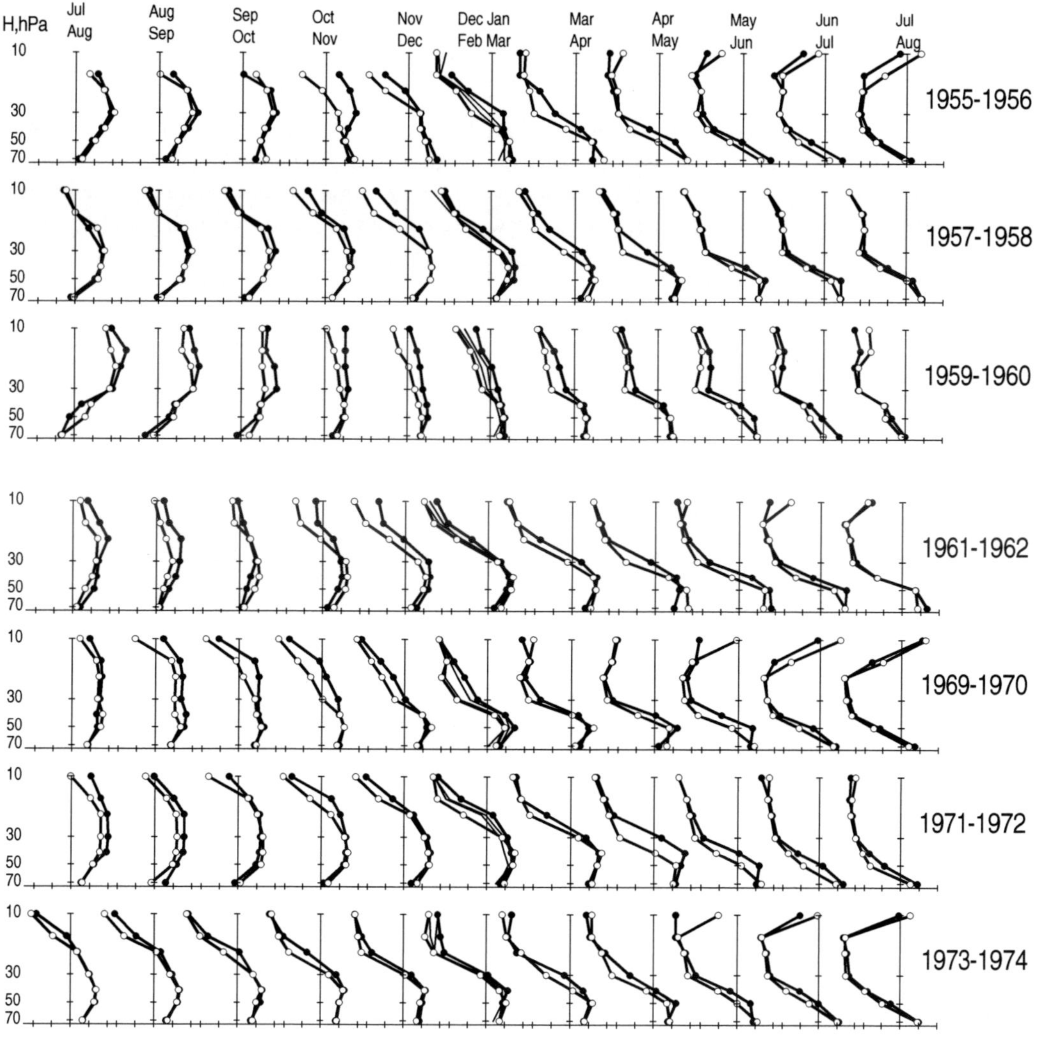

Figure 2. Continued on next page.

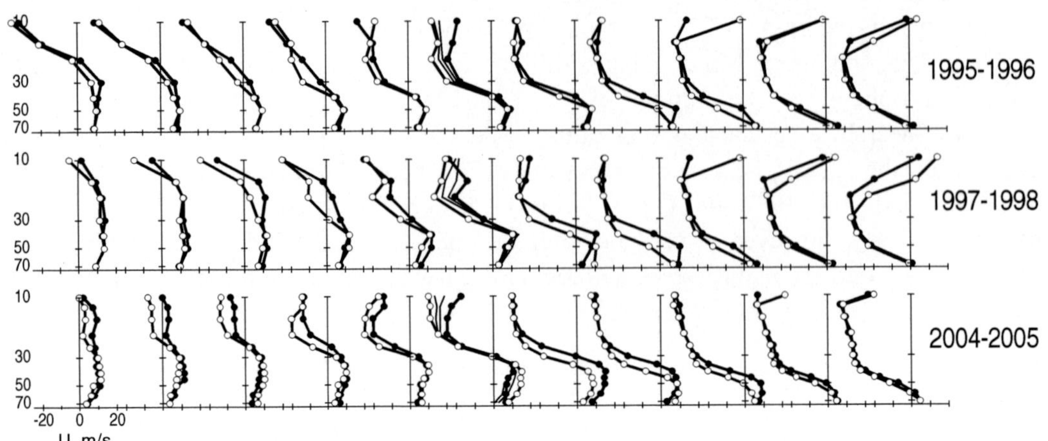

Figure 2. Transformation of height wind profiles nearby the "short" stagnation stage for the 24-months scenarios of QBO-cycle. Shown are ten one year periods from July-August of preceding year to July-August of the subsequent year. The proper months and years are remarked at the upper and right panels of figure, respectively. Each plot in the horizontal row, except the central one, contains profiles for two successive months (the current and previous months being marked by open and solid circles, respectively), the central plot being contained four profiles for interval from December of the preceding year to March of the subsequent year. The December and March profiles are marked by solid and open circles, respectively, January and February profiles are shown by thin lines.

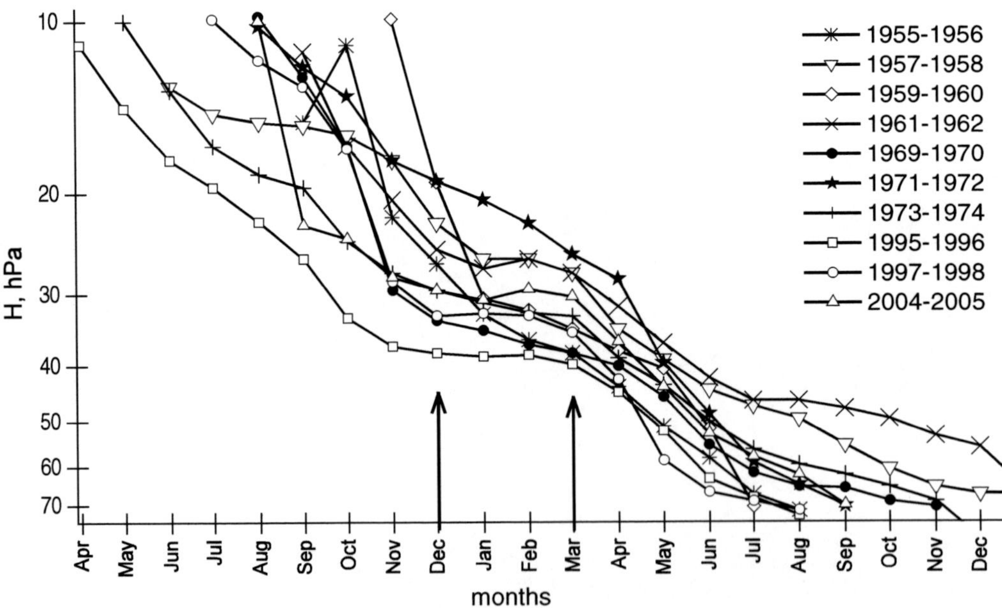

Figure 3. Time-height relationships for ten events of the easterly shear zone descending presented in Figure 2. Each curve illustrates the time position of the zero wind speed altitude for certain E-shear descent, the height interpolation between measurements at the levels, denoted in Figure 2 by solid or open circles, being used. The vertical arrows confine the stalling periods during December-March.

Figure 3 shows the time-height relationship for all ten cases of the easterly shear zone descending presented in Figure 2. Each curve illustrates the time position of the zero wind speed altitude for certain E-shear descent, the height interpolation between adjacent measurements being used. One can see that the shear descent onset, or the 10 hPa level crossing, may be observed within the time interval from April to November. The shear descent end, or the approach to 70 hPa level, may occur within the time interval from July to December of the next year, or later. During the periods from the December of the preceding years to the March of the subsequent years the E-shear is placed at height range 20-40 hPa and demonstrated insignificant altitude variations in each particular case, as mentioned above.

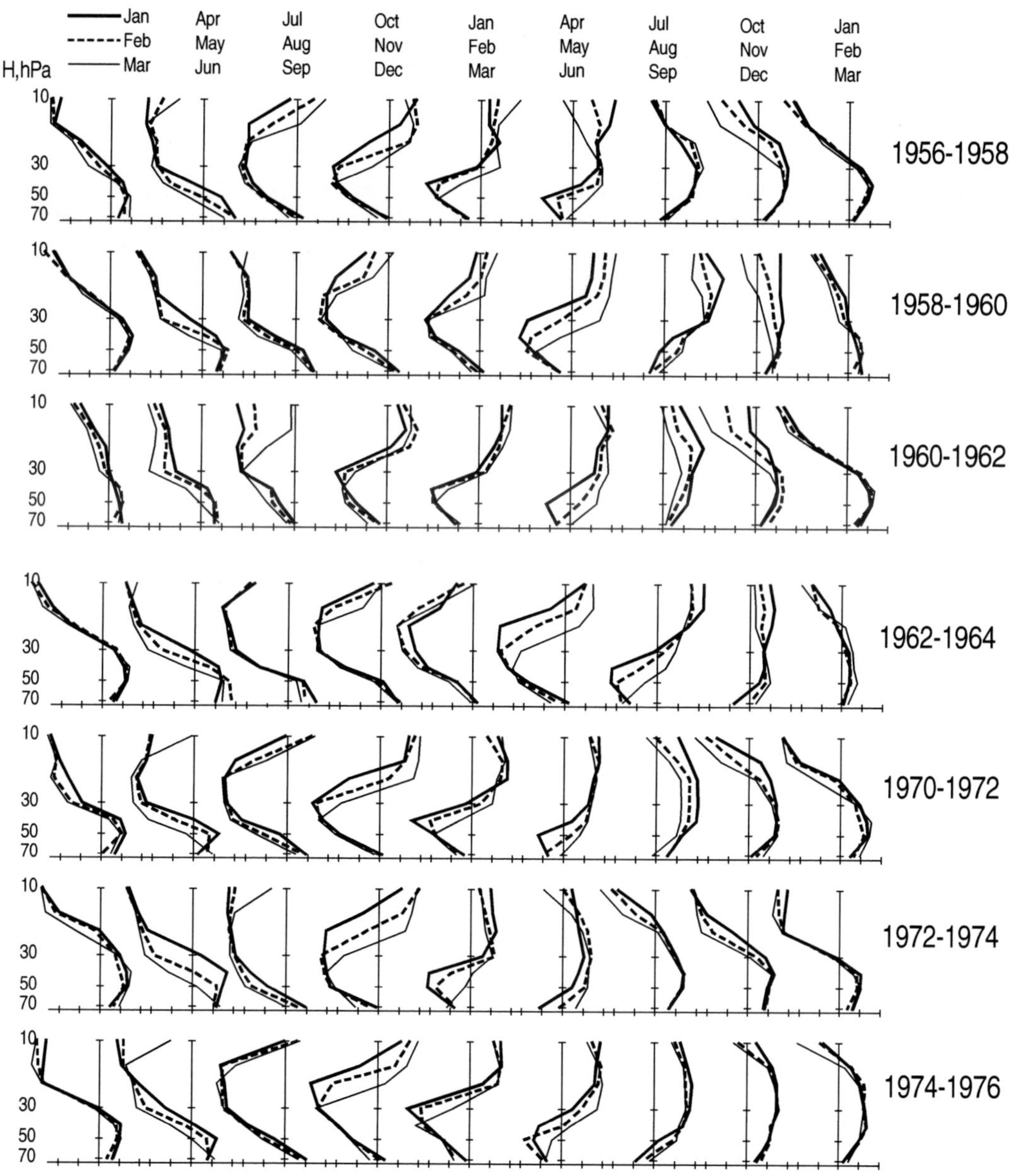

Figure 4. Continued on next page.

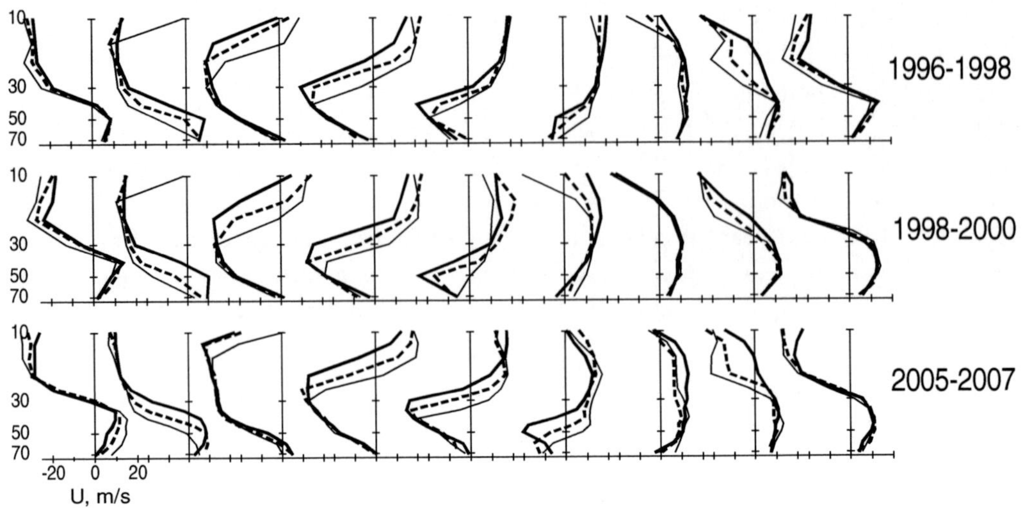

Figure 4. Transformation of height wind profiles for the 24-months scenarios of QBO-cycle. Shown are ten 27-month periods from the stagnation stage onset in January to the stagnation stage of subsequent QBO-cycle. The proper months and years are remarked at the upper and right panels of figure, respectively. Each plot contains three monthly profiles. The first, middle and last profiles on each plot are shown by the hard, dash, and thin line, respectively.

Nevertheless, the zero wind speed altitude in the certain month varies for different E-shear cases, as a result, the zonal wind reversal at the fixed altitude (in range 20-40 hPa) falls on different months for different E-shear cases, depending on the exact height for stagnation stage. It is the reason that the distinct seasonal effect is not revealed in the standard method of the QBO phase derivation, when E-phase or W-phase is determined by the wind direction on fixed level. Figure 3 provides the following very important result: in spite of difference in onsets of E-shear descent at 10 hPa, the stagnation stage starts in solstice (December-January) in all cases, and a new descent resumes in the first succeeding equinox (March-April).

Ten events displayed in Figure 2 are reconstructed in Figure 4 in such a manner that every event starts from the stagnation stage in January-March. Each plot contains three monthly profiles and hence demonstrates the changes of height wind structure during one seasons. The first, middle and last profiles on each plot are shown by the hard, dash, and thin line, respectively. One can see that the stagnation stage in all ten events was observed during one season (three-four months), from the December-January solstice to the nearest equinox in March-April, as it was already noted earlier. After this equinox the wind profile changes are observed lower of the stalling region and the easterly shear descent is finished near 70 hPa within the time interval from July to December in different months for different cases. It is significant that simultaneously with the E-shear arrival to the lower layer the easterly wind speed in higher layer, near 10 hPa, decreases, evidencing the westerly acceleration. The W-shear at 10 hPa starts within the time interval from July to November-December during the first year, in some events before E-shear arrival to the 70 hPa level. The descent of W-shear is completed during four-five seasons, from the onset in June-July of the first year to nearly September-October equinox of the second year. At the same time new E-shear starts to move down from level near 10 hPa in the June-July solstice of the second year and approaches, in all but one case, to nearly the 20-35 hPa level in December-January of the second year. The wind profile changes after the December show insignificant variations near altitude of zero wind speed during some months. Hence the stagnation stage occurred again. Consequently, the stagnation stage was observed during one season, and the full cycle of wind profile changes between the beginnings of two consecutive stagnation stages continues eight seasons. We identify this variant as "24-month scenario" of QBO-cycle evolution.

3.1.2. 30-Month Scenarios

As Figure 4 shows, sometimes the 24-month QBO-scenarios are observed one after other successively. The series of four, three and two repetition of 24-months scenario occurred in 1956-1963, 1970-1975 and 1996-2000, respectively. In other periods (1953-1955, 1964-1969, 1976-1995 and 2000-2004) the wind profile changes demonstrated the larger duration of the stalling period than in 24-month scenario. Figure 5 represents five cases of the QBO-cycles with the stagnation stage which started in December-January solstice. As in the Figure 4 every plot contains three profiles, showing the of height wind structure changes during one season, the first, middle and last profiles being marked on each plot by the hard, dash, and thin line, respectively.

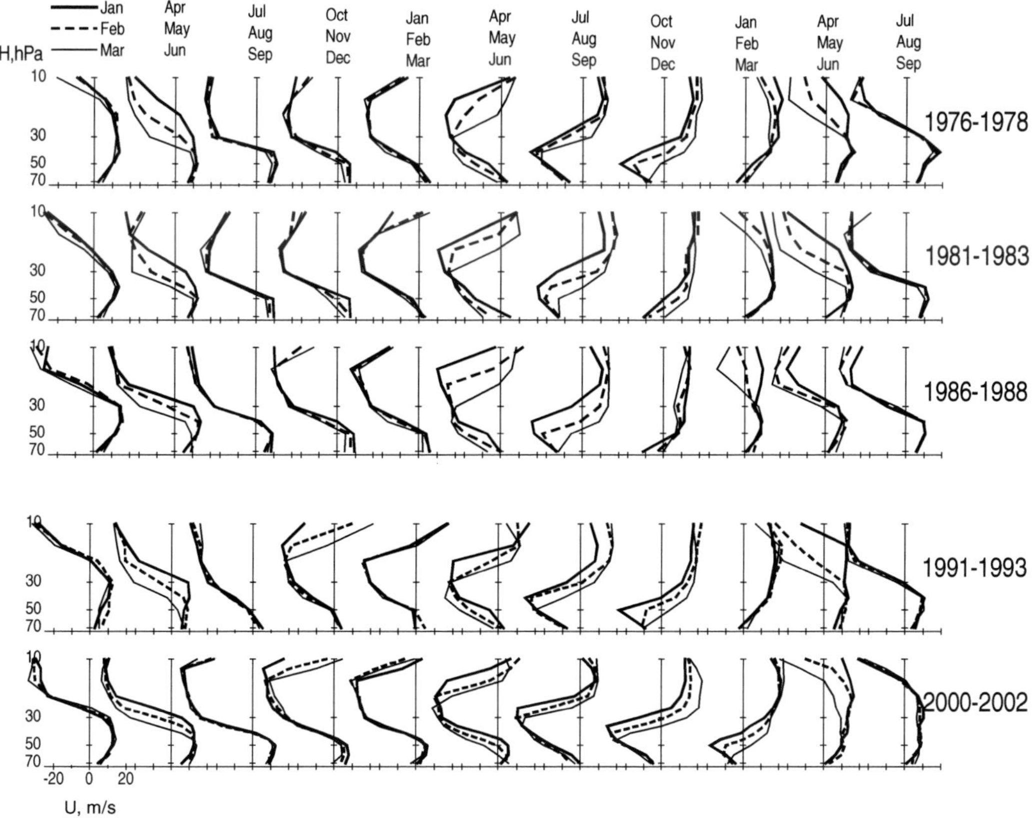

Figure 5. The same as Figure 4, but for January 30-month scenarios. Shown are five 33-month periods from the stagnation stage onset in January to the stagnation stage of subsequent QBO-cycle.

The first plot in each row illustrates insignificant changes in the wind profiles near the easterly shear zone. In the first event (1976-1978) the altitude of zero wind speed was abnormally high (about 13 hPa), but other four cases the zero wind altitude was rather typical (20-28 hPa). The first equinox (March–April) was followed by the slight descent of the easterly shear zone during three-four months. In the next solstice the stopping period occurs again: changes of the wind profile were absent during July-September and the easterly shear zone remained at the constant altitude. However, this altitude was different for different events: 38, 45, 40, 55 and 40 hPa for September 1976, 1981, 1986, 1991 and 2002, respectively. The full period from December-January solstice to the second equinox in September-October turned out to be equal three seasons (or nine – ten months). This period can be regarded as an elongated stagnation stage divided by the intermediate slight easterly shear descent on two steps.

After the second equinox (September-October) the wind profile changes are observed in the lower stratosphere and later the E-shear reaches 70 hPa level. Almost simultaneously the easterly wind starts to decrease in the upper layer, near 10 hPa, and then the westerly shear appears at the 10 hPa level. The subsequent descent of the westerly shear is quite similar to that during 24-month scenarios shown in Figure 4. Throughout four-five seasons, from equinox of the first year (September-October) to the end of the second year, or even later, the W-shear arrives at the 70 hPa level. In the next (third) year the intense wind profile changes occur in the layer higher ~30 hPa. So, it takes about two seasons for the easterly shear descending from 10 to ~30 hPa. As a result in June-July solstice the easterly (westerly) wind are displayed in the middle (lower) stratosphere. This height structures do not change during July-September period, as it seen in Figure 5. So, in each case the beginning of a new stagnation stage is observed in solstice again. The full cycle of the wind profile changes from the initial moment of the first stagnation stage to the initial moment of the stagnation stage in the next QBO-cycle takes ten seasons. The cycle consists of the elongated two-step stagnation stage observed during three seasons and subsequent wind evolution completed within seven seasons. This variant of QBO-cycle evolution we denote as "January 30-month scenario" in conformity with time of the QBO-cycle beginning.

It is obviously that the 30-month January scenario of QBO-cycle with onset in December-January solstice is always followed by the QBO-cycle with beginning in June-July solstice. Figure 6 presents six QBO-cycles, five of which being immediately adjacent to cycles of "January 30-month" scenarios shown in Figure 5. One can see here that in all but one case the wind profile changes were negligible to a great extent during three seasons from June-July solstice to March-April equinox of the next year. In July-September 2002 the altitude of the stagnation stage was higher than in the rest five cases, and the additional E-shear descent took place during period from September to November. However, the shear altitude remained unchanged from December of the first year to the April of the second year. Therefore, in all six events the stagnation stages were observed during three seasons (or nine – ten months), from June-July solstice to March-April equinox of the next year. The subsequent wind profile changes in each case were quite similar to those observed in 24-month and 30-month January scenarios. Seven seasons (or 20-21 months) later the following stagnation stage is observed in December-January solstice. Thus, the full length of QBO-cycles, defined as a period between the successive stagnation stages, turned out to be equal to ten seasons, or 30 months. So, it is a "July 30-month scenario".

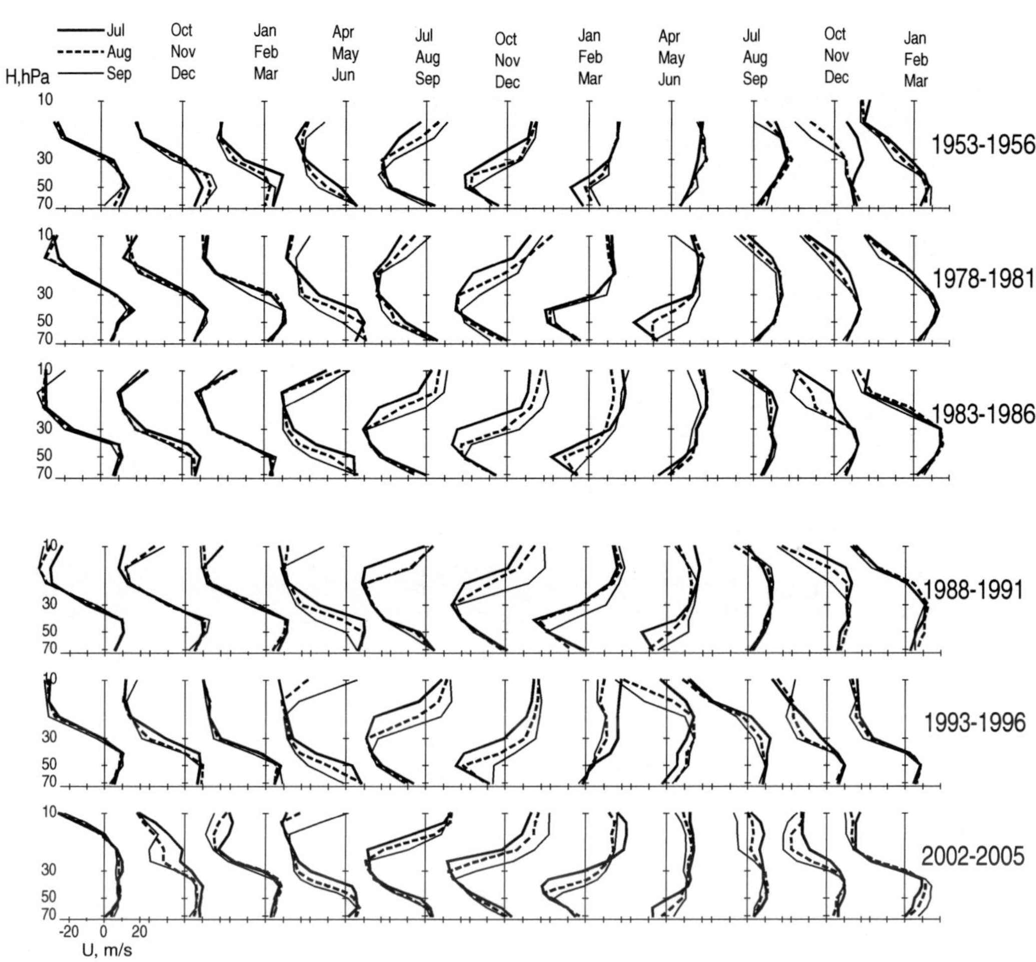

Figure 6. The same as Figure 4, but for July 30-month scenarios. Shown are six 33-month periods from the stagnation stage onset in July to the stagnation stage of subsequent QBO-cycle.

3.1.3. 36-Month Scenarios

Two anomalously long QBO-cycles occurred in 1960s. It was believed [*Dunkerton and Delisi, 1985*] that the unusually slow easterly shear zone descent was a primary cause of lengthened westerly (easterly) QBO-phases at 50 (20) hPa. Figure 7 shows the wind profile changes during two intervals by three years long each: from January 1964 to December 1966 (the upper two rows) and from January 1967 to December 1969 (the lower two rows). Again, like Figures 4, 5 and 6, the plots display the profiles for one season (three months), the first, middle and last profiles on each plot being marked by the hard, dash, and thin line, respectively. As was shown above the series of 24-month QBO-cycles took place in 1956-1963. At the end of the last 24-month QBO-cycles (from January 1962 to December 1963) the decrease of the westerly speed started at the higher layers, however the wind direction remained unchanged. Then westerly was reversed to easterly simultaneously at all altitudes from 24 to 10 hPa in a matter of one month, from December 1963 to January 1964. Later the easterly wind speed in layer higher 29 hPa increases during some months (till July 1964) without essential variations of zero wind speed altitude. From January to June 1964 the altitude changes amount to 5 hPa. Thus, the stagnation stage starts near solstice in December 1963 – January 1964. One can see from Figure 7 that the zero wind speed altitude was very stable near the 30 hPa during the first year. Furthermore, after June 1964 the wind profiles structure at height from 10 to 70 hPa remained practically the same till March-April equinox of the next 1965 year. The easterly shear descent is resumed only 15-16 months, or five seasons, after the stalling period beginning.

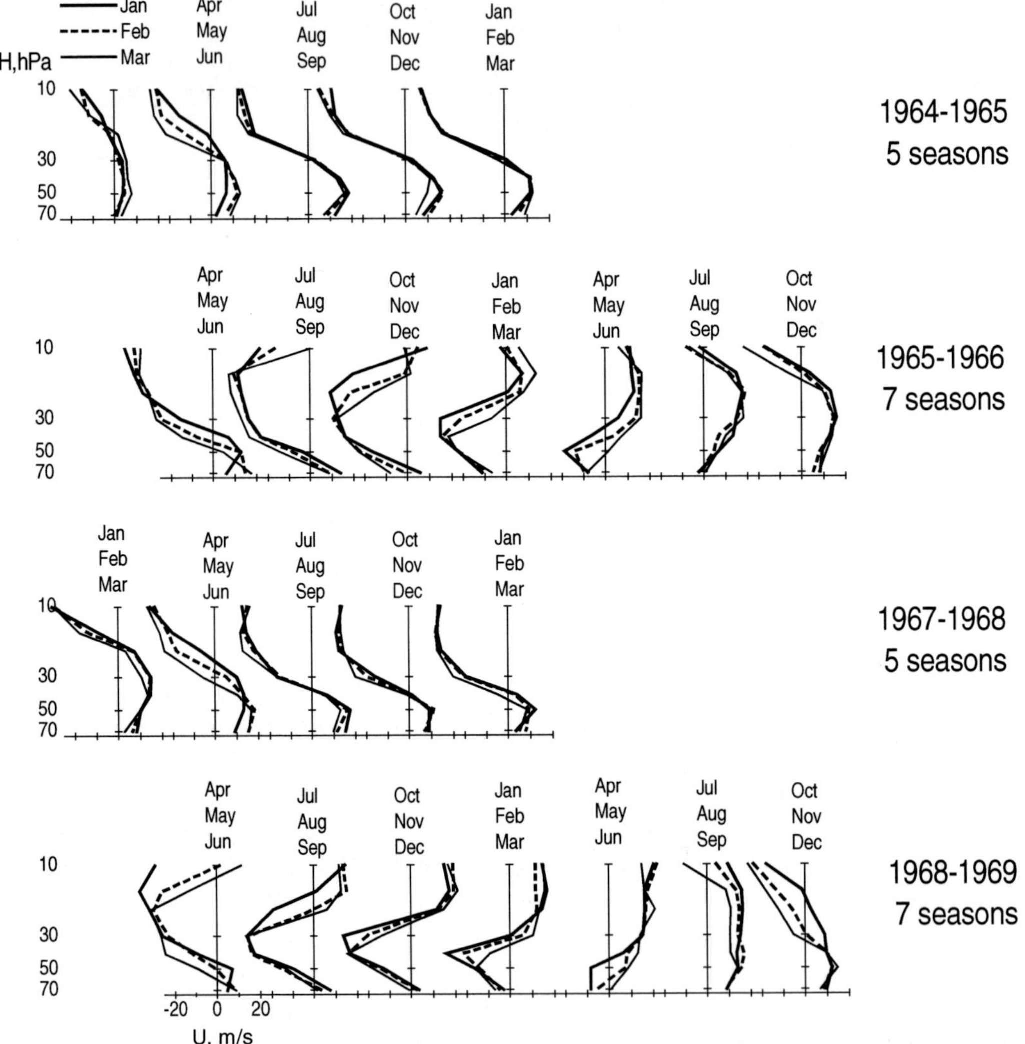

Figure 7. The same as Figure 4, but for 36-month scenarios. Shown are two 39-month periods from the stagnation stage onset in January to the stagnation stage of subsequent QBO-cycle. The first (second) 36-month QBO-cycle is shown in two upper (lower) rows.

As for period till the beginning of a new stagnation stage (marking the following QBO-cycle), the subsequent behavior of easterly and westerly shear descending zones in 1965-1966 was the same as in 24- and 30-month scenarios In such a manner, the cycle was completed in seven seasons (20-21 months) after the end of the stalling period. Thus, the full QBO-cycle length was equal to twelve seasons.

By next solstice (December 1966 – January 1967) the easterly (westerly) were observed again higher (lower) the ~20 hPa level. The altitude of the zero wind speed was kept in the subsequent months, the insignificant changes of the height profiles during December 1966 – March 1967 being indicative of the beginning of subsequent delayed E-shear zone. Behavior of the wind profiles during 1967-1969 was similar to that in 1964-1966 (meaning the durations of the stagnation stage and the full cycle length). The differences between two stagnation stages are concerned only the altitude of halted shear zone but not the moments of the stagnation beginning and end. We denoted the variant of the QBO-cycle evolution, when the total length for each cycle is equal to full three years or twelve seasons, as "the 36-month scenario".

The Table presents the summary of all above examined types of the QBO-cycle evolution observed during period from July 1953 to December 2006.

Table. The QBO-cycles observed during interval from July 1953 to December 2006.

The time intervals	QBO-cycle numbers	24-month scenarios	30-month scenarios		36-month scenarios
			January	July	
2.5 years 1953-1955	1			Jul 53 – Dec 55	
8 years 1956-1963	2	Jan 56 - Dec 57			
	3	Jan 58 - Dec 59			
	4	Jan 60 - Dec 61			
	5	Jan 62 - Dec 63			
6 years 1964-1969	6				Jan 64 – Dec 66
	7				Jan 67 – Dec 69
6 years 1970-1975	8	Jan 70 –Dec 71			
	9	Jan 72 –Dec 73			
	10	Jan 74 –Dec 75			
20 years 1976-1995	11		Jan 76 – Jun 78		
	12			Jul 78 – Dec 80	
	13		Jan 81 – Jun 83		
	14			Jul 83 – Dec 85	
	15		Jan 86 – Jun 88		
	16			Jul 88 – Dec 90	
	17		Jan 91 – Jun 93		
	18			Jul 93 – Dec 95	
4 years 1996-1999	19	Jan 96 –Dec 97			
	20	Jan 98 –Dec 99			
5 years 2000-2004	21		Jan 00 – Jun 02		
	22			Jul 02 – Dec 04	
2 years 2005-2006	23	Jan 05 –Dec 06			

3.2. Seasonal Regularities of the QBO Cycle Evolution

As was noted earlier, the tendency for QBO-phase transitions to be related to annual or seasonal cycles is typical for the lower stratosphere (about 45-50 hPa). Experimental results indicate that the QBO-phase onsets show a preference to occur during northern spring-summer season (April-July) [*Dunkerton and Delisi, 1985; Dunkerton, 1990; Baldwin, et al., 2001; Pascoe et al., 2005*]. However, as was argued in [*Dunkerton, 1990*], the exact seasonal synchronization of the QBO implies that "QBO cycles begin within the limits of data

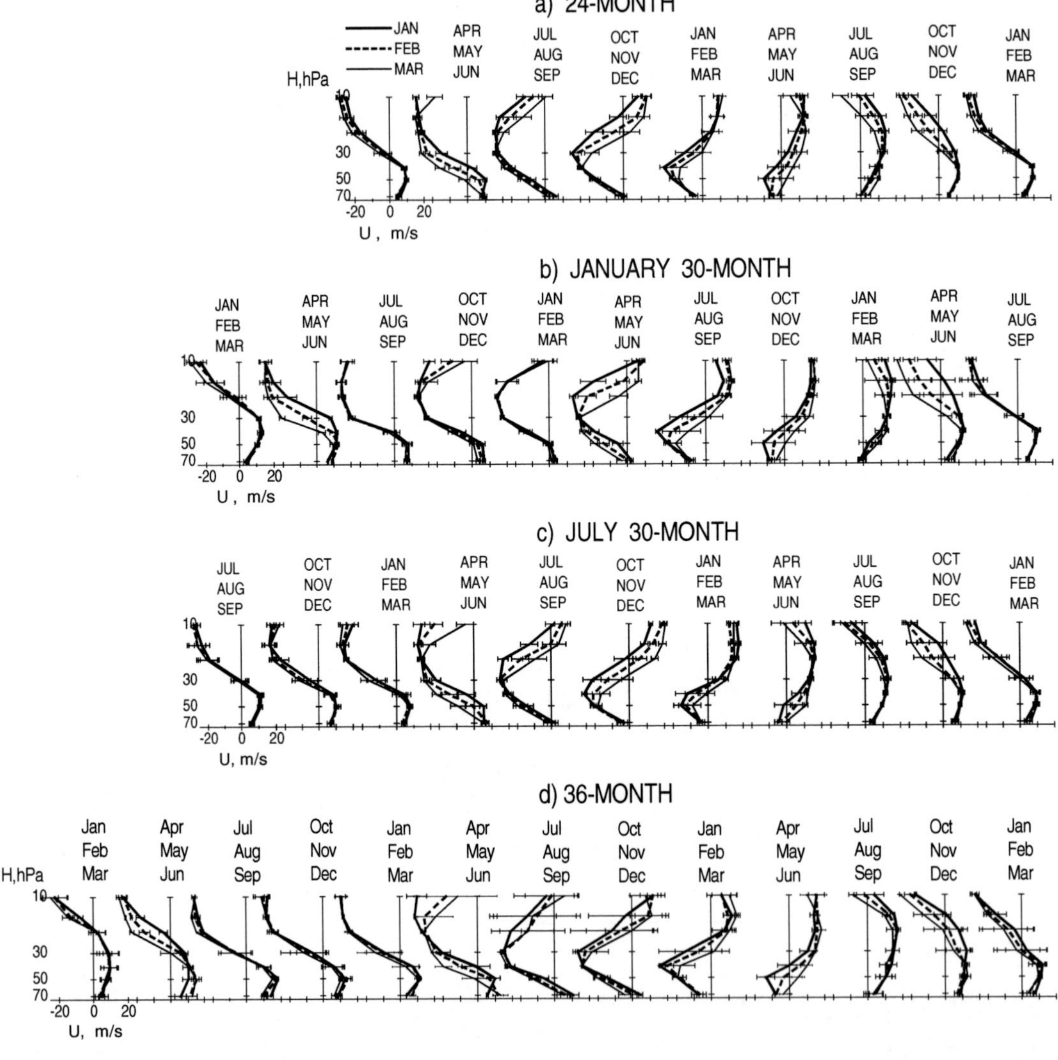

Figure 8. Mean profiles of the wind evolution for QBO cycles with different length: (a) 24-month scenarios, (b) 30-month scenarios, (c) 36-month scenarios. Every plot contains three monthly profiles. Each profile was calculated by averaging the wind speeds at the appropriate heights for the proper months of all QBO-cycles in consistent with dates shown in Table. The first, middle and last profiles on each plot are shown by the hard, dash, and thin line, respectively. The horizontal bars show the standard deviations at each altitude.

accuracy, at exactly the same time or times of the year". But if the QBO-phase (E or W) is determined in the common manner in conformity with the wind direction at a certain fixed pressure level, the phase reversal time can fall at any month of year. In such a case, one can say only about the seasonal modulation in the distribution of the QBO-phase onsets throughout the year.

As it was shown above, analysis of height wind profiles reveals a pronounced seasonal regularity of the QBO-cycle evolution inconspicuous under traditional methods of the QBO examination. The QBO of zonal wind in equatorial stratosphere turned out to be synchronized with the seasonal cycle if the beginning of the easterly shear stagnation stage with westerly (easterly) winds lower (higher) of the stalling region is regarded as the onset for each QBO-cycle. In such a case it is not significant to know the exact altitude of the shear zone during stagnation stage, this altitude varies from one QBO-cycle to other. Figure 8 demonstrates the total scheme of the wind profiles for different scenarios of the QBO evolution. Each profile in Figure 8 was calculated by averaging the wind speeds at the appropriate heights for the proper months of all QBO-cycles in consistent with data shown in the Table. Number of the used QBO cycles at Figures 8a, 8b, 8c and 8d is, respectively, ten, five, six and two. The horizontal bars show the standard deviations for different altitudes. The first plot for each type of scenario shows the profiles observed during first three months after the stagnation stage beginning in solstice. The 24-month, January 30-month and 36-month scenarios begin near the December-January solstice and July 30-month scenarios start near the June-July solstice.

It will be useful to remember that the easterly shear descent from 10 hPa to the stalling region (at about 20-30 hPa)) is observed before the onset of stagnation stage in any type of scenario (see Figures 4-6). Independently from the time of the easterly shear location at 10 hPa level (changing from April to November), the stagnation stage at altitudes 20-30 hPa is formed in solstice (December-January or June-July). The 24-month, both 30-month and 36-month types of QBO scenario are different one season later the stagnation stage beginning. The main feature that defined this difference is the duration (length) of stagnation stage.

One can see from Figure 8a that the stagnation stage in 24-month scenario is finished near the March-April equinox since the easterly shear resumes to descent to the lower stratosphere after this time. It means that the stagnation stage took place during one season. As Figure 8b shows, the two step stagnation stage with slight small intermediate descent during the first equinox is typical of the January 30-month scenario. The following E-shear descent lower the stalling region begins after the second equinox (September-October). Consequently the full duration of stalling period is equals to 9-10 months or three seasons. The July 30-month scenario (Figure 8c) demonstrates the very stable stagnation stage without changes in the zero wind altitude also during 9-10 months or three seasons. And finally the duration of stagnation stage reaches 15-16 month (or five seasons), in 36-month scenario (Figure 8d). Hence the descent of easterly shear zone is stopped at 20-30 hPa layer for over the one, three or five seasons, after that the descent to the lower stratosphere is resumed always in equinox. All cases of the stagnation stage with one season duration were traditionally regarded in research papers as "no stalling events". It was shown in section 3.1 that the easterly shear stalling period is always observed, but, apparently, the one season duration of the stagnation stage is examined as unessential in comparison with three or five seasons, typical of 30 or 36 months scenarios.

The E-shear descent renewal at altitudes lower the stalling region determines the end of the stagnation stage. As Figure 8 shows, the stagnation stage is finished either in March-April

(24-month, July 30-month and 36-month scenarios) or in September-October (January 30-month scenario). The subsequent wind evolution before the next QBO-cycle onset always completed within seven seasons (20-21 month), irrespective of the QBO-scenario. About one season after the stagnation stage end the westerly shear descent starts near the 10 hPa level and reaches the 70 hPa in four-five seasons. Approximately two seasons before a new stagnation stage (identifying the next QBO-cycle) the subsequent easterly shear zone appears in the upper layer and its descent is observed up to solstice, either till December-January (in case of 24-month, July 30-month and 36-month scenarios) or till June-July (in case of January 30-month scenario). Structure of the wind speed changes during these seven seasons varies from cycle to cycle: the westerly and then easterly shear zones appear at 10 hPa level in different time and rate of their succeeding descent to the lower layer varies in large limits. As a result, the wind reversal or QBO-phase transitions times do not reveal the seasonal dependence. At the same time, the tendency for the seasonal modulation of the QBO-phase transitions, mentioned in some investigations, can be easily explained.

The behavior of E-shear descent before the stagnation stage is easily derived from last three-four plots in each row of Figure 8. The westerly speed in layer higher the stalling region starts to decrease about two-three seasons ahead of the stagnation stage beginning that occurs always in solstice (December-January or June-July). Consequently, the transition from W to E QBO-phase at higher levels can be observed either during the second half of year (24-month, July 30-month and 36-month scenarios) or in the first half of year (January 30-month scenario), i.e. throughout the whole year. Nevertheless, there were only five cases of January 30-month scenario over period 1953-2006 in comparison with other eighteens. This fact explains the enlarged probability to observe the W-E transition at higher levels during second half of year.

At 20 hPa (near and slightly higher of the stalling region) the QBO-phase changes from W to E are observed immediately before the December-January solstice that is in last months of the year (October-December). As to the W-E transition at lower layer it is connected with end of the stalling period. Stagnation stage always ends by equinox at altitude about 30-40 hPa. The starts of easterly shear descent from this altitude are observed either in March-April (18 cases) or in September-October (5 cases). Taking into account the mean rate of E-shear descent (~1 km/month) it may be concluded that W-E transition at 45-50 hPa level must be observed more frequently in April-July. The similar explanation is valid also concerning E-W phase transition. The westerly shear appears at 10 hPa about one season after the stagnation stage end. As a result, it occurs at higher layer during first half of year in five cases and during second half of year in eighteen cases. On the contrary, the E-W phase transition in the lower layer takes place during first half of year in eighteen cases and during second half of year in five cases. This circumstance defines the greater probability for E-W transition to be observed at 45-50 hPa in April-June. All above examined regularities in the QBO-phase transitions are in a good consistent with results presented in [*Baldwin, et al., 2001; Pascoe et al., 2005*].

Therefore, it is apparently possible to observe the QBO-phase in any month of year, and the larger or smaller probability of the certain QBO transitions during fixed months or seasons is explained only by the relative number of the realized QBO scenarios. Indeed, the revealed seasonal tendency in distribution of QBO-phase transitions is determined by the larger total number of the QBO events with 24-month, July 30-month and 36-month scenarios in comparison to the events with January 30-month scenario. Since number of July 30-month scenarios is balanced by number of January 30-month scenarios, it is clear that the appointed probability is determined just by group of the 24-month scenarios. Thus, the question of reasons for regular realization of the 24-month QBO scenario is bringing up. It is clear that choice between the 24-month and January 30-month scenarios (i.e. duration of the stalling period) is determined by halt of the E-shear descent during the first equinox after the stagnation stage beginning.

According to [*Pascoe et al., 2005*] whether or not an easterly shear zone becomes stalled depends on the time of year, when the transition reaches the stalling region. If the easterly shear zone arrives at 30 hPa just before the May–July period, the descent to the lower stratosphere occurs without hindrance, because the Brewer-Dobson upwelling at the equator is weakest during this period. It is supposed that the Brewer-Dobson upwelling would act to reinforce the residual upward circulation associated with the easterly QBO shear zone and thereby to stall the descent. This explanation seems us to be unjustified. One can see from Figure 6 that in July 30-month QBO scenarios the shear zone stops in June-July solstice just at ~30 hPa. The monthly wind profiles show that descent in the region higher ~30 hPa proceeded steadily before June-July when the stalling period started. But then in the most of these events only the negligible changes in wind profiles were seen over the subsequent 9-10 months up to the next equinox (March-April), when the easterly shear zone resumed descent to the lower stratosphere. It means that the easterly shear during the first year did not descend in "favored May–July period" (according to [*Pascoe et al., 2005*]), but it moved downward at the same period of next year. Consequently, the enlarged probability (37%) of appearance of the W-E phase transition at 44 hPa in June is determined by occurrence of the easterly shear descent from ~30 hPa altitude to lower stratosphere in March-April in 18 cases from total 23 QBO-cycle observed in 1953-2006.

Thus, the analysis of the height profiles for monthly mean zonal wind has revealed that the process of easterly wind descent to the lower stratosphere always includes the quiet period (not shorter than three months), characterizing by the retardation of the wind speed shear zone in the stalling region at altitudes ~ 20-40 hPa. The stalling period or stagnation stage always starts in the solstice months (December–January or June–July) and ends in one of subsequent equinox, the first, second, or third one. Then seven seasons (20-21 months) are required to complete the current QBO cycle, irrespective of the scenario type. After that a new QBO-cycle starts. Hence, just duration of the stagnation stage is the determining factor for length of the QBO wind rotation cycle. Being dependent on the stagnation stage duration the full QBO cycle can lasts, correspondingly, 24, 30 or 36 months. The minimum length of 24 months is the most likely period for QBO cycle. The longer duration of the cycle (30 months and, sometimes, 36 months) is realized if the beginning of the easterly wind descent to lower layer (or the stagnation stage end) is delayed by no less than two seasons. The aim of the next section is to discover the possible cause of this retardation.

3.3. The Relation of QBO Phases Duration to the Solar UV Irradiance

The distinct seasonal dependence in the zonal wind reversal implies that this process is guided by the internal atmospheric mechanisms. The course of the easterly shear zone descent, being varied for different QBO-cycles, determines the shorter or longer retardation (the stagnation stage) at height near 20-40 hPa. It may be suggested that process of the east wind descent to the lower stratosphere is affected by external conditions, for instance by changes in the solar UV irradiance. As a result, the length of the full QBO cycle may be influenced by the solar UV flux variations through alternation of the of easterly shear zone descent duration.

We examined relations between evolution of the mean height wind profiles and the solar UV irradiance while passing two seasons after the stagnation stage beginning. Intensity of the solar UV irradiance was estimated from the $F_{10.7}$ index, presenting the solar 10.7-sm radio flux. The $F_{10.7}$ index is a standard characteristic of solar activity used in research concerning the solar influence on the Earth's climate. Usage of the $F_{10.7}$ index as the solar activity proxy was established when comparing the time series of the index and UV/EUV measurements [*Donnelly, 1988; Rottman, 2000; Floyd et al., 2005*]. According to [*Soukharev and Hood, 2001; Kane, 2003*] the $F_{10.7}$ index undergoes persistent medium-term variations with quasi-period in the range of 1-3 years, associated with the creation and decay of active regions on the Sun. In our analysis we used the daily $F_{10.7}$ index received from the Space Physics Interactive Data Resource (SPIDR) (http://spidr.ngdc.noaa.gov). To investigate the effects of the solar UV irradiance on quasi-biennial oscillation the medium-term component in $F_{10.7}$ index was derived by subtracting the long- and short-term fluctuations from the index time series. For that the 27-day running mean and the 26-month running mean (the 780 point smoothing) of the index have been evaluated. Then deviation of the 27-day running mean from 780 days smoothed series has been calculated and normalized. The resulting version is denoted as $\Delta F_{10.7}$:

$$\Delta F_{10.7} = \frac{(\langle F_{10.7} \rangle_{27days} - \langle F_{10.7} \rangle_{780days})*100\%}{\langle F_{10.7} \rangle_{780days}}$$

The statistical treatment of the relations between the $F_{10.7}$ index and the wind profiles changes has been performed for period 1953-2005 with use of the epoch superposition method. Results are shown in Figure 9. Three types of the QBO cycles have been examined: the 24-month scenarios, with stagnation stage finished in the nearest equinox and two types of 30-month scenarios ("January" and "July"), with the stagnation stage continued further.

In case of 24-month scenario stagnation stage was always started in December-January, and ended in March-April. Behavior of the $F_{10.7}$ index in these cases was characterized either by the pronounced increase while approaching the stagnation stage end or by the index fluctuations occurring against the high value of $F_{10.7}$. This regularity is demonstrated in Figure 9a, the 1st January within the stagnation stage being taken as "zero date" for the $F_{10.7}$ index. One can see that in case of the 24 months QBO cycle (with minimal duration of stagnation stage) the $F_{10.7}$ index grows, in general, in going from December-January solstice to June-July solstice. The average index $\Delta F_{10.7}$ increases from December to May by 12%, the statistical significance level being better than 0.001. Stagnation stage finishes in March–April against the background of high level of the $F_{10.7}$ index.

In the 30 months scenarios the occurrence of stagnation stage is observed in December-January or in June-July. The appropriate completions of stagnation stage in these scenarios started only during the second equinox (in September-October or in March-April) and stagnation stage lasts about nine months. The low value of the $F_{10.7}$ index in time of the first equinox or pronounced decrease of $F_{10.7}$ was typical of the QBO cycles with 30 months

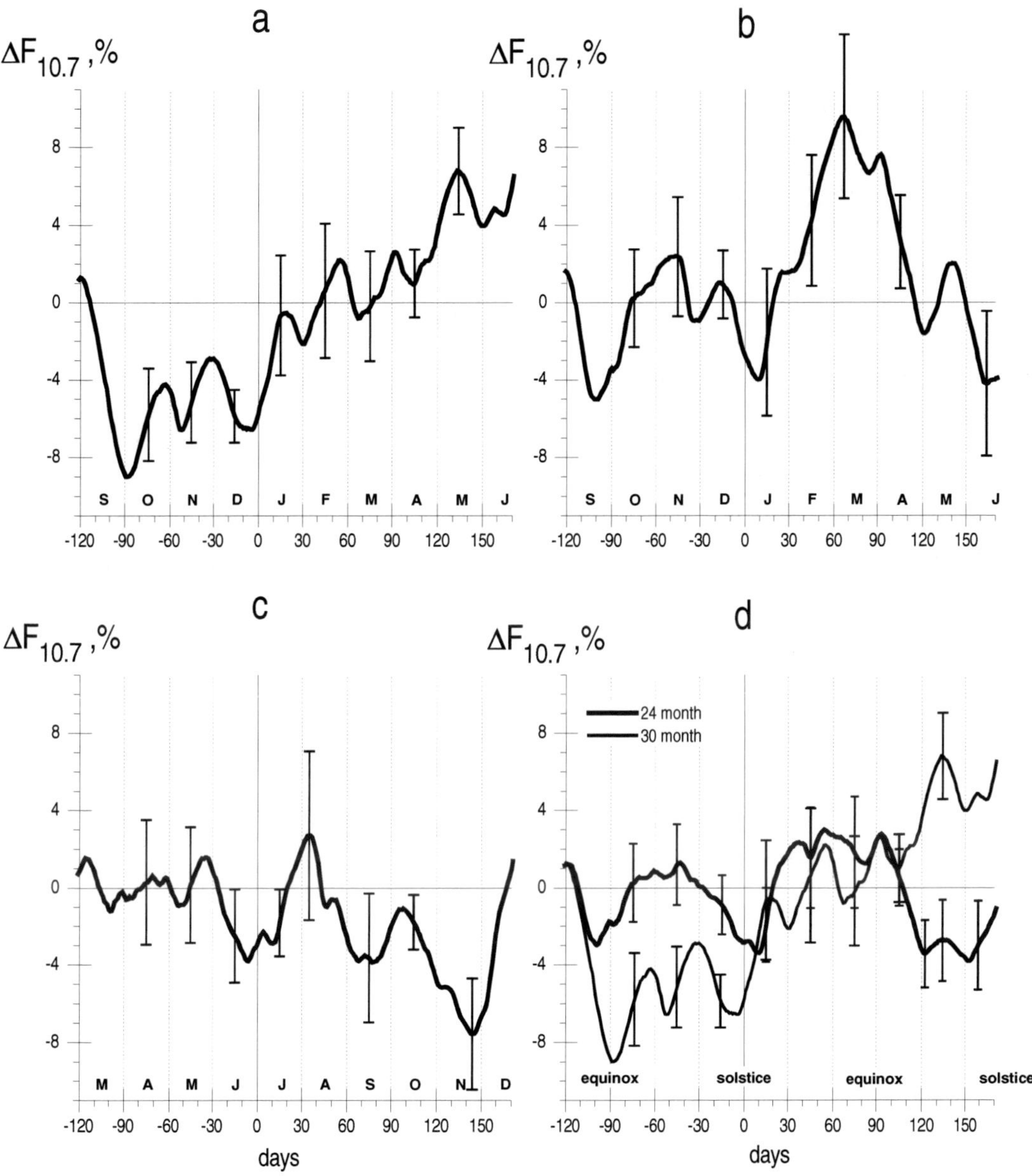

Figure 9. a) Behavior of the mean $\Delta F_{10.7}$ index in case of 24-month scenario of the QBO cycle. The 1st January within the stagnation stage is taken as "zero date" for the $\Delta F_{10.7}$ index. The index $\Delta F_{10.7}$ increases from December to May by 12%, with the statistical significance level better than 0.001.
b) Behavior of the $\Delta F_{10.7}$ index in case of January 30-month scenario of the QBO cycle. The 1st January within the stagnation stage is taken as "zero date" for the $F_{10.7}$ index. The index $\Delta F_{10.7}$ decreases from March to June by 13.8%, the statistical significance level being equal to 0.03.
c) Behavior of the $\Delta F_{10.7}$ index in case of July 30-month scenario of the QBO cycle. The 1st July within the stagnation stage is taken as "zero date" for the $\Delta F_{10.7}$ index. The index $\Delta F_{10.7}$ decreases from August to November by 10%, the statistical significance level being equal to 0.07.
d) Comparison of the mean $\Delta F_{10.7}$ index variations in cases of the 24-month (light curve) and 30-month scenarios (heavy curve), the curve for 30-month scenario being summary of January and July 30-month scenarios shown in Figures 9b and 9c. The difference in the $\Delta F_{10.7}$ index observed at the fifth month after the stagnation stage beginning is 7–9 %, with level of statistical significance equal to 0.008–0.01.

scenarios. The relationship between the wind profile evolution and the $F_{10.7}$ index is shown in Figure 9b for the January 30-month scenario and in Figure 9c for July 30-month scenario, the "zero date" for the $F_{10.7}$ index being taken on the 1st January or the 1st July, respectively. Results of the statistical analysis show that the $\Delta F_{10.7}$ index for January 30-month scenario reduced from March to June by 13.8% (the statistical significance equals to 0.03), the $\Delta F_{10.7}$ index for July 30-month scenario reduced from August to November by 10% (statistical significance equals to 0.07). As a result of the UV decline, the descent of easterly shear zone do not develop below ~ 30 hPa; the finish is delayed for half a year, and total length of the QBO cycle increases up to 30 months.

The quite different character of run of the mean $\Delta F_{10.7}$ index for the 24-month and 30-month QBO scenarios is clearly seen in Figure 9d, where the heavy curve is summary of the both, January and July, types of the 30-month scenario. The difference in the $\Delta F_{10.7}$ index observed at the fifth month after "zero date" in these two scenario types is 7–9 %, with level of statistical significance about 0.01. In case of 36-month scenarios (not shown in the Figure) the $F_{10.7}$ index in course of the first and second equinox either continued to be low ($F_{10.7} \approx 70$-75 during 1964) or decreased (from mean monthly $F_{10.7} \approx 159$ in March 1967 to ≈124 in June 1967 and from ≈156 in August 1967 to ≈135 in October 1967).

Results of the statistical analysis make it possible to conclude that the easterly winds descend below ~30 hPa only under condition of the high level or steady increase of the UV irradiance occurring just in time of the first equinox in the QBO cycle beginning. If the UV irradiance is low or decreases during the first equinox, the easterly winds typical of the upper layer, do not descend below 30 hPa. The decrease of the UV irradiance seems to be conducive to halt the easterly shear in stalling region (at ~30 hPa). Consequently, whether or not an easterly shear zone is retarded at height ~20-30 hPa depends on level of the UV irradiance in time of the first equinox for the partial QBO cycle.

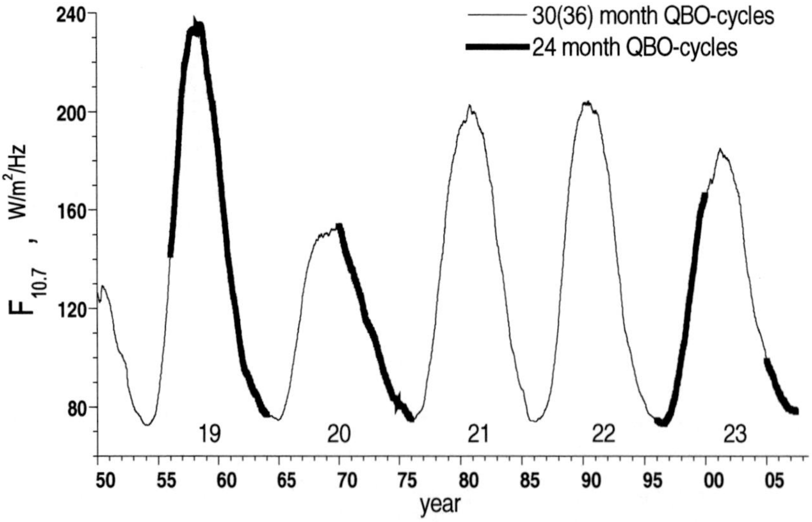

Figure 10. Course of solar activity index $F_{10.7}$ for the period from January 1950 to May 2007. The 26-month running mean (the 780 point smoothing) of the daily $F_{10.7}$ index measurements is presented. The records from 19 to 23 indicate the numbers of solar cycles. Heavy and thing lines correspond to series of short (24 month scenarios) and long (30 or 36 month scenarios) QBO-cycles, respectively.

The QBO-period, duration of the West QBO phase in the lower stratosphere as well as the descent rate of the easterly shear zone demonstrate the periodic variations with a mean period of 10-12 years [*Dunkerton and Delisi, 1985, 1997; Soukharev and Hood, 2001*]. This circumstance explains the attempts to correlate the QBO-period with the 11-year cyclicity of solar activity. In period of solar maxima the rapid descent of easterly shear zone shortens the West QBO phase in the lower stratosphere. During the solar minimum a slow or stalled descent of the easterly shear zone coincides with the lengthy West QBO phase in the lower stratosphere. As suggested in [*Soukharev and Hood, 2001*], this implies more stalling cases in years of solar minimum. Results of our study are not in line with later conclusion. Indeed, the number of lengthy stalling periods resulting in 30 or 36 months scenarios does not depend on phase of the 11-year solar cycle. As Figure 10 shows, the shorter stalling periods (relating to the 24-months scenarios) were dominant in years of solar maximum only during 19th solar cycle (1956-1960). This order broke down in course of the next 20th solar cycle: very long stalling period gave rise to the 36 months QBO scenario observed just in the solar maximum years (1967-1969). Only 30-months scenarios were observed during phases of maximum for the subsequent solar cycles (1978–1983, 1988–1993 and 1999–2003). And vice versa, almost half of the all 24-months scenarios (1962-1963, 1974-1975, 1996-1997 and 2005-2006) occur near solar 11-year minimums. Analysis of all these events provides the basis for our conclusion that delay in the easterly shear zone descending at 30 hPa depends on the intensity of solar UV flux in the proper season. It is well known, that amplitudes of the solar variations at the QBO time-scale (1-3 years) tend to be larger under the solar maximum than under solar minimum conditions [*Donnelly, 1988; Soukharev and Hood, 2001; Kane, 2003; Floyd et al., 2005*]. It makes possible to conclude that observed 11-year variations of the W-QBO phase duration in the lower stratosphere are caused by the 11-year modulation of the shorter-term solar UV oscillations.

According to the QBO numerical models the easterly (westerly) shear zones are related to a cold (warm) temperature anomaly at the equator, and the circulation is characterized by a sinking motion at equator in the westerly shear zones and rising motion in the easterly shear areas [*Reed, 1965; Baldwin et al., 200*]. Based on NCEP reanalysis (1958-2001) *Ribera et al.* [*2004*] reconstructed the spatial and temporal evolution of the several atmospheric variables through a QBO cycle. As they show, the circulation associated with the quasi-biennial oscillation is linked with negative (positive) temperature anomalies caused by the adiabatic rising (sinking) motions over zones of easterly (westerly) wind shear at the equator. Consequently, the solar influence on QBO wind evolution may be related to alterations in distribution and magnitude of the temperature anomalies. The later can be resulted in the additional heating caused by absorption of the increased solar UV radiation at 120-300 nm by ozone and molecular oxygen in the Earth's outer atmosphere.

3.4. Prediction of the QBO Cycle Length

The stagnation stage of the descending easterlies always starts in solstice, either in December-January or in June-July, and having the quantized duration (about 3, 9, or 15 months) always ends by the equinox irrespective of the scenario type. Since the subsequent wind evolution of the particular QBO-cycle is always completed within 21 months, the length of the full QBO-cycle, defined as a period between the beginnings of the successive stagnation stages, will be equal to 24, 30, or 36 months. Consequently, the full length of the current QBO cycle and the appropriate onset of the next one can be predicted while observing the zonal wind transformation during first equinox after the stalling period beginning. As Figure 8 demonstrates, if the westerly winds are observed from 70 to 30-20 hPa and easterly winds occur from 30-20 to 10 hPa in December (or June), this profile will persist without significant changes at least until March of the next year (or until September of the current year). Then, if the easterly shear descent to lower layer has been manifested near the first equinox (in April-May), the length of the current QBO cycle is predictable with certainty.

The previous QBO cycle, starting in December 2004-January 2005, was predicted in 2005 as the cycle developing according to 24-months scenario (see [*Gabis and Troshichev, 2006*]). Prediction was based on experimental data that the easterly shear zone descent resumed in April 2005. It was predicted also that a following QBO-cycle should start in December-January solstice. As a matter of fact, a new QBO-cycle began in December 2006-January 2007 (see Figure 1).

The experimental data available today up to May 2007 make it possible to do the next prediction. Indeed, the analysis of the wind profile changes showed that the easterly shear zone descent from 30 to lower stratosphere is resumed in the first equinox: the zero wind speed altitudes in March, April and May 2007 were equal to 30, 34 and 43 hPa, respectively. It means that stagnation stage starting in December-January solstice lasted only one season and already ended. Therefore, the present QBO cycle (2007-2008) is developed in accordance with 24-months scenario, and the next stagnation stage will start around January 2009.

In the pioneer studies concerning the modulating QBO influence on the solar activity effect on the stratosphere and troposphere [*Holton and Tan, 1980; Labitzke, 1987; Labitzke and van Loon, 1988; Chanin et al., 1989*] the phase of QBO was usually determined by the wind direction at a particular altitude (about 45-50 hPa). To define the QBO-phase during any certain month the different determinations were used in various studying: one of the levels in the range 50-20 hPa, or some combination of levels [*Lait, 1989*], or representation of the wind QBO with use of the first two EOFs (empirical orthogonal functions) of monthly wind observations [*Wallace, et al., 1993; Baldwin and Dunkerton, 1998*]. It is turned out that the correlations involving extratropical parameters depend on the exact level chosen to define the QBO-phase. Furthermore, this level was different in the opposite hemispheres: the strongest QBO-signal was obtained for QBO level of ~40 hPa in the northern hemisphere and ~25 hPa in the southern one [*Baldwin and Dunkerton, 1998*]. According to *Baldwin et al.,* [*2001*] the level for identification of the QBO phase would be chosen in the range from 50 to 25 hPa to maximize the extratropical response in atmospheric processes. Evidently the reason of such uncertainty is the lack of the clear physical mechanisms providing the QBO-influence on the extratropical processes. In addition, the QBO-phase identification by wind direction at a fixed level corresponds to the considerably different height structure of equatorial circulation during various moments of the same QBO-phase.

Our analyses of the height wind profile changes in course the QBO-cycle evolution revealed a clear seasonal regularity. The occurrence of the stagnation stage in a solstice season (December-January or June-July) specifies a quasi-biennial period: 24, 30, or 36 months, depending on the stalling period duration. The entire cycle develops according to certain temporal scales and it is quite insignificant to know the actual altitude of the shear zone during stagnation stage which can vary in different QBO-cycles. As soon as we identify the stagnation stage beginning as the QBO-cycle onset, the QBO transformation of zonal wind in the equatorial stratosphere is synchronized with seasonal cycle.

Firstly QBO cycle was named as "26-months cycle". This length was determined on the basis of data for 4-5 cycles (1953-1962) available in that time [*Reed, et al., 1961; Reed, 1965; Veryard and Ebdon, 1961*]. Accumulation of the observational data led to the increase of mean period to 27.7 months for 1953-1985 [*Naujokat, 1986*], and to 28.2 months for 1953-1995 (Baldwin et al., 2001). The positive trend of the QBO length for last 10 years has been explained by *Mokhov and Eliseev* [*1998*] as due to the influence of the climatic changes. As a matter of fact, the real length of QBO cycle constitutes 24, 30 or 36 months, and "the gradual growth of QBO cycle" is explained by the varying combination of these three lengths in the examined periods. Indeed, as Table shows, there were 3 cycles with 24 months length and only 1 cycle with 30 months length in 1953-1963. The succeeded 20 years included ten cycles and only three of them had 24 months length. As a result, the mean length of QBO cycle essentially increased. After next ten years with four cycles of 30 months, the mean QBO length increased still further. Thus, the 24-month scenario was more common before 1976, and 30-month scenario was more prevalent after 1976. However, the ratio of numbers of the 24- and 30-month cycles level off after 1991, and conclusion about the growth of QBO length can not be regarded as convincing.

3.5. Antarctic Ozone "Hole" and Seasonal Regularities of QBO-Cycle

Earlier it was shown that the level of total ozone (TOZ) over Antarctica during "ozone hole" periods displays significant interannual fluctuations, associated with QBO-phases of zonal wind in the equatorial stratosphere [*Bojkov, 1986; Garcia and Solomon, 1987*]. Later during gathering of experimental data it was revealed the ambiguity concerning relationship of the intensity of TOZ depletion on the QBO-phases [*Lait, et al., 1989; Angell, 1988; 1993; Herman, et al., 1995*]. At present time there are several hypotheses that explain the QBO-modulation at the polar regions by the equatorial circulation influence on meridional exchange and on the strength and isolation of the antarctic winter-spring circumpolar vortex [*Bowman, 1989; O`Sullivan, 1992; O`Sullivan and Salby, 1990; Kinnersley and Tung, 1999; Baldwin, 2001*].

In this section a comparative analysis between monthly mean October TOZ at station Amundsen-Scott (89.98°S; 24.8°W) and direction of zonal circulation in the equatorial stratosphere is presented. TOZ-data are based on ground Dobson-spectrophotometer (1962-2005) [*WOUDC Data Archive: http://www.msc-smc.ec.gc.ca/woudc/index.html*] and satellite TOMS (Total Ozone Mapping Spectrometer) measurements (1978-2005) [*http://toms.gsfc.nasa.gov/*]. The Figure 11 presents the course of TOZ.

It is evidently that for the correct examination of equatorial QBO-circulation influence on "ozone hole" phenomenon it is necessary to take into account the season regularities of the QBO-cycle evolution. During the 24-month and January 30-month scenarios of the QBO-cycles two "ozone holes" are observed. The lasted type of the QBO-cycles is always followed by the July 30-month scenarios starting in June-July solstice. Three "ozone holes" are observed during such QBO-cycles as well as during 36-month scenarios. Here the first "ozone hole" is formed during stagnation stage, i.e. under conditions of the stable circulation in equatorial region with westerly (easterly) in lower (middle) stratosphere. 44 "ozone holes" were observed in Antarctica for the period 1962-2005, and only seven of them occur during stagnation stage. For the most of "ozone holes" either the formation of polar vortex (in May-June) or ozone destruction (in September), or both happen under varying circulation in the equatorial stratosphere.

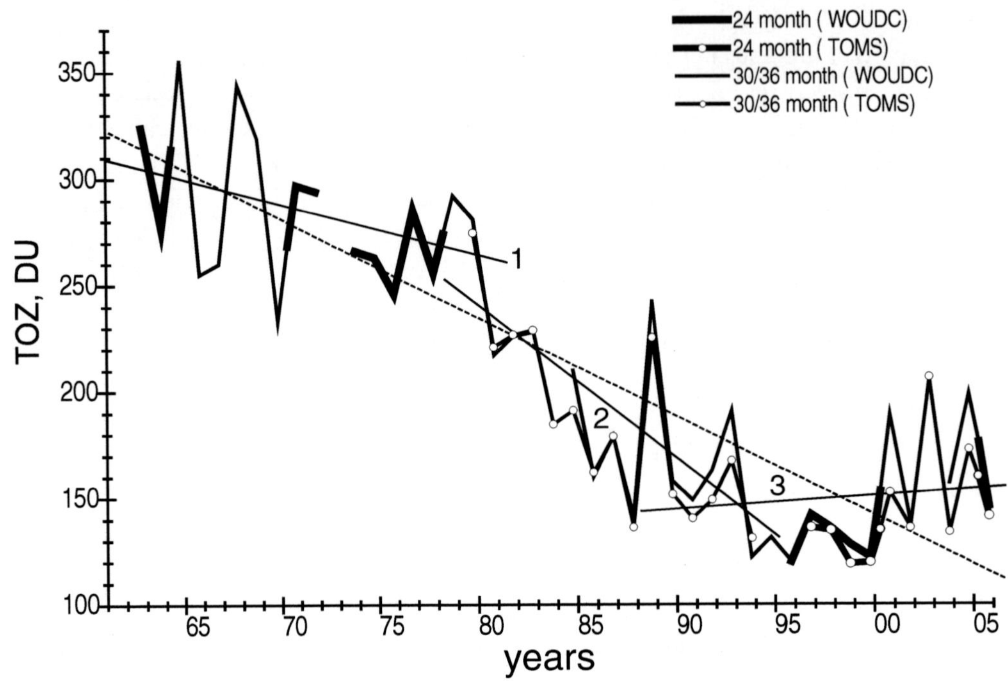

Figure 11. Changes of total ozone (TOZ) during the period 1962-2005. Dash line is trend for period 1962-2005. Thin lines with numbers: 1 – trend for period 1962-1979, 2 – trend for period 1979-1993, 3 – trend for period 1989-2005. Heavy and thing parts of TOZ- curve correspond to series of short (24 month scenarios) and long (30 or 36 month scenarios) QBO-cycles, respectively.

In overwhelming majority cases during the first "ozone hole" in each 24- and January 30-months QBO-cycles TOZ has a relative maximum, while during the second "ozone hole" TOZ has a relative minimum. Nearly always during the third "ozone hole" in July 30-months QBO-cycles TOZ shows a relative minimum. Hence if series of 24-month QBO-cycles is observed in equatorial stratosphere the course of TOZ demonstrates the "true" alternation of relative maxima and minima. Switch from series of 24-month to series of 30-month QBO-cycles results in breakdown of the "true" alternation, because during the July 30-month QBO-cycles the relative TOZ maximum is followed by two minima successively. Figure 11 illustrates that 24-month scenarios were more frequent until the year 1976, and 30-month scenarios more often were observed during 1976-1995. Minor descending trend of TOZ before about 1979 and after 1993 coincides with mixed 24- and 30-months QBO-cycles in the equatorial stratosphere wind. But very large descending trend of TOZ in 1979–1993 was observed when series of 30-months QBO-cycles occurs only. Also one can note the tendency of small amplitude of TOZ-variations during the 24-months QBO-cycles as compared to the amplitude during the 30-month or 36-month QBO-cycles.

4. Conclusion

The changes in equatorial stratospheric wind during QBO-cycle evolution were analyzed with use of the wind speed profiles in height range 70-10 hPa during period from 1953 to 2007. The results showed that the stage of QBO-cycle with easterly above 20-40 hPa and westerly below this height always begins in a solstice and persists for 3-4, 9-10, or 15-16 months without significant changes in the wind structure. This stalling period or stagnation stage is ended by the first, second, or third equinox, respectively, as the descent of the easterly shear to the lower stratosphere is resumed. Seven seasons (20-21 months) after the stagnation stage, the profile with easterly and westerly above and below 20-40 hPa, respectively, is observed again. The full cycle of the wind evolution lasts 24, 30, or 36 months, depending on the stagnation stage duration.

Thus, the reversal in the direction of the equatorial stratospheric zonal mean wind or quasi-biennial oscillations (QBO) have a pronounced seasonal regularity, which is detected when analyzing the height wind speed profiles. The zonal wind transformation turned out to be synchronized with the seasonal cycle if the beginning of the easterly shear stagnation stage is regarded as the onset for each QBO-cycle. In such a case the exact altitude of the shear zone during stagnation stage varies from one QBO-cycle to other, but it is significant that the stalling period always starts in solstice (December-January or June-July), and always ended by the equinox.

The occurrence of the stagnation stage in a solstice season and its different, but quantized duration (about 3, 9, or 15 months) specifies a quasi-biennial period: 24, 30, or 36 months. The minimum length of 24 months is the most likely period for QBO cycle. The longer duration of the cycle (30 months and, sometimes, 36 months) is realized if the easterly wind descending to lower layer (or the stagnation stage end) is delayed by no less than two seasons. The seasonal dependence in the zonal wind reversal implies that this process is guided by the internal atmospheric mechanisms. The start of the easterly shear zone descent lower of the stalling region is varied for different QBO-cycles, suggesting that the process is affected by external conditions, for instance by changes in the solar UV irradiance. As a result, the length of the full QBO cycle may be influenced by the solar UV flux variations through alternation of the start time of the easterly shear zone descending lower stalling region.

If examining the solar UV flux features for different duration of stagnation stage, the following results are obtained. After the short stalling (three–four months) the easterly winds descend below ~30 hPa only under condition of the high level or steady increase of the UV irradiance happening near and after the first equinox in course of the certain QBO cycle. If the UV flux is low or decreases near and after the proper equinox, the easterly winds typical of the upper layer do not descend below 30 hPa, and the stalling period length increases to 9-10 or 15-16 months.

Consequently, the solar UV irradiance influences the QBO cycle length through alternative duration of the easterly shear zone descending from the middle to lower stratosphere. Since the length of the QBO cycle (24, 30 or 36 months) is determined by duration of stagnation stage, the full QBO cycle length can be predicted while observing the zonal wind transformation during the first equinox after the stalling period beginning.

Acknowledgment

The authors are grateful to Barbara Naujokat (Meteorological institute, The Berlin Free University) for kindly providing them the data on monthly mean stratospheric zonal wind components. The work was supported by the Russian Foundation for Basic Research under the grant RFBR-05-05-64633.

References

Angell, J. K. Relation of Antarctic 100 mb temperature and total ozone to equatorial QBO, equatorial SST, and sunspot number, 1958-1987. *Geophys. Res. Lett.* 1988, 15, 915-918.

Angell, J. K. Reexamination of the relation between depth of the Antarctic ozone hole, and equatorial QBO and SST, 1962-1992. *Geophys. Res. Lett.* 1993, 20, 1559-1562.

Baldwin, M. P.; Dunkerton, T. J. Quasi-biennial modulation of the southern hemisphere stratospheric polar vortex. *Geophys. Res. Lett.* 1998, 25(17), 3343-3346.

Baldwin, M. P.; Gray L. J.; Dunkerton, T. J.; Hamilton, K.; Haynes, P. H.; Randel, W. J.; Holton, J. R.; Alexander, M. J.; Hirota, I.; Horinouchi, T.; Jones, D. B. A.; Kinnersley, J. S.; Marquardt, C.; Sato, K.; and Takahashi, M. The Quasi-biennial Oscillation. *Rev. Geophys.* 2001, 39(2), 179-229.

Besprozvannaya, A. S.; Ohl G. I.; Sazonov B. I.; Scherba I. A.; Schuka T. I.; and Troshichev O. A. Influence of short-term changes in solar activity on baric field perturbations in the stratosphere. *J. Atmos. Terr. Phys.* 1997, 59, 1233.

Bojkov, R. D.; The 1979-1985 ozone decline in Antarctic, as reflected in ground level observations. *Geophys. Res. Lett.* 1986, 13, 1236-1239.

Bowman K. P. Global patterns of quasi-biennial oscillation in total ozone. *J. Atmos. Sci.* 1989, 46(21), 3328-3343.

Chanin, M-L.; Keckhut, P.; Hauchecorne, A.; Labitzke, K. The solar activity – QBO effect in the lower thermosphere. *Ann. Geophys.* 1989, 7(5), 463-470.

Donnelly, R. F.. Uniformity in solar UV flux variations important to the stratosphere. *Ann. Geophys.* 1988, 6(4), 417-424.

Dunkerton, T. J. Annual variation of deseasonalized mean flow acceleration in the equatorial lower stratosphere. *J. Meteorol. Soc. Jpn.* 1990, 68(4), 499-508.

Dunkerton, T. J.; Delisi, D. P. Climatology of the equatorial lower stratosphere. *J. Atmosp. Sci.* 1985, 42(4), 376-396.

Dunkerton, T. J.; Delisi, D. P. Interaction of the quasi-biennial oscillation and stratopause semiannual oscillation. *J. Geophys. Res.* 1997, 102, 26107-26116.

Floyd, L.; Newmark, J.; Cook, J.; Herring, L.; McMullin, D. Solar EUV and UV spectral irradiances and solar indices. *J. Atmos. Sol. Terr. Phys.* 2005, 67, 3-15.

Gabis, I. P.; Troshichev, O. A. Influence of solar UV irradiance on quasi-biennial oscillations in the Earth's atmosphere. *Adv. Space Res.* 2004, 34(2), 355-360.

Gabis, I. P.; Troshichev, O. A. QBO cycle identified by changes in height profile of the zonal winds: new regularities. J. *Atmos. Sol. Terr. Phys.* 2005, 67, 33-44.

Gabis, I. P.; Troshichev, O. A. Influence of solar UV irradiance on the quasi-biennial oscillation of zonal winds in the equatorial stratosphere. *J. Atmos. Sol. Terr. Phys.* 2006, 68, 1987-1999.

Garcia, R.; Solomon S. A possible relationship between interannual variability in Antarctic ozone and quasi-biennial oscillation. *Geophys. Res. Lett.* 1987, 14, 848-851.

Georgieva, K.; Kirov, B.; Atanassov, D.; Boneva, A. Impact of magnetic clouds on the middle atmosphere and geomagnetic disturbances. *J. Atmos. Sol. Terr. Phys*. 2005, 67, 163-176.

Herman, J. R.; Newman P. A.; Larko D. Meteor 3/TOMS observations of the 1994 ozone hole. *Geophys. Res. Lett.* 1995, 22(23), 3227-3229.

Holton, J. R.; Lindzen, R. S. An updated theory for the quasi-biennial cycle of the tropical stratosphere. *J. Atmos. Sci.* 1972, 29, 1076-1080.

Holton, J. R.; and Tan, H. C. The influence of the equatorial quasi-biennial oscillation on the global ciculation at 50 mb. *J. Atmosp. Sci.* 1980, 37(10), 2200-2210.

Kinnersley, J. S.; Pawson, S. The descent rates of the shear zone of the equatorial QBO. *J. Atmosp. Sci.* 1996, 53(14), 1937-1949.

Labitzke, K., 1987. Sunspots, the QBO and the stratospheric temperature in the north polar region. *Geophys. Res. Lett.* 14(5), 535-537.

Labitzke, K. On the solar cycle – QBO relationship: a summary. *J. Atmos. Sol. Terr. Phys.* 2005, 67, 45-54.

Labitzke, K.; van Loon, H. Associations Between the 11-year Solar Cycle, the QBO and the Atmosphere, Part I, The Troposphere and Stratosphere in the Northern Hemisphere in winter. *J. Atmos. Terr. Phys*. 1988, 50, 197-206.

Labitzke, K.; van Loon, H. The QBO effect on the solar signal in the global stratosphere in the winter of the Northern Hemisphere. J. *Atmos. Sol. Terr. Phys*. 2000, 62(8), 621-628.

Lait, L. R.; Schoeberl M. R.; Newman P. A. Quasi-biennial modulation of the Antarctic ozone depletion. *J. Geophys. Res*. 1989, 94, 11559-11571.

Kane, R. P. Periodicities of few months in solar indices. *J. Atmos. Sol. Terr. Phys*. 2003, 65, 979-986.

McCormack, J. P. The influence of the 11-year solar cycle on the quasi-biennial oscillation. *Geophys. Res. Lett.* 2003, 30(22), 2162-2165.

Mokhov I. I.; Eliseev A. V. Changes in the characteristics of the quasi-biennial oscillation of zonal wind and temperature in the equatorial lower stratosphere. *Izv. Atmos. Ocean.* Phys. 1998, 34, 291-299.

Naujokat, B. An update of the observed quasi-biennial oscillation of the stratospheric winds over the tropics. *J. Atmos. Sci.* 1986, 43, 1873-1877.

O`Sullivan D.; Salby L. M. Coupling of the quasi-biennial oscillation and extratropical circulation in the stratosphere through planetary wave transport. *J. Atmos. Sci.* 1990, 47(5), 650-673.

O`Sullivan D. Modeling the quasi-biennial oscillation's effect on the winter stratospheric circulation. *J. Atmos. Sci.* 1992, 49(24), 2437-24-48.

Pascoe, C. L.; Gray, L. J.; Crooks, S. A.; Juckes, M. N.; Baldwin, M. P. The quasi-biennial oscillation: analysis using ERA-40 data. *J. Geophys. Res*. 2005, 110, D08105, doi:10.1029/2004JD004941.

Reed, R. J. The present status of the 26-month oscillation. *Bull. Am. Meteorol. Soc.* 1965, 46(7), 374–385.

Reed R. J.; Campbell W. J.; Rasmussen L. A.; Rogers D. G. Evidence of a downward-propagating, annual wind reversal in the equatorial stratosphere. *J. Geophys. Res.* 1961, 66(3), 813-818.

Ribera, P.; Pena-Ortiz, C.; Garcia-Herrera, R.; Gallego, D.; Gimeno, L.; Hernandez, E. Detection of the secondary meridional circulation associated with the quasi-biennial oscillation. *J. Geophys. Res*. 2004. 109, D18112, doi:10.1029/2003JD004363.

Rottman, G. Variations of solar ultraviolet irradiance observed by the UARS SOLSTICE – 1991 to 1999. *Space Sci. Rev*. 2000, 94, 83-91.

Soukharev, B. E.; Hood, L. L. Possible solar modulation of the equatorial quasi-biennial oscillation: additional statistic evidence. J. Geophys. Res. 2001, 106, (D14), 14855-14868.

van Loon, H.; Labitzke, K. The signal of the 11-year solar cycle in the global stratosphere. *J. Atmos. Sol. Terr. Phys*. 1999, 61, 53-61.

van Loon, H.; Labitzke, K. The influence of the 11-year solar cycle on the stratosphere below 30 km: a review. *Space Sci. Rev*. 2000, 94, 259-278.

Veryard, R. G, Ebdon, R. A. Fluctuations in tropical stratospheric winds. *Meteorol. Mag*. 1961, 90, 125-143.

Wallace, J. M.; Panetta R. L.; Estberg J. Representation of the equatorial stratospheric quasi-biennial oscillation in EOF phase space. *J. Atmos. Sci.* 1993, 50(12), 1751-1762.

In: Ozone Depletion, Chemistry and Impacts
Editor: Sem H. Bakker, pp. 269-314
ISBN: 978-1-60692-007-7

Chapter 12

OZONE/ACTIVATED CARBON: A NEW ADVANCED OXIDATION PROCESS TO REMOVE POLLUTANTS FROM WATER

J. Rivera-Utrilla and M. Sánchez-Polo
Departamento de Química Inorgánica, Facultad de Ciencias,
Universidad de Granada, 18071 Granada, España

Abstract

Rising concerns about the quality of drinking water have led both public and private bodies to invest considerable human and economic resources in the development of novel water treatment processes to more effectively remove organic micropollutants, highly toxic for human health (pesticides, herbicides, microtoxins) and sometimes responsible for altering organoleptic properties of the water. Processes based on the simultaneous use of ozone and activated carbon have proven to be very effective for removing contaminants from drinking waters. The results obtained by using naphthalenesulphonic acids as model pollutants have shown that O_3/activated carbon systems have a great efficiency in drinking water treatment because i) it is possible to remove micropollutants due to the high adsorption capacity of activated carbon, and/or ii) they oxidise polar micropollutants, characterised by a low reactivity with ozone, due to the enhancement of ozone transformation into OH· radicals catalysed by activated carbon. Moreover the presence of activated carbon during water ozonation processes reduces the concentration of dissolved organic carbon (TOC) increasing the benefits of the process. Two processes are involved in TOC decrease: i) adsorption of oxidation by-products on activated carbon and ii) mineralization of organic matter by hydroxyl radicals generated in the interaction between ozone and activated carbon. New carbon materials have been developed in our lab in order to potentiate ozone transformation into OH· radicals (activated coke, nitrogen enriched activated carbons and metal doped carbon aerogels) obtaining interesting results especially with metal doped carbon aerogels. The mechanism and the influence of operational parameters in the efficacy of O_3/activated carbon system have also been studied.

Objective

The present work aimed to outline results obtained during the development of a wide project to design a new process, based on the combined use of ozone and activated carbon, for the depuration of waters contaminated with non-biodegradable organic micropollutants, considering naphthalenesulphonic acids as model compounds.

1. Introduction

Effluents from the textile industry present elevated concentrations of sulphonated polyphenols (dyestuffs), ranging between 10 and 80 mg/L and with a mean concentration of 31 mg/L [1]. These effluents are composed of a great variety of chemical compounds, including naphthalenesulphonic acids. The sulphonic groups endow these compounds with high water-solubility and low biodegradability, so that they have been detected in both surface and subterranean waters, at concentrations ranging from 0.5 mg/L for di- and tri-sulphonated rings to 1.5 mg/L for monosulphonated rings [2, 3]. Despite the high concentrations of these contaminants found in the environment, there are scant data on their toxicity. Greihm et al. investigated the toxicity of some sulphonic aromatic compounds in 1994 but could establish no general conclusions about this family of organic compounds [4].

Adsorption on activated carbon is recognized by US Environmental Protection Agency (EPA) as a very effective option for the purification of waters contaminated with aromatic compounds. However, the influence of experimental conditions on the adsorption of aromatic compounds has yet to be fully defined. It is of great interest to know the mechanism and influence of operational variables in this adsorption process, in order to optimise the industrial-scale use of activated carbon in water treatments.

The use of energetic oxidants is recommended for the degradation of organic compounds, which are refractory to biological treatment. Ozone is one of the most widely used oxidants for the depuration of waters containing toxic organic compounds. In the last two decades, many researchers have studied the reaction between ozone and organic compounds [5-8]. Ozone is very costly, therefore, the kinetic parameters of the reactions of ozone to oxidate naphthalenesulphonic acids need to be determined in order to establish the feasibility of its use in water treatment and optimize the parameters required in the design of reactors.

Advanced oxidation processes (AOPs) have been developed to increase the efficacy of ozone in purification, all based on the generation of $HO\cdot$ radicals [9, 10]. The $HO\cdot$ radical is a highly reactive specie and attacks most organic molecules, with reaction rate constants ranging from 10^6 to 10^9 M^{-1} s^{-1}. The use of solid catalyst has been proposed to increase the extent of ozonation [11-13]. Activated carbon has been signalled as a promising catalyst of ozonation because of its chemical and textural properties and low cost [14]. However, there are no data on the physical and chemical processes that take place to enable the correct design of a reactor for this process. Thus, it is not known which properties of activated carbon are involved in increasing the extension of the ozonation process or what reactions are involved. Moreover, the influence of experimental and instrumental operational parameters on the adsorption/ozonation process is unknown.

The present work aimed to outline results obtained during the development of a wider project to design a new process based on the combined use of ozone and activated carbon for the depuration of waters contaminated with non-biodegradable organic micropollutants, considering naphthalenesulphonic acids as a model compounds. Some of these results have already been published [14-29].

Knowledge of the ozonation process of naphthalenesulphonic acids in the absence of activated carbon (reaction mechanism, kinetics, influence of operational variables, oxidation by-products) and of the interactions involved in the adsorption of these contaminants on activated carbon will contribute to explain the behaviour of activated carbon both as catalyst and adsorbent in the treatment system based on the simoultaneous use of ozone and activated carbon.

2. Experimental

2.1. Materials

1-Naphthalenesulphonic acid (NS), 1,5-naphthalenedisulphonic acid (NDS) and 1,3,6-naphthalenetrisulphonic acid (NTS) were supplied by Fluka. Sodium p-chlorobenzoate (pCBA) was obtained from Merck. In the liquid chromatography, tetrabutyl ammonium bromide (TBABr) (Fluka) was used as ion exchanger and NaH_2PO_4 (Merck) as pH regulator in the mobile phase. Tert-butyl alcohol (t-BuOH) (Merck) was used as the compound to scavenge radicals. All of the aqueous solutions were prepared with ultrapure water obtained from a Millipore Milli-Q device.

The commercial activated carbons used in the study were Filtrasorb 400, Sorbo, Merck, Ceca GAC, Ceca AC40, Norit y Witco.

2.2. Ozonation Experiments

Ozone was generated by an OZOKAV ozone generator with a maximum capacity of 76 mg/min. The experimental system was described in detail elsewhere [20].

In each experiment the reactor was filled with 1 L of pH 2.3 buffer solution (H_3PO_4/NaH_2PO_4), with NS, NDS or NTS. After stabilization of the partial pressure of the ozone, the gas mixture O_3/O_2 was fed to the reactor. Samples were taken regularly for chemical assay; $NaNO_2$ was used to stop the ozonation reaction. Naphthalenesulphonic acid concentration, dissolved ozone, genotoxic activity and total organic carbon (TOC) were determined after different time periods. In experiments where activated carbon was used, the proper dose of activated carbon was fed to the reactor simultaneously with the ozone.

2.3. Adsorption Experiments

Adsorption isotherms of NS, NDS and NTS on activated carbon samples were determined at 298 K in aqueous solutions. Typically, a 0.1 L aqueous solution containing 25-200 mg/L of naphthalenesulphonic acid was mixed with 0.1 g activated carbon in stirred Erlenmeyer flasks and maintained at 298 K for 72 h. This time was longer than that needed to reach constant solution concentrations. These experiments were carried out in the absence of any pH setting; the starting solutions had a pH of 5.5-6. The pH was monitored and it did not vary during these experiments. The naphthalenesulphonic acid concentration was determined using a Spectronic Genesys 5 spectrophotometer. The wavelengths used were 223 nm, 286 nm and 223 nm for NS, NDS, and NTS, respectively.

Adsorption of the three naphthalenesulphonic acids (NS, NDS, and NTS) was also studied in a dynamic regime using columns of activated carbon. For this purpose, a solution of naphthalenesulphonic acid (50 mg/L) was passed through an 8 x 1 cm column at a flow of 1.5 mL/min. As in the determination of isotherms, the solutions were prepared in the absence of any pH-controlling compound. Samples of the solution were collected at the column outflow until saturation was reached. The results of these experiments were used to determine the breakthrough curves and characteristics of the carbon columns.

2.4. Pretreatment of Activated Carbon

To determine changes in activated carbon activity due to ozonation, it was pre-treated with aqueous and gaseous ozone. 0.5 g of activated carbon Filtrasorb 400 (F400) was suspended in 0.5 L of a 1 mg/L ozone solution (pH 7) for 1 h. This activated carbon was then dried at 100 °C for 24 h and again subjected to the same ozone treatment. This cycle was repeated three times, obtaining the samples designated F400-1, F400-2 and F400-3. The activated carbon F400 was also treated with gaseous ozone in a fixed bed reactor loaded with 2 g of carbon under a constant ozone flow of 76 mg O_3/min, (25 °C, 1 atm). Ozone was fed to the reactor for different exposure times (10 and 120 min). The samples obtained were designated F400-10 and F400-120.

2.5. Preparation of Activated Carbons from Petroleum Coke

A series of activated carbons were prepared from calcinated petroleum coke supplied by REPSOL-YPF (Alicante, Spain). This coke will be designated as C, and its particle size ranged from 0.5 to 0.8 mm. KOH was used as activating agent for the petroleum coke activation. Different activated cokes were prepared, with a KOH/coke mass ratio ranging from 1 to 4. The samples will be designated with the letter C followed by the KOH/coke mass ratio used in the activation process. Briefly, 25 g of petroleum coke was mixed with 100 mL of a KOH solution at the appropriate concentration for each sample. This mixture was first heated at 60 °C for 48 h and was then treated in N_2 atmosphere (flow 300 mL/min) at 400 °C for 2 h. Finally, it was carbonized by heating the mixture at 700 °C for 1 h under an N_2 flow of 300 mL/min. In all heat treatments, the oven heating rate was 10 °C/min. Finally, the samples were washed with deionized water to constant conductivity.

2.6. Preparation of Nitrogen Enriched Activated Carbons

Activated carbon Witco, with a particle size of 0.5-0.8 mm, was treated with ammonia (W-A), ammonium carbonate (W-C), or urea (W-U) in an autoclave at 135 °C and 2 atm of pressure. For this purpose, 25 g of activated carbon was placed in an Erlenmeyer flask with 20 g of nitrogenating agent dissolved in 100 mL of distilled water. The contact was maintained during 1 h. Then, the carbon residue was filtered and washed with distilled water to reach a constant pH. Finally, the samples were dried in an oven at 110 °C and stored in a dessicator until their use. Details of this procedure are given elsewhere [24].

2.7. Preparation of Metal Doped Carbon Aerogels

Preparation of metal doped carbon aerogels was reported elsewhere [30, 31]. Briefly, resorcinol (R) and formaldehyde (F) were dissolved in water (W) containing cobalt acetate, manganese acetate or titanium tetra butyl ammonium as catalyst (C). The stoichiometric R/F and R/W molar ratios were 0.15 and 0.13, respectively. The stoichiometric R/C ranged from 15 to 200. Another aerogel was prepared in the same way but adding sodium carbonate for use as a blank, referred to in the text as A. The mixtures were stirred to obtain homogeneous solutions that were cast into glass molds (25 cm length x 0.5 cm internal diameter) and cured for a certain period of time. The gel rods were then cut into 5 mm pellets and supercritically dried with carbon dioxide to form the corresponding aerogels. Samples will be referred to in the text as A followed by the metal present and the R/C ratio used.

To determine changes in metal doped carbon aerogel activity due to ozonation process, aerogels were pre-treated with ozone; 0.5 g of aerogel was suspended in 0.5 L of 1 mg/L ozone aqueous solution (pH 7) for 1 h. The aerogel was then dried at 100 °C for 24 h; Mn-doped aerogel was subjected to the same ozone treatment three times. Samples were designated A-Co(II)-15-1, A-Ti(IV)-15-1, A-Mn(II)-15-1, A-Mn(II)-15-2, A-Mn(II)-15-3.

2.8. Activated Carbon Characterization

The surface area of carbon samples was determined from the BET equation applied to the N_2 adsorption isotherms at 77 K, which were obtained using a Micromeritics Gemini 2370 adsorption unit. The volumes of macropores (V_3) and wider mesopores (V_2) were determined by mercury porosimetry using a Quantachrome Autoscan 60 apparatus.

Carbon samples were chemically characterized using selective neutralization [16, 32], as well as determining their pH_{PZC} [33], ash content and elemental analysis.

2.9. Analytical Methods

The concentration of ozone in the gas mixture was analyzed spectrophotometrically [34] with a Genesis 5 model Spectrophotometer. The ozone dissolved in aqueous solution was determined colorimetrically using the Karman-Indigo method [35].

Concentrations of NS, NDS, NTS and pCBA were followed up by means of a Merck-Hitachi apparatus with UV detection, using a 250 mm LiChrosphere 100 RP-18 (5 mm) column. The mobile phase consisted of a solution of methanol-water (35/65) containing $5 \cdot 10^{-3}$ M TBABr as ionic exchanger and $1 \cdot 10^{-2}$ M NaH_2PO_4 as pH regulator at a flow of 1.3 mL/min.

The total organic carbon (TOC) was measured using a Shimadzu TOC-5000A apparatus.

The formation of $O_2^{\cdot -}$ radicals was detected by its fast reaction with tetranitromethane (TNM, $k = 2 \cdot 10^9$ $M^{-1}s^{-1}$) by measuring the formation of nitroform anion, $C(NO_2)_3^-$ (NF^-, $\varepsilon_{(350nm)} = 15.000$ $M^{-1}cm^{-1}$), a common technique in peroxyl radical chemistry but also applied to ozone chemistry [36] (Equations 1 and 2). The TNM concentration used in these experiments was determined to minimize $O_2\cdot^-$ reaction with ozone (Equation 3).

$$O_2\cdot^- + C(NO_2)_4 \longrightarrow O_2 + C(NO_2)_3^- + NO_2\cdot \qquad k_1 = 2\cdot 10^9\ M^{-1}\ s^{-1} \tag{1}$$

$$O_2\cdot^- + O_3 \longrightarrow O_2 + O_3\cdot^- \qquad k_2 = 1.6\cdot 10^9\ M^{-1}\ s^{-1} \tag{2}$$

The fraction of $O_2\cdot^-$ reacting with ozone was determined according to Equation 3. It was 0.01 % for a $C(NO_2)_4$ concentration of $1.6\cdot 10^{-3}$ M.

$$f_{O3} = \frac{k_2\,[O_3]}{k_1\,[C(NO_2)_4] + k_2\,[O_3]} \tag{3}$$

2.10. Bacillus Subtilis "Rec Assay": Test with Isogenic Strains

This test was done following the procedure described by Mazza in 1982 [37]. 100 mL of different concentrations of each contaminant under study were added to two series of tubes containing 2 mL of soft agar. To each of the series was then added 100 mL of a 10^{-4} dilution of *B. Subtilis 1652* or *B. Subtilis 1791* strains. Each tube was agitated and its contents emptied onto a plate of soft agar. The plates were incubated at 310 K for 24 h. A similar procedure was used to determine the evolution of genotoxicity during treatments with ozone and ozone/activated carbon.

The quantitative measurement of DNA damage was expressed as the plate efficiency (N/N_0) of the *B. Subtilis* 1791 strain, where N is the total bacteria count in presence of the contaminant and N_0 is the total bacteria count in its absence.

The results can also be expressed as the relative affinity (A), defined as the relationship between the plate efficiency of the *B. Subtilis 1791* and *B. Subtilis 1652* strains.

3. Results and Discussion

3.1. Adsorption of Naphthalenesulphonic Acids on Activated Carbon

In the present work, a new alternative is proposed for the purification of waters contaminated with these organic compounds, based on the simultaneous use of ozone and activated carbon. In order to better understand the behaviour of the ozone/activated carbon purification system, a study was conducted on the adsorption process of the naphthalenesulphonic acids on activated carbons. The aim was to determine the adsorption capacity and investigate the predominant interactions in the adsorption processes, as well as to measure the adsorption effect when activated carbon is used as catalyst of the ozonation of these sulphonic acids.

3.1.1. Adsorption in Static Conditions

The adsorption capacity (Xm) of activated carbon F400 for the three naphthalenesulphonic acids was determined by applying the Langmuir equation to the adsorption isotherms. Table 1 lists these data and the relative affinities (BXm) of these compounds to be adsorbed on the carbon sample. The adsorption capacity (Xm) and relative affinity (BXm) decreased as the number of sulphonic groups in the aromatic rings of the acids increased. Under these experimental conditions, i) activated carbon presents positive surface charge (pHsolution (6) < pH point of zero charge) and ii) NS and NDS molecules are totally dissociated, whereas NTS presents two dissociated sulphonic acid groups (ie.,$pK_1 \cong 2$, $pK_2 \cong 4$, $pK_3 \cong 10$). Thus, at a pH of 6 there would be two negative charges on both NDS and NTS molecules, as compared with only one in the case of NS. This suggests that there should be a greater adsorption affinity for NDS and NTS than for NS. However, the actual adsorptive behavior, as shown in Table 1 contradicts this logic, because the experimental results show that the adsorption affinities decrease in the following order:

$$NS > NDS > NTS$$

Thus, electrostatic interactions alone cannot explain the adsorption behaviour of these aromatic sulphonic acids on untreated carbon at an initial pH around 6. The adsorption mechanism of organic compounds on activated carbon is controversial. According to Leon y Leon et al. [38], basic carbons are characterized by a high content of electron rich sites on their basal planes, and a low concentration of electron-withdrawing groups. Thus, aromatic ring deactivation due to the increase in the number of electron-withdrawing sulphonic groups seems to be the main cause for the reduction in adsorption affinity of activated carbon F400 towards to the sulphonic acids in the above order. Therefore, the present results suggest that the adsorption of naphthalenesulphonic acids on basic activated carbon, such as carbon F400, occurs via π-π dispersion interactions between aromatic ring electrons in the acids and those present in the carbon basal plane, which is consistent with the mechanism suggested by *Coughlin et al., 1968* [39]. More details about the adsorption process of naphthalenesulphonic acids on activated carbon can be found elsewhere [21].

Table 1. Results obtained by applying the Langmuir equation to naphthalenesulphonic acid adsorption isotherms on untreated activated carbon

Sample	NS		NDS		NTS	
	Xm (mg/g)	BXm (L/g)	Xm (mg/g)	BXm (L/g)	Xm (mg/g)	BXm (L/g)
F400	84.03	69.44	38.46	8.13	34.97	7.63

3.1.2. Adsorption in Dynamic Regime

In order to establish the real applicability of activated carbon in the removal of naphthalenesulphonic acids from water, their adsorption in dynamic regime was investigated. In this study, the breakthrough curves of carbon F400 columns were determined for the three organic acids. The characteristics of the columns were determined from these curves and the values are shown in Table 2. It can be seen how the breakthrough volume (V_B) and the amount adsorbed at 0.8 of breakthrough ($X_{0.8}$) were much greater for NS than for the other compounds studied, due to the smaller number of sulphonic groups in its molecule. The differences in adsorption capacity between NS and the other two acids (NDS and NTS) were more marked in the dynamic than in the static regime. The height of the mass transfer zone (H_{MTZ}) was very elevated, especially for NDS and NTS, whereas the fractional capacity was very low. The degree of utilization of the columns (D_U) was particularly low in the case of NTS. These results indicate that a system exclusively based on activated carbon is not fully effective in removing these contaminants from waters, above all in the case of NDS and NTS.

Table 2. Characteristics of activated carbon columns (carbon F400).

Compound	V_B (mL)	X_B (mg/g)	$X_{0.8}$ (mg/g)	H_{MTZ} (cm)	ϕ	D_U (%)
NS	800.79	51.74	81.61	7.31	0.56	63.39
NDS	63.12	5.14	13.19	22.36	0.29	39.96
NTS	36.98	2.62	10.34	21.90	0.30	25.33

VB: Volume of effluent treated to initial breakthrough
XB: Amount of naphthalenesulphonic acids adsorbed to initial breakthrough
X0.8: Amount of naphthalenesulphonic acids adsorbed at 0.8 breakthrough value
HMTZ: Height of the mass transfer zone
ϕ : Fractional capacity of naphthalenesulphonic acid within the mass transfer zone
DU : Utility degree

3.2. Ozonation of Naphthalenesulphonic Acids

3.2.1. Determination of Both Stoichiometry and Mechanism of Reaction

Study of the kinetics of oxidation with ozone requires determination of the stoichiometry and rate constants of the direct and radical reaction. Calculation of the stoichiometry is hampered by the presence of competitive reactions that also consume ozone. The stoichiometry of the oxidation reaction can be calculated from these experiments by means of equation 4:

$$z = \frac{(C_{O_3})_0 - (C_{O_3})_f}{(C_M)_0 - (C_M)_f} \tag{4}$$

where $(C_{O3})_o$ is the initial concentration of ozone, $(C_{O3})_f$ is the final concentration of ozone, $(C_M)_o$ is the initial concentration of sulphonic acid and $(C_M)_f$ is the final concentration of sulphonic acid.

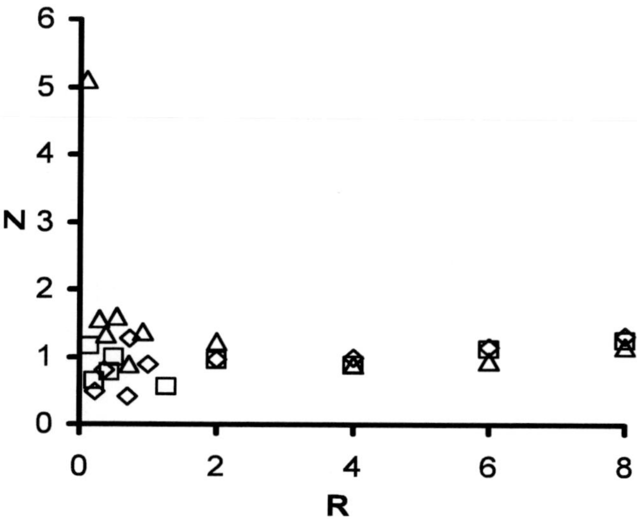

Figure 1. Stoichiometry of the direct ozonation reaction. (◇), NS; (□) NDS, (△) NTS.

Figure 1 shows the stoichiometric coefficient values (Z) as a function of the initial ratio (R) of sulphonic acid and ozone concentrations. The stoichiometric factor was determined to be one mol of ozone consumed per mol of sulphonic acid consumed in all cases.

Bailey in 1982 [40] studied the oxidation of aromatic compounds and concluded that when the stoichiometry is one, the main mechanism is 1,3-dipolar cycloaddition at the double bond of highest electron density. Beltran et al. in 1995 [41] corroborated these results after analyzing the ozonization of polycyclic aromatic hydrocarbons. Pryor et al. [42] observed that the ozonization of alkenes with electron-withdrawing substituents takes place mainly via 1,3-dipolar cycloaddition. Moreover, olefins in which the double bond is connected to electron-donating groups react many times faster than those in which it is connected to electron-withdrawing groups.

Table 3. Electron density at the bond critical point, ρ (ea0-3)

Bond (Structure)	NS	NDS	NTS
1-2	0.339	0.340	0.340
2-3	0.304	0.304	0.306
3-4	0.345	0.346	0.343
4-10	0.303	0.304	0.303
10-5	0.302	0.302	0.304
5-6	0.344	0.340	0.344
6-7	0.305	0.304	0.302
7-8	0.342	0.346	0.343
8-9	0.302	0.303	0.303
9-1	0.297	0.296	0.297
9-10	0.318	0.317	0.319

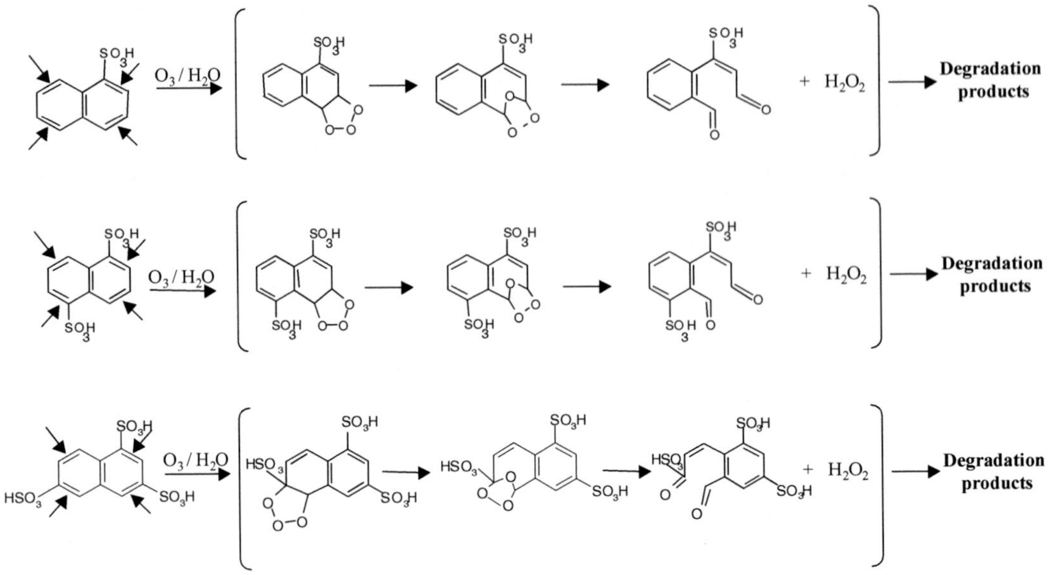

Figure 2. Direct ozonization reactions.

In the light of the above observations and the present results, it can be proposed that the ozone attack on aromatic sulphonic acids proceeds mostly by 1,3-dipolar cycloaddition to the carbon-carbon double bond of highest electron density. We have carried out a theoretical study of the distribution of the electron density of these molecules using the Gaussian-98 series of programs. The electron density was analyzed using the topological formalism Atoms in Molecules (AIM). This theory postulates the bond critical point as a fundamental location to characterize the bond properties. The electron density values at the bond critical point are displayed in Table 3: in all three acids, the double bonds of highest electron density are those at positions 1-2, 3-4, 5-6 and 7-8, with values of around 0.34 ea^{-3}. Thus, these bonds will be the most sensitive to ozone attack and, therefore, according to the above observations, we propose the ozonation of these acids takes place following the mechanism shown in Figure 2.

3.2.2. Kinetic Parameters of Naphthalenesulphonic Acids Ozonization

The oxidation of an organic compound by ozone in water may have two causes: direct attack by ozone and attack by hydroxyl radicals [43]. The attack by hydroxyl radicals to organic molecules is considered the basis of Advanced Oxidation Processes. Hydroxyl radicals are produced by the decomposition of the ozone, initiated by the hydroxyl ion or by other trace substances in solution. Thus, the ozonization rate of an organic compound can be expressed by equation 5:

$$\frac{dC_M}{dt} = r_D + r_{OH} = k_D C_M C_{O_3} + k_{OH} C_M C_{OH} \tag{5}$$

where r_D represents the contribution of direct ozonation and r_{OH} the radical oxidation reaction.

It can be observed in Figure 3 that as the number of sulphonic groups in the aromatic ring increased, there was a greater resistance of the organic compound to ozonation. This results from a reduction in the electronic density in the aromatic ring because of the electron-withdrawing sulphonic groups in it. These functional groups withdraw electronic density from the aromatic ring, deactivating the aromatic ring against electrophilic attacks [44].

Table 4 lists the values of the direct reaction constants for the sulphonic acids studied, showing that the constant decreased considerably with the increase of the number of sulphonic groups in the aromatic ring. The values were much lower than the value of the reaction constant of naphthalene with ozone ($k_D = 1500\ M^{-1}\ s^{-1}$) [45], supporting the influence of electron-withdrawing functional groups in the aromatic ring on its deactivation against ozone attack. Comparison with findings published by Hoigné et al. [8] on the oxidation of aromatic compounds shows the low reactivity of the sulphonic acids to the ozone reaction. The reaction constant we found for NTS is very similar to that reported by other researchers for aromatic compounds with electron-withdrawing groups in their ring [42].

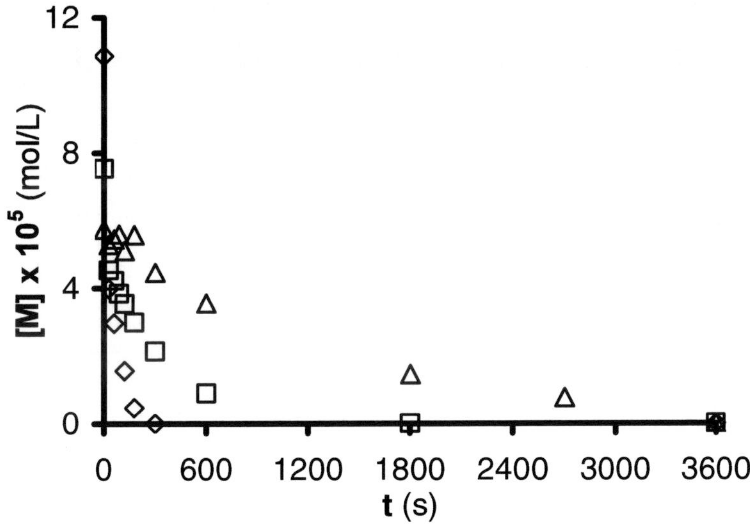

Figure 3. Variation of concentration of sulphonic acids with ozone treatment time. (◇), NS; (□) NDS, (△) NTS.

Table 4. Kinetic parameters of the ozonization of the naphthalenesulphonic acids

Compound	k_D ($M^{-1}\ s^{-1}$)	Ea ($kJ\ mol^{-1}$)	$k_{OH}\times10^{-9}$ ($M^{-1}\ s^{-1}$)
NS	111.8	37.9	5.73
NDS	23.6	38.4	4.92
NTS	6.7	37.2	3.68

kD =Rate constant of direct reaction.

Ea = Activation energy of the direct reaction.

kOH = Rate constant of indirect reaction.

Table 4 also shows the activation energy values of the ozonization reaction for each sulphonic acid, calculated using the Arrhenius equation. The values were all similar and the number of sulphonic groups had no appreciable influence on the activation energy of the reaction. These results confirm that the ozonization reaction of naphthalenesulphonic acids proceeds by the same mechanism in all cases, as explained above. The activation energy values shown in Table 4 are similar to those reported by other authors in the oxidation of aromatic compounds with electron-withdrawing groups in their ring [42].

The radical reaction constant was determined in order to complete the kinetic study of the ozonation reaction of the naphthalenesulphonic acids. Table 4 lists the values obtained for the rate constant of radical reaction for each sulphonic acid. There was a slight reduction in this constant with an increase in sulphonic groups in the aromatic ring, because of deactivation of the ring induced by the electron-withdrawing character of the sulphonic groups. Thus, the radical constant of NS was 1.5-fold that of NTS. The values obtained for the radical constant of the sulphonic acids are similar to those reported by other authors who studied the ozonization of organic compounds in aqueous solution [46].

3.2.3. Evolution of TOC and Genotoxicity during Ozone Treatment of Naphthalenesulphonic Acids

In order to evaluate the applicability of ozone in the purification of waters contaminated with naphthalenesulphonic acids, it is necessary to know the evolution of the amount of dissolved organic compounds with treatment. For this reason, the amount of total organic carbon dissolved (TOC) was determined at different times during the ozonation process.

Figure 4 plots the TOC values versus the ozonation time of the naphthalenesulphonic acids. It can be seen that the amount of TOC remains virtually constant during the treatment time. According to these results, ozone oxidizes naphthalenesulphonic acid, successively degrading it into compounds of lower molecular weight, but has inadequate oxidizing capability to produce its mineralization and transformation into CO_2.

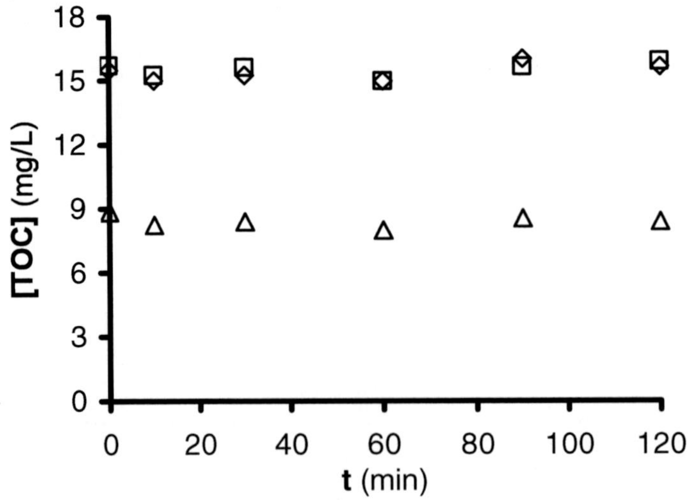

Figure 4. TOC evolution during naphthalenesulphonic acids ozonation. (◇), NS; (□) NDS, (△) NTS.

The evolution of oxidation by-products genotoxicity during naphthalensulphonic acids ozonation treatment was evaluated using B. Subtilis "rec assay". The quantitative measurement of DNA damage was expressed as the plate efficiency (N/N_0) of the *B. Subtilis* 1791 strain, where N is the total bacteria count in presence of the contaminant and N_0 is the total bacteria count in its absence. The results can also be expressed as the relative affinity (A), defined as the relationship between the plate efficiency of the *B. Subtilis 1791* and *B. Subtilis 1652* strains.

The genotoxic activity of the oxidation by-products decreased in the order NS > NDS > NTS (Figure 5). Analysis of the naphthalenesulphonic acid degradation by-products confirmed the presence of oxidation by-products that contained the sulphonic group in their organic molecule. It was observed that after 10 min of treatment, 57% of the sulphonic groups of NS but only 41% and 22% of those of NDS and NTS, respectively, were transformed into sulphate anions.

As mentioned above the presence of the sulphonic group, would prevent organic compounds from penetrating the interior of the bacteria and preclude their action on the DNA. This would explain the greater genotoxicity of the oxidation by-products of NS with respect to those of NDS or NTS. Figure 5 shows that genotoxic activity was always eliminated in high-time treatments, because the main compounds present under these conditions are highly oxidized linear short chain compounds (oxalic acid, formic acid, acetic acid, etc.). More data about the effect of ozone on genotoxic activity can be found elsewhere [18].

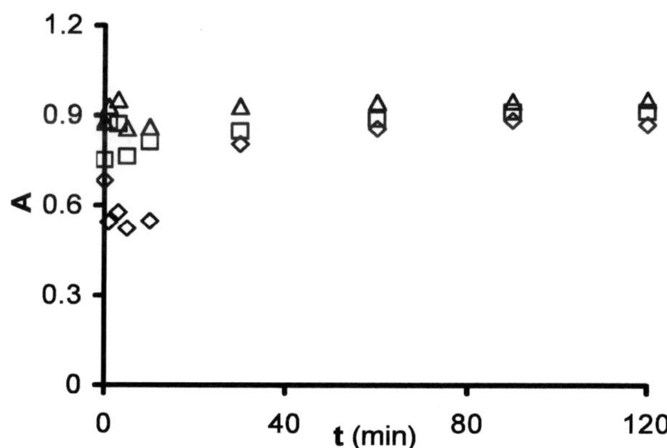

Figure 5. Genotoxic activity evolution during naphthalenesulphonic acids ozonation. (◇), NS; (□) NDS, (△) NTS.

3.3. Ozonation of Naphthalenesulphonic Acids in Presence of Activated Carbon

3.3.1. Influence of Chemical and Textural Characteristics of Activated Carbon on the NTS Oxidation Rate: Mechanism Involved

Table 5 summarises the results obtained from the textural and chemical characterisation of the commercial activated carbons used in this study. Greater surface areas, in the range 1200-

1300 m^2/g, were shown by Sorbo, Merck and Ceca AC40 activated carbons, whereas Witco carbon presented the smallest surface area (808 m^2/g). According to the data presented in Table 5, the Sorbo and Norit carbons had the most marked macroporosity, while Ceca GAC carbon had the greatest mesoporosity ($V_2 = 0.13$ cm^3/g). Witco carbon had the lowest V_2 and V_3 values.

Regarding the surface chemistry of the activated carbons, Sorbo ($pH_{PZC} = 9.42$) and Norit ($pH_{PZC} = 9.18$) carbons showed the greatest concentration of surface basic groups (1713 and 2050 μeq/g, respectively), whereas Ceca AC40 carbon ($pH_{PZC} = 5.29$) had the greatest concentration of surface acid groups.

With regard to the mineral matter, Ceca GAC carbon featured the highest ash content (12 %) and Witco carbon the lowest (0.3%) (Table 5). Filtrasorb 400 ash presented a high Fe (6.32 %) and Al (8.41 %) content, whereas Norit and Sorbo had a low content of Fe and Al and a significant fraction of Mg and Ca (around 9-10 % each). Unlike other samples, ashes from Ceca GAC showed a high P (4.11 %) content. A substantial concentration of Ti (≅ 1 %) was detected in Filtrasorb 400 and Merck ash samples, while Mn contents were significant in Norit and Sorbo ashes (0.25% and 0.13%, respectively). These metals are frequently used as catalysts in oxidation processes. Indeed, Ti is favoured as catalyst in photocatalysis [47], and Ma and Graham [48] used low concentrations of MnO_2 to enhance chlorobenzene degradation by ozone.

Table 5. Characterization of activated carbons

Activated Carbon	S_{N2} (m^2/g)	V_2 [a] (cm^3/g)	V_3 [b] (cm^3/g)	pH_{PZC}	Acid Groups[c] (μeq/g)	Basic Groups[d] (μeq/g)	% ash
F400	1075	0.11	0.26	7.91	234	570	6.6
Sorbo	1295	0.06	0.37	9.42	88	1713	5.9
Merck	1301	0.09	0.26	7.89	114	582	5.2
Ceca GAC	966	0.13	0.16	6.83	323	99	12.0
Ceca AC40	1201	0.07	0.32	5.29	438	102	8.3
Norit	968	0.10	0.42	9.18	139	2050	4.8
Witco	808	0.04	0.05	6.85	183	253	0.3

a) Volume of pores with diameter of 50 to 6.6 nm.
b) Volume of pores with diameter above 50 nm.
c) Determined by NaOH (0.1 N) neutralization.
d) Determined by HCl (0.02 N) neutralization.

Figure 6 shows experimental results of NTS ozonation in the presence of the different activated carbons. All of the carbons increased the ozonation rate. Sorbo, Norit, and Ceca GAC carbons greatly enhanced NTS degradation rates, whereas Witco activated carbon had a lower effect on the NTS degradation rate. These results were obtained at pH 2.3, at which there is poor NTS reactivity with ozone in the absence of activated carbon. The greater rate of NTS ozonation in the presence of these carbons could, therefore, be explained by an increase in HO· concentration. As shown above, the NTS direct reaction constant was 6.72 $M^{-1}s^{-1}$, whereas the indirect reaction (free radical reaction) constant was $3.7 \cdot 10^9$ $M^{-1}s^{-1}$. Thus, the radical reaction proved more efficacious in oxidizing NTS.

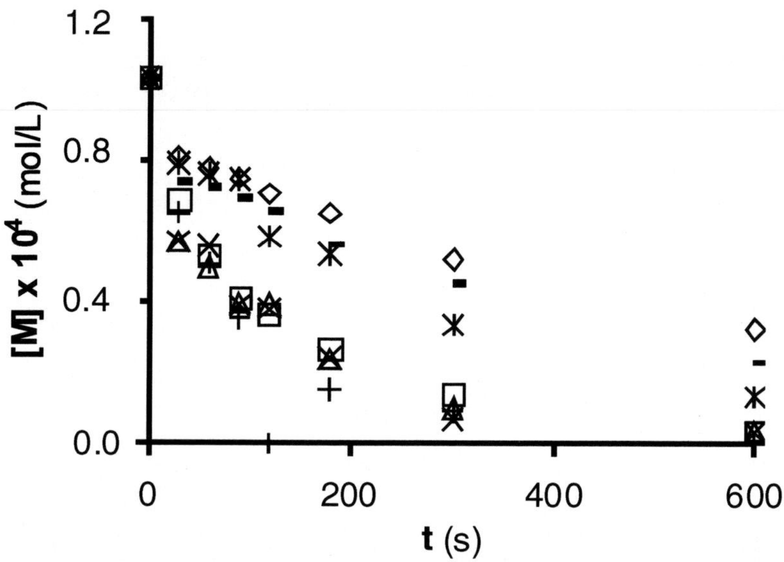

Figure 6. NTS ozonation in the presence of commercial activated carbon. (◇), without carbon; (○), Filtrasorb 400; (△), Merck; (□), Ceca GAC; (*), Ceca AC40; (x), Norit; (+), Sorbo; (-) Witco.

The carbons that most enhanced the NTS ozonation rate were those with greatest pH_{PZC} values and highest concentrations of surface basic groups (Table 5). However, no clear relationship was observed between the NTS ozonation rate and the S_{N2} of the activated carbon. Thus, Merck carbon had the largest surface area (S_{N2} = 1301 m^2/g) but did not show the highest rate of NTS oxidation. These results indicate that the process of ozonation catalysed by activated carbon is not affected by its microporosity. On the other hand, the carbons that most favoured the removal of NTS from the medium (Sorbo and Norit) were those with greatest macropore volumes (Table 5). These pores act as transport pores, facilitating the access of ozone to the carbon surface and reducing diffusion problems. Thus, the low catalytic activity showed by Witco carbon could be related, in part, to its small V_2 and V_3 values. However, there was no close relationship between macropore volume and catalytic activity in the remaining carbons under study.

Although activated carbon is a heterogeneous material with a large number of surface groups and different pore size distributions, the above results suggest that the catalytic activity of these activated carbons in NTS ozonation is mainly a function of the carbon basicity. Thus, the catalytic activity seems to be enhanced by increased carbon basicity.

The basicity of an activated carbon is due to the presence of basic oxygen-containing functional groups (e.g. pyrones or chromenes) and/or graphene layers acting as Lewis bases and forming electron donor-acceptor (EDA) complexes with H_2O molecules. These latter basic sites are located at π electron-rich regions within the basal planes of carbon crystallites away from the crystallite edges (equation 6). This delocalised π electron system can act as a Lewis base in aqueous solution:

$$-C\pi + 2\,H_2O \longrightarrow C\pi\text{--}H_3O^+ + OH^- \qquad (6)$$

The delocalised π electron system of basic carbons and oxygenated basic groups (chromene and pyrone) would, therefore, act as catalytic centres of reaction, reducing the ozone molecules to hydroxyl ion and hydrogen peroxide following the reactions:

$$O_3 + H_2O + 2e^- \longrightarrow O_2 + 2\,OH^- \tag{7}$$

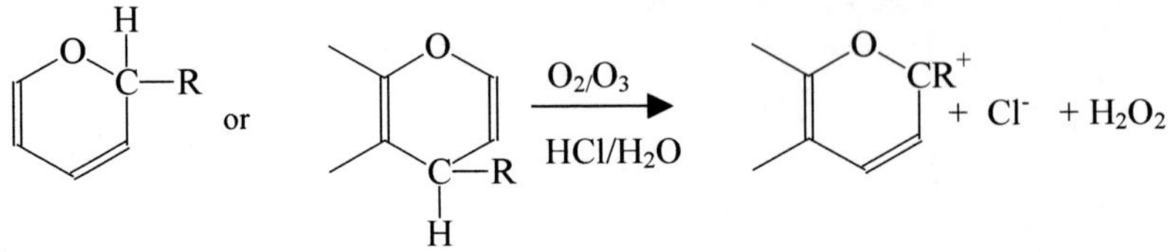

$$\text{(chromene: O–CH(R)) or (pyrone: C(H)–R)} \xrightarrow[HCl/H_2O]{O_2/O_3} \text{(O–}CR^+\text{)} + Cl^- + H_2O_2 \tag{8}$$

It is widely known [44] that both hydroxyl ion and hydrogen peroxide act as initiators of the ozone decomposition process in aqueous phase. Thus, the higher degradation rate of NTS in the presence of Sorbo (pH_{PZC} = 9.42) and Norit (pH_{PZC} = 9.18) carbons is because these carbons have greater reducing properties, favouring reactions 7 and 8 and, therefore, increasing the extent of ozone decomposition into highly oxidative radicals.

3.3.2. NTS Ozonation in Presence of Different Commercial Activated Carbons. Kinetic Study

The total NTS degradation rate of ozonation in presence of activated carbon can be defined as the sum of the homogeneous reaction rate, $(-r_M)_{homo}$, calculated in absence of activated carbon, and the heterogeneous reaction rate, $(-r_M)_{hetero}$, due to the presence of activated carbon. Then, the total NTS degradation rate can be mathematically expressed as:

$$(-r_{total}) = (-r_{homo}) + (-r_{hetero}) = \left(-\frac{dC_M}{dt}\right)_{homo} + \left(-\frac{dC_M}{dt}\right)_{hetero} \tag{9}$$

In accordance with Hoigné et al. [8], the homogeneous reaction rate can be represented by Equation 10:

$$\left(-\frac{dC_M}{dt}\right)_{homo} = (-r_D) + (-r_{OH}) = k_D C_{O_3} C_M + k_{OH} C_{OH} C_M \tag{10}$$

where r_D represents the contribution of the direct ozonation reaction and r_{OH} the contribution of the radical oxidation, and k_D and k_{OH} are the corresponding reaction constants. The values of the constants were $k_D = 6.72\ M^{-1}\ s^{-1}$ and $k_{OH} = 3.68 \cdot 10^9\ M^{-1}\ s^{-1}$ and a good fit was observed between the proposed model and the experimental results.

The NTS concentration decay data shown in Figure 6 were obtained under the same experimental conditions except that the type of activated carbon was different for each run. Figure 6 indicates that the NTS degradation rate decreased considerably with increased treatment time. The NTS ozonation rate in presence of Filtrasorb 400 carbon, as an example, was approximately $3 \cdot 10^{-7}$ mol/L s in the first 5 min of treatment, whereas this rate was reduced by 84% in the next 5 min. This behavior can be explained by considering that the heterogeneous degradation rate is dependent on the concentration of NTS.

Likewise, experiments were conducted to determine whether the dissolved ozone concentration affects the heterogeneous reaction rate of NTS. For this purpose, NTS ozonation was carried out in presence of Filtrasorb 400 and the supply of ozone to the reactor was interrupted after 60 s of treatment (Figure 7). Figure 7 shows that the NTS degradation rate was reduced when the dissolved ozone concentration in the system diminished. These results indicate that, as in the case of the kinetic equation of the homogeneous reaction rate, the concentration of dissolved ozone is a factor to be included in the kinetic equation that represents the heterogeneous degradation rate of NTS.

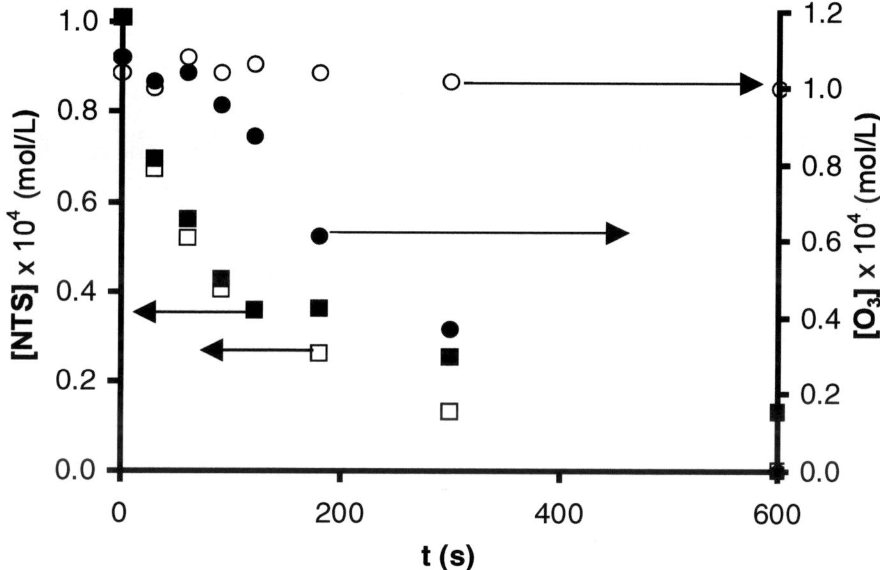

Figure 7. Influence of the ozone concentration on the NTS degradation in presence of Filtrasorb 400 carbon. (□), NTS concentration without interruption of ozone supply to reactor; (■), NTS concentration with interruption of ozone supply to reactor; (○), Ozone concentration without interruption of ozone supply to reactor; (●), Ozone concentration with interruption of ozone supply to reactor.

According to the results depicted in Figures 6 and 7, it may be inferred that the heterogeneous reaction rate should be of the following type:

$$\left(-\frac{dC_M}{dt}\right)_{hetero} = k_{hetero}\ f\left(C_M, C_{O3}\right) \tag{11}$$

where k_{hetero} represents the heterogeneous reaction constant when activated carbon catalyzes the reaction.

The order of this reaction with respect to the NTS and ozone were determined using the method of initial rates of reaction. For this purpose, NTS ozonation experiments were conducted in presence of Filtrasorb 400 activated carbon, varying the NTS concentration and keeping the ozone concentration constant and vice versa. The initial concentration of the NTS was varied from $5.76\cdot10^{-5}$ to $1.03\cdot10^{-4}$ mol /L and that of the ozone from $4.1\cdot10^{-5}$ to $1.04\cdot10^{-4}$ mol/L. The results indicate that the reaction order with respect to both NTS and ozone is 1; therefore the overall reaction order is 2. Equation 11 can be written as:

$$(-r_{hetero})=\left(-\frac{dC_M}{dt}\right)_{hetero}=k_{hetero}\,C_M C_{O_3} \tag{12}$$

The k_{hetero} value for each activated carbon was determined by using Equation 12, which is expressed as follows at a constant concentration of dissolved ozone:

$$(-r_{hetero})=\left(-\frac{dC_M}{dt}\right)_{hetero}=k_{obs}\,C_M \tag{13}$$

Table 6. Values of the heterogeneous reaction constants for the original and demineralized activated carbons evaluated according to the proposed model

Activated Carbon	k_{obs} (s^{-1})	$(k_{obs})_{real}$ (s^{-1})	k_{hetero} ($(mol/L)^{-1}s^{-1}$)	$(k_{hetero})_{demi}$ ($(mol/L)^{-1}s^{-1}$)
F400	0.0115	0.0140	134.6	106.1
Sorbo	0.0152	0.0197	189.4	142.3
Merck	0.1005	0.0119	114.4	105.7
Ceca GAC	0.0155	0.0203	195.2	90.4
Ceca AC40	0.0086	0.0099	95.2	88.5
Norit	0.0653	0.0219	210.5	190.4
Witco	0.0085	0.0098	94.2	94.2

The $(-r_{hetero})$ values were calculated as the difference between $(-r_{total})$ and $(-r_{homo})$ (Equation 9). The experimental data of $(-r_{hetero})$ vs. C_M were interpreted with Equation 13, yielding the value of k_{obs}. The values of k_{obs} are given in Table 6.

The Thiele modulus, φ, and effectiveness factor, η, were applied in order to confirm that diffusion limitations were not important to determine k_{obs} under the conditions used. For a pseudo-first order reaction (Equation 13) and spherical particles, these parameters are defined as:

$$\varphi=R_p\sqrt{\frac{(k_{obs})_{real}}{D_E}} \tag{14}$$

$$\eta = \frac{3}{\varphi^2}\left(\varphi \operatorname{Coth}\varphi - 1\right) \tag{15}$$

where the $(k_{obs})_{real}$ is the k_{obs} without diffusion effects.

Intraparticle diffusion effects are not important when the effectiveness factor equals 1. Assuming that k_{obs} had no diffusion effects ($(k_{obs})_{real} = k_{obs}$) and using the numerical values for F400 carbon in these Equations (Table 6), it was calculated that $\varphi = 1.73$ and $\eta = 0.84$, indicating that the k_{obs} value included diffusion effects of little importance. The k_{obs} must be corrected to eliminate the diffusion effects, and this can be done on the basis of the definition of the effectiveness factor (Equation 16).

$$\left(-r_{hetero}\right) = \eta\ \left(-r_{hetero}\right)_{real} \tag{16}$$

where $(-r_{hetero})_{real}$ represents the heterogeneous reaction rate without diffusion limitations, which is defined as Equation 17:

$$\left(-r_{hetero}\right)_{real} = (k_{obs})_{real}\, C_M \tag{17}$$

Replacing Equations 16 and 17 in Equation 15, the following Equation is obtained:

$$k_{obs} = \eta\ \left(k_{obs}\right)_{real} \tag{18}$$

The $(k_{obs})_{real}$ value was estimated by an iterative method using the Equations 16, 17, and 18 and the k_{obs} value. The final results for F400 carbon were $\varphi = 1.91$, $\eta = 0.818$ and $(k_{obs})_{real} = 0.0140\ s^{-1}$. Table 6 lists the $(k_{obs})_{real}$ values for the different carbons.

The relationship between k_{hetero} and $(k_{obs})_{real}$ is $(k_{obs})_{real} = k_{hetero}\, C_{O3}$. The k_{hetero} value for each carbon was obtained by substituting the dissolved ozone concentration value ($1.04 \cdot 10^{-4}$ mol/L), and the results are shown in Table 6. The k_{hetero} values were higher than the k_D value but much lower than the k_{OH} value.

Therefore, the global ozonation rate in presence of activated carbon can be expressed as follows:

$$-\frac{dC_M}{dt} = \left(k_D C_{O_3} + k_{OH} C_{OH} + \eta\ k_{hetero} C_{O_3}\right) C_M \tag{19}$$

This is a general kinetic model because it includes the homogeneous and heterogeneous reaction rates as well as the diffusion effects. In the ozonation catalyzed by activated carbon the NTS concentration decay data for different particle diameters can be predicted by using the Equation 19 and calculating the effectiveness factor with Equations 17 and 18. The results are depicted in Figure 8, demonstrating that the proposed model, shown by the discontinuous line, adequately represented the experimental results. The k_{hetero} value can be used to quantify the catalytic activity of activated carbon in the NTS ozonation.

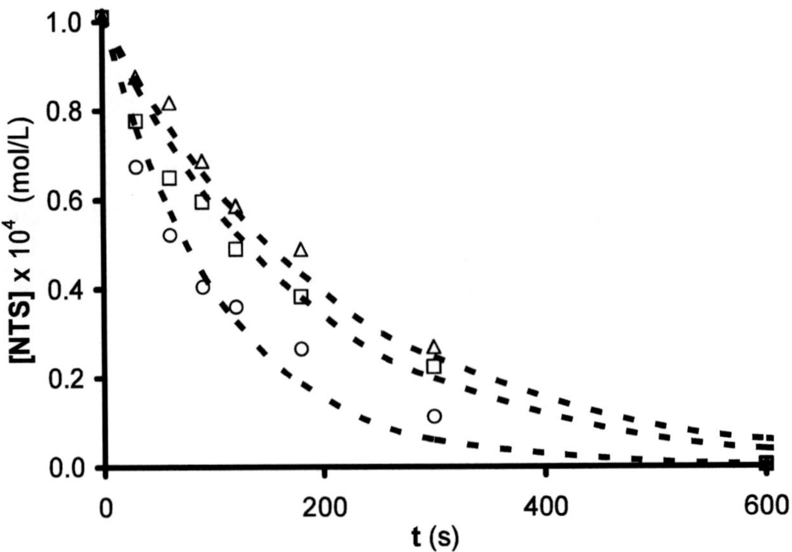

Figure 8. Prediction of the proposed general kinetic model (discontinuous line) for the NTS ozonation in presence of Filtrasorb 400 activated carbon as a function of the particle size. pH = 2.3, T = 298 K. (○), 200-500 μm, $\eta = 0.818$ y $\varphi = 1.91$; (□), 500-800 μm, $\eta = 0.64$ y $\varphi = 3.26$; (Δ), 800-1000 μm, $\eta = 0.58$ y $\varphi = 3.74$.

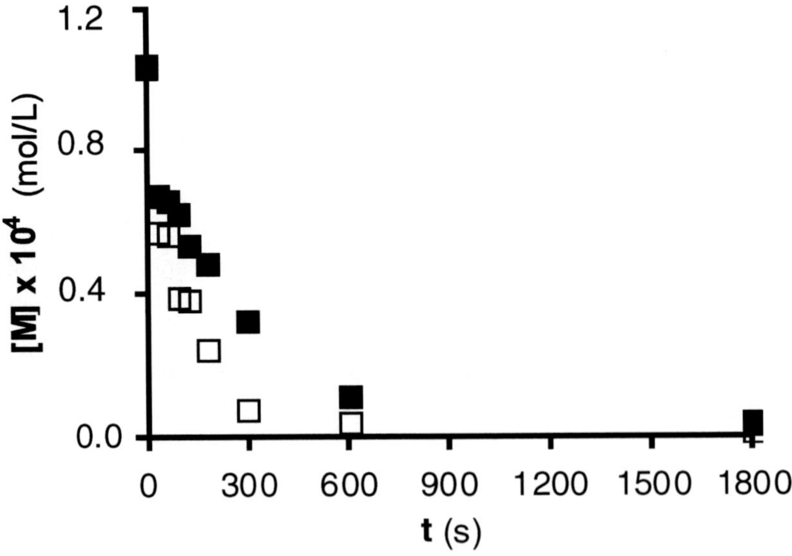

Figure 9. Effect of demineralization on the catalytic capacity of Ceca GAC carbon in the NTS ozonation. pH = 2.3, T = 298 K. (□), Untreated; (■), Demineralized.

The relationship between the k_{hetero} value (Table 6) and the chemical characteristics (Table 5) of each activated carbon was further studied. It was noticed that the NTS degradation was especially enhanced in the presence of carbons with a high content of ashes and a high concentration of basic groups. Both are catalytic centers that can decompose the ozone into highly reactive species. Consequently, the Norit, Sorbo, and Ceca GAC activated carbons presented the highest k_{hetero} values.

In order to determine the contribution of the ash to the heterogeneous ozonation rate of NTS, ozonation experiments were conducted in presence of demineralized carbons. Demineralization of activated carbons with HCl and HF was carried out using the procedure described in detail elsewhere [49]. Figure 9 depicts, as an example, the results obtained for the original and demineralized Ceca GAC carbon. In all of the carbons except for Witco, due to its very low ash content (Table 5), the degradation rate was reduced when the activated carbon was demineralized.

These results corroborate that the mineral matter present in activated carbon contributes positively to the catalytic effect of activated carbon in the NTS ozonation. All the metals that have been found to show catalytic activity in the ozonation of organic compounds are present in the mineral matter of the activated carbons used. However, it is difficult to ascertain the role of each metal in the catalyzed ozonation of NTS.

Previous studies demonstrated that the demineralization of carbons has no major effect on the concentration of surface basic groups [50]. In addition, the pH_{PZC} values of the demineralized carbons were very similar to those of the original carbons. Therefore, in the case of demineralized carbons, the NTS degradation rate was enhanced due to the presence of basic groups on the activated carbon surface. A similar method was applied to determine the heterogeneous reaction constant in presence of the demineralized carbons, $(k_{hetero})_{demi}$. (Table 6).

Once the $(k_{hetero})_{demi}$ value is known, the contribution of the basic groups to the catalytic activity of the activated carbons in the NTS ozonation process can be determined by dividing the $(k_{hetero})_{demi}$ value by the k_{hetero} value. Except in the case of Ceca GAC carbon, the $(k_{hetero})_{demi}$ contributed more than 50 % of the k_{hetero} value (Table 6). These results indicate that, although the mineral matter contributed positively to the catalytic activity of the activated carbon, especially in the case of Ceca GAC carbon, the basicity of the carbon was mainly responsible for the enhancement of the NTS degradation rate.

Furthermore, the heterogeneous reaction constant obtained for the demineralized carbons can be related to the concentration of basic groups in each carbon (Figure 10) or to the surface concentration of these groups (Figure 10), a linear relationship between them was observed. On the other hand, the fact that the ordinate in the origin was not zero indicates that other aspects of the surface chemistry of the activated carbon contributed positively to the catalytic effect of the carbon in the NTS ozonation process.

Another aspect of major interest was observed when the difference between the k_{hetero} and $(k_{hetero})_{demi}$ values for each carbon was divided by its ash content (Table 5, 6). The results indicate that the contribution of the mineral matter of the activated carbon to its catalytic activity in NTS ozonation does not depend on the amount but rather on the type of mineral present in each carbon. This is due to the fact that some components present in the carbon do not contribute to its catalytic activity in the ozonation process (Mg, Na, K, Si). The concentrations of these components vary between one activated carbon and another. Interestingly, the mineral matter with the greatest catalytic activity in the NTS ozonation process was that present in the Ceca GAC, Sorbo and Filtrasorb 400 carbons. Because of the chemical complexity of the mineral matter of activated carbon, much greater efforts must be applied to the identification of the specific metal compounds that participate in catalytic ozonation.

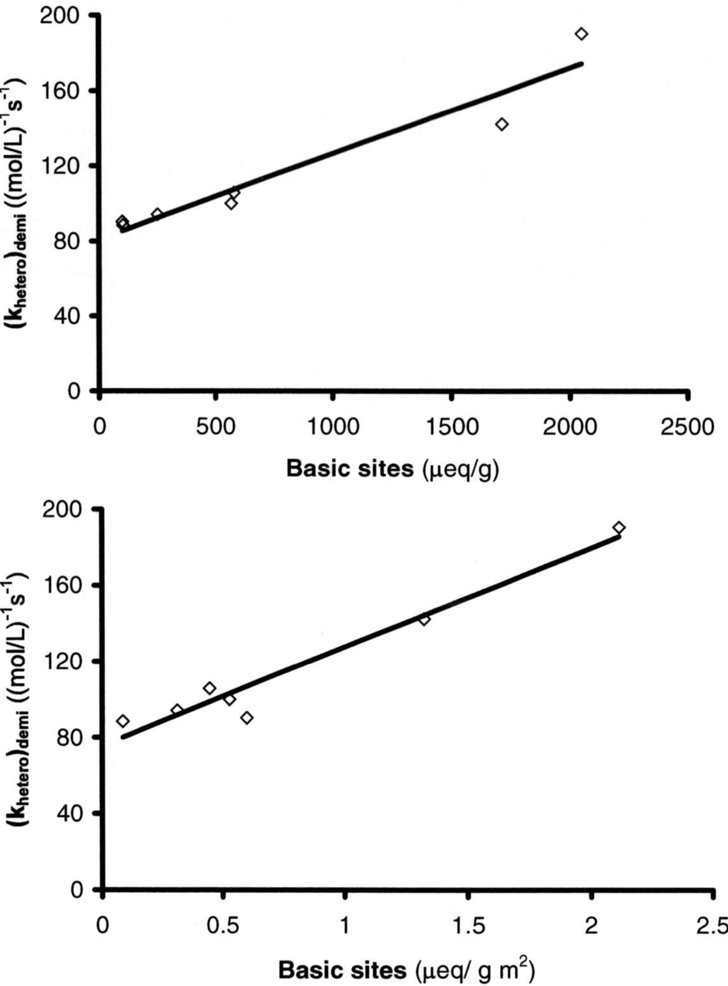

Figure 10. Relationship between the $(k_{hetero})_{demi}$ value and the concentration of basic groups in the activated carbons.

3.3.3. Ozonation of Micropollutants in Presence of Activated Carbon. Influence of Operational Parameters

As indicated above, ozonation in presence of activated carbon can lead to oxidation of micropollutants either by a direct reaction of the compounds with ozone or by ·OH radicals that are produced in the interaction of ozone with the surface of activated carbon. The concentration of both oxidants must be known to define and calibrate this process with respect to its oxidation capacity. An experimental approach to determine the concentration of ozone and ·OH radicals during conventional ozonation or the AOP O_3/H_2O_2 has been developed in previous studies [51]. As described by Elovitz and von Gunten, a R_{ct} value can be defined as the ratio of the exposure of ·OH radicals and ozone (i.e., concentration of oxidant integrated over the reaction time (Equation 20)).

$$R_{ct} = \frac{\int [\cdot OH]dt}{\int [O_3]dt} \tag{20}$$

The ratio R_{ct} indicates the efficiency of the transformation of ozone into ·OH radicals in a particular system. The R_{ct} can be calculated from the measurement of the decrease of a probe compound, which reacts fast with ·OH but not with ozone, and a simultaneous determination of the ozone concentration. To investigate the efficiency of ·OH radical formation, sodium para-chlorobenzoate (pCBA) was added as a probe compound for ·OH radicals. This compound present a low reactivity with ozone ($k_{O3} = 0.15\ M^{-1}s^{-1}$) and a high affinity by HO· radicals ($k_{OH} = 5.2 \cdot 10^9\ M^{-1}s^{-1}$).

3.3.3.1. Influence of Ozone Dose

The concentration of dissolved ozone present in the system is one of the main variables to be studied in AOPs, because while an increase in ozone dose increases the potential to oxidise micropollutants it also leads to an increased formation of oxidation by-products such as bromate [52]. In addition, if disinfection is an issue, a high O_3 exposure is required which might not be the case in a system optimised for ·OH radical production.

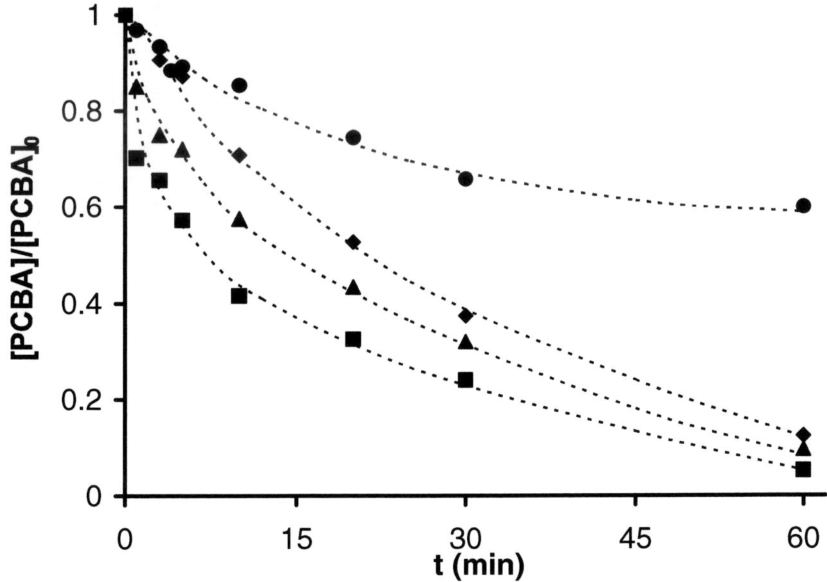

Figure 11. Influence of ozone concentration on oxidation of pCBA in presence of activated carbon. $[pCBA]_0 = 5 \cdot 10^{-7}$ M, pH 7 $[NaH_2PO_4] = 5 \cdot 10^{-3}$ M, $[t\text{-}BuOH]_0 = 8 \cdot 10^{-5}$ M, [Activated carbon] = 0.5 g/L, T 295 K. (●), Ozone only, $[O_3]_0 = 2 \cdot 10^{-5}$ M; (◆), O_3/F400, $[O_3]_0 = 2 \cdot 10^{-5}$ M;, (▲), O_3/F400, $[O_3]_0 = 4 \cdot 10^{-5}$ M; (■), O_3/F400, $[O_3]_0 = 6 \cdot 10^{-5}$ M.

Figure 11 depicts pCBA oxidation as a function of the ozonation time in presence and absence of activated carbon. It can be observed that the presence of activated carbon increased the pCBA oxidation rate. Because this probe compound has a low reactivity with ozone and is not adsorbed on activated carbon, the enhanced rate of oxidation can be explained by an increased rate of ·OH radical formation. Figure 11 also shows that an increase in the ozone dose in the combined O_3/activated carbon system produces an increase in the pCBA oxidation rate.

The R_{ct} values were determined for each experiment (Table 7, experiments 1-4), using the model proposed by Elovitz and von Gunten [52]. Table 7 shows that the presence of activated carbon F400 during pCBA ozonation leads to an increase in the R_{ct} value. This increase is mainly due to an increase in the rate constant (k_D) for ozone decomposition, determined by a first-order kinetic model. Therefore, these results confirm that the presence of activated carbon enhances the transformation of ozone into ·OH radicals. Moreover, the results listed in Table 7 show that an increase in the concentration of dissolved ozone in the combined O_3/activated carbon system produced a marked increase in the R_{ct} value.

Table 7. Rct values for the different experiments at pH 7 in Milli-Q water ([NaH2PO4] = 5·10-3 M, [t-BuOH] = 8·10-5 M)

Experiment	Sample	Carbon dose (g/L)	$[O_3]$ (M)	k_D (s^{-1})	R_{ct}
1	Without carbon	0.00	$2·10^{-5}$	$6.0·10^{-4}$	$2.7·10^{-9}$
2	F400	0.50	$2·10^{-5}$	$3.2·10^{-3}$	$1.2·10^{-8}$
3	F400	0.50	$4·10^{-5}$	$3.6·10^{-3}$	$1.6·10^{-8}$
4	F400	0.50	$6·10^{-5}$	$4.0·10^{-3}$	$4.7·10^{-8}$
5	F400	0.01	$2·10^{-5}$	$6.1·10^{-4}$	$3.0·10^{-9}$
6	F400	0.25	$2·10^{-5}$	$9.0·10^{-4}$	$6.0·10^{-9}$
7	F400	0.85	$2·10^{-5}$	$8.0·10^{-3}$	$5.7·10^{-8}$
8	F400-1	0.50	$2·10^{-5}$	$2.9·10^{-3}$	$1.4·10^{-8}$
9	F400-2	0.50	$2·10^{-5}$	$2.6·10^{-3}$	$1.5·10^{-8}$
10	F400-3	0.50	$2·10^{-5}$	$2.4·10^{-3}$	$1.5·10^{-8}$
11	F400-10	0.50	$2·10^{-5}$	$1.0·10^{-3}$	$5.6·10^{-9}$
12	F400-120	0.50	$2·10^{-5}$	$4.0·10^{-4}$	$3.8·10^{-9}$

3.3.3.2. Influence of Activated Carbon Dose

The minimum dose of activated carbon required to produce an ozone transformation into ·OH radicals is an essential parameter for the application of this process. It was observed (results not shown) that when the dose of activated carbon in the system was increased, the pCBA oxidation rate also increased. A higher rate of ozone decomposition was observed when de dose of activated carbon was increased (Table 7, Experiments 2, 5-7). If the k_D or Rct value is plotted relative to the activated carbon dose a strong increase can be observed for higher doses (Figure 12). These results indicate that activated carbon dose is a decisive factor for the ozone transformation into ·OH radicals.

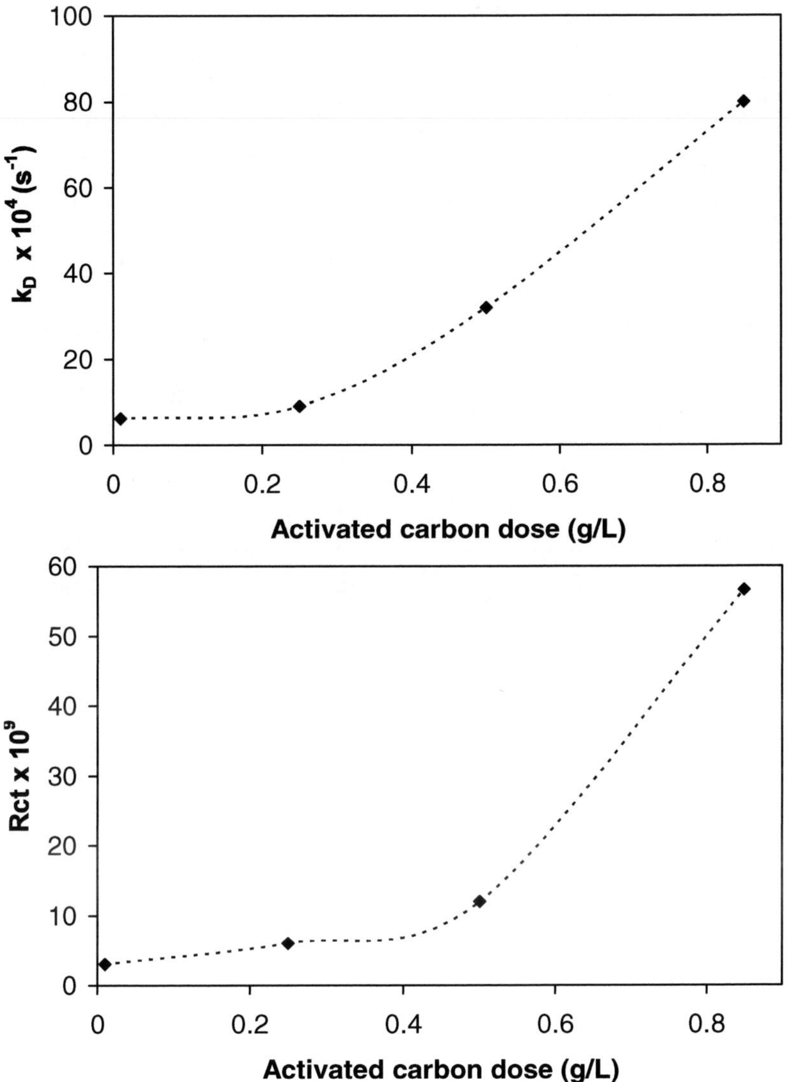

Figure 12. Influence of activated carbon dose on a) pseudo first-order rate constant for ozone decomposition (k_D) and b) $R_{ct} = [\cdot OH]/[O_3]$.

3.3.3.3. Influence of Activated Carbon Pre-ozonation on Ozone Transformation

To determine changes in the activated carbon activity during ozonation process, oxidation experiments were carried out with pre-ozonated activated carbons (F400-1, F400-2 and F400-3). The obtained results indicate that the carbon activity for the transformation of ozone into ·OH radicals was not affected by the ozone treatment applied, and very similar R_{ct} values were found for the three samples studied (Table 7, Experiments 2, 8-10). To assess the behaviour of the activated carbon for extended ozone treatment, some activated carbon samples were subjected to a much more drastic ozonation treatment (F400-10, F400-120). These activated carbon samples showed a marked reduction in their activity to transform ozone into ·OH radicals (Table 7, Experiments 2, 11, 12) with increasing treatment time. The R_{ct} value decreased from $1.2 \cdot 10^{-8}$ (untreated) to $3.8 \cdot 10^{-9}$ when the activated carbon was subjected to 120 min gas phase ozonation. These results indicate that the oxidation of the activated carbon by ozone reduces its catalytic properties due to an increase in the number of acidic oxygenated surface functional groups and a decrease in the surface area as shown in Table 8. These functional groups (electron-withdrawing groups) reduce the electronic density of the activated carbon graphenic layers, thereby diminishing the reductive properties of the activated carbon and hence its reactivity with ozone. From these observations we concluded that the activated carbon is not really a catalyst for ozone transformation but rather acts as a conventional initiator or promoter in the ozone transformation process into ·OH radicals.

Table 8. Textural and chemical characterisation of the activated carbons

Sample	S_{N2} (m^2/g)	V_2 [a] (cm^3/g)	V_3 [b] (cm^3/g)	pH_{PZC}	Acid groups [c] (μeq/g)	Basic groups [d] (μeq/g)	Ash (%)
F400	1075	0.11	0.26	8.8	234	570	6.6
F400-1	1062	0.11	0.25	8.9	239	562	6.6
F400-2	1053	0.11	0.25	8.7	265	548	6.6
F400-3	1025	0.12	0.26	8.6	285	529	6.6
F400-10	1023	0.12	0.27	5.9	702	437	6.6
F400-120	632	0.14	0.32	2.6	3370	< d.l.	6.6

[a] Volume of pores with diameter of 50 to 6.6 nm;
[b] Volume of pores with diameter above 50 nm.
[c] Determined by NaOH (0.1 N) titration;
[d] Determined by HCl (0.02 N) titration.

3.3.3.4. Influence of Carbon Particle Size on NTS Ozonation Rate

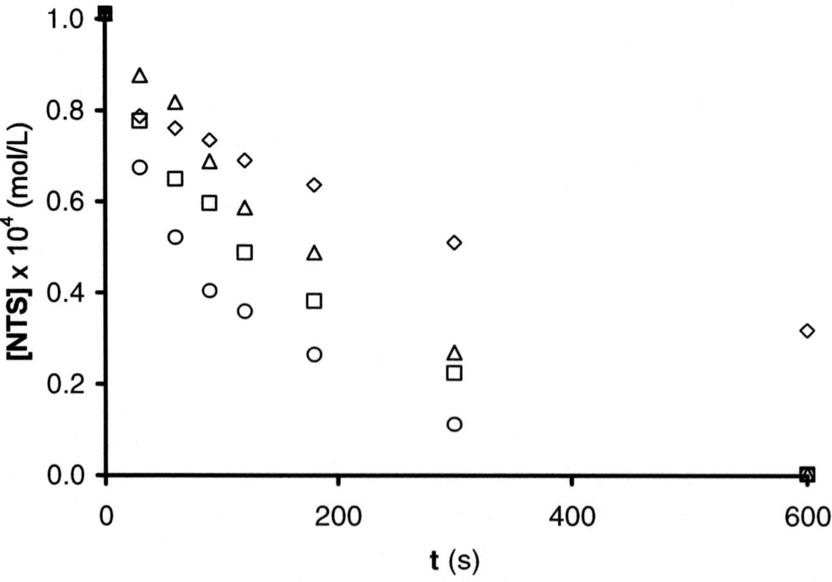

Figure 13. Influence of the particle size of Filtrasorb 400 carbon on the NTS ozonation. pH = 2.3, T = 298 K. (◇), without carbon; (○), 200-500 mm; (□), 500-800 mm; (Δ), 800-1000 mm.

Figure 13 depicts, as an example, the results of the NTS ozonation in presence of Filtrasorb 400 activated carbon for different particle sizes. The NTS degradation was greater in the presence of this carbon than in its absence, and the NTS ozonation rate increased with decreasing particle size. Moreover, at short contact times the ozonation rate of NTS was slightly decreased by the presence of the activated carbon with particle sizes ranging 800-1000 mm; however, the ozonation rate was substantially raised when adding activated carbon with particle sizes ranging 500-800 and 200-500 mm. These results could indicate that the degradation rate of NTS in the presence of activated carbon is influenced by diffusional processes.

In order to determine whether the increase in the ozonation rate by decreasing particle size was due to intraparticular diffusion limitations in the carbon, the Weisz-Prater parameter (Equation 21) was calculated (Table 9) for each particle size studied. The Weisz-Prater criterion indicates that the NTS ozonation rate in presence of activated carbon would not be influenced by pore diffusion when $C_{WP} << 1$ [53].

$$C_{WP} = \frac{(-r_M)R_P^2}{D_E C_M} \tag{21}$$

where $(-r_M)$ is the NTS degradation rate (mol/L s),, R_p is the particle radius (cm), D_E is the effective diffusivity (cm^2/s), and C_M is the concentration of NTS (mol/L).

The effective diffusivity of NTS was estimated using Equation 22 [54]:

$$D_E = \frac{\varepsilon_P D_{AB}}{\tau_P} \tag{22}$$

where ε_P is the void fraction of the carbon particle, D_{AB} is the molecular diffusivity, and τ_P is the tortuosity factor of the activated carbon. The void fraction was calculated from the experimental values of pore volume and particle density ($\varepsilon_P=0.63$ for Filtrasorb 400). The tortuosity factor for activated carbon can be considered approximately 3 [55] and the molecular diffusivity of the NTS was calculated by using the Wilke-Chang Equation (Equation 23).

$$D_{AB} = \frac{7.4 \times 10^{-8} (\phi M_B)^{1/2}\, T}{\eta_{BV} V_A^{0.6}} \tag{23}$$

In this Equation, ϕ is the association parameter of water and has a value of 2.6, M_B is the molecular weight of water (18 g/mol), T is temperature (298 K), η_{BV} is viscosity of water (0.8904 cp), and V_A is the molar volume of the solute at its boiling temperature (196 cm^3/ mol). The value of V_A was calculated using Schroeder's method [55].

Table 9. Values of the Weisz-Prater parameter (CWP) for each particle size of activated carbon Filtrasorb 400

Particle size (μm)	C_{WP}
200-500	0.41
500-800	1.42
800-1000	2.73

According to results in Table 9, it was determined that for particle sizes above 200-500 μm the NTS degradation rate is controlled by intraparticular diffusion.

3.3.4. Activated Carbons Prepared in Our Laboratory to Enhance Ozone Decomposition into HO·

As shown above, the combined use of ozone and activated carbon in a single process is an attractive option to eliminate toxic organic compounds from waters. Thus, it was observed that the capacity of activated carbons to produce ozone transformation into ·OH radicals, which have a higher oxidant power than ozone, is directly related to their porous texture, surface chemistry (basic groups), and mineral matter content. With this background, some carbon materials were prepared in our laboratory to enhance ozone transformation into HO· radicals (activated carbon from petroleum coke, nitrogen enriched activated carbons and metal doped carbon aerogels).

3.3.4.1. Activated Carbons from Petroleum Coke

Petroleum coke is a residue of the petrochemical industry, which generates around 4 tons of carbon for every 100 tons of crude oil refined. Because of its high concentration of heavy metals (Ni, V, Fe), this residue cannot be used in any productive process. However, these same characteristics could make petroleum coke a very promising material in the ozonation of aromatic pollutants.

Table 10 shows the results of the textural characterization of the activated carbons prepared with different KOH/coke mass ratios. These results indicate that the activation process considerably developed the porosity in all of the samples studied, increasing the volume of micro- (V_{mic}), meso- (V_2) and macropores (V_3). Thus, the surface area of the original coke (C) markedly increased after the activation, with sample C-1 showing the highest value. Moreover, it was observed that the micropore volume reduced with an increase in the amount of KOH added to the coke. Regarding to the surface chemistry of the samples, it was observed that activation of the coke modified their chemical nature (Table 10). Thus, whereas the original coke was a mildly acid material (pH_{PZC} = 6.5), the pH_{PZC} of the KOH-activated coke ranged from 8.4 for sample C-1 to 9.7 for sample C-4. This change is largely due to the generation of surface basic groups during the activation process which increases with the KOH/coke ratio.

Figure 14 shows the results of NTS ozonation in the presence of the activated coke samples. The NTS oxidation rate was observed to increase when the ozonation was performed in their presence. The NTS adsorption rate on the activated coke samples was very slow; the adsorption kinetics of NTS on sample C-1 is shown in Figure 14 as an example. Thus, practically no NTS was adsorbed after 60 min of contact. Therefore, the increased removal rate of this pollutant in the presence of activated carbons is solely due to an increase in the ozone decomposition rate into highly oxidant species in the system.

In order to quantify the increase in NTS removal rate due to the presence of activated coke in the system, the value of k_{hetero} was calculated for each sample (Table 11). These results show that the chemical activation process increases the activity of petroleum coke in the NTS ozonation process. Thus, the activation of the coke produces an increase in the k_{hetero} value ranging from 83 % for sample C-1 to 16 % for sample C-4, mainly due to: i) development of the porosity of the coke, allowing greater accessibility of the ozone to its surface active sites and mineral matter (Table 10); and ii) an increase in the surface basicity of the original coke (Tables 10) which favours ozone reduction on its surface and thereby enhances its transformation into highly oxidant species

Table 10. Textural characterization of the original and activated cokes

Sample	KOH/Coke	S_{N2} (m^2/g)	V_{mic} (cm^3/g)	V_2[a)] (cm^3/g)	V_3[b)] (cm^3/g)	pH_{PZC}
C	0	< 30	0.02	Nil	0.011	6.5
C-1	1	1619	0.55	0.063	0.132	8.4
C-2	2	1261	0.41	0.061	0.154	8.8
C-3	3	1021	0.25	0.058	0.176	9.3
C-4	4	970	0.20	0.051	0.263	9.7

a) Volume of pores with diameter of 50 to 6.8 nm.
b) Volume of pores with diameter above 50 nm.

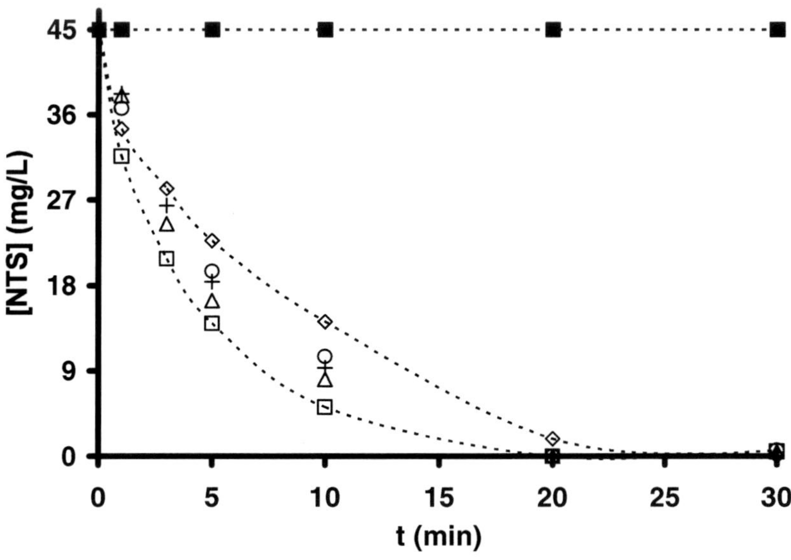

Figure 14. NTS ozonation in the presence of the activated cokes. pH 2, T 25 °C. (◇),Without coke; (□), C-1; (△), C-2; (+), C-3; (○), C-4. Black symbol (■) represents NTS adsorption kinetic on activated coke C-1.

Interestingly, the activity of the activated coke diminished with an increase in the amount of KOH in the coke during the activation process (Figure 14, Table 11). This may in part be due to its reduced surface area (Table 10) and therefore the reduced number of active sites capable of decomposing ozone into highly oxidant species.

In order to determine the participation of the different functional groups and of the mineral matter in the activity of the activated coke in NTS ozonation process, analyses were conducted on the transformations of the surface oxygenated groups and of the metals Fe, Ni and V, the main mineral matter components in the coke samples, which take place during the ozonation due to the interaction of the ozone with the activated coke surface. Table 12 shows the results of deconvoluting the XPS O1s spectrum of sample C-1 after its use as catalyst in the NTS ozonation process (sample designated C-1-ozonated). Thus, there was a marked increase in the concentration of oxygenated phenolic and carboxylic groups, whereas the concentration of carbonyl groups was reduced. Moreover, the percentage of surface oxygen of sample C-1 showed a 15 % increase after its use as catalyst. These results indicate that oxidation of the carbon surface takes place during the NTS ozonation process.

Table 11. Values of rate constants of the heterogeneous reaction for the activated coke samples in the NTS ozonation

Sample	k_{hetero} ((mol/L)$^{-1}$s^{-1})
C	57.7
C-1	105.8
C-2	86.5
C-3	76.9
C-4	67.3

Table 12. Results of the deconvolution of the XPS O1s spectrum of the activated coke samples

Sample	(-C=O) (%) 530.7±0.2 eV	(C-OH, C-O-C) (%) 532.1±0.2 eV	(-COOH) (%) 533.3±0.2 eV	H_2O and/or O_2 (%) 535.3±0.2 eV
C-1	76	14	6	4
C-1-ozonated	50	22	23	5

Ozonation of the carbon affects not only its surface oxygenated groups but also its metal sites, which can undergo changes in their oxidation state. Analysis of the Fe $2p_{3/2}$ spectrum for samples C-1 and C-1-ozonated showed a 14 % increase in the surface concentration of Fe_2O_3. These results appear to indicate that, during NTS ozonation, the ozone may attack Fe(II) metal sites on the carbon surface, generating Fe(III) (equations 24 and 25). These reactions may also contribute favourably to the NTS ozonation process by increasing the ozone transformation rate into ·OH radicals. The results of the study of the Ni $2p_{3/2}$ and V $2p_{3/2}$ spectra (Ni, and V are the major components of mineral matter) indicated that no changes were produced in the initial oxidation state, with Ni remaining in oxidation state +2 and V in oxidation state +5.

$O_3 + Fe^{2+}\ FeO^{2+} + O_2 \longrightarrow$ (24)

$FeO^{2+} + H_2O\ Fe^{3+} + \cdot OH + OH^- \longrightarrow$ (25)

Some components of the mineral matter of carbons behave as active sites to enhance ozone transformation process into ·OH radicals and, therefore, NTS oxidation rate (Figure 14). XPS analysis shown that the presence of metals susceptible to oxidation by ozone on the surface of activated cokes is potential promoters of NTS ozonation (Reactions 24, 25). The results shown in Table 11 could also indicate that other components of the mineral matter besides Fe(II) have to contribute positively to coke activity in the ozone transformation process into OH radicals (the value of heterogeneous constant for the original coke is very high considering the low extension of the reactions 24 and 25), but due to the chemical complexity of this mineral matter, it is very difficult to determine the components involved and the mechanism followed in the ozonation process. Greater efforts have to be done in this research area to dilucidate these aspects.

3.3.4.2. Activated Carbons Enriched in Basic Superficial Groups

Activated carbons enriched in basic superficial groups were prepared to explore the possibility of increasing the purification efficacy of the ozone/activated carboy system in the removal of naphthalenesulphonic acids.

Table 13 exhibits the results of the textural characterization of activated carbon Witco (W) and the nitrogenating agent-treated samples. The surface area of carbon W (S_{N2}) was increased by treatment with nitrogenating agents, especially urea (W-U). Carbon W-U presented an increased microporosity because the urea attacks the carbon and gasifies it, with the resulting development of its porosity. The pore volume values determined by mercury porosimetry (V_2 and V_3) are included in Table 13. It can also be deduced from these results that treatment with ammonia (W-A) or ammonium carbonate (W-C) develops to a small degree the meso- and macroporosity of the carbon, whereas treatment with urea developed to a large degree the porosity of the carbon across the whole range, i.e., the micro-, meso- and macropores.

Table 13. Textural characterization of activated carbon samples

Sample	S_{N2} (m^2/g)	V_{micro} (cm^3/g)	V_2 (cm^3/g)	V_3 (cm^3/g)
W	812	0.238	0.040	0.050
W-A	904	0.206	0.047	0.091
W-C	825	0.222	0.042	0.102
W-U	1057	0.289	0.064	0.122

SN2= Apparent surface area determined applying BET equation to N2 adsorption isotherm.

Vmicro= Micropore volume determined applying Dubinin-Raduskevich equation to CO2 adsorption isotherm.

V2= Volume of pores with diameter of 50 -6.6 nm.

V3= Volume of pores with diameter above 50 nm.

Tables 14 and 15 exhibit the results of the characterization of the surface chemistry of the carbons. The aim was to determine the modifications in the carbon surface chemistry that result from the nitrogenating agent treatments and to identify the groups that may act as catalytic centers in the ozonation process. The results of the elemental analysis of the samples (Table 14) indicate that treatment of the carbon with the nitrogenating agents reduced their carbon content because they introduced groups composed of N, H or O. Thus, the percentage of these heteroatoms in the samples increased with the treatments, especially when urea was the treatment agent. The deconvolution of the XPS spectra corresponding to the N1s region of the activated carbon samples (Table 15) indicated that the concentration of nitrogenated functional groups introduced into the carbon depends on the treatment used. Thus, carbon W-A was characterized by a high concentration of pyridone groups, carbon W-C by a high concentration of pyridine and pyrrol-type groups, whereas the urea-treated sample was characterized by high concentrations of pyrrol and pyridone groups.

After characterization of the activated carbon samples, NTS ozonation processes were studied to determine their possible catalytic activity. Figure 15 depicts the results of NTS ozonation in presence of the activated carbon samples under study. The NTS degradation rate increased when activated carbon was added to the system and was especially enhanced in presence of carbon W-U.

Table 14. Elemental analysis of activated carbon samples (% dry weight)

Sample	C	N	H	O (by difference)
W	92.55	0.00	0.17	4.24
W-A	91.30	0.50	0.41	4.78
W-C	88.22	0.51	0.31	8.13
W-U	76.88	1.64	0.37	19.30

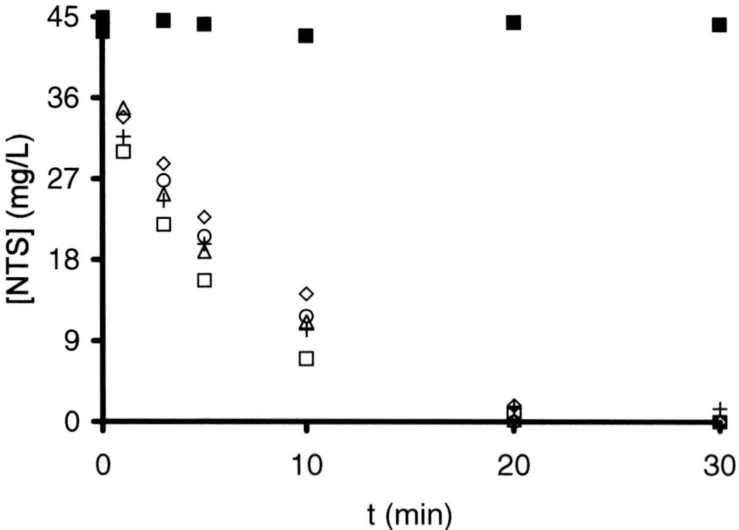

Figure 15. NTS ozonation in presence of activated carbon. pH 2, T 298 K. (∩), Without activated carbon; (○), W; (△), W-A; (+), W-C; (ϒ), W-U. The black symbols correspond to adsorption kinetics of NTS on activated carbons used.

Table 15. XPS analysis of N1s region of activated carbon samples (%)

Sample	Pyridine (398.5±0.2 eV)	Pyridone (399.5±0.2 eV)	Pyrrol (400.5±0.2 ev)
W-A	11	79	10
W-C	56	--	44
W-U	8	42	50

When the catalytic activity was considered in relation to the chemical properties of the samples, it was observed to increase in the basic activated carbon samples. Interestingly, in the sample with the greatest catalytic activity in the NTS ozonation process (W-U), a high proportion of the surface nitrogenated groups were pyrrol groups, with a low proportion of pyridine groups (Table 15). These results appear to indicate that pyrrol-type groups increase the catalytic activity of the carbon. This hypothesis could be explained as follows: in the pyrrol group, the pair of nitrogen electrons forms part of the electronic cloud of the ring and is, therefore, delocalized among the five atoms that form the molecule. As a result, pyrrol has 6 π electrons on 5 centers, so that these are π-excessive aromatic rings. Therefore, the presence of pyrrol groups on the activated carbon surface increases the electronic density of its basal plane, and this carbon will consequently have a greater capacity to produce a reduction of the ozone dissolved on its surface (equation 26).

$$O_3 + H_2O + 2e^- \quad O_2 + 2\,OH^- \longrightarrow \qquad (26)$$

Moreover, the increase in the π electron system of the carbon due to the presence of pyrrol groups produces a greater degree of interaction with the water molecules. In both processes, OH^- ions are generated in the medium, which act as initiators of the ozone decomposition process into·OH radicals, which are highly reactive against NTS, increasing the degradation rate (Reactions 27-29).

$$O_3 + OH^- \quad O_2\cdot^- + HO_2\cdot \longrightarrow \qquad (27)$$

$$O_2\cdot^- + O_3 \quad O_2 + O_3\cdot^- \longrightarrow \qquad (28)$$

$$O_3\cdot^- + H^+ \quad OH\cdot + O_2 \longrightarrow \qquad (29)$$

This reasoning may explain, in part, the high catalytic activity of carbon W-U due to its elevated concentration of pyrrolic groups. In addition, the greater catalytic activity of carbon W-C versus carbon W-A, despite the larger surface area of the latter, may be due to the higher concentration of pyrrol groups in carbon W-C. Thus, the lower nitrogen percentage of carbon W-C compared with carbon W-U and, therefore, its much lower pyrrol concentration (Table 14 and 15) explains that carbon W-U presents a higher catalytic activity than carbon W-C.

In contrast, the presence of pyridine and pyridone groups on the activated carbon surface does not favour the above reactions, because in pyridine, the nitrogen does not donate an excess of electronic density to the aromatic ring. In this heterocycle, the single pair of electrons of N are localized in a sp^2 hybridization orbital, and they do not form part of the electronic cloud of the aromatic ring. On the contrary, because the nitrogen is more electronegative than the carbon, it attracts electronic density from the ring by both induction and resonance. Therefore, in the pyridine and pyridone groups, the π cloud is partially anchored on the N, being p-deficient heterocycles. Hence, the presence of these groups in the carbon surface reduces the electronic density of its graphene layers, which does not potentiate the decomposition of the ozone into radicals, in accordance with the above reactions. For this reason, carbons W-A and W-C, with their high concentration of pyridone and pyridine rings, respectively, presented a low reactivity in NTS ozonation in comparison with carbon W-U.

In order to determine the participation of the different functional groups in the catalytic activity of the activated carbon in NTS ozonation, we analyzed the transformations of the carbon surface chemical groups during ozonation, which result from interaction of the ozone with the carbon surface. Table 16 displays some of the chemical characteristics of carbon W and carbon W-U after their use as catalysts in the ozonation process. The acidity of these carbons considerably increased with their ozonation to pH_{PZC} values of 1.7 and 2.3, much lower than the values of pH_{PZC} presented by these carbons before the ozonation. This is because the ozonation of the carbon considerably increases the concentration of carboxylic, lactonic, and phenolic groups.

Table 16. Ozonation effect on oxygenated surface groups concentration (μeq/g).

Sample	pH_{PZC}	Carboxyl	Lactone	Phenol
W (Ozonated)	1.7	600	1520	1130
W-U (Ozonated)	2.3	520	960	966

Table 17. Ozonation effect on nitrogen surface groups concentration (%)

Sample	Pyridine (398.5±0.2 eV)	Pyridone (399.5±0.2 eV)	Pyrrol (400.5±0.2 eV)	N-Oxide (402.5±0.2 eV)
W-U	8	42	50	--
W-U (ozonated)	7	47	16	30

The ozonation of carbon affected not only the oxygenated groups but also the nitrogenated groups. Thus, Table 17 shows the changes observed in sample W-U as an example. It can be observed that the pyridone groups were not affected by the ozonation, whereas a large number of the pyrrol-type groups were oxidized and transformed into N-oxide-type groups.

Therefore, these results appear to indicate that the ozone may attack the pyrrolic groups of the activated carbon graphene planes during ozonation, yielding N-oxide type groups and the hydroperoxide radical (Reaction 30).

$$\text{NH} + O_3 \longrightarrow \overset{+}{\text{N}}=\overset{-}{\text{O}} + HO_2^{\bullet} \qquad (30)$$

This reaction, which explains the transformations in the pyrrol group observed during ozonation, may always contribute favourably to the NTS ozonation process, given that the hydroperoxide radical attacks the ozone, enhancing its decomposition into radicals that are highly effective in NTS degradation. This might contribute to the greater catalytic activity of pyrrol groups present in the carbon surface.

The generation of $HO_2\cdot$ radicals during ozonation processes in presence of sample W-U (Equation 30) was confirmed by measuring the concentration of $O_2\cdot^-$ radicals with tetranitromethane [37]. In Figure 16 it is shown that the concentration of NF^- for the W-U carbon is a factor of 4 higher after 10 μM of ozone are consumed, relative to the blank (O_3 only) and ozonation in presence of W activated carbon. The results shown in Figure 16 would also indicate that the concentration of surface pyrrol groups present in sample W-U is exhausted rapidly because the NF^- concentration is constant after 12 μM of ozone consumed. This is further emphasized by their observed relative decrease after ozonation by XPS (Table 17).

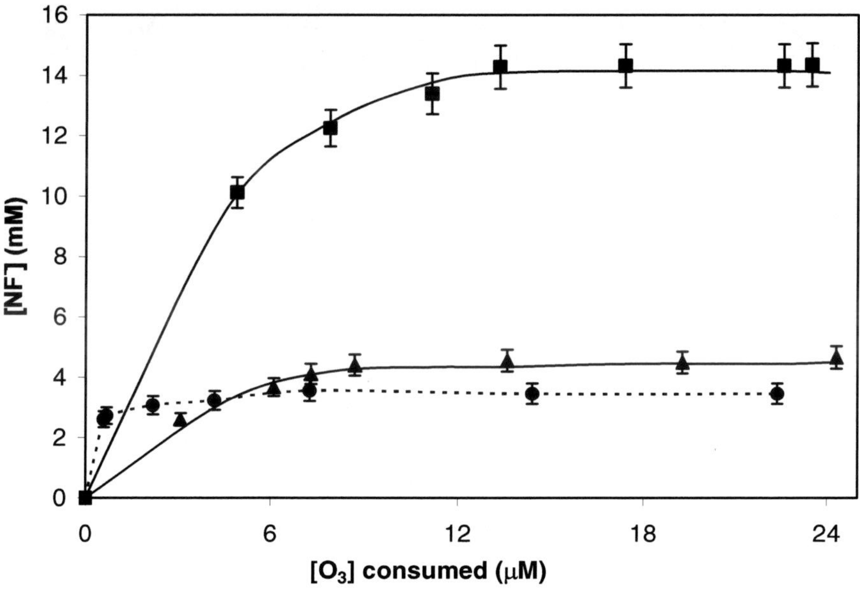

Figure 16. Concentration of NF^- anion versus ozone consumed. $[pCBA]_0 = 5\cdot10^{-7}$ M, pH 7, $[t\text{-}BuOH]_0 = 8\cdot10^{-5}$ M, T 295 K, $[O_3]_0 = 2\cdot10^{-5}$ M, $[TNM]_0 = 1.6\cdot10^{-3}$ M. (●), Ozone only; (▲), O_3/W; (■), O_3/W-U.

3.3.4.3. Metal Doped Carbon Aerogels

Carbon aerogels, first developed at the end of the 1980s [31, 32], are now used as the basis of numerous industrial applications because of their chemical and textural properties and easy preparation [56, 57, 58]. They are prepared from resorcinol/formaldehyde gels by supercritical drying methods. The skeletal structure of wet aerogels is maintained by supercritical drying, obtaining solids with high porosity and specific surface areas. A potentially important feature of these materials is that metal-doped aerogels can be easily prepared by adding a soluble metal salt to the initial resorcinol/formaldehyde mixture [59,

60]. After gelation, the metal salt is trapped within the gel structure and the metal ions are chelated by functional groups of the polymer matrix. These properties make carbon aerogels highly promising materials to enhance ozone transformation process into ·OH radicals, because advantage can be taken of the properties of transition metals with contrasting catalytic activities in the ozonation process, avoiding their dissolution in and the separation from the system.

The textural characteristics of the original aerogels (A, A-Co(II)-15, A-Mn(II)-15 and A-Ti(IV)-15) are compiled in Table 18; all are mainly meso- and macroporous materials. Thus, both the volume of pores with diameter in between 6.6 nm and 50 nm (V_2) and the macropore volume (V_3) are much greater than the micropore volume (V_{mic}). Moreover, their pore size distributions showed the maxima located at around 12 nm in diameter. N_2 surface area (S_{N2}), CO_2 surface area (S_{CO2}) and V_{mic} values were relatively low for sample A (blank) and were slightly higher in aerogels doped with a transition metal (Co(II), Mn(II) or Ti(IV)).

Table 18. Textural characterization of the aerogel samples

Sample	S_{N2} (m^2/g)	S_{CO2} (m^2/g)	V_{mic} (cm^3/g)	V_2 (cm^3/g)	V_3 (cm^3/g)
A	500	200	0.07	0.36	0.68
A-Co(II)-15	562	206	0.07	0.43	0.97
A-Ti(IV)-15	550	203	0.07	0.40	0.92
A-Mn(II)-15	554	210	0.07	0.41	0.95
A-Mn(II)-15-1	540	200	0.07	0.41	0.94
A-Mn(II)-15-2	534	204	0.07	0.40	0.93
A-Mn(II)-15-3	546	206	0.07	0.41	0.95
A-Co(II)-15-1	560	210	0.07	0.41	0.92

SN2 = Apparent surface area determined applying BET equation to N2 adsorption isotherm.

SCO2 = Apparent surface area determined applying Dubinin-Raduskevich equation to CO2 adsorption isotherm.

Vmic= Micropore volume determined applying Dubinin-Raduskevich equation to CO2 adsorption isotherm.

V2= Volume of pores with diameter of 50 -6.6 nm.

V3= Volume of pores with diameter above 50 nm.

Comparisons between results for A-Co(II)-15, A-Mn(II)-15, and A-Ti(IV)-15 (Table 18) indicated that the textural properties of the samples did not differ according to the metal added, with all of them showing values close to 550 m^2/g, 0.4 cm^3/g, and 0.9 cm^3/g for S_{N2}, V_2, and V_3, respectively. The value of S_{N2} was higher than that of S_{CO2} in these samples indicating that a large fraction of the surface of these samples corresponded to meso- and macropores [61].

The carbon aerogels (A, A-Co(II)-15, A-Ti(IV)-15 and A-Mn(II)-15) were chemically characterized by determination of the pH_{PZC}, XRD, and XPS. Results obtained are shown in Tables 19 and 20. The pH_{PZC} values (Table 19) showed the aerogels to have a high surface acidity (pH_{PZC} = 3.5- 4.3).

Table 19. Chemical characterization of the aerogel samples

Sample	pH_{PZC}	C (%)	O (%)	Co(II) (%)	Mn(II) (%)	Mn(III) (%)	Mn(IV) (%)	Ti(IV) (%)
A	3.5	68	32	--	--	--	--	--
A-Co(II)-15	3.8	64	22	14	--	--	--	--
A-Ti(IV)-15	4.3	64	21	--	--	--	--	15
A-Mn(II)-15	4.2	62	22	--	16	--	--	--
A-Mn(II)-15-1	4.0	54	30	--	10	4	2	--
A-Mn(II)-15-2	4.1	49	35	--	8	4	4	--
A-Mn(II)-15-3	3.9	42	42	--	6	4	6	--
A-Co(II)-15-1	3.9	55	31	14	--	--	--	--
A-Ti(IV)-15-1	4.2	53	32	--	--	--	--	15

Table 20. Deconvolution of O1s spectra of the aerogel samples

Sample	(-C=O) 532±0.2eV (%)	(-C-O) 533.9±0.2eV (%)
A	52	48
A-Co(II)-15	55	45
A-Ti(IV)-15	49	51
A-Mn(II)-15	76	24
A-Mn(II)-15-1	83	27
A-Mn(II)-15-2	91	9
A-Mn(II)-15-3	96	4
A-Co(II)-15-1	68	32
A-Ti(IV)-15-1	65	35

The XPS results (Table 19) showed a high concentration of surface oxygen with values greater than 20%. To determine the nature of the oxygenated surface groups, deconvolution of the O1s XPS spectrum was performed for each sample, observing that the surface oxygen was distributed almost equally between ether (–C-O-C) and carbonyl (>C=O) groups, samples A, A-Co(II)-15 and A-Ti(IV)-15. However, in sample A-Mn(II)-15 the amount of >C=O (76%) is much higher than that of –C-O-C (24%).

XRD results for samples A-Co(II)-15, A-Mn(II)-15, and A-Ti(IV)-15 showed a wide dispersion of the metal on the aerogel surface, since no diffraction peaks were observed in any case (results not shown). Importantly, XPS analysis of the same samples revealed that the metals on the aerogel surface were in oxidation state +2 in the case of Mn and Co, and +4 in the case of Ti, and that the surface percentage of each metal was around 15% in all cases (Table 19).

Textural analyses of the aerogel samples treated with ozone (A-Mn(II)-15-1, A-Mn(II)-15-2, A-Mn(II)-15-3, A-Co(II)-15-1 and A-Ti(IV)-15-1) showed that this treatment did not significantly affect the S_{N2}, S_{CO2}, V_2, and V_3 values obtained, which were similar to those in the non-pretreated samples (Table 18). XPS results (Table 20) showed that, regardless of the sample considered, an increase was produced in the percentage of surface oxygen with an increase in the number of ozonation cycles to which the samples were subjected. No such increase was detected in the pH_{PZC} values of these samples, which remained close to initial values (pH_{PZC} = 3-4). In all aerogel samples the surface oxygen created during their ozonation was mainly in the form of –C=O (Table 20). Interestingly, in the O_3-pretreated Mn-doped aerogel samples Mn(III) and Mn(IV) are formed; the percentage of Mn in oxidation state +4 increased with a larger number of ozonation cycles in the pretreatment. However, the results for samples A-Co(II)-15-1 and A-Ti(IV)-15-1 showed that the oxidation state of the Co and Ti remained at +2 and +4, respectively (Table 19) after the ozonation pretreatment.

Figure 17 depicts the results of pCBA ozonation in presence of the original carbon aerogels and the corresponding adsorption kinetics of this compound on the samples. The adsorption kinetics of pCBA on the aerogel samples studied was very slow, and no adsorption of pCBA was observed during 60 min of contact. The Rct values obtained in the different experiments and the corresponding ozone decomposition constants (k_D), determined by a first-order kinetic model, are listed in Table 21.

The presence of the aerogel prepared as blank (sample A) did not increase the pCBA removal rate (Figure 17), and the Rct and k_D values were close to those observed in absence of aerogel (Table 21). These findings indicate that the organic matrix used to generate the aerogel does not positively contribute to the transformation of ozone into ·OH radicals.

Table 21. Determination of Rct value in the different ozonation experiments performed pH 7, T 25°C, [O3] = 2 10-5 M, [t-BuOH] = 8 10-5 M)

Experiment	Sample	Carbon dose (mg)	k_D (s^{-1})	Rct
1	Without aerogel	0	$6.0\cdot10^{-4}$	$2.68\cdot10^{-9}$
2	A	2.5	$6.2\cdot10^{-4}$	$2.74\cdot10^{-9}$
3	A-Co(II)-15	2.5	$5.8\cdot10^{-4}$	$2.56\cdot10^{-9}$
4	A-Ti(IV)-15	2.5	$6.1\cdot10^{-4}$	$2.73\cdot10^{-9}$
5	A-Mn(II)-15	2.5	$4.2\cdot10^{-3}$	$5.36\cdot10^{-8}$
6	A-Mn(II)-15-1	2.5	$2.6\cdot10^{-3}$	$3.35\cdot10^{-8}$
7	A-Mn(II)-15-2	2.5	$2.1\cdot10^{-3}$	$2.68\cdot10^{-8}$
8	A-Mn(II)-15-3	2.5	$1.4\cdot10^{-3}$	$1.78\cdot10^{-8}$

Results presented in Figure 17 also show that the pCBA removal rate was increased in presence of Mn aerogel, whereas the presence of Co- or Ti-doped carbon aerogels had virtually no effect on this rate. Because of the slow reactivity of pCBA against ozone [7] and the slow adsorption kinetic of pCBA on the aerogel samples studied (Figure 17), the increase in its oxidation rate in presence of Mn aerogel would mainly result from the generation of ·OH radicals in the system. Moreover, the fact that the presence of aerogel A (blank) did not increase the rate of removal of pCBA from the medium (Figure 17) confirms that the activity of the aerogels in the transformation of ozone into ·OH radicals is directly related to the presence of the metal on their surface.

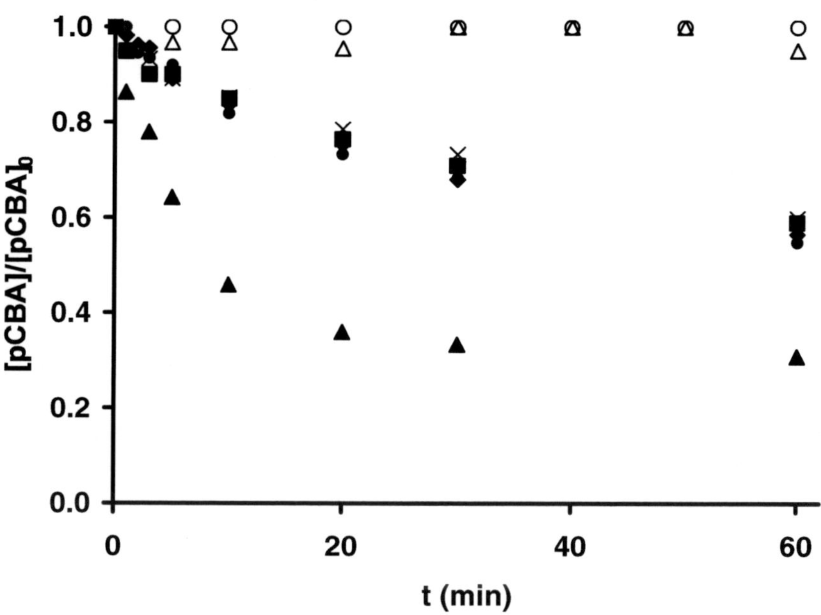

Figure 17. Ozonation of pCBA in presence of the aerogels. pH 7, T 25°C, [O3] = 2·10-5 M, [t-BuOH] = 8·10-5 M, [Aerogel] = 2.5 mg/L. (x), Without aerogel; (●), A; (■), A-Co(II)-15; (◆), A-Ti(IV)-15; (▲), A-Mn(II)-15. Open symbols represent the adsorption kinetics on samples A (○) and A-Mn(II)-15 (△).

The presence of Mn aerogel during pCBA ozonation increased Rct values 20-fold, whereas the Rct values obtained in presence of Co(II) and Ti(IV) were very similar to those obtained in their absence. Likewise, the presence of Mn(II) aerogel during pCBA ozonation increased the k_D value, whereas the presence of Co(II) or Ti(IV) aerogels had practically no effect on this parameter (Table 21, Experiments 2 - 5). These findings confirm that the presence of Mn doped carbon aerogel during this ozonation process accelerates ozone transformation into ·OH radicals.

In order to determine the mechanism by which the Mn(II) aerogel accelerates the pCBA removal rate, samples A-Co(II)-15, A-Mn(II)-15, and A-Ti(IV)-15 were studied by XPS after their prior ozonation treatment as described in the Experimental section (samples A-Co(II)-15-1, A-Mn(II)-15-1, and A-Ti(IV)-15-1). Results of the textural and chemical characterization of these samples were discussed in detail previously. However, it is of interest to note that the aerogel sample that enhanced ozone transformation into ·OH radicals (A-Mn(II)-15) is also the one in which the oxidation state increased after the ozonation process (Table 19). Thus, 10% of the surface Mn was present in Mn(II) form, 4% in Mn(III) form, and 2% in Mn(IV) form on sample A-Mn(II)-15-1. For Co and Ti (samples A-Co(II)-15 and A-Ti(IV)-15), the oxidation state of the metal was unchanged at +2 and +4, respectively. These results indicate that the mechanism by which A-Mn(II)-15 aerogel enhances transformation of the ozone into ·OH radicals is based on oxidation-reduction reactions.

Hence, and in agreement with the results presented in Figure 17 and Tables 18, 19, 20 and 21 possible reactions responsible for accelerating the transformation of ozone into ·OH radicals due to the presence of A-Mn(II)-15 aerogel during pCBA ozonation may be:

$$O_3 + Mn^{2+} \longrightarrow MnO^{2+} + O_2 \quad (31)$$

$$MnO^{2+} + H_2O \longrightarrow Mn^{3+} + \cdot OH + OH^- \quad (32)$$

$$2Mn^{+3} + 2\,H_2O \longrightarrow MnO_2 + Mn^{2+} + 4\,H^+ \quad (33)$$

In this way, oxidation of surface Mn(II) to Mn(III) and Mn(IV) during the oxidation process results in a transformation of ozone into ·OH radicals, explaining the increase in the pCBA removal rate (Figure 17) and the Rct and k_D values observed (Table 21). This mechanism is in agreement with the proposed by other authors [62, 63] where the ozonation of atrazine and oxalic acid in presence of dissolved Mn(II) and MnO_2 were studied. More information related to the use of metal-doped aerogels in micropollutant ozonation process can be found elsewhere [30].

3.3.5. Evolution of Both TOC and the Genotoxicity of Naphthalenesulphonic Acids during Ozone/Activated Carbon Treatment

Figure 18 depicts, as an example, the variations in TOC during the treatment of NTS; similar results were obtained for NS and NDS. In the absence of activated carbon the TOC keeps constant during ozonation; these results indicate that the oxidant power of ozone is low to mineralize the degradation by-products from naphthalene sulphonic acids ozonation; however in the presence of activated carbon, the TOC decreased with increased treatment time and after 120 min of treatment, 80% of the initial TOC was eliminated. Nevertheless, during the first 10 min, an increase in TOC is observed due to the attack of ozone to activated carbon. Two processes are involved in the TOC decrease: i) adsorption of oxidation by-products on activated carbon and ii) mineralization of dissolved organic carbon by hydroxyl radicals generated in the interaction between ozone and activated carbon.

The evolution of the genotoxicity of the naphthalene sulphonic acids was also studied during this combined treatment. In the case of NTS, the addition of activated carbon had no great influence on the elimination of the genotoxicity of the oxidation by-products. This is because the oxidation by-products of these contaminants do not have a high genotoxicity, due to the presence of sulphonic groups in the organic molecule. As explained above, the sulphonic group prevents penetration into the bacterian cell. Similar results were obtained for NDS. However, in the case of NS, the addition of a small dose of activated carbon to the system did enhance the elimination of the genotoxicity of the oxidation by-products of NS (Figure 19). This may be because the presence of activated carbon produces the adsorption of the more hydrophobic compounds and increases the extent of ozonation through the generation of hydroxyl radicals.

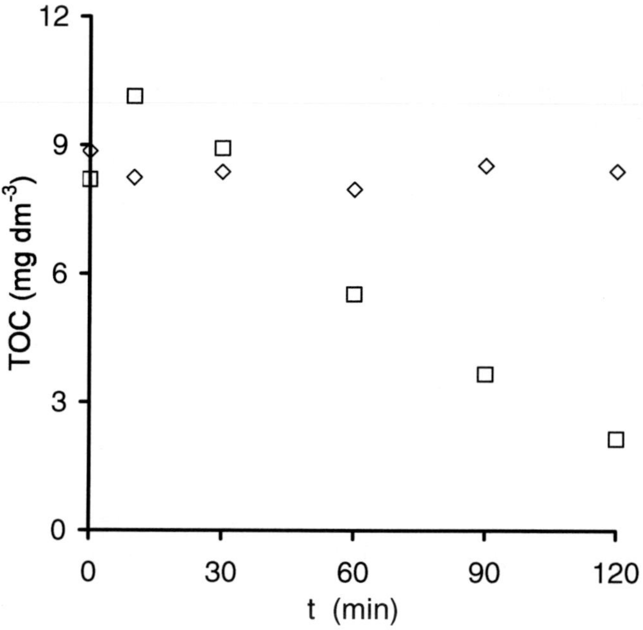

Figure 18. Evolution of Total Organic Carbon during treatment of NTS. pH 2, T 298 K, O_3 flow 76 mg/min. (◇), ozone; (□), ozone/activated carbon.

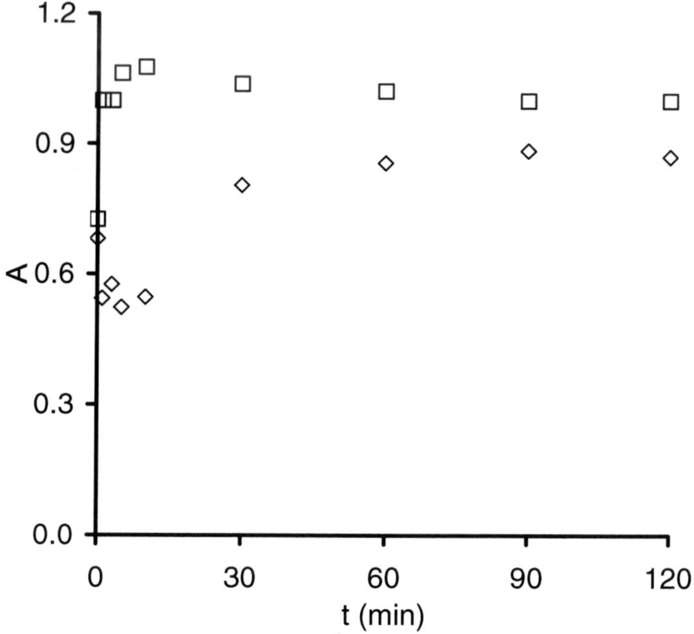

Figure 19. Evolution of genotoxicity of NS. pH 2, T 298 K, O_3 flow 76 mg/min. (◇), ozone; (□), ozone/activated carbon.

4. Conclusion

The adsorption of naphthalenesulphonic acids on activated carbons is regulated both by π-π dispersion interactions and by electrostatic interactions. However, the π-π interactions have a more determining influence on this process. The extremely low breakthrough volumes of the activated carbon columns and the excessively high mass transference zone indicate that a system based exclusively on activated carbon is not completely efficacious in the removal of these contaminants from waters.

The aromatic acids studied showed a low reactivity to ozone. The greater the number of sulphonic groups in the molecule, the lower was the reactivity. The electron-withdrawing character of the sulphonic groups deactivated the aromatic ring against the reaction with ozone. The stoichiometry of the ozonization reaction in all cases was one mol of ozone consumed per one mol of aromatic sulphonic acid degraded. The activation energy of the reaction ranged from 37-42 kJ/mol in all cases. These results indicate that the reaction of aromatic sulphonic acids occurs by the same mechanism in all cases, most probably by 1,3 dipolar cycloaddition at the carbon-carbon bond of greatest electronic density (positions 1-2, 3-4, 5-6 and 7-8).

The presence of activated carbon during the ozonation of NTS increases the rate of degradation and produces the removal of part of the dissolved organic carbon by its mineralization. Basic activated carbons have greatest catalytic activity in the ozonation process. The basal plane electrons and oxygenated surface groups of basic nature (chromene and pyrone) in activated carbons are mostly responsible for ozone decomposition in aqueous phase. The ozone reduction on the surface of activated carbon generated OH^- ions or H_2O_2 that initiated the decomposition of ozone in aqueous phase into highly oxidative species, which are responsible for the increase in the NTS ozonation rate. Furthermore, these species are able to mineralize dissolved organic carbon, decreasing the TOC. A high macroporosity in carbon also enhances the ozonation process, reducing diffusion problems and, therefore, favoring access of the ozone to the active centers of the carbon surface. The study has also quantified the activated carbon activity for the transformation of aqueous ozone into ·OH radicals. The transformation rate increases when the concentration of dissolved ozone and activated carbon in the system is increased.

In accordance with the Weisz-Prater criterion, it was determined that for particle sizes above 200-500 μm the NTS degradation rate is controlled by intraparticular diffusion limitations. The kinetic equation that represents the heterogeneous reaction rate depends on the concentrations of contaminant and dissolved ozone. The results indicated that the reaction order with respect to both NTS and ozone is 1. Furthermore, the proposed general kinetic model adequately represents the experimental results, allowing quantification of the catalytic activity of the carbon in the NTS ozonation process. The k_{hetero} constant values determined ranged from 94.2 $(mol/L)^{-1}$ s^{-1} for Witco to 210.5 $(mol/L)^{-1}$ s^{-1} for Norit carbon.

Chemical activation of the petroleum coke with KOH produces an increase in its catalytic activity in the NTS ozonation process, due to the increased basicity of the carbon and to the development of its porosity, allowing greater access by the ozone to active sites on the carbon surface. The catalytic activity of activated carbons in NTS ozonation is increased after their treatment with nitrogenating agent due to i) an increase in the meso- and macropore volume and ii) an increase in the basicity, largely due to the creation of pyrrol groups on the carbon surface. The capacity of metal doped carbon aerogels to accelerate the transformation of ozone into ·OH radicals is related to the metal on their surface. Only the presence of metals susceptible to oxidation by ozone is effective. The presence of Mn(II) doped carbon aerogels during ozonation accelerates ozone transformation into ·OH radicals. The capacity to accelerate this process depends on the dose of aerogel and the concentration of Mn(II) on its surface. The capacity of the Mn doped carbon aerogel to accelerate ozone transformation into ·OH radicals decreases with longer exposure of the aerogel to ozone. The oxidation of surface Mn(II) to higher oxidation states is responsible for this behavior and leads to an inactivation of this material.

The presence of activated carbon and ozone in the system produces a reduction in the genotoxicity of the naphthalenesulphonic acid oxidation by-products. These results indicate the great efficacy of this novel purification system.

References

[1] Reemtsma, T., Jochimsen, J., Jekel M. Persistence of sulphonated polyphenols in the biological treatment of industrial wastewater. *Vom Wasser*, 1993 81, 353-363.

[2] Brouwer, E.R., Slobodnik, J., Lingeman, H., Brinkman U.A. Determination by reverse-phase ion-pair chromatography of aromatic sulphonic acids in surface water, *Analysis*, 1992 20, 121-126.

[3] Zerbinatti, O., Ostacoli, G., Gastaldi D., Zelano V. Determination and identification by high-performance liquid chromatography and spectrofluorimetry of twenty.three aromatic sulphonates in natural waters. *J. Chromatogr.,* 1993 640, 231-239.

[4] Greihm, H., Ahlers, J., Bias, R., Broeker, B., Hollander, H., Gelbke, H.P., Klimisch, H.J., Mangelsdorf, I., Payz, A., Schon, N., Stroop, G., Vogel, R., Weber, C., Ziegler Skylakakis, K., Bayer E. Toxicity and ecotoxicity of sulfonic acids: Structure-activity relationship. *Chemosphere*, 1994 28, 2203-2236.

[5] Doré, M. Mechanism de l´ozonation des herbicides de 1 ácide phenoxyacetique: 2,4-D and MCPA. *Wat. Res.*, 1979 14, 767-773.

[6] Yao, C.C.D., Haag, W.R., Rate constants for direct reactions of ozone with several driking water contaminants, *Wat. Res.*, 1991 25, 761-773..

[7] Hoigné, J., Bader, H. Rate constants of reactions of ozone with organic and inorganic compounds in water. 1. Non-dissociated organic compounds. *Wat. Res.*, 1983 17, 173-183.

[8] Beltrán, F.J., Ovejero, G., Rivas, J. Oxidation of polynuclear aromatic hydrocarbons in water. 4. Ozone combined with hydrogen peroxide. Ind. Eng. Chem. Res., 1996 35, 891-898.

[9] Bruny, R., Bourbigot, M.M., Doré, M. Oxidation of organic compounds through the combination ozone-hydrogen peroxide. *Ozone Sci. and Eng.*, 1985 7, 241-257.

[10] Beltrán, F.J., García Araya, J.F., Acedo, B. Advanced oxidation of atrazine in water. II Ozonation combined with ultraviaolet radiation. *Wat. Res.*, 1994 28, 2615-2174.

[11] Andreozzi, R., Insola, A., Caprio, V., Marotta, R., Tufano V. The use of manganeso dioxide as a heterogeneous catalyst for oxalic ozonation in aqueous solution. *App. Catal. A: General*, 1996 138, 75-81.

[12] Logemann, F.P., Anne J.H.J. Water treatment with a fexed bed catalytic ozonation process. *Wat. Sci. Tech.*, 1997 35, 353-360.

[13] Jans, U., Hoigné, J. Activated carbon and carbon black catalyzed transformation of aqueous ozone into OH-radicals. *Ozone Sci. and Eng*., 1998 20, 67-90.

[14] Sánchez-Polo, M., Rivera-Utrilla, J., Zaror, C.A. Advanced oxidation with ozone of 1,3,6-naphthalenetrisulphonic acid in aqueous solution. *J. Chem. Tech. Biotech.*, 2002 77, 148-154.

[15] Valdés, H., Sánchez-Polo, M., Rivera-Utrilla, J., Zaror, C.A. Effect of ozone treatment on surface properties of activated carbon. *Langmuir*, 2002 18, 2111-2116, 2002.

[16] Rivera-Utrilla, J., Sánchez-Polo, M., Zaror, C.A. Degradation of naphthalenesulphonic acids by oxidation with ozone in aqueous phase. *Phys. Chem. Chem. Phys.* 2002 4, 1129-1134.

[17] Rivera-Utrilla, J., Sánchez-Polo, M., Mondaca, M.A., Zaror, C.A. Effect of ozone and ozone/activated carbon treatments on genotoxic activity of naphthalenesulphonic acids. *J. Chem. Tech. and Biotech.*, 2002 77, 883-890.

[18] Sánchez-Polo, M., Rivera-Utrilla, J. Adsorbent-adsorbate interactions in the adsorption of Cd(II) and Hg(II) on ozonised activated carbons. *Env. Sci. Tech.*, 2002 36, 3850-3854.

[19] Rivera-Utrilla, J., Sánchez-Polo, M. Ozonation of 1,3,6-naphthalenetrisulphonic acid catalysed by activated carbon in aqueous phase. *Appl. Catalysis B: Environmental*, 2002 39, 319-329.

[20] Rivera-Utrilla, J., Sánchez-Polo, M. The role of dispersive and electrostatic interactions in the aqueous phase adsorption of naphthalenesulphonic acids on ozone-treated activated carbons. *Carbon*, 2002 40, 2685-2691..

[21] Sánchez-Polo, M., Rivera-Utrilla, J. Effect of O_3/carbon reaction on the catalytic activity of activated carbon during degradation of 1,3,6-naphthalenetrisulphonic acid with ozone, *Carbon,* 2003 41, 303-307.

[22] Rivera-Utrilla, J., Sánchez-Polo, M. Degradation and removal of naphthalenesulphonic acids by means of adsorption and ozonation catalyzed by activated carbon in aqueous phase. *Water Resources Research*, 2003 39, 1232-1244,

[23] Sánchez-Polo, M., Rivera-Utrilla, J. Ozonation of naphthalenesulphonic acid in aqueous phase in the presence of basic activated carbons. *Langmuir*, 2004 20, 9217-9222.

[24] Sánchez-Polo, M., Leyva-Ramos, R., Rivera-Utrilla, J. Kinetics of 1,3,6-naphthalenetrisulphonic acid ozonation in presence of activated carbon. *Carbon* 2005 43, 962-969.

[25] Sánchez-Polo, M., von Gunten, U., Rivera-Utrilla, J. Efficiency of activated carbon to transform ozone into ·OH radicals: influence of operational parameters. *Water Research*, 2005 39, 3189-3198.

[26] Sánchez-Polo, M., Rivera-Utrilla, J. Ozonation of naphthalenetrisulphonic acid in the presence of activated carbons prepared from petroleum coke. *Appl. Catalysis B: Environmental*, 2006 67, 113-120.

[27] Sánchez-Polo, M., von Gunten, U., Rivera-Utrilla, J. Combination of ozone with activated carbon as an alternative to conventional advanced oxidation processes. *Ozone Sci. and Eng.*, 2006 28, 237-245.

[28] Sánchez-Polo, M., Rivera-Utrilla, J., J. Méndez-Díaz, Canonica, S., von Gunten, U. Photooxidation of naphthalenesulfonic acids. Comparison between processes based on O_3, O_3/activated carbon and UV/H_2O_2. *Chemosphere*, 2007 68, 1814-1820.

[29] Sánchez-Polo, M., Rivera-Utrilla, J., von Gunten, U. Metal-doped carbon aerogels as catalysts during ozonation processes in aqueous solutions. *Water Research*, 2006 40, 93-100.

[30] Pekala R.W. Organic aerogels from polycondensation of resorcinol with formaldehyde. J. Mater Sci. 1986 24, 3221-3227.

[31] Pekala, R.W., Alviso C.T., Lemey, J.D. Organic aerogels-Microestructural dependence of mechanical properties in compression. *J. of Non-Cristalline Solids* 1990 125, 37-75.

[32] Boehm, H.P. Chemical identification of surface groups. *Advances in Catalysis* 1966 16, 179-274.

[33] Ferro-García, M.A., Utrera-Hidalgo, E., Rivera-Utrilla, J., Moreno-Castilla, C., Joly, J.P. Regeneration of activated carbons exhausted with chlorophenols. *Carbon* 1993 31, 857-863.

[34] Masschelein, W.J. Measurement of high ozone concentrations in gases by KI titration and monitoring by UV-absorption. *Ozone Sci. and Eng.*, 1998 20, 489-49.

[35] Bader H., Hoigné J., Determination of ozone in water by the Indigo method, *Wat. Res.*, **15**, 449-456, 1981.

[36] Flyunt, R., Leitzke, A., Mark, G., Mvula, E., Reisz, E., Schick, R. and von Sonntag C. Determination of ·OH, $O_2^{\cdot -}$ and hydroperoxide yields in ozone reactions in aqueous solution. *J. Chem. Phys. B.* 2003 107, 7242-7253.

[37] Mazza G., Bacillus Subtilis "rec assay" test with isogenic strains. *Appl. Env. Microb.*, 1982 43, 177-184.

[38] Leon y Leon C.A., Solar J.M., Calemma V., Radovic L.R. Evidence of the protonation of basal plane sites on carbon. *Carbon* 1992 30, 3797-3805.

[39] Couglhin, R.W., Ezra, F.S. Role of surface acidity in the adsorption of organic pollutants on the surface of carbon, *Env. Sci. Tech.*, 1968 2, 291-298.

[40] Bayley, P.S., Ozonation in organic chemistry, Academia Press, Inc. New York, 1982.

[41] Beltrán, F.J., Encinar, J.M. Alonso, M.A. Nitroaromatic hydrocarbon ozonation in water. 1. Single ozonation. *Ind. Eng. Chem. Res.*, 1998 37, 25-31.

[42] Pryor, W.A., Giamalva, D. Church D.F. Kinetics of ozonation. 1. Electron-deficient alkenes. *J. Am. Chem. Soc.*, 1983 105, 6858-6861.

[43] Staehelin, J., Hoigné, J. Decomposition of ozone in water in the presence of organic solutes acting as promoters and inhibitors of radical chain reactions. *Env. Sci. Tech.* 1985 19, 1206-1213.

[44] Vollhardt, Peter C., Organic Chemistry, Neil E. Schore and W.H. Freeman Company, USA, 1994.

[45] Legube, B., Sugimitsu, H., Guyon, S., Doré, M. Ozonation of naphthalene in aqueous solution II. Kinetic studies of the inicial reaction step. *Wat. Res.*, 1986 20, 209-214.

[46] Hagg, W.R., David, Yao, C.C. Rate constants for reaction of hydroxyl radicals with several drinking water contaminants. *Env. Sci. Tech.*, 1992 26, 1005-1013.

[47] Yeber, M.C., Rodríguez, J., Freer, J., Baeza, J., Durán, N., Mansilla, H.D. Toxicity abatement and biodegradability by advanced chemical oxidation. *Wat. Sci. Tech.* 1999 40, 337-342.

[48] Ma, J., Graham, N.J.D. Preliminary investigation of manganese-catalyzed ozonation for the destruction of Atrazine. *Ozone Science and Engineering,* 1997 19, 227-236.

[49] Morgan, M.E., Jenkins, R.G., Walter P.L. Inorganic constituents in American lignites, *Fuel* 1981 60, 189-193.

[50] Moreno-Castilla, C., Carrasco-Marín, F., Maldonado-Hódar, F.J., Rivera-Utrilla, J. Effects of non-oxidant and oxidant acid treatments on the surface properties of an activated carbon with very low ash content. *Carbon* 1998 36, 145-151.

[51] Elovitz, M.S., von Gunten, U. Hydroxyl radical/ozone ratios during ozonation processes. I. The Rct concept. *Ozone Science and Engineering* 1999 21, 239-245.

[52] Von Gunten, U. Ozonation of drinking water: Part II. Disinfection and by-product formation in presence of bromide, iodide or chlorine. *Water Research* 2003 37, 1469-1487.

[53] Fogler, H, Scott, L, Elements of chemical reaction engineering,", (3rd Ed.), Prentice Hall, New Jersey, 1999.

[54] Leyva-Ramos, R., Geankoplis, C.J., Diffusion in liquid filled pores of activated carbon. I: Pore volume diffusion. *Can. J. Chem. Eng.* 1994 72, 262-271.

[55] Reid, R.C., Prausnitz, J.M., Poling, B.E. Properties of gases and liquids. (4th Ed.), McGraw- Hill, New York, 1987.

[56] Wencui, L., Reichenauer, G. Frick, J. Carbon aerogels derived from cresol-resorcinol-formaldehyde for supercapacitors. *Carbon* 2002 40, 2955-2959.

[57] Moreno-Castilla, C., Maldonado-Hódar, F.J., Rivera-Utrilla, J. Rodríguez-Castellón E. Group 6 metal oxide carbon aerogels. Their synthesis, characterization and catalytic activity in the skeletal isomerization of 1-butene. *Appl. Catal. A: General* 1999 183, 345-356.

[58] Rotter, H., Landau, V., Carrera, M., Goldfarb, D. Herskowitz, M. High surface area chromia aerogel efficient catalyst and catalyst support for ethlyacetate combustion. *Appl. Catalysis B: Environmental*, 2004 47, 111-126.

[59] Maldonado-Hódar, F.J., Ferro-García, M.A., Rivera-Utrilla, J. Moreno-Castilla, C. Synthesis and textural characteristics of organic aerogels, transition-metal containing organic aerogels and their caronized derivates. *Carbon* 1999 37, 1199-1205.

[60] Maldonado-Hódar, F.J., Moreno-Castilla, C., Rivera-Utrilla, J., Hanzawa Y., Yamada, Y. Catalytic graphitizacion of carbon aerogels by transition metals. *Langmuir,* 2000 16, 4367-4373.

[61] Rodríguez-Reinoso, F., Linares-Solano, A. Microporous structure of activated carbons as revealed by adsorption methods. *Chemistry and Physics of Carbon*, vol. 21, Ed. P.L. Walker, Marcel Dekker, New York. 1989.

[62] Andreozzi, R., Insola, A., Caprio, V. D´amore M.G. The kinetics of Mn(II)-catalysed ozonation of oxalic acid in aqueous solution. *Wat. Res.*, 1992 26, 917-925.

[63] Ma, J., Graham, N.J.D. Degradation of atrazine by manganese-catalysed ozonation—influence of radical scavengers. *Water Research*, 2000 34, 3822-3828.

INDEX

A

Aβ, 130
abatement, 45, 313
absorption, 2, 8, 9, 54, 56, 61, 64, 65, 76, 77, 78, 123, 261
acceptor, 113, 283
access, 118, 283, 310
accessibility, 297
accidents, 142
accuracy, 7, 200, 255
acetaldehyde, 6
acetate, 37, 273
acetic acid, 281
acetone, 6, 56, 57
acetonitrile, ix, 97
acetophenone, 61, 64
acetylation, 102
acid, viii, ix, xi, 19, 22, 24, 25, 26, 27, 28, 29, 31, 32, 33, 34, 35, 37, 38, 42, 43, 44, 45, 46, 47, 48, 49, 50, 53, 67, 68, 73, 82, 88, 89, 93, 94, 95, 97, 99, 101, 104, 107, 108, 128, 148, 150, 151, 153, 155, 156, 158, 187, 271, 272, 275, 276, 277, 280, 281, 282, 296, 308, 310, 311, 312, 314
acidic, viii, 41, 45, 53, 54, 59, 60, 92, 93, 107, 148, 151, 153, 294
acidity, 32, 153, 302, 304, 313
ACS, 44
actin, 121
activated carbon, xii, 18, 20, 32, 35, 36, 37, 38, 39, 49, 50, 51, 153, 154, 155, 269, 270, 271, 272, 274, 275, 276, 281, 282, 283, 284, 285, 286, 287, 288, 289, 290, 291, 292, 293, 294, 295, 296, 297, 299, 300, 301, 302, 303, 308, 309, 310, 311, 312, 313, 314
activation, 35, 44, 121, 122, 124, 129, 141, 148, 216, 272, 280, 296, 297, 298, 310
activation energy, 35, 44, 280, 310
active centers, 32, 310
active site, 33, 153, 155, 297, 298, 299, 310
acute, ix, 123, 128, 236
adaptation, 211, 214, 215, 226
additives, 136, 145
adhesion, 121, 122, 125, 127, 129, 130
adiabatic, 261
adjustment, 198, 236, 237
adsorption, xii, 20, 26, 30, 31, 32, 33, 35, 36, 37, 39, 46, 49, 153, 155, 156, 157, 269, 270, 271, 273, 274, 275, 276, 296, 297, 299, 300, 304, 306, 307, 308, 310, 312, 313, 314
adsorption isotherms, 273, 275
aerobic, 18, 100, 101, 140
aerogels, xii, 38, 51, 269, 273, 296, 303, 304, 306, 307, 308, 311, 313, 314
aerosol, xi, 187, 188
aerosols, 13
afternoon, 202, 204
agar, 274
age, 236
agent, 20, 27, 136, 139, 143, 148, 157, 158, 272, 299, 300, 310
agents, 18, 101, 113, 123, 125, 299, 300
aggregation, 158
agricultural, 98, 100
agriculture, 139, 140, 144
agrochemicals, 140
aid, 94, 138
AIM, 278
air, vii, viii, x, xi, 1, 2, 3, 5, 6, 7, 11, 16, 39, 49, 111, 118, 135, 139, 140, 141, 142, 148, 149, 163, 164, 185, 186, 187, 188, 192, 195, 196, 197, 198, 199, 200, 201, 202, 217, 219, 221, 225, 226, 227, 231, 234, 235, 236, 237
air pollutant, vii, 7
air quality, 164, 185, 186, 196, 197, 198, 200, 202, 219, 221, 225, 226, 231, 235

aircraft, 14
airway epithelial cells, 124, 127, 128, 130, 131
airway hyperresponsiveness, 127, 131, 132, 133
airway inflammation, ix, x, 117, 118, 120, 122, 124, 127, 128
airway responsiveness, ix, 117, 118, 119, 120, 126
airway surface, 123
airways, 118, 122, 123, 126, 128, 130, 131
Alberta, 237
alcohol, 75, 76, 77, 98, 99, 271
alcohols, 6
aldehydes, 6, 34, 99, 152
algae, 20
algorithm, 172, 210, 213, 214, 215, 216, 221, 228, 233, 237
alkaline, 21, 54, 104, 153
alkalinity, 22
alkanes, 3, 4
alkenes, 3, 277, 313
allergen challenge, 129
allergic asthma, 125
allergic reaction, 127
allergy, 124, 141
alpha, 128, 130, 132
alternative, xi, 50, 148, 164, 195, 215, 226, 227, 231, 236, 265, 274, 312
alternatives, 27
aluminum, 47, 48
aluminum oxide, 47, 48
alveolar macrophages, 120
ambient air, vii, viii, 1, 2, 3, 5, 6, 11, 197, 201
ambiguity, 263
amiloride, 131
amine, 88
amines, 126
amino, 74, 101
amino acids, 101
ammonia, 272, 299
ammonium, 49, 119, 271, 272, 273, 299
amphoteric, 153
amplitude, 228, 264
Amsterdam, 185
anaerobic, 18, 25, 43, 100, 101, 140
analytical tools, 10
anatase, 31, 158
aniline, 74, 158
animal care, 139
animals, vii, 138, 140
anionic surfactant, 21, 44
anode, x, 135
Antarctic, xii, 239, 240, 263, 266, 267
Antarctic polar vortex, xii, 240
anthracene, 75
anthropogenic, 2, 140, 188
antibacterial, 145
antibiotics, 139, 141, 144
antigen, x, 117, 124, 125, 127, 131, 132
antigen presenting cells, 124
antioxidant, 120, 127, 128
AOC, 14
apoptosis, 129, 132
APP, 163
appendix, 219, 229, 230
application, viii, x, 2, 6, 7, 10, 12, 17, 18, 19, 20, 23, 27, 39, 100, 113, 135, 136, 138, 140, 141, 143, 144, 147, 148, 151, 152, 156, 157, 158, 159, 164, 185, 186, 214, 292
aquaculture, 139
aquaporin, 123, 131
aqueous solution, viii, x, 21, 22, 35, 36, 40, 41, 42, 43, 44, 45, 46, 47, 48, 49, 51, 53, 56, 57, 75, 94, 95, 104, 137, 138, 147, 148, 153, 271, 273, 280, 283, 311, 312, 313, 314
aqueous solutions, viii, 40, 41, 44, 45, 47, 49, 51, 53, 56, 75, 137, 138, 153, 271, 313
aqueous suspension, 152
Arctic, 239
aromatic compounds, 25, 41, 76, 270, 277, 279, 280
aromatic hydrocarbons, 3, 41, 311
aromatic rings, 76, 99, 102, 103, 275, 301
Arrhenius equation, 280
artery, 119
aseptic, 123
ash, 273, 282, 289, 314
assessment, 13, 186, 229, 236
assets, 16
assumptions, 166, 214
asthma, ix, 117, 118, 120, 121, 122, 123, 125, 126, 127, 128, 129, 130, 131, 132, 141
Athens, 233
Atlantic, 199
atmosphere, vii, x, xi, 6, 135, 188, 193, 196, 239, 240, 261, 266, 267, 272
atoms, 11, 19, 67, 70, 71, 75, 76, 82, 93, 103, 190, 192, 301
atopic asthma, 130, 132
attacks, 54, 270, 279, 299, 303
Austria, 14
automation, 202
automobiles, 196
availability, 107
averaging, 254, 255
awareness, 142
axon, 126
azo dye, 82, 83, 93, 94, 95, 158

B

B cells, 124
BAC, 50
Bacillus, 274, 313
backscattered, 8
bacteria, 18, 136, 137, 138, 139, 143, 274, 281
bacterial, 137, 140, 141
bacterial strains, 141
BAL, 119, 124
barrier, ix, 117, 123, 127, 138
barriers, 118
basement membrane, 121
basic fibroblast growth factor, 126
basicity, 32, 153, 283, 289, 297, 310
Bayesian, x, 163, 164, 167, 168, 184, 186
Bayesian analysis, x, 163, 186
Bayesian estimation, 186
bcl-2, 120, 127, 128
beating, 123
behavior, 29, 36, 48, 99, 106, 184, 240, 244, 252, 256, 275, 285, 311
Beijing, 187, 189
bending, 76
benefits, viii, x, xii, 17, 18, 135, 136, 140, 159, 197, 227, 269
benzene, 48, 49, 61, 68, 70, 74, 75, 76, 77, 82, 88, 93
beverages, 136
bicarbonate, 22
binding, 120, 126, 128
bioaccumulation, 18
biodegrability, 140
biodegradability, 270, 313
biodegradable, 18, 41, 48, 54, 148, 158, 270, 271
biodegradation, 21
biofilm formation, 145
biofilms, 139
biomass, 98, 99, 107
bioremediation, 144
biosphere, 6
bipolar, 137
black-box, 198
bleaching, 98, 99, 139
bleeding, 119
blood, 122, 140, 141
blood stream, 141
blot, 124
boiling, 3, 10, 295
bonding, 93
bonds, 54, 58, 59, 71, 72, 74, 76, 98, 278
bounds, 209, 218
bovine, 128
Brazil, 147, 161
breakdown, 264
breathing, 118, 141
British Columbia, 235
bronchial epithelial cells, 118, 120, 121, 122, 124, 125, 127, 128, 129, 130, 132
bronchial epithelium, ix, 117, 118, 121, 127
bronchitis, 123, 124
bronchoalveolar lavage, 119, 131
bronchoconstriction, 120
Brussels, 233
bubble, 144
buffer, 11, 271
buildings, 138
burn, 173
burning, 140, 186
buses, 196, 205
butyl ether, 34
by-products, viii, xii, 17, 25, 34, 38, 48, 100, 113, 136, 148, 269, 271, 281, 291, 308, 311

C

cadherins, 121
calcitonin, 125, 126, 133
calcium, 123
calving, 138
CAM, 122
Canada, 237
capacity, xii, 10, 21, 23, 26, 27, 32, 36, 37, 39, 56, 124, 126, 140, 141, 153, 156, 269, 271, 274, 275, 276, 288, 290, 296, 301, 311
capillary, 57
carbon atoms, 11, 103
carbon dioxide, 10, 273
carbon materials, xii, 49, 269, 296
carbon monoxide, 158
carbonates, 26, 32, 37
carbonyl groups, 298
carboxylic, 34, 46, 150, 152, 298, 302
carboxylic acids, 34, 46, 150, 152
carboxylic groups, 298
carpets, 140
case study, 95, 228
cast, 273
catalase, 120, 128
catalysis, 27, 30, 33, 45, 48, 158
catalyst, 24, 25, 26, 27, 28, 29, 30, 31, 32, 33, 34, 35, 36, 37, 38, 39, 43, 45, 46, 47, 48, 49, 50, 112, 113, 148, 152, 153, 155, 156, 157, 158, 270, 271, 273, 274, 282, 294, 298, 311, 314
catalytic activity, 24, 30, 31, 32, 33, 35, 36, 37, 38, 148, 153, 158, 283, 287, 289, 300, 301, 303, 310
catalytic effect, 28, 155, 289

catalytic properties, 153, 158, 294
catalytic system, viii, x, 17, 19, 147, 148
catechol, 43
cation, 30, 153, 155
causality, 166, 167, 169, 170, 178, 182
cell, 6, 98, 101, 118, 119, 121, 122, 123, 124, 125, 126, 128, 129, 130, 131, 132, 138, 141, 308
cell adhesion, 122, 129
cell cycle, 121
cell death, 121, 125
cell differentiation, 121
cell growth, 101
cell line, 124, 125
cellulose, 98, 99, 108
cereals, 98
cerium, 148, 149
CH4, 25
channels, 123, 131
chemical agents, 140
chemical approach, 21
chemical oxidation, 27, 144, 313
chemical properties, 301
chemical reactions, 2, 196
chemical stability, 32, 156
chemical structures, 54
chemicals, 98, 99, 108, 139
chemiluminescence, 190
chemisorption, 156
chemoattractant, 122
chemokine, 130
chemokines, 118, 122, 130, 131
chemometrics, 7, 14
chemotaxis, 131, 132
children, 234, 236
Chile, 235
China, 53, 56, 117, 140, 187, 193
chloride, 31, 119, 123
chlorination, 21
chlorine, 20, 39, 136, 138, 139, 143, 314
chlorobenzene, 33, 47, 282
chlorophenol, 26, 29, 33, 45, 47
chromatograms, 10, 11, 83, 88
chromatography, 6, 7, 13, 15, 42, 54, 57, 74, 189, 311
chronic obstructive pulmonary disease, 121, 127, 132
cigarette smoke, 128
cilia, 123
Cincinnati, 143
circulation, xi, xii, 122, 141, 183, 239, 240, 241, 242, 244, 257, 261, 262, 263, 264, 267, 268
classes, 165
classical, 150, 164, 212
classification, 211, 229, 244
clay, 31
clean technology, x, 135
cleaning, 42, 139, 140
cleavage, 98, 99
Clostridium botulinum, 137
clouds, 267
cluster analysis, 233
clusters, 158
CO2, viii, 6, 17, 18, 77, 82, 83, 93, 150, 280, 299, 304
coagulation, 20, 54, 93, 94, 144
coagulation process, 20
coastal areas, 192
cobalt, 46, 158, 273
coffee, 188
coke, xii, 269, 272, 296, 297, 298, 299, 310, 312
colors, 57
combined effect, 95, 210
combustion, 2, 314
communication, 122
communities, 184
community, 39
complex systems, viii, 2
complexity, 20, 36, 39, 186, 289, 299
compliance, 13, 119
components, vii, viii, 1, 3, 8, 53, 98, 100, 105, 107, 109, 123, 126, 138, 165, 167, 169, 170, 196, 200, 203, 266, 289, 298, 299
composition, 6, 22, 54, 57, 98, 123, 125, 159
compounds, xi, 4, 5, 6, 11, 12, 18, 20, 21, 24, 27, 30, 31, 35, 36, 38, 39, 40, 44, 47, 49, 54, 74, 76, 77, 78, 82, 93, 98, 99, 107, 111, 112, 118, 138, 139, 140, 143, 144, 148, 157, 187, 188, 270, 271, 275, 276, 280, 281, 289, 290, 308, 311
computing, xi, 195, 199
concentration, xii, 5, 6, 7, 12, 35, 38, 54, 56, 57, 65, 67, 73, 74, 75, 82, 83, 93, 100, 111, 113, 119, 138, 142, 149, 151, 164, 189, 190, 191, 192, 193, 196, 202, 203, 204, 207, 209, 221, 226, 231, 234, 269, 270, 271, 272, 273, 274, 275, 276, 279, 282, 284, 285, 286, 287, 288, 289, 290, 291, 292, 295, 296, 298, 300, 301, 302, 303, 305, 310, 311
concrete, 138
conductance, 131
conduction, 113
conductive, 123
conductivity, viii, 53, 66, 67, 93, 272
confidence, 213, 217, 225
configuration, 3, 11, 20
conformity, 250, 255
Congress, 46, 94, 143
conjugation, 71, 75

connective tissue, 126
consensus, 159
conservation, 137
constraints, 228
construction, 142
consumption, 22, 33, 37, 111, 145, 147, 148
contact time, 137, 138, 295
contaminant, 18, 29, 30, 31, 34, 38, 39, 159, 274, 281, 310
contaminants, x, xii, 18, 23, 31, 32, 34, 35, 40, 139, 143, 147, 269, 270, 271, 276, 308, 310, 311, 313
contaminated soils, 144
contamination, 18, 100, 137
contractions, 137
control, 13, 43, 101, 119, 124, 132, 190, 197, 209, 215, 233, 235
control group, 119
convergence, 213, 214
conversion, x, 147
cooking, 140
cooling, 10, 140
COPD, 118, 121, 123, 124
corn, 99, 102, 103, 104, 107, 108
corona discharge, x, 135
correlation, 10, 166, 167, 169, 170, 182, 190, 202, 204, 205, 215, 221, 240
correlation analysis, 202
correlation coefficient, 221
correlations, 207, 241, 262
cosine, 203
cost saving, 136
cost-effective, 24
costimulatory molecules, 124
costs, 20, 143
coughing, 141
cracking, 101
Crete, 11
crops, 6, 98, 101, 140
cross-linking, 98
cross-validation, 216, 229
crude oil, 296
cryogenic, 10
crystallites, 283
CSF, 122, 125, 130
culture, 120, 132
CXC, 130
cyanide, 43
cyanosis, 140
cycles, xii, 5, 35, 38, 239, 249, 250, 251, 253, 254, 255, 257, 258, 260, 261, 262, 263, 264, 265, 306
cyclohexane, 48, 189, 190, 193
cystic fibrosis, 123
cysts, 138, 143
cytokine, 124
cytokines, 118, 122, 126, 127, 130
cytoprotective, 128
cytoskeleton, 121
cytosolic, 130

D

dairy, 136
danger, 140
data analysis, 12, 14, 185, 235
data processing, 13
data set, 165, 180, 226, 227, 235, 237
database, 192, 216
dating, 152
death, 124
decay, xi, 187, 188, 189, 190, 191, 192, 193, 258, 284, 287
decisions, 202
decomposition, viii, ix, 17, 18, 20, 22, 23, 29, 30, 32, 34, 35, 36, 37, 38, 40, 41, 43, 44, 46, 47, 48, 49, 50, 54, 97, 111, 136, 138, 148, 149, 151, 152, 153, 154, 155, 156, 157, 158, 278, 284, 292, 293, 297, 301, 302, 303, 306, 310
decompression, 98
deconvolution, 298, 300, 305
defects, ix, 117, 118, 127, 129
defense, 118, 120
definition, 10, 165, 166, 211, 221, 229, 287
degradation, viii, ix, x, 17, 18, 21, 22, 23, 24, 25, 30, 32, 36, 42, 43, 46, 47, 48, 50, 53, 54, 57, 59, 64, 65, 69, 70, 72, 73, 74, 75, 76, 77, 78, 82, 83, 88, 92, 93, 94, 97, 98, 99, 100, 101, 104, 105, 106, 107, 108, 111, 112, 113, 139, 144, 147, 148, 149, 150, 151, 152, 153, 155, 156, 158, 270, 281, 282, 284, 285, 288, 289, 295, 296, 300, 301, 303, 308, 310, 312
degradation mechanism, 69
degradation pathway, ix, 53, 112
degradation process, viii, 53, 59, 82, 92, 101, 105, 106, 107
degradation rate, 73, 112, 282, 284, 285, 289, 295, 296, 300, 301, 310
degrading, 54, 280
Degussa, 24, 25, 26
demand, 17, 18, 139, 196, 198, 236
demineralized, 286, 289
dendritic cell, 124
denitrifying, 18
density, 64, 65, 92, 150, 166, 167, 168, 169, 170, 171, 201, 208, 277, 278, 279, 294, 295, 301, 302, 310
density values, 278

dental plaque, 141
dentistry, 141
deodorizing, 137
derivatives, viii, ix, 27, 53, 54, 88, 89, 99, 212, 213, 214
dermatitis, 141
desorption, 2, 11, 155
destruction, xii, 45, 46, 47, 99, 112, 136, 205, 240, 264, 314
detachment, 129
detection, vii, 1, 4, 6, 7, 9, 10, 24, 25, 142, 186, 210, 215, 216, 217, 219, 220, 229, 232, 235, 237, 273
detoxification, 113, 139, 144
Deviance Information Criterion, 172, 182
deviation, 258
diaphragm, 11
differentiation, 10, 121, 124, 125, 126, 129
diffraction, 305
diffusion, 28, 32, 137, 157, 283, 286, 287, 295, 296, 310, 314
diffusivity, 295
digestion, 25, 43, 120, 127
dimensionality, 164, 185
diseases, 123, 127, 236
disinfection, viii, 17, 18, 45, 138, 139, 141, 143, 144, 291
disorder, 125, 127
dispersion, 35, 198, 199, 275, 305, 310
distilled water, 137, 272
distribution, xi, 4, 7, 13, 43, 101, 102, 125, 132, 133, 165, 166, 167, 168, 169, 170, 171, 172, 173, 177, 178, 180, 227, 239, 255, 256, 261, 278
diuron, 26, 152
DNA, 101, 274, 281
DNA damage, 274, 281
donor, 283
doped, xii, 42, 51, 269, 273, 296, 303, 304, 306, 307, 308, 311, 313
double bonds, 40, 54, 70, 75, 278
drinking, viii, xii, 17, 19, 20, 39, 40, 41, 47, 138, 139, 140, 147, 158, 159, 269, 313, 314
drinking water, viii, xii, 19, 20, 39, 40, 41, 47, 140, 147, 158, 159, 269, 313, 314
drying, 303
duration, xii, 205, 209, 215, 216, 219, 221, 225, 239, 240, 241, 244, 249, 255, 257, 258, 261, 262, 265
dust, 201
dyeing, 93, 94
dyes, viii, 42, 44, 45, 53, 54, 56, 57, 58, 59, 60, 61, 65, 66, 67, 68, 69, 70, 71, 72, 73, 74, 75, 76, 77, 78, 81, 82, 83, 87, 88, 92, 93, 94, 95
dynamic systems, 235

E

E.coli, 137, 138
earth, 98, 205
E-cadherin, 121
ECM, 126
economic resources, xii, 269
ecosystems, 196
edema, 120
effluent, 21, 24, 25, 26, 28, 42, 54, 94, 139, 276
effluents, 17, 18, 20, 23, 24, 27, 36, 42, 48, 49, 93, 94, 95, 139, 143, 270
egg, 137
elaboration, 70
elastin, 126
elderly, 164
electric field, 6
electrical conductivity, 66, 93
electrolysis, x, 135
electrolytes, 94
electron, 65, 92, 113, 150, 152, 156, 157, 275, 277, 278, 279, 280, 283, 284, 294, 301, 310
electron density, 150, 277, 278
electrons, 5, 25, 75, 113, 152, 201, 275, 301, 310
electron-transfer, 157
electrostatic interactions, 49, 275, 310, 312
emission, vii, 1, 2, 4, 9, 13, 198, 204
employment, 138
endocrine, 127, 140, 144
endothelin-1, 126
energy, 98, 111, 113, 136, 201, 205, 227, 228, 279
energy consumption, 111
engagement, 124
environment, 11, 13, 27, 95, 98, 101, 113, 118, 123, 136, 137, 140, 159, 211, 236, 270
environmental effects, xi, 98, 195, 197, 225
environmental factors, 136
environmental impact, 18
environmental issues, 184
environmental protection, 198
Environmental Protection Agency, 1, 140, 270
enzymatic, 100
enzymes, 120, 136, 138
eosinophils, 119, 120, 122, 131
EPA, vii, 1, 2, 3, 6, 7, 13, 14, 15, 140, 143, 145, 270
epidermal growth factor, 120, 128, 131, 132
epithelia, 118, 121, 122, 126, 128, 132
epithelial cells, x, 117, 118, 120, 121, 122, 123, 124, 125, 126, 127, 128, 129, 130, 131, 132
epithelial lining fluid, 120
epithelium, ix, x, 117, 118, 120, 121, 122, 123, 124, 125, 126, 127, 129, 131
ERA, 267

Escherichia coli, 143
ester, 77, 88, 89
esters, 6, 103
estimating, 212, 231
estimator, 216, 227
estimators, 214, 228
ethylene, 137
Europe, 145, 147, 197, 233, 235
European Commission, 233
evening, 208
evolution, x, xii, 105, 106, 107, 135, 224, 239, 241, 242, 244, 249, 250, 253, 254, 255, 256, 258, 260, 261, 262, 263, 265, 274, 280, 281, 308
examinations, 144
exceedance, 214
exclusion, 119
experimental condition, 27, 32, 100, 107, 140, 270, 275, 284
exposure, ix, xi, 117, 118, 119, 121, 123, 125, 127, 128, 129, 131, 133, 141, 142, 184, 185, 195, 196, 198, 216, 235, 272, 290, 291, 311
extracellular matrix, 120, 121
extraction, 18, 99
extrapolation, 138

F

fabrication, 11
failure, viii, 17, 18, 237
FAK, 121
false alarms, 219, 229, 232
family, 6, 121, 124, 270
FDA, 136, 142
February, 242, 244, 246
Federal Register, 145
ferrous ion, 45, 54
fiber, 38, 51
fibers, 98
fibroblast, 123, 126
fibroblast growth factor, 126
fibroblasts, 122, 125, 126
fibronectin, 125, 126
fibrosis, 123, 131
filters, vii, 198, 199, 237
filtration, 37, 54
finance, 236
financial support, 184, 233
fire, 9
fish, 136
fixation, 120
flame ionization detector (FID), vii, 1, 3, 8, 12, 189
flexibility, 200, 214
flocculation, 20
flood, 237
flow, 11, 24, 25, 26, 28, 29, 31, 33, 34, 35, 37, 38, 56, 118, 266, 272, 273, 309
flow rate, 56
fluctuations, xi, xii, 198, 239, 240, 242, 258, 263
fluid, 51, 119, 120, 123, 131
fluid transport, 131
focusing, viii, ix, 17, 117, 118, 125
food, 10, 18, 118, 136, 137, 138, 139, 143, 144, 145
food additives, 145
Food and Drug Administration, 145
food industry, 136, 139, 144
food processing industry, 136
food products, 136
forecasting, 195, 198, 199, 232, 233, 235, 236, 237
formaldehyde, 273, 303, 313
fossil, 186
Fourier, vii, 1, 6, 9
Fourier transformation, 9
fragmentation, 6
France, xi, 20, 147, 195, 197, 205, 237
free radical, 21, 23, 30, 41, 152, 156, 282
freedom, 167, 170
freezing, 6
Friday, 207
fruits, 137, 138
FTIR, vii, 1, 6, 7, 9, 13, 102, 103
fuel, 186
fugitive, vii, 1, 4, 9
Fulvic acid, 34, 35
fungal, 43, 137
fungi, 26, 136
fungicide, 26
fungicides, 139

G

Gamma, 167, 168, 169, 170, 171, 180
gas, x, xi, 8, 13, 15, 19, 24, 26, 30, 35, 37, 56, 72, 135, 136, 137, 140, 142, 144, 145, 187, 189, 190, 193, 271, 273, 294
gas chromatograph, 13, 15, 189
gas phase, 30, 137, 140, 145, 193, 294
gases, vii, 1, 2, 7, 120, 187, 188, 313, 314
Gaussian, 212, 217, 227, 278
gel, 152, 273, 304
gelation, 304
gels, 303
gene, 120, 121, 123, 125, 126, 128, 129, 132, 133
gene expression, 120, 128, 129
generation, xii, 21, 22, 23, 25, 30, 38, 148, 149, 150, 152, 156, 158, 239, 241, 270, 296, 303, 306, 308
generators, 142

genotoxic, 50, 271, 281, 312
Germany, 57, 159
germination, 26
Gibbs, 172, 173, 186
gland, 123
glass, 273
glow discharge, 6
glucocorticosteroids, 129
glutathione peroxidase, 120, 128
glycoproteins, 121, 123
GM-CSF, 122, 125, 130, 131
goals, xi, 195, 198
goodness of fit, 221
GPC, 101, 102
grasses, 98, 101
grassland, 6
grazing, 6
Greece, 11, 135
groundwater, ix, 40, 41, 97, 100, 101
groups, ix, 32, 36, 37, 53, 61, 64, 65, 71, 72, 73, 75, 77, 78, 92, 93, 94, 95, 98, 101, 104, 138, 153, 158, 196, 270, 275, 276, 277, 279, 280, 281, 282, 283, 284, 288, 289, 290, 294, 296, 298, 299, 300, 301, 302, 303, 304, 305, 308, 310, 313
growth, 100, 101, 118, 120, 122, 124, 126, 128, 129, 130, 131, 132, 137, 234, 263
growth factor, 118, 120, 122, 126, 128, 130, 131, 132
Guangzhou, 53
guidance, 142, 197, 226
guidelines, 136, 145, 219

H

half-life, 137, 138, 191, 193
halogen, 4
handling, 144
hands, 98
hardwoods, 98
harmful effects, vii, 1
harvest, 140
harvesting, 6
hazardous materials, 41
hazardous substances, 139
hazards, xi, 195, 197, 225
healing, 141
health, vii, xi, xii, 1, 2, 13, 140, 141, 142, 164, 184, 195, 196, 197, 225, 236, 269
health effects, 2, 13, 196, 236
health problems, 164
heart, 10, 196
heat, 120, 128, 272
heating, 137, 140, 261, 272
heating rate, 272
heavy metals, 27
height, xii, 239, 240, 241, 242, 243, 244, 245, 246, 247, 248, 249, 250, 252, 253, 255, 257, 258, 260, 262, 265, 266, 276
hemicellulose, 98
hemidesmosome, 121
hemisphere, 262, 266
hemp, 98
herbicide, ix, xii, 94, 97, 98, 101, 111, 112, 113, 139, 149, 269, 311
heterocycles, 302
heterogeneity, 98, 201
heterogeneous, viii, x, 17, 19, 27, 30, 32, 36, 37, 39, 43, 46, 47, 50, 95, 100, 147, 148, 151, 152, 153, 154, 155, 156, 157, 158, 283, 284, 285, 286, 287, 289, 298, 299, 310, 311
heterogeneous catalysis, 27
heterogeneous systems, 156
heteroscedasticity, 185
high pressure, 98
high resolution, 3
high temperature, 2, 35, 219
higher quality, 139
high-performance liquid chromatography, 311
histogram, 217, 223
histological, 93
HLA, 124, 125, 132
Holland, 185
homeostasis, 122, 127
horizon, 159
hormone, 126
hospital, 144, 236
hospitality, 141
hospitalization, 236
hospitals, 139
host, 138
hot water, 104
human, vii, ix, xi, xii, 1, 13, 97, 100, 101, 117, 120, 121, 124, 125, 126, 128, 129, 130, 131, 139, 141, 145, 159, 195, 196, 197, 225, 269
human exposure, 141
humans, 138, 140
humic acid, 35, 43
Humic acid, 26, 29, 35
humic substances, 31, 45, 46, 148
humidity, 137, 202, 204, 205, 211
hybrid, 148, 237
hybridization, 301
hydro, 3, 4, 18, 37, 41, 148, 196, 277, 311
hydrocarbon, 4, 76, 77, 144, 233, 313
hydrocarbons, 4, 18, 37, 196

hydrogen, 18, 21, 22, 23, 40, 41, 44, 47, 48, 49, 54, 75, 76, 95, 99, 100, 109, 111, 113, 139, 153, 284, 311
hydrogen atoms, 76
hydrogen peroxide, 18, 21, 22, 23, 40, 41, 44, 47, 48, 49, 54, 95, 99, 100, 109, 111, 113, 139, 153, 284, 311
hydrolysis, 49, 101
hydrolyzed, 95
hydrophilic, 148
hydrophobic, 308
hydroponics, 140, 145
hydroquinone, 34, 99
hydrosphere, 100
hydroxide, 40, 94
hydroxyl, viii, xii, 17, 18, 21, 22, 24, 27, 30, 31, 36, 40, 54, 75, 76, 113, 148, 149, 150, 151, 152, 153, 154, 155, 156, 157, 158, 269, 278, 284, 308, 313
hydroxyl groups, 153, 158
hydroxylation, 49
hygiene, 137
hyperbolic, 211, 216
hyperplasia, 123
hypersensitivity, 125
hypothesis, 105, 118, 124, 154, 166, 167, 301

I

ICAM, 121, 122, 127, 129, 130, 132
ice, 147
identification, ix, 12, 24, 25, 26, 95, 97, 111, 196, 210, 226, 235, 240, 262, 289, 311, 313
IFN, 125
IgE, 118
IGF, 126
IGF-1, 126
IL-1, 122, 127
IL-6, 122, 131
IL-8, 122, 127, 130, 131
illumination, 8
images, 8, 9
imaging, vii, 1, 8, 13, 15
imaging techniques, vii, 8, 13
imidazolinone, ix, 97, 98, 101, 113
immune response, 122, 124
immune system, 141
immunity, 118
immunocompromised, 140
immunofluorescence, 124
immunomodulatory, 126
implementation, 16, 36, 39, 136, 140, 198, 214, 225
impregnation, 31
IMS, 195
in situ, 11, 38, 100, 152
in vitro, 130, 132, 138, 141
in vivo, 118, 138
inactivation, 42, 136, 138, 141, 143, 311
inclusion, 202
income, 99
independence, 10, 168
indicators, 140, 185, 199, 217, 219, 221, 222, 229, 230
indices, 12, 219, 266, 267
induction, 128, 302
industrial, vii, viii, 2, 4, 5, 7, 9, 13, 17, 20, 23, 37, 94, 136, 137, 139, 147, 202, 270, 303, 311
industrial application, 303
industry, x, 8, 36, 54, 99, 135, 136, 137, 139, 143, 144, 196, 270, 296
infection, 123, 124, 125, 132
infections, 141
inflammation, ix, x, 117, 118, 120, 122, 123, 124, 126, 127, 128, 130, 131
inflammatory, ix, x, 117, 118, 120, 121, 122, 124, 125, 127, 130, 132
inflammatory cells, 118, 121, 122, 125
inflammatory response, ix, 117, 120, 125, 130
inflation, 185
infrared, vii, 1, 6, 8, 9, 54, 57, 75, 76, 77, 78, 88, 93, 205
infrared light, 6, 8
infrared spectroscopy, vii, 1, 9
inhibition, 101, 125, 126, 131
inhibitors, 41, 100, 143, 313
initial state, 221, 226
initiation, 40, 122, 142, 156
injury, ix, 117, 120, 122, 123, 125, 126, 127, 128, 131
innervation, x, 117
innovation, 193
inorganic, 18, 20, 21, 32, 38, 40, 41, 59, 60, 67, 78, 83, 92, 93, 143, 311
insight, 219
inspection, 183
instruments, vii, 1, 12, 189
insurance, 236
integration, 103
integrin, 120, 121, 125, 129
integrins, 121, 122, 125
integrity, ix, x, 117, 118, 121, 123, 127
intensity, xii, 64, 202, 205, 211, 240, 241, 242, 261, 263
interaction, x, xii, 25, 31, 113, 122, 124, 125, 135, 149, 153, 156, 157, 269, 290, 298, 301, 302, 308
interactions, 40, 45, 49, 124, 130, 132, 210, 271, 274, 275, 310, 312

intercellular adhesion molecule, 121
interface, 65
interference, 7, 101, 142
interferon, 122, 130, 132, 141
interleukin, 129, 130, 141
interleukin-1, 129, 130
interleukins, 122, 130
interpretation, 7
interval, 13, 241, 244, 246, 247, 248, 253
inventories, 13
iodine, 142
ionic, 18, 21, 30, 40, 68, 273
ionization, 5, 189
ions, ix, 5, 22, 40, 43, 45, 53, 54, 66, 68, 70, 71, 73, 78, 92, 93, 123, 142, 148, 149, 150, 151, 153, 155, 158, 301, 304, 310
iron, 20, 45, 46, 48
irradiation, ix, 44, 95, 97, 99, 104, 105, 106, 111, 150, 158
irritation, 141
Islam, 233
isolation, 263
isoleucine, 101
isomerization, 314
isoprene, 6
isotherms, 271, 272, 273, 275
Israel, 129
Italy, 97, 100, 108

J

January, xii, 172, 174, 175, 235, 239, 242, 244, 246, 248, 249, 250, 251, 252, 253, 255, 256, 257, 258, 259, 260, 261, 262, 263, 264, 265
Japan, 43, 56, 57
Jordan, 14
Jun, 45, 245, 246, 247, 249, 251, 252, 253, 254
Jung, 43, 158, 161

K

Kalman filter, 198, 199, 204, 212, 214, 236, 237
keratinocytes, 124
ketones, 6
killing, 141
kinases, 130, 132
kinetic constants, 38, 113
kinetic model, 41, 42, 287, 288, 292, 306
kinetic parameters, 111, 270
kinetics, ix, xi, 22, 25, 28, 39, 44, 47, 48, 65, 92, 94, 95, 97, 107, 110, 124, 138, 187, 188, 271, 276, 297, 300, 306, 307, 314
KOH, 272, 296, 297, 298, 310

L

labor, vii, 1, 8
labor-intensive, 8
Lagrangian, 228
laminin, 126
landfill, 140, 144
Langmuir, 275, 312, 314
large-scale, 20, 36, 39
laser, 8, 9, 10
laundry, 44
law, 6, 15, 21, 140
leachate, 140, 144
leaching, 32, 33, 35
lead, 9, 124, 153, 196, 226, 231, 290
leaks, vii, 1, 8, 9, 13
learning, 216, 233
legislation, 20, 141, 198
legislative, 197
lens, 9
lettuce, 143
leucine, 101
leukocyte, ix, 117, 118, 122, 130, 132
leukocytes, 121, 122, 127, 130
Lewis acids, 32
lifetime, 192, 193
ligand, 36, 48, 121, 126, 130
ligands, 121, 124, 128
lignin, ix, 42, 97, 98, 99, 100, 101, 102, 103, 104, 105, 106, 107, 108, 109
likelihood, 164, 168, 169, 170, 171, 172, 185, 202
limitation, 11
limitations, vii, 1, 138, 210, 219, 286, 287, 295, 310
linear, 23, 41, 65, 109, 111, 119, 196, 198, 199, 201, 210, 211, 214, 215, 216, 220, 225, 226, 227, 228, 229, 230, 231, 281, 289
linear function, 210
linear model, 198, 214, 227, 228, 229, 230
linear programming, 228
linear regression, 119, 215, 220, 228
linkage, 98
links, 68, 69
lipid, 118, 141
lipids, 136
lipoproteins, 141
liquid chromatography, 42, 271, 311
liquid nitrogen, 11
liquid phase, 30
liquids, 314
liquor, 75
Listeria monocytogenes, 137, 143

liver, 93
localization, 121
location, 9, 244, 255, 278
London, 95, 235, 236
long distance, 9
losses, 31, 137, 189
Louisiana, 14, 15
Louisiana State University, 15
low molecular weight, 44, 105
lumen, 120, 122
lung, 119, 120, 123, 126, 127, 131, 133, 196
lungs, 13, 119
lymphocytes, 122, 124

M

machine learning, 237
machinery, 124
macromolecules, 123
macrophages, 119, 124
macropores, 273, 296, 299, 304
magnetic, 267
maintenance, 127, 159
major histocompatibility complex, 132
malic, 99
malignant, 141
malignant cells, 141
management, vii, 1, 2, 100, 136, 198
manganese, 20, 45, 46, 48, 155, 157, 158, 273, 314
Manganese, 45
manifold, 112
manufacturing, 94
MAPK, 122, 125
mapping, 185
marine environment, 235
maritime, 199
market, 140, 186
Markov, x, 163, 164, 186
Markov chain, 164
mass spectrometry, 13
mass transfer, 276, 310
mast cells, 122
matrix, 22, 120, 121, 125, 165, 166, 167, 171, 200, 212, 213, 214, 304, 306
maturation, 137
MCP, 122, 130
MCP-1, 122
measurement, vii, 1, 3, 5, 6, 7, 9, 13, 119, 142, 145, 172, 173, 201, 212, 214, 234, 274, 281, 291
measures, 142, 163, 164, 183, 186, 197, 221
meat, 136, 137
mechanical properties, 313
media, 41, 144
mediators, ix, x, 117, 118, 128, 129
Mediterranean, 99
memory, 186
Merck, 271, 273, 282, 283, 286
mercury, 273, 299
mesoporous materials, 30, 157
metabolic, 10
metabolism, 141
metabolites, 139
metal ions, 148, 149, 150, 151, 304
metal oxide, 30, 32, 33, 34, 46, 100, 147, 148, 151, 152, 153, 155, 156, 157, 158, 159, 314
metals, x, 18, 30, 31, 32, 34, 35, 37, 39, 46, 137, 147, 148, 151, 152, 153, 156, 157, 282, 289, 296, 298, 299, 304, 305, 311, 314
meteorological, vii, 1, 4, 5, 13, 196, 198, 199, 200, 201, 202, 204, 205, 209, 210, 211, 215, 216, 220, 226, 228, 229, 233, 236, 237
methane, 4, 5, 9, 13
methanol, ix, 6, 37, 97, 273
metropolitan area, 226, 228
Mexico, x, 163, 164, 165, 172, 173, 182, 183, 184, 185, 235
Mexico City, x, 163, 164, 172, 173, 182, 183, 184, 185
MHC, 124, 131
mice, 118, 119, 121, 126, 129
microbes, 138
microbial, 98, 100, 101, 136, 137, 138, 141, 143
microbiota, 143
microelectronics, 139, 144
microenvironment, 127
microflora, 137, 139
microorganism, 138, 140, 141
microorganisms, 18, 20, 100, 138, 139, 141, 143
micropollutants, 290
microscope, 119
microscopy, 120
migration, 121, 123, 126, 127
mimicking, ix, 117
mineralization, ix, xii, 24, 26, 27, 31, 35, 38, 57, 66, 93, 97, 100, 101, 113, 140, 148, 149, 150, 151, 152, 153, 158, 269, 280, 308, 310
mineralized, 150
minerals, 158
mirror, 7
misidentified, 6
missions, 226
MIT, 235
mixing, 142, 198, 229
modeling, 13, 45, 164, 185, 236
models, x, xi, 120, 129, 163, 164, 165, 166, 167, 168, 169, 170, 171, 172, 183, 184, 185, 186, 195,

196, 197, 198, 199, 225, 226, 229, 230, 231, 233, 234, 235, 236, 237, 241, 261
modulation, 10, 11, 126, 129, 132, 255, 256, 261, 263, 266, 267, 268
modules, 209
modulus, 286
moisture content, 137
molar ratios, 273
molar volume, 295
molecular mass, viii, 6, 53, 57, 88, 92, 93
molecular oxygen, 261
molecular structure, 70, 75, 92
molecular weight, ix, x, 6, 44, 97, 101, 102, 105, 106, 107, 135, 148, 280, 295
molecular weight distribution, 101
molecules, ix, x, 5, 6, 31, 43, 53, 59, 60, 64, 65, 66, 70, 71, 72, 74, 75, 76, 77, 78, 82, 83, 88, 92, 93, 97, 105, 111, 113, 117, 120, 121, 122, 123, 124, 125, 150, 152, 156, 157, 158, 188, 270, 275, 278, 283, 284, 301
monocyte, 124
monocytes, 122
monomeric, 98
monomers, 98
Monte Carlo, x, 163, 164, 186
Moon, 128, 159
morbidity, 196
morning, 192, 208
morphogenesis, 121
morphological, 241
mortality, 138, 184, 185, 196, 234, 235
motion, 261
mouse, 128
mouse model, 128
movement, 122, 123
MTBE, 47
mucin, 123
mucosa, 123
mucus, 118, 120, 123, 127, 131
mucus hypersecretion, 123
multidimensional, 10
multivariate, 164, 165, 185, 210
muscle, 118, 126
muscle cells, 126
mutations, 138
myofibroblasts, 122, 125, 126, 132

N

naphthalene, 37, 68, 75, 76, 279, 308, 313
national, 13, 185, 198
natural, 2, 7, 8, 20, 33, 35, 38, 47, 48, 111, 113, 157, 158, 159, 188, 192, 311
natural resources, 159
neoplastic, 125
nerve, x, 117, 126
Netherlands, 236
network, vii, x, xi, 1, 3, 163, 164, 172, 184, 185, 195, 197, 199, 209, 216, 225, 233, 234, 235
neural network, xi, 164, 185, 195, 197, 199, 209, 225, 233, 234
neural networks, 164, 199, 233, 234
neuroendocrine cells, 126
neurogenic, 118, 126
neurons, 216
neuropeptides, 125, 127, 133
neurosecretory, 126
neurotransmitter, 126
neutralization, 273, 282
neutrophils, 119, 120, 122, 130
New England, 234
New York, 39, 43, 48, 93, 184, 235, 313, 314
Newton, 221, 233
nicotinic acid, ix, 97
Nielsen, 194
NiO, 33
nitrate, 18, 25, 31, 36, 49
nitric acid, 148
nitric oxide, 128
nitric oxide synthase, 128
nitrobenzene, 48, 155, 158
nitrogen, xii, 2, 10, 11, 18, 40, 42, 54, 56, 70, 72, 75, 77, 93, 196, 201, 207, 269, 296, 301, 302
nitrogen dioxide, 207
nitrogen gas, 56, 72
NMR, 101, 103, 104
noise, 200, 210, 212, 214, 215, 227, 228, 229
nonlinear, xi, 195, 196, 197, 198, 199, 209, 210, 211, 212, 214, 219, 221, 225, 226, 227, 228, 229, 230, 231, 233, 234, 235, 237
nonlinear dynamic systems, 210
nonlinear systems, 214, 235
nonparametric, 233
nonsmokers, 235
normal, 8, 121, 122, 125, 126, 204
norms, 227
Northeast, 172
Northern Hemisphere, 267
Nrf2, 123
NTS, 271, 272, 273, 275, 276, 277, 279, 280, 281, 282, 283, 284, 285, 286, 287, 288, 289, 294, 295, 296, 297, 298, 299, 300, 301, 302, 303, 308, 309, 310
nuclear power plant, 44
nucleic acid, 143
nutrients, 145

O

observations, 32, 164, 199, 202, 212, 240, 241, 262, 266, 267, 278, 294
obstruction, ix, 117, 120, 123
OCT, 254
odors, 141
Ohio, 143
oil, ix, 10, 97, 98, 99, 100, 101, 108, 109, 110, 111, 296
olefins, 277
olive, ix, 97, 98, 99, 100, 108, 109, 110, 111
olive oil, ix, 97, 98, 99, 100, 108, 109, 110, 111
one dimension, viii, 2, 11, 12, 13
online, vii, 1, 2, 3, 4, 5, 6, 13, 198, 199, 226, 234, 235
operator, 8, 9
optical, 8
optimization, xi, 32, 36, 39, 195, 210, 214, 221, 226, 227, 228, 229, 230, 231, 236
optimization method, 236
oral, 141
organic, viii, ix, x, xi, xii, 5, 17, 18, 19, 20, 21, 22, 23, 26, 27, 30, 31, 32, 34, 36, 39, 40, 41, 42, 43, 44, 47, 48, 49, 50, 53, 59, 60, 65, 73, 75, 77, 78, 82, 83, 88, 92, 93, 94, 95, 100, 101, 109, 113, 138, 139, 140, 143, 144, 147, 148, 149, 150, 151, 152, 153, 155, 156, 157, 158, 159, 187, 188, 196, 205, 269, 270, 271, 274, 275, 278, 279, 280, 281, 289, 296, 306, 308, 310, 311, 313, 314
organic compounds, viii, ix, x, 17, 18, 22, 26, 27, 32, 40, 41, 42, 43, 44, 49, 53, 73, 77, 78, 82, 83, 93, 95, 100, 113, 138, 139, 147, 148, 152, 155, 156, 158, 159, 205, 270, 274, 275, 280, 289, 296, 311
organic matter, xii, 22, 23, 26, 27, 32, 36, 47, 48, 109, 148, 158, 269
organochlorinated, 20
organoleptic, xii, 269
orthogonal functions, 241, 262
orthogonality, 10
oscillation, 240, 258, 261, 266, 267, 268
oscillations, 240, 261, 265, 266
outliers, 227
overproduction, 123
oxalate, 150, 151
oxalic, 28, 29, 31, 32, 35, 44, 46, 47, 48, 51, 99, 107, 108, 150, 151, 158, 281, 308, 311, 314
oxalic acid, 28, 29, 31, 32, 35, 44, 46, 47, 48, 51, 107, 108, 150, 151, 158, 281, 308, 314
oxidants, viii, 17, 18, 19, 39, 98, 120, 145, 192, 270, 290
oxidation, viii, ix, x, xii, 12, 17, 18, 20, 21, 22, 23, 27, 30, 34, 35, 36, 38, 39, 40, 41, 42, 43, 44, 45, 46, 47, 48, 49, 50, 65, 67, 94, 95, 97, 98, 99, 100, 107, 108, 109, 111, 112, 113, 117, 127, 138, 140, 141, 142, 144, 149, 151, 152, 153, 156, 158, 188, 192, 193, 196, 269, 270, 271, 276, 277, 278, 279, 280, 281, 282, 283, 284, 290, 291, 292, 293, 294, 296, 298, 299, 305, 306, 307, 308, 311, 312, 313
oxidation products, 100, 107, 156
oxidation rate, 36, 65, 291, 292, 296, 299, 306
oxidative, 43, 54, 57, 99, 104, 120, 123, 126, 128, 147, 150, 151, 152, 153, 284, 310
oxidative damage, 126
oxidative destruction, 99
oxidative stress, 120, 123, 128
oxide, x, 32, 47, 48, 49, 128, 147, 152, 153, 155, 156, 157, 158, 207, 302
oxides, 30, 31, 33, 35, 48, 74, 148, 153, 155, 196, 201
oxidizability, 61, 92
oxygen, ix, x, 2, 19, 20, 47, 54, 56, 97, 99, 100, 104, 105, 106, 120, 128, 135, 136, 139, 140, 141, 145, 153, 157, 158, 261, 283, 298, 305, 306
ozonation, viii, x, xii, 17, 18, 19, 21, 22, 23, 27, 28, 30, 31, 32, 33, 36, 37, 38, 39, 40, 41, 42, 43, 44, 45, 46, 47, 48, 49, 50, 51, 54, 56, 57, 60, 64, 65, 66, 67, 68, 70, 71, 72, 73, 75, 76, 83, 93, 95, 100, 107, 112, 113, 138, 144, 147, 148, 149, 150, 151, 152, 153, 155, 156, 157, 158, 159, 269, 270, 271, 272, 273, 274, 277, 278, 279, 280, 281, 282, 283, 284, 285, 287, 288, 289, 290, 291, 292, 293, 294, 295, 296, 297, 298, 299, 300, 301, 302, 303, 304, 306, 307, 308, 310, 311, 312, 313, 314
ozone hole, xii, 240, 263, 264, 266
ozone reaction, 50, 188, 190, 192, 193, 279, 313
ozonolysis, 42, 95, 107, 108

P

p38, 122, 125, 130
packaging, 137
PAHs, 140
pain, 141
palladium, 49
paper, 48, 98, 99, 139, 144, 163, 164, 165, 166, 172, 184, 186, 216, 219
paracrine, 122
parameter, 32, 99, 105, 178, 197, 198, 205, 212, 214, 219, 224, 225, 226, 227, 228, 229, 230, 231, 240, 292, 295, 296, 307
parameter estimates, 214
parameter estimation, 197, 226, 227
parasites, 136
parenchyma, 120
Paris, 237

particle density, 295
particles, 123, 126, 158, 196, 286
particulate matter, 130, 235
passive, ix, 2, 7, 8, 117
pathogenesis, x, 117, 118, 121, 122, 132
pathogenic, 123, 139
pathogens, 136, 137, 141
pathology, 120, 121
pathways, ix, 22, 38, 53, 112, 121, 122, 125, 148, 157
patients, 122, 123, 124, 131, 132, 140
pattern recognition, 13
peat, 46
peptide, 31, 120, 124, 125, 126, 128, 132, 133
peptides, 46, 124, 125, 126, 129, 130
performance, 5, 6, 20, 24, 25, 26, 30, 33, 38, 153, 157, 211, 215, 216, 217, 221, 228, 229, 230, 231, 237
performance indicator, 217, 221, 229, 230, 231
perinatal, 131
periodic, 183, 201, 261
periodicity, xi, 239
peripheral blood, 122
permit, ix, 97, 111, 184
peroxidation, 141
peroxide, 19, 21, 22, 34, 41, 44, 49, 113
perturbation, 126
perturbations, 266
pesticides, xii, 18, 27, 43, 100, 139, 152, 153, 269
petrochemical, vii, 1, 7, 8, 13, 296
petroleum, 272, 296, 297, 310, 312
pH, viii, 22, 24, 25, 26, 27, 28, 29, 30, 31, 32, 33, 34, 35, 37, 38, 45, 46, 48, 53, 54, 56, 57, 59, 60, 66, 75, 76, 77, 82, 92, 93, 94, 101, 104, 119, 136, 137, 138, 149, 150, 151, 153, 154, 155, 158, 159, 271, 272, 273, 275, 282, 288, 291, 292, 294, 297, 300, 303, 306, 307, 309
pH values, 27, 59, 92, 138
pharmaceutical, 139, 144
phase space, 268
phase transitions, 254, 256
phenol, 31, 33, 34, 38, 43, 46, 47, 48, 49, 98, 99, 100, 109, 148, 149, 152, 157
phenolic, 27, 36, 40, 42, 44, 49, 51, 99, 100, 109, 298, 302
phenolic compounds, 27, 36, 44, 49
phenotype, 126
phosphates, 32
phospholipids, 141
phosphorus, 18
phosphorylation, 130
photocatalysis, 21, 26, 42, 43, 95, 113, 152, 153, 282
photocatalytic, 43, 144
photochemical, 2, 13, 21, 99, 142, 196, 198, 199, 229, 233, 235
photochemical degradation, 99
photodegradation, 41, 42, 157
photoirradiation, 105, 106
photolysis, 23, 41, 111, 113, 150
physical mechanisms, 262
physical treatments, 139
physicochemical, 54, 201
physiological, 118, 123, 124
pipelines, 9
planetary, 267
planning, xi, 195, 197
plants, 20, 37, 98, 100, 137
plasma, 21
platelet, 126
platinum, 49
play, ix, 18, 21, 36, 117, 118, 120, 121, 122, 123, 124, 127, 187, 188, 199
plethysmography, 118
PMS, xi, 187, 188, 189, 190, 191, 192, 193
poison, 3
Poisson, 164, 184
polarity, 10
policy makers, 164
pollutant, vii, x, xi, 1, 2, 7, 13, 18, 24, 26, 27, 35, 36, 39, 47, 140, 147, 148, 149, 150, 164, 195, 196, 198, 200, 201, 202, 204, 207, 237, 297
pollutants, xii, 7, 9, 10, 13, 18, 20, 21, 22, 24, 27, 29, 32, 34, 41, 44, 45, 50, 94, 100, 125, 148, 152, 155, 163, 196, 200, 205, 207, 211, 225, 226, 269, 296, 313
pollution, x, xi, 54, 109, 140, 142, 163, 164, 182, 184, 185, 186, 195, 196, 197, 198, 199, 201, 202, 204, 219, 225, 226, 227, 231, 233, 234, 235, 236, 237
polycondensation, 313
polycyclic aromatic hydrocarbon, 144, 277
polymer, 98, 100, 304
polymer matrix, 304
polymerization, ix, 97, 100, 105, 106, 107
polymerization process, ix, 105, 106, 107
polymorphonuclear, 130
polynomial, 211, 214
polynomials, 212
polyphenols, 38, 100, 114, 270, 311
polysaccharides, 99
poor, 213, 214, 226, 229, 231, 282
population, 158, 164, 197, 234
pore, 283, 295, 299, 304
pores, 282, 283, 294, 297, 299, 304, 314
porosity, 156, 296, 297, 299, 303, 310
porous, 36, 144, 159, 296

porous media, 144
portability, 214
ports, 11
poultry, 136
power, 7, 9, 24, 36, 44, 208, 296, 308
power plants, 44
precipitation, 18, 34, 105
prediction, xi, 195, 196, 197, 198, 200, 202, 204, 209, 213, 214, 215, 216, 217, 218, 222, 223, 224, 225, 226, 228, 229, 233, 234, 235, 237, 262
predictive models, xi, 195, 196, 198
preference, 254
pressure, 24, 25, 26, 98, 101, 118, 119, 137, 202, 205, 208, 211, 240, 241, 255, 271, 272
prevention, xi, 138, 141, 195, 198
preventive, 164
prices, 185
primary data, 172
private, xii, 269
probability, 83, 215, 227, 256, 257
probe, 291
production, x, 8, 11, 12, 25, 41, 60, 67, 68, 69, 70, 71, 72, 73, 74, 75, 98, 100, 109, 120, 124, 130, 135, 137, 141, 148, 152, 153, 183, 291
program, vii, 1, 7, 12, 118
proinflammatory, 122
proliferation, 121, 122, 124, 125, 126, 129, 130, 132
promote, ix, 30, 39, 97, 101, 141, 158
promoter, 37, 156, 294
propagation, 154, 155, 216
property, xi, 55, 239, 240
propionic acid, 26, 47
protection, 138, 198
protein, 101, 119, 120, 123, 124, 125, 130
protein synthesis, 101
proteins, 34, 121, 129, 136
protocol, 7, 14
protocols, 138
protons, 103
prototype, 225
protozoa, 136
proxy, 258
PSD, 208
pseudo, ix, xi, 65, 66, 97, 111, 187, 190, 191, 286, 293
Pseudomonas, 140
p-type, 157
public, xi, xii, 13, 16, 17, 139, 140, 141, 143, 195, 196, 198, 269
public health, 139, 140, 143
pulmonologist, 127
pulp mill, 98
pulse, 40, 41
purification, 45, 100, 147, 189, 270, 274, 280, 299, 311
pyridine ring, 302
pyruvic, 29, 45

Q

QBO, xi, xii, 239, 240, 241, 242, 244, 246, 248, 249, 250, 251, 252, 253, 254, 255, 256, 257, 258, 259, 260, 261, 262, 263, 264, 265, 266, 267
quadratic programming, 227, 233
quantum, 24
quinones, 99

R

radiation, x, 7, 18, 24, 27, 44, 95, 135, 196, 201, 202, 205, 211, 261, 311
radical, 19, 21, 22, 24, 25, 30, 31, 32, 33, 36, 38, 39, 40, 41, 95, 100, 104, 113, 136, 143, 148, 149, 150, 151, 152, 153, 154, 156, 157, 159, 189, 270, 273, 276, 278, 280, 282, 284, 291, 292, 302, 303, 313, 314
radical formation, 291, 292
radical mechanism, 33
radical reactions, 40
radio, 126, 258
radius, 295
rain, xi, 187, 199
range, ix, 6, 37, 100, 109, 117, 140, 190, 193, 196, 199, 217, 232, 240, 241, 242, 247, 248, 258, 262, 265, 281, 299
RANTES, 122, 130
rat, 131
RAW, 94
reactant, 188
reactants, 189, 193
reaction mechanism, 12, 24, 25, 28, 30, 32, 47, 82, 148, 198, 271
reaction medium, 149, 152, 153, 154, 155, 157, 158, 159
reaction order, 111, 286
reaction rate, 21, 24, 25, 35, 37, 57, 59, 64, 65, 92, 138, 148, 190, 270, 284, 285, 287, 310
reaction rate constants, 21, 24, 37, 190, 270
reaction temperature, 158
reaction time, viii, 53, 59, 60, 65, 66, 82, 92, 93, 290
reactive groups, 73
reactive oxygen, 120
reactivity, viii, xii, 17, 18, 21, 99, 136, 139, 150, 192, 269, 279, 282, 291, 294, 302, 306, 310
reagent, ix, 21, 45, 94, 97, 100, 108, 109, 111

reagents, ix, 27, 30, 97, 104, 151
real time, 8
reasoning, 301
receptor sites, vii, 1
receptors, 124, 132
recognition, 13, 136
recombination, 152
recovery, 18, 99, 108
recruiting, x, 117
recycling, 139
red blood cells, 119
redox, 148, 149
reduction, ix, xi, 18, 21, 24, 28, 31, 37, 42, 59, 82, 83, 97, 100, 105, 109, 110, 111, 139, 140, 148, 151, 153, 187, 195, 198, 210, 275, 279, 280, 293, 297, 301, 307, 310, 311
refineries, vii, 1
reflection, 7
refractory, 44, 47, 50, 100, 148, 152, 270
regeneration, 35, 38, 54, 94, 129, 150
regional, 199
regression, 190, 196, 198, 199, 215, 220, 226, 228, 233, 234
regression analysis, 196
regressions, 211
regrowth, 141
regular, xi, 239, 240, 257
regulation, 100, 121, 123, 125, 126, 127, 129, 130, 132
regulations, 140, 184
reinforcement, 113
relationship, 8, 15, 36, 196, 216, 226, 247, 260, 263, 267, 274, 281, 283, 287, 288, 289, 311
relationships, 82, 207, 210, 211, 216, 225, 246
relevance, 192
reliability, 3, 216
remediation, 140, 144, 159
remodeling, ix, x, 117, 118, 120, 122, 125, 126, 127, 132
remodelling, 127
remote sensing, 7, 13
repair, 98, 121, 122, 125, 126, 127, 129
research, vii, x, 4, 20, 32, 33, 36, 117, 126, 127, 141, 159, 199, 228, 235, 255, 258, 299
researchers, viii, 17, 19, 21, 22, 23, 27, 30, 36, 157, 270, 279
residuals, 136, 230
residues, 101, 136
resin, 43
resistance, x, 98, 117, 119, 120, 141, 279
resolution, vii, 1, 3, 4, 7, 10
resorcinol, 273, 303, 313
resources, xii, 20, 159, 269
respiratory, vii, 13, 118, 123, 124, 125, 131, 132, 196, 236
response time, 6
responsiveness, ix, x, 117, 118, 119, 123, 127
retardation, 257
retention, 3, 7, 10, 12, 13, 123
returns, 165, 173, 175
reusability, 54
Reynolds, 40, 236
rhinovirus infection, 124
rings, 54, 76, 270, 302
risk, 98, 121, 141, 184
risks, 142
road dust, 236
rods, 273
rolling, 121
room temperature, xi, 9, 19, 27, 151, 152, 187, 192, 193
ROS, 120
Rosseau, 130
Russia, 239
Russian, 266

S

sacrifice, 7
safety, 139, 141
saline, 119
Salmonella, 137
salt, x, 54, 147, 303, 304
salts, 27
sample, 4, 5, 57, 66, 76, 82, 99, 100, 101, 102, 104, 108, 109, 111, 168, 173, 272, 275, 296, 297, 298, 300, 301, 302, 303, 304, 305, 306, 307
sample design, 298
sampling, 2, 3, 5, 6, 7, 172, 189, 201, 213
sanitation, 46
SARS, 140
satellite, 263
saturated hydrocarbons, 4
saturation, 272
Saturday, 207, 208
scavenger, 41, 151, 155
scheduling, xi, 195, 197, 209, 215, 219, 220, 225
school, 234
scientists, 159
scripts, 15
sea level, 199
seafood, 139
secretion, x, 117, 118, 120, 123, 126, 127, 129, 130, 131, 132
sediment, 101
sedimentation, 20

selecting, 164, 202
selectivity, 9, 37, 148
semiconductor, 9, 10, 139, 153
semiconductor lasers, 9
semiconductors, 157
sensing, 7, 13
sensitivity, viii, x, 2, 6, 9, 13, 117
separation, viii, 2, 3, 7, 10, 304
series, x, xi, 9, 122, 163, 164, 165, 183, 185, 186, 195, 197, 198, 199, 201, 202, 204, 205, 208, 210, 211, 218, 219, 222, 225, 226, 227, 231, 235, 237, 240, 249, 251, 258, 260, 264, 272, 274, 278
serum, 123
severity, 99
Shanghai, 56
shear, 241, 242, 243, 244, 245, 246, 247, 248, 250, 251, 252, 253, 255, 256, 257, 258, 260, 261, 262, 265, 267
shock, 124
short period, 4, 100
short-term, 184, 195, 196, 198, 202, 258, 266
Siemens, x, 135
sign, 229
signal transduction, 127
signaling pathways, 121, 132
signals, 103, 122, 124, 125, 216, 228
significance level, 258, 259
silica, 32, 152, 153, 158
silicates, 157
siloxane, 57
silver, 47
simulation, 172, 199
sine, 203
Singapore, 240, 242
singular, 215
sintering, 31
SiO2, 35
sites, vii, 1, 32, 33, 131, 139, 152, 153, 155, 156, 200, 214, 275, 283, 290, 297, 298, 299, 310, 313
sludge, 34, 48, 94, 100, 139, 144
smog, xi, 187, 188, 193, 195, 196, 197, 209, 216, 221, 225, 232, 233
smooth muscle, 118, 126
smooth muscle cells, 126
smoothing, 237, 258, 260
SO2, 190, 200
SOD, 120, 128
sodium, 44, 50, 99, 119, 123, 131, 142, 273, 291
software, 7, 172, 214, 225
soil, 47, 100, 101, 139, 140, 144
soils, 98, 100, 101, 140, 144
solar, xi, xii, 202, 205, 211, 239, 240, 241, 242, 258, 260, 261, 262, 265, 266, 267, 268
sol-gel, 157
solid phase, 18, 30
solubility, viii, 17, 18, 35, 59, 101, 136, 138, 140, 270
solutions, vii, viii, ix, 1, 41, 45, 48, 53, 54, 57, 60, 66, 70, 73, 74, 75, 76, 77, 82, 83, 88, 92, 93, 94, 148, 152, 228, 271, 272, 273
solvents, 196
somatostatin, 126
soybean, 101
Spain, 17, 272
spatial, 7, 9, 10, 13, 125, 127, 132, 133, 236, 261
spatial information, 13
species, 3, 5, 6, 9, 11, 12, 13, 14, 21, 22, 27, 30, 32, 104, 107, 113, 120, 128, 138, 149, 151, 152, 153, 156, 157, 188, 196, 198, 289, 297, 298, 310
specific surface, 303
spectral analysis, 208
spectrophotometer, 273
spectrophotometric, 142
spectroscopy, 10, 13, 103
spectrum, ix, 5, 7, 9, 12, 60, 61, 62, 63, 64, 75, 76, 77, 78, 81, 87, 92, 97, 101, 102, 103, 104, 110, 111, 112, 122, 136, 298, 305
speculation, 122
speed, 11, 202, 203, 204, 205, 206, 208, 211, 215, 219, 224, 240, 242, 243, 244, 246, 247, 248, 249, 250, 251, 252, 253, 256, 257, 262, 265
St. Petersburg, 239
stability, viii, 17, 18, 22, 32, 36, 37, 46, 136, 138, 156
stabilization, 271
staffing, 198
stages, xii, 139, 153, 240, 241, 245, 249, 250, 253, 261
stagnation stage, xii, 239, 240, 245, 246, 248, 249, 250, 251, 252, 253, 255, 256, 257, 258, 259, 261, 262, 263, 264, 265
standard deviation, 172, 176, 177, 178, 179, 181, 255
standards, 139, 142, 197, 217, 219, 221, 226, 231
starch, 142
statistical analysis, 260
statistics, 186, 227
stem cells, 121
stochastic, x, 163, 164, 165, 167, 169, 170, 171, 173, 175, 177, 179, 181, 182, 183, 184, 185, 186, 210, 227, 235
stochastic model, 235
stochastic processes, 227
stock, 186
stoichiometry, 28, 29, 45, 74, 276, 277, 310
storage, 137
strain, 274, 281

strains, 138, 141, 274, 281, 313
strategies, 13, 118, 126
stratosphere, xi, xii, 239, 240, 244, 245, 250, 254, 255, 257, 258, 261, 262, 263, 264, 265, 266, 267, 268
strawberries, 143
strength, 30, 76, 98, 263
stress, ix, 117, 118, 120, 123, 125, 127, 128, 130
stretching, 102, 113
substances, 21, 27, 45, 65, 66, 67, 88, 92, 100, 118, 120, 126, 136, 139, 148, 153, 278
substitution, 21, 36, 48, 76
substitution reaction, 48
Succinic, 31, 38
sugar, 137
sulfate, 36, 100
sulfur, xi, 54, 187, 188
sulfur dioxide, 188
sulphate, 45, 281
summer, 197, 254
Sun, 42, 128, 129, 132, 133, 144, 258
Sunday, 207, 208
sunspot, 266
supercritical, 21, 100, 303
superiority, 136, 139
superoxide, ix, 34, 97, 99, 104, 107, 120
superposition, 61, 258
supply, 285
surface area, 32, 38, 158, 273, 281, 282, 283, 294, 296, 298, 299, 301, 304, 314
surface chemistry, 36, 38, 282, 289, 296, 300
surface properties, 153, 312, 314
surface water, 139, 147, 311
surfactant, 40, 50
surfactants, 21, 23, 27, 40
surplus, 125
surveillance, 9
survival, 125, 128, 145, 216
survival signals, 125
susceptibility, 120, 121, 122, 125
suspensions, 43
Sweden, 143
symbols, 300, 307
symptoms, 141
synchronization, 254
synchronous, 76
synergistic, 42, 43
synergistic effect, 42, 43
synthesis, 32, 36, 101, 120, 126, 155, 158, 314
systems, vii, viii, xii, 2, 9, 10, 18, 20, 21, 24, 27, 30, 34, 36, 41, 45, 48, 50, 100, 120, 128, 137, 139, 140, 142, 149, 158, 198, 205, 233, 269

T

T cells, 124, 125, 132
T lymphocyte, 122
tachykinins, 126
targets, 127
taste, 18, 20, 39
TCE, 40
TCR, 124
technology, viii, 6, 7, 15, 17, 18, 46, 50, 98, 136, 139, 140, 141, 143, 152, 159, 183, 197
Teflon, 56, 189
telephone, 236
temperature, xi, 7, 11, 12, 30, 49, 57, 94, 98, 99, 136, 137, 158, 159, 188, 192, 199, 202, 203, 204, 205, 211, 215, 219, 239, 240, 261, 266, 267, 295
temporal, vii, 1, 13, 127, 133, 197, 209, 216, 221, 236, 261, 262
tetrachloroethylene, 44
tetracycline, 42
tetroxide, 41
Texas, 6, 15, 16, 235
textile, 27, 36, 42, 50, 93, 94, 95, 144, 270
textile industry, 36, 270
TGF, 122, 123, 126
thawing, 6
theory, 19, 118, 164, 172, 197, 215, 227, 236, 267, 278
therapy, 127, 138, 141
thermal plasma, 21
thin films, 43
three-dimensional, 98, 99
threshold, 5, 142, 164, 185, 197, 209, 215, 216, 217, 219, 220, 221, 226, 229, 230, 231
threshold level, 217
thresholds, 197, 209, 215, 219, 220, 233, 236
throat, 141
tics, 130
tight junction, 121
time consuming, vii, 1, 8
time periods, 141, 271
time resolution, 4
time series, x, 9, 163, 164, 185, 186, 197, 198, 199, 201, 202, 204, 205, 208, 210, 211, 218, 219, 222, 225, 226, 227, 235, 237, 258
timing, 241
TiO_2, ix, 24, 25, 26, 27, 30, 31, 32, 33, 34, 35, 42, 43, 46, 47, 48, 97, 111, 112, 113, 152, 153, 157, 158
tissue, 6, 120, 122, 125, 126
titania, 153, 157
titanium, 43, 46, 157, 273
titanium dioxide, 43, 46, 157

titration, 142, 294, 313
TNF, 122
tobacco, 140
toluene, 6
topological, 278
total organic carbon, viii, 53, 148, 271, 273, 280
toxic, xii, 18, 20, 36, 49, 54, 113, 120, 151, 269, 270, 296
toxicity, viii, 17, 18, 24, 25, 33, 42, 88, 93, 270
toxins, 18
trace elements, 145
traffic, 142, 202, 226
training, 221
trans, 99
transcription, 123
transfer, 11, 20, 37, 215
transference, 23, 32, 310
transformation, viii, xii, 17, 18, 19, 30, 32, 36, 39, 41, 49, 69, 125, 153, 158, 198, 240, 242, 262, 263, 265, 269, 280, 291, 292, 293, 294, 296, 297, 298, 299, 304, 306, 307, 308, 310, 311, 312
transformations, 9, 210, 298, 302, 303
transforming growth factor, 126, 130
transition, 30, 46, 61, 148, 203, 212, 256, 257, 304, 314
transition metal, 30, 46, 148, 304, 314
transitions, 256
translocation, 140
transmembrane, 121, 123
transmembrane glycoprotein, 121
transmission, 123
transport, 2, 123, 131, 137, 138, 198, 199, 229, 267, 283
transportation, 127
transpose, 165
transverse section, 120
travel, 141, 226
trend, 59, 66, 78, 93, 164, 173, 186, 204, 205, 242, 263, 264
troposphere, xi, 188, 239, 240, 241, 262
trucks, 196, 205
trypsin, 120
tubular, 33
tumor, 129
turbulence, 137
Turkey, 235
two-dimensional, 10, 11, 12

U

ultrasound, 95
ultraviolet, vii, x, 43, 44, 60, 61, 92, 95, 135, 139, 196, 268
ultraviolet light, vii, 43
uncertainty, 30, 209, 214, 218, 219, 227, 262
uniform, 7, 141, 168
United States, 145
upper airways, 123
urban areas, 185, 196, 205, 208, 231, 233, 236
urea, 272, 299, 300
UV irradiation, 44, 95, 150, 158
UV light, 27, 54, 149
UV radiation, 18, 24, 27, 44, 261
UV spectrum, 101, 102, 110
UV-irradiation, 97, 111

V

vaccination, 138
vaccine, 138
vacuum, 6, 21
valence, 19
Valencia, 130, 144, 145
validation, 5, 197, 217, 226
valine, 101
values, xi, 21, 27, 59, 65, 83, 92, 105, 106, 111, 113, 138, 164, 168, 172, 173, 182, 185, 186, 191, 192, 195, 196, 198, 200, 211, 214, 215, 216, 217, 219, 221, 226, 227, 228, 229, 231, 235, 276, 277, 278, 279, 280, 282, 283, 284, 286, 287, 289, 292, 293, 295, 299, 302, 304, 305, 306, 307, 308, 310
vapor, 6, 98
variability, xii, 184, 185, 229, 239, 240, 241, 267
variable, 12, 166, 167, 169, 198, 203, 204, 206, 207, 208, 210, 211, 229, 237
variables, x, 24, 25, 30, 33, 37, 38, 39, 135, 165, 166, 168, 170, 196, 199, 200, 201, 202, 204, 205, 208, 209, 210, 211, 212, 213, 215, 216, 226, 228, 229, 237, 261, 270, 271, 291
variance, 165, 166, 167, 171, 180, 185, 214
variance-covariance matrix, 165, 166, 167, 171
variation, ix, 53, 54, 60, 66, 82, 204, 205, 207, 213, 214, 215, 219, 233, 266
vascular wall, 121
vasoactive intestinal peptide, 120, 125, 128, 133
vasodilatation, 126
vector, 165, 166, 167, 168, 169, 170, 171, 209, 210, 212, 229
vegetables, 137, 143
vegetation, 6, 196
velocity, 112
ventilation, 142
versatility, 21
vessels, 98
vibration, 75, 76
video clips, 8

video surveillance, 9
vinasse, 43
VIP, 125, 126, 127, 132
virus, 31, 122, 136, 140
viruses, 136, 138
viscosity, 123, 295
visible, 9, 60, 61, 78, 92, 99
volatility, x, 163, 164, 165, 166, 167, 169, 170, 173, 176, 177, 179, 180, 182, 183, 184, 185, 186
vortex, 263, 264, 266

W

waste treatment, 30, 39
wastes, 40
wastewater, viii, ix, x, 17, 19, 40, 42,43, 44, 47, 50, 51, 54, 93, 94, 97, 98, 100, 108, 109, 110, 111, 139, 140, 143, 144, 145, 147, 148, 159, 311
wastewater treatment, 44, 50, 93
wastewaters, 24, 37, 54, 144
water, viii, ix, x, xii, 6, 17, 18, 19, 20, 21, 22, 26, 27, 33, 34, 35, 36, 38, 39, 40, 41, 42, 43, 44, 45, 46, 47, 48, 49, 50, 54, 94, 95, 97, 98, 100, 101, 108, 109, 110, 111, 113, 118, 123, 131, 135, 136, 137, 138, 139, 140, 142, 143, 145, 147, 148, 151, 158, 159, 235, 269, 270, 271, 272, 273, 275, 278, 292, 295, 301, 311, 313, 314
water quality, 138, 139, 235
water supplies, 39, 139, 143
water vapor, 6
wavelengths, 9, 13, 54, 60, 64, 272
weather prediction, 234
web, 234
Weibull, 215, 227
WHO, 142, 145
wind, xi, xii, 172, 199, 202, 203, 204, 205, 206, 207, 211, 215, 219, 224, 239, 240, 241, 242, 243, 244, 246, 247, 248, 249, 250, 251, 252, 253, 254, 255, 256, 257, 258, 260, 261, 262, 263, 264, 265, 266, 267
wind speeds, 242, 254, 255
wine, 140
winter, 263, 267
wood, 98

X

XPS, 298, 299, 300, 301, 303, 304, 305, 306, 307
XRD, 304, 305

Y

yield, 98, 99, 188

Z

zeolites, 32
zinc oxide, 158
zirconia, 153
ZnO, 24, 42, 158
zoonotic, 138